T0212929

Lecture Notes in Computer Science　　9251

Commenced Publication in 1973
Founding and Former Series Editors:
Gerhard Goos, Juris Hartmanis, and Jan van Leeuwen

More information about this series at http://www.springer.com/series/7407

Victor Malyshkin (Ed.)

Parallel Computing Technologies

13th International Conference, PaCT 2015
Petrozavodsk, Russia, August 31 – September 4, 2015
Proceedings

 Springer

Editor
Victor Malyshkin
Russian Academy of Sciences
Novosibirsk
Russia

ISSN 0302-9743   ISSN 1611-3349 (electronic)
Lecture Notes in Computer Science
ISBN 978-3-319-21908-0   ISBN 978-3-319-21909-7 (eBook)
DOI 10.1007/978-3-319-21909-7

Library of Congress Control Number: 2015944720

LNCS Sublibrary: SL1 – Theoretical Computer Science and General Issues

Springer Cham Heidelberg New York Dordrecht London

Printed on acid-free paper

Springer International Publishing AG Switzerland is part of Springer Science+Business Media
(www.springer.com)

Preface

The PaCT 2015 (Parallel Computing Technologies) conference was a four-day conference held in Petrozavodsk (Russia). This was the 13th international conference in the PaCT series. The conferences are held in Russia every odd year. The first conference, PaCT 1991, was held in Novosibirsk (Academgorodok), September 7–11, 1991. The next PaCT conferences were held in Obninsk (near Moscow), August 30 to September 4, 1993; in St. Petersburg, September 12–15, 1995; in Yaroslavl, September, 9–12, 1997; in Pushkin (near St. Petersburg), September, 6–10, 1999; in Academgorodok (Novosibirsk), September 3–7, 2001; in Nizhni Novgorod, September, 15–19, 2003; in Krasnoyarsk, September 5–9, 2005; in Pereslavl-Zalessky, September 3–7, 2007; in Novosibirsk, August 31 – September 4, 2009; in Kazan, September 19–23, 2011, and in St. Petersburg, September 30 to October 4, 2013. Since 1995 all the PaCT proceedings are published by Springer in the LNCS series. PaCT 2015 was jointly organized by the Institute of Computational Mathematics and Mathematical Geophysics (Russian Academy of Sciences), Novosibirsk State University, Novosibirsk State Technical University, Institute of Applied Mathematical Research (Karelian Research Centre of Russian Academy of Sciences), and Petrozavodsk State University. The aim of the conference is to give an overview of new developments, applications, and trends in parallel computing technologies. We sincerely hope that the conference will help our community to deepen its understanding of parallel computing technologies by providing a forum for an exchange of views between scientists and specialists from all over the world. The conference attracted 87 participants from around the world, with authors from 13 countries submitting papers. Of these, 53 papers were selected for the conference as regular ones; there was also an invited speaker. All the papers were reviewed by at least three referees. Many thanks to our sponsors: the Ministry of Education and Science, Russian Academy of Sciences, and Russian Fund for Basic Research.

September 2015   Victor Malyshkin

Organization

PaCT 2015 was organized by the Supercomputer Software Department, Institute of Computational Mathematics and Mathematical Geophysics (ICM&MG), Siberian Branch, Russian Academy of Science, Novosibirsk State University (NSU), Novosibirsk State Technical University (NSTU), Institute of Applied Mathematical Research (IAMR), Karelian Research Centre of Russian Academy of Sciences, and Petrozavodsk State University (PetrSU).

Organizing Committee

Conference Co-chairs

Victor Malyshkin	ICM&MG, Novosibirsk
Evgeny Ivashko	IAMR, Petrozavodsk

Conference Secretary

Maxim Gorodnichev	ICM&MG

Contact Volume Editor

Olga Bandman	ICM&MG

Organizing Committee Members

Valentina Markova	ICM&MG
Mikhail Ostapkevich	ICM&MG
Yuri Medvedev	ICM&MG
Georgy Schukin	ICM&MG
Sergey Kireev	ICM&MG
Vladimir Mazalov	IAMR
Andrey Pechnikov	IAMR
Evgeny Ivashko	IAMR
Alexander Rumiantsev	IAMR
Natalia Nikitina	IAMR
Alexander Golovin	IAMR
Vladimir Timofeev	NTSU

Sponsoring Institutions

Ministry of Education and Science, Russia
Russian Academy of Sciences
Russian Fund for Basic Research

Program Committee

Victor Malyshkin	Russian Academy of Sciences, Chair
Farid Ablaev	Kazan Federal University, Russia
Sergey Abramov	Russian Academy of Sciences
Farhad Arbab	Leiden Institute for Advanced Computer Science, The Netherlands
Stefania Bandini	University of Milano-Bicocca, Italy
Olga Bandman	Russian Academy of Sciences
Thomas Casavant	University of Iowa, USA
Pierpaolo Degano	University of Pisa, Italy
Dominique Désérable	National Institute for Applied Sciences, Rennes, France
Sergei Gorlatch	University of Münster, Germany
Bernard Goossens	University of Perpignan, France
Yuri G. Karpov	St. Petersburg State Polytechnical University, Russia
Alexey Lastovetsky	University College Dublin, Ireland
Jie Li	University of Tsukuba, Japan
Thomas Ludwig	University of Hamburg, Germany
Mikhail Marchenko	Russian Academy of Sciences
Giancarlo Mauri	University of Milano-Bicocca, Italy
Nikolay Mirenkov	University of Aizu, Japan
Dana Petcu	West University of Timisoara, Romania
Viktor Prasanna	University of Southern California, USA
Michel Raynal	Research Institute in Computer Science and Random Systems, Rennes, France
Bernard Roux	National Center for Scientific Research, France
Mitsuhisa Sato	University of Tsukuba, Japan
Carsten Trinitis	University of Bedfordshire, UK and Technical University of Munich, Germany
Mateo Valero	Technical University of Catalonia, Spain
Roman Wyrzykowski	Czestochowa University of Technology, Poland

Additional Reviewers

S. Abramov	M. Gorodnichev	I. Menshov
D. Akhmed-Zaki	K. Kalgin	A. Nepomniaschaya
O. Bandman	Y. Karpov	V. Perepelkin
T. Casavant	S. Kireev	D. Petcu
D. Deserable	A. Kireeva	V.K. Prasanna
B. Goossens	Y. Klimov	N. Shilov
S. Gorlatch	A. Lastovetsky	V. Toporkov
H. Dirks	V. Malyshkin	R. Wyrzykowski
F. Wuebbeling	M. Marchenko	Y. Zagorulko
A. Rasch	V. Markova	
M. Haidl	G. Mauri	

Contents

Unconventional Computing - Cellular Automata

Distributed Computing

Special Processors Programming Techniques

Applications

Parallel Models, Algorithms and Programming Methods

Software System for Maximal Parallelization of Algorithms on the Base of the Conception of Q-determinant

Valentina N. Aleeva$^{(\boxtimes)}$, Ilya S. Sharabura, and Denis E. Suleymanov

South Ural State University, Chelyabinsk 454080, Russia
aleevavn@susu.ac.ru, run174@yandex.ru, kingmidas1992@gmail.com

Abstract. The development and the usage of parallel computing systems make it necessary to research parallelization resource of algorithms for search of the most rapid implementation. The algorithm representation as Q-determinant is one of the approaches that can be applied for that case. Such representation allows getting the most rapid possible implementation of the algorithm evaluates its performance complexity. Our work is to develop software system QStudio, which presents algorithm in the form of Q-determinant using the flowchart, finds the most rapid implementation of that one and builds an execution plan. The obtained results are oriented to ideal model of parallel computer system. However they can be a basis for automated execution of the most rapid algorithm implementations for real parallel computing systems.

Keywords: Algorithm representation as Q-determinant · Parallelization algorithm · Most rapid implementation of the algorithm · Execution plan of the most rapid implementation · Parallel computing system

1 Introduction

The performance increase of parallel computing systems is one of the priorities of the development of computer technology. One of the ways of solving that problem is the most rapid implementation of algorithms due to maximal use parallelization resource. Our work is devoted to the same research area.

The paper presents the results of work on the project Maximum Parallelization of Algorithms, briefly, MPA. The aim of the project is to automate execution of the most rapid implementations of algorithms for parallel computing systems. To achieve that it is necessary to solve the following problems:

1. software engineering QStudio for the analysis of resource parallelization of any algorithm including its most rapid implementation and constructing an execution plan of the most rapid implementation of the algorithm or the most effective implementation of the algorithm for parallel computer systems;
2. development and implementation of technology to execute the most rapid parallel implementation of the algorithm for the parallel computer systems using its execution plan.

© Springer International Publishing Switzerland 2015
V. Malyshkin (Ed.): PaCT 2015, LNCS 9251, pp. 3–9, 2015.
DOI: 10.1007/978-3-319-21909-7_1

The present article includes the results of the project MPA to solve the problem No 1 and those results will be used for solving the second problem.

The concept of Q-determinant is the basis of the studies conducted in the framework of MPA.

2 The Conception of Q-determinant

First we describe the conception of Q-determinant [1]. The representation of algorithm as Q-determinant allows obtaining all possible implementations of the algorithm including detection of the most rapid implementation meaning that every algorithm operation is performed if its operand values are ready. There are some algorithms for those their most rapid implementation cannot be performed because we need infinite number of operations need to perform at the same time. Hence an infinite number of processors (compute nodes) are required. In [1] there are analysis of the most rapid algorithm implementation and so it is possible to obtain estimates of complexity: the number of processors and the number of work cycles of computer system to accomplish one. As the result we can compare the resources of algorithm parallelization for solving the maximal parallelization problem and choose the algorithm with maximal resource of parallelization.

For a long time there were many investigations of algorithm parallelization, but just the conception of Q-determinant allows to detect of possibility of the most rapid algorithm implementation. There is no doubt that the investigation parallel structure program is very important and advanced for their realizations of parallel computer systems. One of the most significant researches in that direction is the system V-Ray [2,3]. However, if you want to solve the task of maximal algorithm parallelization then it is may be not advisable of the usage of the program that realizes the algorithm because it may not contain all algorithm realizations, particularly, the most rapid implementation. That's why, in our opinion the more correct approach is the algorithm parallelization of the algorithm itself.

Now we give a brief description of the Q-determinant conception. Let α be an algorithm to solve the algorithmic problem $\bar{y} = F(N, B)$ where N is a parameter dimension set of the problem, B is a set of input data, $\bar{y}$ is a set of output data. Let Q be a basic set of arithmetic and logical type operations. *The expression* is called the set of operands of arithmetic or logical type that use operations from Q. *Q-term* is the map from the problem dimension to a structured set of expressions that we need to calculate one of the output variables of the problem. The set of Q-terms can be unconditional, conditional and conditionally infinite according to the structure of expression set.

Q-determinant is the set of Q-terms that we need to calculate each of the problem output data. Let an algorithm α be in the form of $y_i = f_i (i = 1, \ldots, m)$ where f_i is Q-term to calculate y_i, m is the number of output data. Then we consider that the algorithm α represents in the form of Q-determinant. If the algorithm has some representation as flowchart then it can be represent in the form of Q-determinant [4].

We consider the Gauss–Jordan solution of a system of linear equations as an example of representation of the algorithm in the form of Q-determinant. Let $A\bar{x} = \bar{b}$ be a system of linear equations, where $A = [a_{ij}]_{ij=1,\ldots,n}$ is a $n \times n$ invertible matrix, $\bar{x} = (x_1, \ldots, x_n)^T$, $\bar{b} = (a_{1,n+1}, \ldots, a_{n,n+1})^T$. At the first step we suppose that the leading element is the first nonzero element of the first row of the original matrix, and at k-th step ($2 \leq k \leq n$) we select the first nonzero element of the k-th row that obtained at $(k-1)$-th step. Then the Q-determinant of Gauss–Jordan method consists of n conditional Q-terms and its representation in the form of Q-determinant has the shape

$$x_j = \left\{ (u_1, w_1^j), \ldots, (u_{n!}, w_{n!}^j) \right\} \ (j = 1, \ldots, n).$$

Unconditional Q-terms $u_1, \ldots, u_{n!}$, $w_1^j, \ldots, w_{n!}^j$ $(j = 1, \ldots, n)$ depends on terms of the matrix A and vector b. In detail their structure is described in [1].

The realization of the algorithm in the form of Q-determinant is called the process of calculating the Q-terms $f_i(i = 1, \ldots, m)$ that are included in the Q-determinant. If the calculation of all Q-terms $f_i(i = 1, \ldots, m)$ is produced at the same time and as rapid as possible, i.e. the operations of the set are executed as soon as they are ready to perform, in this case we have *the most rapid implementation of the algorithm*. Generally speaking, the set Q may include any operation, not only which are considered in the paper [1]. So the software system QStudio allows specifying any operation, as well as if it is necessary to redefine the basic ones. All operations both basic and asked again are the functions of variable collections, so we will call them functions. All functions are separate projects in the framework of using system QStudio.

3   The Software System QStudio

The software system QStudio makes possible to calculate Q-determinant of any algorithm (if the algorithm has some representation as flowchart), to find the most rapid possible implementation (in the sense mentioned above) and to build its execution plan for the parallel system. The system was developed with the help of the technology .Net in C# and the system for building client application Window Presentation Foundation (WPF).

That software technology allows separating the application logic from the client interface and for interface scaling. WPF with compiled markup language XAML allows to redefine the appearance and behavior of default controls and to create custom controls. That feature has allowed developing the library of controls that can be used in development of other applications. The appearance is stylized to look like Modern UL interfaces beginning from redefined Window component to standard text labels. When we have developed the custom elements two classes were modified: class TabControl (now it allows to close tabs) and class TreeViewControl (icons of elements were added). Besides redefining default elements new custom elements were created such as console, file explorer and text processor. They were allocated into a new library. By WPF utilization of routed events most components are noninteracting. In other words, the system has weak connectivity.

The core of software system contains such basic structure as graph and Q-determinant. Using this structure one can build the logic of most components and due to object-oriented architecture it is also possible to extend the functionality. In addition, the core contains the object converter of different data formats, which allows standardizing input and output for all plugins. All project files are contained in serialized format, so that a user cannot unintentionally change the files. The basis of system is the common class `Adapter<IDeterminant,IPlan>`. As Q-determinant and execution plan modules should support certain interfaces their implementation was united in one class that allows changing the basic logic without changing external modules and even other core classes.

All mentioned above is the system basis. All the rest can be implemented as plugins. For example, the view of the execution plan in the form of visual graph is possible only with the plugin `ImplementationPlanViewer`. Otherwise QStudio will open it as text files in JSON. Similarly, QStudio will continue to operate without components that implement `IDeterminant` or `IPlan` interfaces, but the message will appear that there is no opportunity to define the resource of parallelism. If several modules are detected then a user will be able to choose what algorithm to use. To create it is necessary as dependents to enable core and possibly the visual component. After that `IPlugin Interface` must be implement. QStudio will enable and instantiate this plugin by itself.

The Compiler acts as separate software. GUI allows editing files, enabling dependents, creating new projects, etc. as any IDE. Then the compiler on the basis of all having processes data produces necessary operations to receive a result. The given separation allows to develop two different projects independently, namely, QStudio and Compiler. Since we have plugin support in QStudio and reflexes support in Compiler the connectivity of project is weak.

The whole system is developed with the use TDD methodology that minimizes unauthorized and undocumented states. Also, different patterns are used such that Abstract factory, Adapter, Observer.

All the code of the project is an open source and is published in repository on Github. That gives opportunity to conduct the remote teamwork and it guarantees the code security.

4 Preparation Q-determinant Algorithm

As the source data for preparation Q-determinant of the algorithm the software system QStudio uses a flowchart as it was proposed in [4]. The flowcharts can be described in different formats, for example, XML or JSON. Furthermore, the supported formats are MessagePack and GPB. QStudio converts the description of the flowchart from one format to another. The methods of Q-determinant preparation are contained in a separate class library that implements the interface `IDeterminant`. The Q-determinant preparation of the algorithm is carried out by using the module `QDeterminant`, which receives the flowchart as an input.

The module analyzes a flowchart to obtain Q-terms that are component of Q-determinant. The analysis is produced at fixed parameters of dimension of the problem. The algorithm starts from the terminal flowchart symbol Start then it tracks all the ways to the terminal flowchart symbol End. If we use the approach of [1] we can be get that every passage in the flowchart at fixed parameters of dimension $\bar{N}$ of the problem determines either an expression $w(\bar{N})$, or a pair of expressions $u(\bar{N}), w(\bar{N})$, relating to one of the Q-terms that makes up the Q-determinant algorithm. The result of the work of the module is the Q-determinant of the algorithm that is the set of Q-terms at fixed parameters of dimension of the problem.

Analysis of flowchart elements depends on the type of elements. The expression written in the element to check the condition falls into unconditional logical Q-term. The examples of such Q-terms are $u_1, \ldots, u_{n!}$ from Q-determinant for the Gauss–Jordan method. If the computable expression is contained in the flowchart symbol then it can be added to unconditional Q-term of any type.

5 The Detection of the Most Rapid Algorithm Realization and the Building of Plan of Its Execution

All processing algorithms of Q-determinant and receiving execution plan are contained in the class library that implements the interface IPlan. The processing of the Q-determinant and obtaining the execution plan are implemented by the module ImplementationPlan.

By the module QDeterminant we obtain data structure that contains a description of Q-determinant at fixed values of parameters of the dimension problem. Then by the program it is divided into separate units, which represent a description of unconditional Q-terms or couples of unconditional Q-terms that enter the Q-determinant. For example, for the Q-determinant of the Gauss–Jordan method the blocks describe pairs $(u_1, w_1^j), \ldots, (u_{n!}, w_{n!}^j)$ $(j = 1, \ldots, n)$.

Next it is made the following analysis: what operations and on which of the work cycle should be performed to calculate the Q-terms simultaneously and as fast as possible. That is the most rapid algorithm implementation. If we have some restrictions of the number of processors or other restrictions then sometimes we cannot receive the most rapid algorithm implementation. Taking into account of the number of processors we get the most effective algorithm implementation [5].

The software system QStudio builds its execution plan of the found realization that is the directed tiered graph where the zero level contains the input data of the algorithm and each non-zero level matches the cycle of the computer system and contains all the operations of the algorithm running in parallel during this cycle [6]. To construct the execution plan we use the principle of the reverse Polish notation at which not string representation is placed into the stack but the structure that describe the vertex of the graph. To evaluate the algorithm parallelization resource we can count the number of functions performed at each step simultaneously calculating oall Q-terms included in the Q-determinant.

If it is necessary we can limit the number of processors and optimize execution plan in order to obtain execution plan of the most effective realization that allows us to determine exactly how many cycles the algorithm will execute. The execution plan can be optimized, i.e. eliminate duplicate input values and repetitive calculations. The flowchart and the execution plan can be converted into different data formats and displayed in the graphical form for the convenience and experience of the user.

The figure below displays the graph of the execution plan of the most rapid implementation of the algorithm Scalar. That algorithm calculates the scalar product of the vectors. In the given example the vector dimension is equal to 9. The figure is obtained with the help of the system Qstudio Fig. 1.

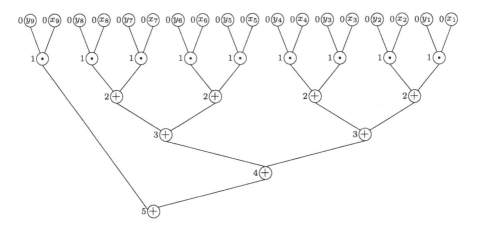

Fig. 1. The example of the execution plan of the algorithm Scalar

6   Conclusion

This article describes the development of the software system QStudio designed to find the most rapid and the most effective algorithm implementations and to build their execution plans on parallel computer systems. In the design we use the approach to parallelization algorithms based on the conception of Q-determinant. The obtained results are the basis of the maximal resource usage of parallelization algorithms at the implementation for parallel computer systems which should lead to greater productivity.

The next step of the investigation is to generate to executable code on the basis of the execution plan of the most rapid algorithm implementation plan. Also we develop the idea how the algorithm execution plan of the most rapid algorithm realization overlays on the architecture of a particular computer system. As the first additional plugin it will be added the possibility of the transformation of the execution plan on the architecture of the supercomputer "Tornado" of South Ural State University.

References

1. Aleeva, V.N.: Analysis of parallel numerical algorithms: Preprint No. 590. Novosibirsk, Computing Center of the Siberian Branch of the Academy of Sciences of the USSR (1985)
2. Voevodin, V.V., Voevodin, V.V.: The V-ray technology of optimizing programs to parallel computers. In: Vulkov, L.G., Yalamov, P., Waśniewski, J. (eds.) WNAA 1996. LNCS, vol. 1196, pp. 546–556. Springer, Heidelberg (1997)
3. Shamakina, A.V., Sokolinsky L.B.: Testing methodology of instrumental complexes for constructing parallel programs. In: Scientific service on the Internet: Multicore Computer World. 15 years RFBR: Proceedings of the all-Russian Scientific Conference (Novorossijsk, Russia, September, 24–29, 2007), pp. 227–230. Publishing of the Moscow State University, Moscow (2007)
4. Ignatyev, S.V.: Definition of parallelism resource of algorithms on the base of the concept of Q-determinant. In: Scientific Service on the Internet: Supercomputer Centers and Tasks: Proceedings of the International Supercomputer Conference (Novorossijsk, Russia, September, 20–25, 2010), pp. 590–595. Publishing of the Moscow State University, Moscow (2010)
5. Svirihin, D.I.: Definition of parallelism resource of algorithm and its effective use for of a finite number of processors. In: Scientific Service on the Internet: the Search for New Solutions: Proceedings of the International Supercomputer Conference (Novorossijsk, Russia, September, 17–22, 2012), pp. 257–260. Publishing of the Moscow State University, Moscow (2012)
6. Svirihin, D.I., Aleeva, V.N.: Definition the maximum effective realization of algorithm on the base of the conception of Q-determinant. In: Parallel Computational Technologies (PCT 2013): Proceedings of the International Scientific Conference (Chelyabinsk, Russia, April, 1–5, 2013), p. 617. Publishing of the South Ural State University, Chelyabinsk (2013)

Highly Parallel Multigrid Solvers for Multicore and Manycore Processors

Oleg Bessonov[⊠]

Institute for Problems in Mechanics of the Russian Academy of Sciences, 101,
Vernadsky Avenue, 119526 Moscow, Russia
bess@ipmnet.ru

Abstract. In this paper we present an analysis of parallelization properties and implementation details of the new Algebraic multigrid solvers. Variants of smoothers and multicolor grid partitionings are discussed. Optimizations for modern throughput-oriented processors are considered together with different storage schemes. Finally, comparative performance results for multicore and manycore processors are presented.

1 Introduction

This paper is devoted to the development of efficient parallel algebraic methods for solving large sparse linear systems arising in discretizations of partial differential equations. Historically, Conjugate Gradient methods have been widely used for solving such linear systems [1]. In order to accelerate the solution, implicit preconditioners of the Incomplete LU-decomposition type (ILU) are applied [2,3]. However, implicit preconditioners have limited parallelization potential and therefore can't be efficiently employed on massively parallel computers.

On the other hand, there exists a separate class of implicit methods, multigrid, which possess very good convergence and parallelization properties. Multigrid solves differential equations using a hierarchy of discretizations. At each level, it uses a simple smoothing procedure to reduce corresponding error components.

In a single multigrid cycle, both short-range and long-range components are smoothed. It means that information is propagated instantly throughout the domain within a such cycle. As a result, this method becomes very efficient for elliptic problems that propagate physical information infinitely fast.

At the same time, multigrid can be efficiently and massively parallelized because processing at each grid level is performed in the explicit manner, and data exchanges are needed only between adjacent subdomains at the end of a cycle.

In this paper we will consider the Algebraic multigrid (AMG) approach [4] which is based on matrix coefficients rather than on geometric parameters of a domain. Parallelization properties and implementation details of this algorithm will be analyzed, and performance results for modern throughput processors will be presented.

V. Malyshkin (Ed.): PaCT 2015, LNCS 9251, pp. 10–20, 2015.
DOI: 10.1007/978-3-319-21909-7_2

2   Iterative Methods and Their Parallelization Properties

The multigrid approach will be presented here in the context of the hierarchy of iterative methods. Iterative methods are used for solving large linear systems arising in discretizations of partial differential equations in many areas (fluid dynamics, semiconductor devices, quantum problems). They can be applied to ill-conditioned linear systems, both symmetric and non-symmetric.

The most popular class of iterative methods is the Conjugate Gradient (CG). In order to accelerate the convergence, this method requires preconditioning. There are two main classes of preconditioners: explicit, that act locally by means of a stencil of limited size and propagate information through the domain with low speed, and implicit, that operate globally and propagate information almost instantly. Due to this implicit preconditioners work much faster and have better than linear dependence of convergence on the geometric size of the problem.

The same applies to stand-alone iterative methods that possess similar properties and may be classified as being explicit or implicit (we consider here the nature of internal iterations rather than the outer properties of a method).

Parallel properties of iterative solvers strongly depend on how information is propagated in the algorithm. For this reason methods with the implicit nature of iterations can't be easily parallelized, and many efforts are needed for finding geometric or algebraic approached of parallelization [3].

However, some methods with explicit iterations are also essentially sequential, and their parallelization may become difficult. In particular, the Gauss-Seidel and Successive Over-Relaxation (SOR) methods in their original form need the sequential processing of grid points in a domain. In order to overcome this problem, multicolored approaches are used when all grid nodes in a domain are partitioned in such a way that each node has no connection with another nodes of the same color. Owing to this, computations for all grid points with the same color can be performed in any order thus allowing arbitrary splitting of the domain. For the 7-point stencil in 3D, it is enough to use 2 colors (Red-black partitioning), while the more general 27-point stencil requires 8 colors.

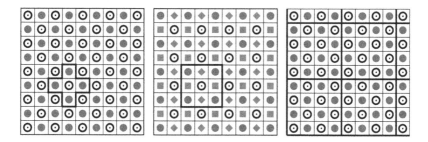

Fig. 1. Red-black (left) and multicolor (center) grid partitionings; splitting of a computational domain with the red-black partitioning for the parallelization (right) (Color figure online)

Figure 1 illustrates grid partitionings for 2 and 4 colors and splitting of a computational domain into subdomains for processing them in parallel.

Properties of different iterative methods are presented in Table 1. Solutions of the Poisson equation on two different uniform grids were used for this test, with the convergence stopping criteria 10^{-10}. The approximate cost of an iteration is presented for each method (relative to the Jacobi method). For the Gauss-Seidel and SOR, Red-black variants are given. Fortunately they have the same convergence properties as their sequential counterparts. For the Conjugate Gradient, several preconditioners were tested: plain diagonal one (CG), polynomial Jacobi explicit preconditioner (CG Jacobi) and Incomplete LU in two variants (CG ILU and MILU) [2,3]. For the multigrid, two implementations were used: a plain AMG solver and an AMG-preconditioned Conjugate Gradient.

Table 1. Convergence of different iterative algorithms for square matrices of size N

Iterative method	$N = 65$	$N = 129$	$f(N)$	Cost of iteration
Jacobi	23650	90850	$O(N^2)$	1
Gauss-Seidel	12210	46940	$O(N^2)$	1
SOR	299	600	$O(N)$	1
CG	258	514	$O(N)$	1.5
CG Jacobi	130	257	$O(N)$	2.25
CG ILU	109	212	$O(N)$	3
CG MILU	48	68	$O(N^{1/2})$	3
AMG	8	8	$O(1)$	5
CG AMG	7	7	$O(1)$	6

The above results confirm that the convergence of the multigrid doesn't depend on the problem size. The next efficient method is the CG MILU with the required number of iterations of the order of $O(N^{1/2})$. This method is still applicable and efficient for many problems that are not convenient for the multigrid (anisotropic grids, systems of equations etc.). However, the ILU methods have very limited parallelization potential [2,3] and, for this reason, simple explicit methods and preconditioners of the $O(N)$ class remain attractive in some cases.

Let's consider now the Red-black Gauss-Seidel and SOR methods. These methods become fully explicit and can be represented as two half-sweeps (Fig. 2). Here parts 1 and 2 of a matrix represent grid nodes of the first and the second color respectively. The principal operation in each half-sweep is the multiplication of a sparse matrix by a corresponding vector. Such kind of operations can be massively and efficiently parallelized. Therefore they can be used as smoothers in highly parallel implementations of the multigrid.

Efficient implementation of the matrix-vector multiplication for multicolor grid partitionings requires reordering and reorganization of the original matrix.

$$(D + L)x_{k+1} = b - L^{\mathrm{T}}x_k$$

Two half-sweeps:

$$D^{(1)}x_{k+1}^{(1)} = b^{(1)} - L^{\mathrm{T}}x_k^{(2)}$$
$$D^{(2)}x_{k+1}^{(2)} = b^{(2)} - Lx_{k+1}^{(1)}$$

Explicit representation:

$$x_{k+1} = D^{-1}(I - LD^{-1})(b - L^{\mathrm{T}}x_k)$$

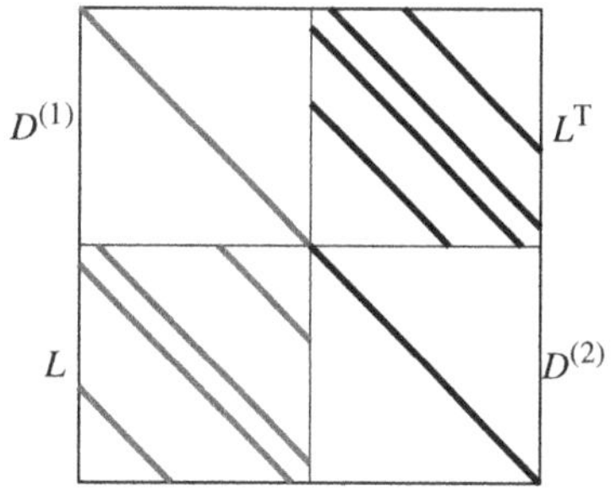

Fig. 2. Iteration of the Red-black Gauss-Seidel method for a symmetric matrix (Color figure online)

3 Throughput-Oriented Processors and Storage Schemes

All modern high-performance microprocessors belong to the class of throughput-oriented processors. This means that their performance is achieved in cooperative work of many processor cores and depends not only on the computational speed of cores but also on the throughput of the memory subsystem. The latter is determined by characteristics of the cache memory hierarchy, number and width of integrated memory controllers, memory access speed and capacity of inter-core or interprocessor communications. Additionally some memory optimization features are presented in the processor such as streamlined prefetch of regularly accessed data, fast access to unaligned data and facilities for efficient indirect accesses (gather, scatter) which are needed for processing sparse matrices.

Another main feature of throughput-oriented processors is vectorization: several elements of data can be packed in a vector and processed simultaneously by a single machine instruction. This feature is supported by the smart vectorizing compilers which can be controlled by auxiliary directives in a source code.

Finally, shared memory organization with coherent caches is needed for parallelization, together with the appropriate software support (OpenMP compilers and other parallel environments).

There are two principal classes of throughput-oriented processors – multicore and manycore. Multicore means just a presence of several traditional processor cores in a single semiconductor chip, with the typical number of cores up to 8–12, clock frequency around 3 GHz and standard integrated memory interfaces (up to 4 64-bit channels of DDR3 or DDR4). Typical vector width is 128 or 256 bits.

Manycore (MIC) is a new class of processors with large number of simple and relatively slow cores. Each core is equipped with a very powerful floating point unit (FPU) that processes 512-bit vectors organized as 8×64-bit or 16×32-bit words. The only current implementation of MIC is Intel Xeon Phi, that have the following characteristics (for the model 5110P): 60 cores, up to 4 threads per core (240 threads total), frequency 1.05 GHz, 8 channels of GDDR5 memory. Thus manycore processors are also throughput-oriented, they possess all necessary properties: parallelization, vectorization and memory optimization.

However, manycore processors differ from their multicore counterparts in the balance between components of performance: they need much more threads to saturate a processor (240 vs. 8–12), rely on slower scalar performance of a single thread (speed ratio about 1:10), benefit from regular vectorized floating point operations (512-bit vectors vs. 256-bit) and from very powerful memory subsystem (150–20 GB/s vs. 40–50 GB/s). As a results they are able to demonstrate the level of performance about 1.5 times higher than bi-processor systems built on multicore processors when running realistic memory-bound applications.

Manycore is often considered just as a superfast arithmetic engine (like GPU). However, it is conceptually a universal parallel processor and, compared to GPU, its performance potential is much wider, as well as the range of applications.

Both multicore and manycore processors need special optimizations of application programs in order to achieve high performance, such as contiguous data placement, avoiding very sparse structures etc. It is a good programming practice if the same source code is developed for both types of processors.

The most important points of these optimizations is a storage format for sparse matrices. Usually, the Compressed Row Storage (CRS) is used, when non-zero elements of a matrix are stored contiguously row-by-row, being accompanied with their column indexes. In this case, addressing elements of a vector by which this matrix is being multiplied is performed indirectly. However, a set of vector elements being processed for a given row can be located very sparsely: for example, in the discretization of a regular 3D domain of the size n^3, the distance between elements will be $O(n^2)$. Such non-regular and non-contiguous data accesses are very non-optimal for modern processors, especially if they are performed using vectorized forms of data load operations (gather).

A good alternative is the Compressed Diagonal Storage (CDS). This format is applicable for more-or-less regular discretizations with a stencil of the limited size – in particular, for Cartesian discretizations in arbitrary domains (e.g. in the Level set method). In this format, a matrix is considered as consisting of the limited number of diagonals (more exactly, quasi-diagonals), and matrix elements are stored contiguously within each diagonal, being accompanied with their column indexes. Each diagonal must contain an element for each row even if it is zero. On the other hand, corresponding vector elements being addressed by the indexes are located almost densely. As a result, vectorized forms of data load operations can be employed very efficiently.

Both storage schemes were evaluated using the Jacobi-preconditioned Conjugate Gradient solvers for two matrix sizes. For the smaller matrix (0.32 M grid points), the performance gain of the CDS scheme was about 1.25 and 1.4 on multicore processors (for the FPU vector size 128 bits and 256 bits, respectively) and about 1.8 on a manycore (vector size 512 bits). For the large matrix (4.8 M grid points), the full memory throughput rate of 80 GB/s was achieved on the tested bi-processor multicore system (Sandy Bridge). On the manycore processor (model 3120P), the achieved level was 120 GB/s that is also close to the limit. Thus, these tests demonstrated the importance of adequate data structures and access patterns for modern high-performance processors.

4   Description of the Algebraic Multigrid

Algebraic multigrid (AMG) is based on matrix coefficients rather than on the geometry of a computational domain. Owing to the strict mathematical foundation, this method works fine and demonstrates excellent convergence on solving ill-conditioned elliptic linear systems. For regular grids, however, the Algebraic multigrid has the natural geometric interpretation.

Here we consider implementations of the AMG for Cartesian discretizations in arbitrary domains. These discretizations produce unified 7-point stencils and, as a result, it becomes possible to use more simple and regular data structures.

An iteration of the multigrid is usually represented as a V-cycle (Fig. 3). Within this cycle, a smoothing procedure is performed on a hierarchy of discretizetions thus reducing high-frequency and low-frequency components of the residual vector.

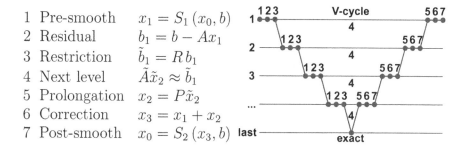

1	Pre-smooth	$x_1 = S_1\left(x_0, b\right)$
2	Residual	$b_1 = b - Ax_1$
3	Restriction	$\tilde{b}_1 = R\,b_1$
4	Next level	$\tilde{A}\tilde{x}_2 \approx \tilde{b}_1$
5	Prolongation	$x_2 = P\tilde{x}_2$
6	Correction	$x_3 = x_1 + x_2$
7	Post-smooth	$x_0 = S_2\left(x_3, b\right)$

Fig. 3. Multigrid algorithm (left) and illustration of V-cycle (right)

Smoothing is the most important part of the algorithm. It is first performed in the beginning of each level reducing corresponding error components. Then the residual vector is restricted (averaged) into the more coarse grid, and the algorithm is recursively executed at the next level. After that, the new coarse residual vector is prolongated (interpolated) into the current fine grid, the result is corrected, and the smoothing procedure is performed again. At the last level, the coarsest grid equation is solved either exactly or by a simple iterative procedure (not necessarily accurate).

At the initialization phase of the algorithm, a hierarchical sequence of grids should be built, together with intergrid transfer operators R and P. Also, matrices for all grid levels should be constructed.

A procedure of building the next (coarse) grid from the current (fine) grid is called coarsening. At the particular level, a subset of variables (grid nodes) is selected for the next level. They are called Coarse nodes (C-nodes), while the remaining ones are Fine nodes (F-nodes). Different coarsening algorithms exist, more or less aggressive. Here, the natural geometric coarsening is applied, when the number of grid points is reduced by 8 at each level in 3D (Fig. 4).

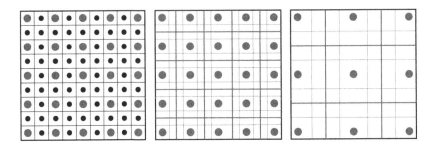

Fig. 4. Fine grid and two levels of coarsening (left to right)

F-nodes can be connected to C-nodes either directly (strong connection) or indirectly (weak connection). These connections are used for building intergrid transfer operators, firstly the prolongation (interpolation) operator P (Fig. 3).

In the AMG, operator-dependent interpolations are used, which are based on matrix coefficients. For strongly connected F-nodes, interpolations are performed in accordance with coefficient weights in the connections to adjacent C-nodes (direct interpolation). For weakly connected nodes, intermediate weights in the connections to adjacent strong-F-nodes are used (standard interpolation). For more weakly connected F-nodes, an additional interpolation step is needed [4].

The restriction (averaging) operator is constructed as a transpose of the prolongation operator: $R = P^T$.

The next step is to build a matrix for the coarser level from the current one by means of the Galerkin operator (Fig. 5). It is related to the restriction (R) and prolongation (P) operators. The Galerkin operator retains the symmetry and some other properties of a matrix.

$$\tilde{x} = Rx = P^T x$$
$$\tilde{A} = RAP = P^T AP$$
$$x = P\tilde{x}$$

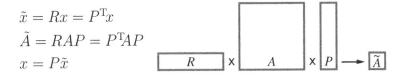

Fig. 5. Building a coarser level matrix by the Galerkin operator

For the above interpolations (in 3D), all constructed matrices have 27 quasi-diagonals, that corresponds to a 27-point discretization stencil at each level except the first one (which has a 7-point stencil for orthogonal grids). Despite this, the first level is the most computationally expensive because of the larger number of grid points. The second level with 8 times less nodes is still expensive because of 4 times larger stencil. Therefore, algorithms for the first two levels should be well optimized while the remaining levels are less important for performance.

Within an AMG iteration, pre- and post-smoothers are the most important parts of the algorithm both for efficiency and convergence. In the current implementation, Red-black Gauss-Seidel or SOR smoothers are used at the first level and 8-color Gauss-Seidel at others. Several iterations of the smoothing algorithm can be applied. In particular, the typical number of iterations is $1\frac{1}{2}$ (3 half-sweeps) or 2 for the first two levels, and up to 3 or 4 for others.

At the first level, the Compressed Diagonal storage scheme is used for the main matrix. The matrix is stored in two parts to separate red and black nodes. This is necessary because at each step of the algorithm, only elements of a particular color are processed, and storing them in the natural order (when red and black elements alternate with each other) would lead to the twofold increase of data being read from the memory.

For the next levels, matrix coefficients are stored densely with 27 elements in a row, being accompanied with their column indexes in another dense array.

For storing the restriction operator R, a similar scheme is used at all levels because this operator also corresponds to a 27-point stencil. On the other hand, stencils of the prolongation operator P are not uniform and may have different number of points, depending on the class of a grid node (C, strong-F, weak-F, weak-weak-F). For this reason, several sparse structures are organized to assist the interpolation.

Additionally, some auxiliary data array are build for storing different characteristics of grid nodes (color, interpolation class, intergrid references etc.).

The above optimizations of sparse data structures allowed to eliminate difficulties in parallelization and vectorization of the algorithm and increase its computational speed.

For the last grid level, a simple iterative solver of the Conjugate Gradient class can be applied. This solver doesn't need to be accurate, it is usually enough to reduce the residual norm by only two orders of magnitude. The solver works fine on multicore computer systems with a moderate number of threads. However, on manycore processors there is too few computational work in each thread (for the typical last-level matrix size about 500). At the same time, several barrier synchronizations for all threads are needed at each iteration of the CG algorithm that leads to large delays and increases the computational time.

In order to avoid this problem, a variant of the solver based on the matrix inversion was developed. In this approach, the original last-level sparse matrix is explicitly inverted by the Gauss-Jordan algorithm at the initialization phase. The resulting full (dense) matrix is used at the execution phase in the very simple algorithm of matrix-by-vector multiplication. This algorithm is perfectly parallelized and doesn't need synchronizations. The only requirement is that the original last-level matrix should not be large, i.e. the sufficient number of multigrid levels should be defined. This algorithm works well on both manycore and multicore processors. Another point is that it is direct and therefore more robust in comparison with iterative CG solvers.

The described multigrid solver for non-symmetric linear systems was implemented in Fortran with OpenMP parallelization. The same source code works both on multicore and manycore processors, demonstrating good parallelization efficiency. The solver has many parameters for fine tuning the AMG algorithm.

5   Multigrid as a Preconditioner

Another approach is to use the multigrid idea for preconditioning in a solver of the Conjugate Gradient class. In this approach, the meaning of a multigrid iteration (V-cycle) is different. While in the plain multigrid V-cycles are repeated until convergence, gradually reducing the residual norm, in the multigrid-preconditioned CG a V-cycle is applied once in each CG iteration in order to reduce the condition number of a matrix.

Surprisingly, the multigrid-preconditioned Conjugate Gradient method often behaves substantially better than the plain multigrid: it is more robust and converges faster.

Additionally, it becomes possible to use the single-precision arithmetic for the multigrid part of the algorithm instead of the traditional double-precision without loosing the overall accuracy. Due to this, the computational cost of the algorithm can be decreased because of reduced sizes of arrays with floating point data and corresponding reduction of the memory traffic.

As a result, the AMG-preconditioned CG solver becomes faster than the plain AMG in a single iteration and at the same time needs less iterations to converge. Currently the AMG CG solver is implemented for symmetric matrices only. Later it will be extended for the nonsymmetric case as an AMG-preconditioned BiCGStab. Typical iteration counts are shown below for the CG AMG solver vs. the plain AMG for several test matrices:

- small spherical domain (0.32 M grid points): 7 vs. 10,
- large spherical domain (2.3 M grid points): 8 vs. 11,
- long cylindrical domain, aspect ratio 5:1 (2 M grid points): 9 vs. 14.

For the last problem, the CG AMG solver is about 1.7 times faster than the plain AMG.

In order to achieve better convergence, some tuning of the algorithm is necessary, such as defining the number of grid levels, the number of iterations (halfsweeps) for smoothing at each level and the value of the over-relaxation factor at the first level. In particular, the SOR factor around 1.15 is usually optimal for convergence.

6   Performance of the Multigrid Solvers

Performance results of the new AMG-preconditioned Conjugate Gradient solver in comparison with the Jacobi-preconditioned (explicit) Conjugate Gradient are shown in Table 2. The following computer systems were used for these tests:

- Bi-processor Xeon E5-2690v2, 3 GHz, 2×10 cores, 20 threads;
- Xeon Phi model 5110P, 1.05 GHz, 59 cores, 236 threads.

Table 2. Performance results for multicore and manycore computer systems

Computer system	CG Jacobi	CG AMG	AMG : CG
multicore bi-Xeon	4.87 s	0.172 s	28 : 1
manycore Xeon Phi	3.15 s	0.185 s	17 : 1
manycore : multicore	1.55 : 1	0.93 : 1	—

The test matrix represents a discretization in a long cylindrical domain with the aspect ratio 5:1 (2 M grid points). Iteration counts are 9 for the CG AMG and 690 for the CG Jacobi (this corresponds to 1380 iterations for the simple diagonally-preconditioned CG [3]).

The above results demonstrate the great superiority of the AMG-preconditioned method over the plain Conjugate Gradient. The important reason is that this particular matrix represents a long domain with the number of grid points about 400 in the longest direction, that determines the iteration count for explicit CG algorithms. This example also confirms that the multigrid iteration count doesn't depend on the problem size.

On shorter domains, superiority of the multigrid is a little bit less. In particular, for a large spherical domain (2.3 M grid points), iteration counts are 8 and 350 respectively, and AMG : CG speed ratio is about 16:1 for the multicore Xeon processor and about 10:1 for the manycore Xeon Phi. These proportions are expected to be typical for most matrices of the similar size.

The comparison of the manycore Xeon Phi to the multicore bi-Xeon system demonstrates its moderate superiority in the simple explicit CG which is a pure memory-bound algorithm. The speed proportion 1.55:1 in favor of Xeon Phi roughly corresponds to the ratio between achievable memory throughput rates for these systems. The multigrid algorithm, in turn, is much more complicated and less regular, especially in memory access patterns and at higher (coarser) grid levels. For this reason, the relative performance of Xeon Phi becomes less. Nevertheless this result should be considered as very reasonable taking into account the highly parallel nature of this processor.

7   Conclusion

In this work we have presented new efficient parallel solvers for Cartesian discretizations in general domains. Two variants of the solvers are built, based on the Algebraic multigrid approach (AMG) and on the Conjugate Gradient method with the AMG preconditioning (CG AMG). Both solvers use the advanced Compressed Diagonal storage format suitable for efficient processing on modern throughput-oriented computer systems. The solvers are targeted on the solution of Poisson-like and other ill-conditioned linear systems, both symmetric and non-symmetric.

The new solvers have been evaluated and tested on multicore (bi-Xeon) and manycore (Xeon Phi) processors using several matrices of different size.

In comparison to the Conjugate Gradient class solvers with explicit preconditioning, they have demonstrated the speed increase up to 10–16 times typically and up to 17–28 for elongated domains.

The obtained results have also demonstrated that manycore processors can be efficiently employed for solving algebraic problems of the general class using standard programming languages and parallel environments (Fortran, OpenMP).

Acknowledgements. This work was supported by the Russian Foundation for Basic Research (project RFBR-15-01-06363) and by the Institute of mathematics (IMATH) of the University of Toulon. Computations have been performed at the BULL's Computing Center, IMATH and Mésocentre of the University of Aix-Marseille, France.

References

1. Shewchuk, J.R.: An Introduction to the Conjugate Gradient Method without the Agonizing Pain. Carnegie Mellon University, School of Computer Science, Pittsburgh (1994)
2. Accary, G., Bessonov, O., Fougère, D., Gavrilov, K., Meradji, S., Morvan, D.: Efficient parallelization of the preconditioned conjugate gradient method. In: Malyshkin, V. (ed.) PaCT 2009. LNCS, vol. 5698, pp. 60–72. Springer, Heidelberg (2009)
3. Bessonov, O.: Parallelization properties of preconditioners for the conjugate gradient methods. In: Malyshkin, V. (ed.) PaCT 2013. LNCS, vol. 7979, pp. 26–36. Springer, Heidelberg (2013)
4. Stüben, K.: A review of algebraic multigrid. J. Comput. Appl. Math. **128**, 281–309 (2001)

Hierarchical Optimization of MPI Reduce Algorithms

Khalid Hasanov$^{(\boxtimes)}$ and Alexey Lastovetsky

University College Dublin, Belfield, Dublin 4, Ireland
khalid.hasanov@ucdconnect.ie, Alexey.Lastovetsky@ucd.ie

Abstract. Optimization of MPI collective communication operations has been an active research topic since the advent of MPI in 1990s. Many general and architecture-specific collective algorithms have been proposed and implemented in the state-of-the-art MPI implementations. Hierarchical topology-oblivious transformation of existing communication algorithms has been recently proposed as a new promising approach to optimization of MPI collective communication algorithms and MPI-based applications. This approach has been successfully applied to the most popular parallel matrix multiplication algorithm, SUMMA, and the state-of-the-art MPI broadcast algorithms, demonstrating significant multi-fold performance gains, especially for large-scale HPC systems. In this paper, we apply this approach to optimization of the MPI reduce operation. Theoretical analysis and experimental results on a cluster of Grid'5000 platform are presented.

Keywords: MPI · Reduce · Grid'5000 · Communication · Hierarchy

1 Introduction

Reduce is important and commonly used collective operation in the Message Passing Interface (MPI) [1]. A five-year profiling study [2] demonstrates that MPI reduction operations are the most used collective operations. In the reduce operation each node i owns a vector x_i of n elements. After completion of the operation all the vectors are reduced element-wise to a single n-element vector which is owned by a specified root process.

Optimization of MPI collective operations has been an active research topic since the advent of MPI in 1990s. Many general and architecture-specific collective algorithms have been proposed and implemented in the state-of-the-art MPI implementations. Hierarchical topology-oblivious transformation of existing communication algorithms has been recently proposed as a new promising approach to optimization of MPI collective communication algorithms and MPI-based applications [3,5]. This approach has been successfully applied to the most popular parallel matrix multiplication algorithm, SUMMA [4], and the state-of-the-art MPI broadcast algorithms, demonstrating significant multi-fold performance gains, especially on large-scale HPC systems. In this paper, we apply this approach to optimization of the MPI reduce operation.

© Springer International Publishing Switzerland 2015
V. Malyshkin (Ed.): PaCT 2015, LNCS 9251, pp. 21–34, 2015.
DOI: 10.1007/978-3-319-21909-7_3

1.1 Contributions

We propose a hierarchical optimization of legacy MPI reduce algorithms without redesigning them. The approach is simple and general, allowing for application of the proposed optimization to any existing reduce algorithm. As by design the original algorithm is a particular case of its hierarchically transformed counterpart, the performance of the algorithm will either improve or stay the same in the worst case scenario. Theoretical study of the hierarchical transformation of six reduce algorithms, which are implemented in Open MPI [7], is presented. The theoretical results have been experimentally validated on a widely used Grid'5000 [8] infrastructure.

1.2 Outline

The rest of the paper is structured as follows. Section 2 discusses related work. The hierarchical optimization of MPI reduce algorithms is introduced in Sect. 3. The experimental results are presented in Sect. 4. Finally, Sect. 5 concludes the presented work and discusses future directions.

2 Related Work

In the early 1990s, the CCL library [9] implemented collective reduce operation as an inverse broadcast operation. Later collective algorithms for wide-area clusters were proposed [10], and automatic tuning for a given system by conducting a series of experiments on the system was discussed [11]. Design and high-performance implementation of collective communication operations and commonly used algorithms, such as minimum-spanning tree reduce algorithm, are discussed in [12]. Five reduction algorithms optimized for different message sizes and number of processes are proposed in [13]. Implementations of MPI collectives, including reduce, in MPICH [15] are discussed in [16]. Algorithms for MPI broadcast, reduce and scatter, where the communication happens concurrently over two binary trees, are presented in [14]. Cheetah framework [17] implements MPI reduction operations in a hierarchical way on multicore systems, which supports multiple communication mechanisms. Unlike that work, our optimization is topology-oblivious, and MPI reduce optimizations in this work do not design new algorithms from scratch, employing the existing reduce algorithms underneath. Therefore, our hierarchical design can be built on top of the algorithms from the Cheetah framework as well. This work focuses on reduce algorithms implemented in Open MPI such as flat, linear/chain, pipeline, binary, binomial and in-order binary tree algorithms.

   We extend our previous studies on parallel matrix multiplication [3] and topology-oblivious optimization of MPI broadcast algorithms on large-scale distributed memory platforms [5,6] to MPI reduce algorithms.

2.1   MPI Reduce Algorithms

We assume that the time to send a message of size m between any two MPI processes is modeled with Hockney model [18] as $\alpha + m \times \beta$, where α is the latency per message and β is the reciprocal bandwidth per byte. It is also assumed that the computation cost per byte in the reduction operation is γ on any MPI process. Unless otherwise noted, in the rest of the paper we will call MPI process just process.

– Flat tree reduce algorithm.
  In this algorithm, the root process sequentially receives and reduces a message of size m from all the processes participating in the reduce operation in $p-1$ steps:

$$(p-1) \times (\alpha + m \times \beta + m \times \gamma). \tag{1}$$

  In a segmented flat tree algorithm, a message of size m is split into X segments, in which case the number of steps is $X \times (p-1)$. Thus, the total execution time will be as follows:

$$X \times (p-1) \times \left(\alpha + \frac{m}{X} \times \beta + \frac{m}{X} \times \gamma\right). \tag{2}$$

– Linear tree reduce algorithm.
  Unlike the flat tree, here each process receives or sends at most one message. Theoretically, its cost is the same as the flat tree algorithm:

$$(p-1) \times (\alpha + m \times \beta + m \times \gamma). \tag{3}$$

– Pipeline reduce algorithm.
  It is assumed that a message of size m is split into X segments and in one step of the algorithm a segment of size $\frac{m}{X}$ is reduced between p processes. If we assume a logically reverse ordered linear array, in the first step of the algorithm the first segment of the message is sent to the next process in the array. Next, while the second process sends the first segment to the third process, the first process sends the second segment to the second process, and the algorithm continues in this way. The first segment takes $p-1$ and the remaining segments take $X-1$ steps to reach the end of the array. If we also consider the computation cost in each step, then overall execution time of the algorithm will be as follows:

$$(p + X - 2) \times \left(\alpha + \frac{m}{X} \times \beta + \frac{m}{X} \times \gamma\right). \tag{4}$$

– Binary tree reduce algorithms.
  If we take a full and complete binary tree of height h, its number of nodes will be $2^{h+1} - 1$. In the reduce operation, a node at the hight h will receive two messages from its children at the height $h+1$. In addition, if we segment a message of size m into X segments, the overall run time will be as follows:

$$2 \left(\log_2 (p+1) + X - 2\right) \times \left(\alpha + \frac{m}{X} \times \beta + \frac{m}{X} \times \gamma\right). \tag{5}$$

Open MPI uses the in-order binary tree algorithm for non-commutative operations. It works similarly to the binary tree algorithm but enforces order in the operations.

- Binomial tree reduce algorithm.
  The binomial tree algorithm takes $\log_2(p)$ steps and the message communicated at each step is m. If the message is divided into X segments, then the number of steps and the message communicated at each step will be $X \times \log_2(p)$ and $\frac{m}{X}$ respectively. Therefore, the overall run time will be as follows:

$$\log_2 (p) \times (\alpha + m \times \beta + m \times \gamma). \tag{6}$$

- Rabenseifner's reduce algorithm.
  The Rabenseifner's algorithm [13] is designed for large message sizes. The algorithm consists of reduce-scatter and gather phases. It has been implemented in MPICH [16] and used for message sizes greater than 2 KB. The reduce-scatter phase is implemented with recursive-halving, and the gather phase is implemented with binomial tree. Therefore, the run time of the algorithm is the sum of these two phases:

$$2 \log_2 (p) \times \alpha + 2 \frac{p-1}{p} \times m \times \beta + \frac{p-1}{p} \times m \times \gamma. \tag{7}$$

The algorithm can be further optimized by recursive vector halving, recursive distance doubling, recursive distance halving, binary blocks, and ring algorithms for non-power-of-two number of processes. An interested reader can consult [13] for more detailed discussion of those algorithms.

3   Hierarchical Optimization of MPI Reduce Algorithms

This section introduces a topology-oblivious optimization of MPI reduce algorithms. The idea is inspired by our previous study on the optimization of the communication cost of parallel matrix multiplication [3] and MPI broadcast [5] on large-scale distributed memory platforms.

The proposed optimization technique is based on the arrangement of the p processes participating in the reduce into logical groups. For simplicity, it is assumed that the number of groups divides the number of MPI processes and can change between one and p. Let G be the number of groups. Then there will be $\frac{p}{G}$ MPI processes per group. Figure 1 shows an arrangement of 8 processes in the original MPI reduce operation, and Fig. 2 shows the arrangement in a hierarchical reduce operation with 2 groups of 4 processes. The hierarchical optimization has two phases: in the first phase, a group leader is selected for each group and the leaders start reduce operation inside their own group in parallel (in this example between 4 processes). In the next phase, the reduce is performed between the group leaders (in this example between 2 processes). The grouping can be done by taking the topology into account as well. However, in this work the grouping is topology-oblivious and the first process in each group is selected as the group leader. In general, different algorithms can be used for reduce operations between group leaders and within each group. This work focuses on the case where the same algorithm is employed at both levels of hierarchy. Algorithm 1 shows the pseudocode of the hierarchically transformed MPI reduce operation. Line 4 calculates the root for the reduce between the groups. Then line 5 creates a sub-communicator of G processes between the groups, and line 6 creates a sub-communicator of $\frac{p}{G}$ processes inside the groups. Our implementation uses the MPI_Comm_split MPI routine to create new sub-communicators.

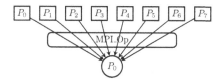

Fig. 1. Logical arrangement of processes in MPI reduce.

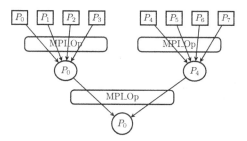

Fig. 2. Logical arrangement of processes in hierarchical MPI reduce.

3.1 Hierarchical Transformation of Flat Tree Reduce Algorithm

After the hierarchical transformation, there will be two steps of the reduce operation: inside the groups and between the groups. The reduce operations inside the groups happen between $\frac{p}{G}$ processes in parallel. Then, the operation continues between G groups. The cost of the reduce operations inside groups and between groups will be $(G-1)\times(\alpha + m\times\beta + m\times\gamma)$ and $(\frac{p}{G} - 1)\times(\alpha + m\times\beta + m\times\gamma)$ respectively. Thus, the overall run time can be seen as a function of G:

$$F(G) = \left(G + \frac{p}{G} - 2\right) \times (\alpha + m\times\beta + m\times\gamma) \tag{8}$$

The derivative of the function is $(1 - \frac{p}{G^2})\times(\alpha + m\times\beta + m\times\gamma)$, it can be shown that $p = \sqrt{G}$ is the minimum point of the function in the interval $(1, p)$. Then the optimal value of the function will be as follows:

$$F(\sqrt{p}) = (2\sqrt{p} - 2) \times (\alpha + m\times\beta + m\times\gamma) \tag{9}$$

Algorithm 1. Hierarchical optimization of MPI reduce operation.

Data: p - Number of processes
Data: G - Number of groups
Data: $sendbuf$ - Send buffer
Data: $recvbuf$ - Receive buffer
Data: $count$ - Number of entries in send buffer (integer)
Data: $datatype$ - Data type of elements in send buffer
Data: op - MPI reduce operation handle
Data: $root$ - Rank of reduce root
Data: $comm$ - MPI communicator handle
Result: The root process has the reduced message
begin

```
1    MPI_Comm comm_outer      /* communicator between the groups */
2    MPI_Comm comm_inner      /* communicator inside the groups */
3    int root_outer           /* root of reduce between the groups */
4    root_outer = Calculate_Root_Outer(G, p, root, comm)
5    comm_outer = Create_Comm_Between_Groups(G, p, root_outer, comm)
6    comm_inner = Create_Comm_Inside_Groups(G, p, root, comm)
7    MPI_Reduce(sendbuf, recvbuf, count, datatype, op, root, comm_inner)
8    MPI_Reduce(sendbuf, recvbuf, count, datatype, op, root_outer, comm_outer)
```

3.2 Hierarchical Transformation of Pipeline Reduce Algorithm

If we sum the costs of reduce inside and between groups with pipeline algorithm, the overall run time will be as follows:

$$F(G) = \left(2X + G + \frac{p}{G} - 4\right) \times \left(\alpha + \frac{m}{X} \times \beta + \frac{m}{X} \times \gamma\right) \qquad (10)$$

In the same way, it can be easily shown that the optimal value of the cost function is as follows:

$$F(\sqrt{p}) = (2X + 2\sqrt{p} - 4) \times \left(\alpha + \frac{m}{X} \times \beta + \frac{m}{X} \times \gamma\right) \qquad (11)$$

3.3 Hierarchical Transformation of Binary Reduce Algorithm

For simplicity, we will take $p + 1 \approx p$ in the formula 5. Then the cost of the reduce operations between the groups and inside the groups will be as follows respectively: $2\log_2(G) \times (\alpha + m \times \beta + m\gamma)$ and $2\log_2(\frac{p}{G}) \times (\alpha + m \times \beta + m\gamma)$. If we add these two terms, the overall cost of the hierarchical transformation of the binary tree algorithm will be equal to the cost of the original algorithm.

3.4 Hierarchical Transformation of Binomial Reduce Algorithm

Similarly to the binary reduce algorithm, the cost function of the binomial tree will not change after hierarchical transformation.

3.5   Hierarchical Transformation of Rabenseifner's Reduce Algorithm

By applying the formula 7 between the groups with G processes and inside the groups with $\frac{p}{G}$ processes, we can find the run time of hierarchical transformation of Rabenseifner's algorithm. Unlike the previous algorithms, now the theoretical cost increases in comparison to the original Rabenseifner's algorithm. Therefore, theoretically the hierarchical reduce implementation should use the number of groups equals to one, in which case the hierarchical algorithm retreats to the original algorithm.

$$2\log_2(p)\times\alpha + 2m\times\beta\times\left(2 - \frac{G}{p} - \frac{1}{G}\right) + m\gamma\left(2 - \frac{G}{p} - \frac{1}{G}\right) \qquad (12)$$

3.6   Possible Overheads in the Hierarchical Design

Our implementation of the hierarchical reduce operation uses MPI_Comm_split operation to create groups of processes. The obvious questions would be to which extent the split operation can affect the scalability of the hierarchical algorithms. Recent research works show different approaches to improve the scalability of MPI communicator creation operations in terms of run time and memory footprint. The research in [20] introduces a new MPI_Comm_split algorithm, which scales well to millions of cores. The memory usage of the algorithm is $O(\frac{p}{g})$ and the time is $O(g\log_2(p) + \log_2^2(p) + \frac{p}{g}\log_2(g))$, where p is the number of MPI processes, g is the number of processes in the group that perform sorting. More recent research work in [21] improves the previous algorithm with two variants. The first one, which uses a bitonic sort, needs $O(\log_2(p))$ memory and $O(\log_2^2(p))$ time. The second one is a hash-based algorithm and requires $O(1)$ memory and $O(\log_2(p))$ time. Having these algorithms, we can utilize MPI_Comm_split operation in our hierarchical design with negligible overhead of creating MPI subcommunicators. There will not be any overhead at all for large messages as the split operation does not depend on the message size.

4   Experiments

The experiments were carried out on the Grid'5000 infrastructure in France. The platform consists of 24 clusters distributed over 9 sites in France and one in Luxembourg which includes 1006 nodes, 8014 cores. Almost all the sites are interconnected by 10 Gb/s high-speed network. We used the Graphene cluster from Nancy site of the infrastructure as our main testbed. The cluster is equipped with 144 nodes and each node has a disk of 320 GB storage, 16 GB of memory and 4-cores of CPU Intel Xeon X3440. The nodes in the Graphene cluster interconnected via 20 Gb/s Infiniband and Gigabyte Ethernet. More comprehensive information about the platform can be found on the Grid'5000 web site (http://www.grid5000.fr).

The experiments have been done with Open MPI 1.4.5, which provides a few reduce implementations. Among those implementations there are several reduce algorithms such as linear, chain, pipeline, binary, binomial, and in-order binary algorithms and platform/architecture specific algorithms, some of which are reduce algorithms for Infiniband networks, and the Cheetah framework for multicore architectures. In this work, we do not consider the platform specific reduce implementations. We used the same approach as described in MPIBlib [19] to benchmark our experiments. During the experiments, the mentioned reduce algorithms were selected by using Open MPI MCA (Modular Component Architecture) coll_tuned_use_dynamic_rules and coll_tuned_reduce_algorithm parameters. MPI_MAX operation has been used in the experiments. We have used Graphene cluster with two experimental settings, one process per core and one process per node with the Infiniband-20G network. A power-of-two number of processes have been used in the experiments.

4.1   Experiments: One Process per Core

The nodes in the Graphene cluster are organized into four groups and connected to four switches. The switches in turn are connected to the main Nancy router. We have used 10 patterns of process to core mappings, but we will show experimental results only with one such mappings where the processes are grouped by their rank in increasing order. The measurements with different groupings showed similar performance.

The theoretical and experimental results showed that the hierarchical approach mainly improves the algorithms which assume flat arrangements of the processes, such as linear, chain and pipeline. On the other hand the native Open MPI reduce operation selects different algorithms depending on the message size, the count and the number of processes sent to the MPI_Reduce function. This means the hierarchical transformation can improve the native reduce operation as well. The algorithms used in the Open MPI decision function are linear, chain, binomial, binary/in-order binary and pipeline reduce algorithms which can be used with different sizes of segmented messages.

Figure 3 shows experiments with default Open MPI reduce operation with a message of size 16 KB where the best performance is achieved when the group size is 1 or p, in which case the hierarchical reduce obviously turns into the original non-hierarchical reduce. Here for different numbers of groups the Open MPI decision function selected different reduce algorithms. Namely, if the number of groups is 8 or 64 then Open MPI selects the binary tree reduce algorithm between the groups and inside the groups respectively. In all other cases the binomial tree reduce algorithm is used. Figure 4 shows similar measurements with a message of size 16 MB where one can see a huge performance improvement up to 30 times. This improvement does not come solely from the hierarchical optimization itself, but also because of the number of groups in the hierarchical reduce resulted in Open MPI decision function to select the pipeline reduce algorithm with different segment sizes for each groups. The selection of the algorithms for different number of groups is described in Table 1.

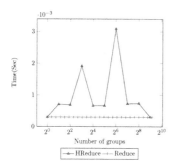

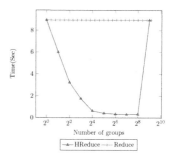

Fig. 3. Hierarchical native reduce. m = 16 KB and p = 512.

Fig. 4. Hierarchical native reduce. m = 16 MB and p = 512.

Table 1. Open MPI algorithm selection in HReduce. m = 16 MB, p = 512.

Groups	Inside groups	Between groups
1	-	Pipeline 32 KB
2	Pipeline 32 KB	Pipeline 64 KB
4	Pipeline 32 KB	Pipeline 64 KB
8	Pipeline 32 KB	Pipeline 64 KB
16	Pipeline 64 KB	Pipeline 64 KB
32	Pipeline 64 KB	Pipeline 64 KB
64	Pipeline 64 KB	Pipeline 32 KB
128	Pipeline 64 KB	Pipeline 32 KB
256	Pipeline 64 KB	Pipeline 32 KB

As mentioned in Sect. 3.6, it is expected that the overhead from the MPI_Comm_split operation should affect only reduce operations with smaller message sizes. Figure 5 validates this with experimental results. The hierarchical reduce operation of 1 KB message with the underlying native reduce achieved its best performance when the number of groups was one as the overhead from the split operation itself was higher than the reduce.

It is interesting to study the pipeline algorithm with different segment sizes as it is used for large message sizes in Open MPI. Figure 6 presents experiments with the hierarchical pipeline reduce with a message size of 16 KB with 1 KB segmentation. We selected the segment sizes using coll_tuned_reduce_algorithm_segmentsize parameter provided by MCA. Figures 7 and 8 shows the performance of the pipeline algorithm with segment sizes of 32 KB and 64 KB respectively. In the first case, we see a 26.5 times improvement, while with the 64 KB the improvement is 18.5 times.

Figure 9 demonstrates speedup of the hierarchical transformation of native Open MPI reduce operation, linear, chain, pipeline, binary, binomial, and in-order binary reduce algorithms with message sizes starting from 16 KB up to

16 MB. Except binary, binomial and in-order binary reduce algorithms, there is a significant performance improvement. In the figure, NT is native Open MPI reduce operation, LN is linear, CH is chain, PL is pipeline with 32 KB segmentation, BR is binary, BL is binomial, and IBR denotes in-order binary tree reduce algorithm. We would like to highlight one important point that Fig. 9 does not compare the performance of different Open MPI reduce algorithms, it rather shows the speedup of their hierarchical transformations. Each of these algorithms can be better than the others in some specific settings depending on the message size, number of processes, underlying network and so on. At the same time, the hierarchical transformation of these algorithms will either improve their performance or be equally fast.

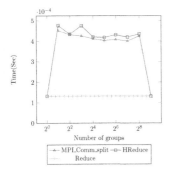

Fig. 5. Time spent on MPI_Comm_split and hierarchical native reduce. m = 1 KB, p = 512.

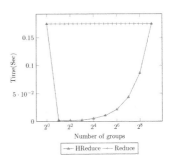

Fig. 6. Hierarchical pipeline reduce. m = 16 KB, segment 1 KB and p = 512.

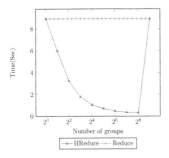

Fig. 7. Hierarchical pipeline reduce. m = 16 MB, segment 32 KB and p = 512.

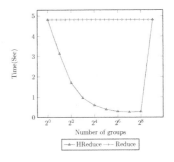

Fig. 8. Hierarchical pipeline reduce. m = 16 MB, segment 64 KB and p = 512.

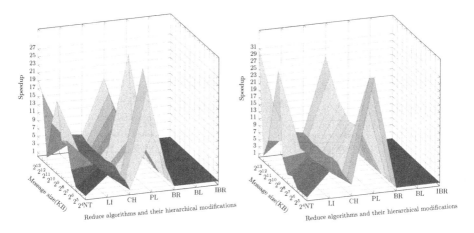

Fig. 9. Speedup on 256 (left) and 512 (right) cores, one process per core.

4.2 Experiments: One Process per Node

The experiments with one process per node showed a similar trend to that of with the one process per core setting. The performance of linear, chain, pipeline and native Open MPI reduce operations can be improved by the hierarchical approach. Figures 10 and 11 show experiments on 128 nodes with message sizes of 16 KB and 16 MB accordingly. In the first setting, the Open MPI decision function uses the binary tree algorithm when the number of processes is 8 between or inside groups, in all other cases the binomial tree is used.

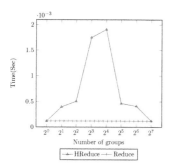

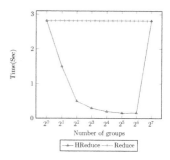

Fig. 10. Hierarchical native reduce. m = 16 KB and p = 128.

Fig. 11. Hierarchical native reduce. m = 16 MB and p = 128.

The pipeline algorithm has similar performance improvement to that of with 512 processes, Fig. 12 shows experiments with a message of size 16 MB segmented by 32 KB and 64 KB sizes. The labels on the x axis has the same meaning as in the previous section.

Figure 13 presents speedup of the hierarchical transformations of all the reduce algorihms from Open MPI "TUNED" component with message sizes

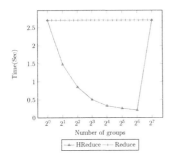
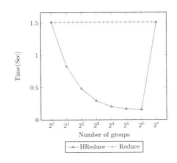

Fig. 12. Hierarchical pipeline reduce. m = 16 MB, segment 32 KB (left) and 64 KB (right). p = 128.

from 16 KB up to 16 MB on 64 (left) and 128 (right) nodes. Again, the reduce algorithms wich has "flat" design and Open MPI default reduce operation have multi-fold performance improvement.

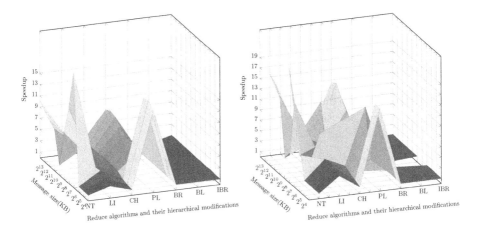

Fig. 13. Speedup on 64(left) and 128(right) cores. 1 process per node.

5   Conclusion

Despite there has been a lot of research in MPI collective communications, this work shows that their performance is far from optimal and there is some room for improvement. Indeed, our simple hierarchical optimization, which transforms existing MPI reduce algorithms into two-level hierarchy, gives significant improvement on small and medium scale platforms. We believe that the idea can be incorporated into Open MPI decision function to improve the performance of reduce algorithms even further. It can also be used as a standalone software on top of MPI based applications.

The key feature of the optimization is that it can never be worse than any other optimized reduce operation. In the worst case, the algorithm can use one group and fall back to the native reduce operation.

As the future work, we plan to investigate if using different reduce algorithms in each phase and different number of processes per group can improve the performance. We would also like to generalize our optimization to other MPI collective operations.

Acknowledgments. This work has emanated from research conducted with the financial support of IRCSET (Irish Research Council for Science, Engineering and Technology) and IBM, grant number EPSPG/2011/188, and Science Foundation Ireland, grant number 08/IN.1/I2054.

The experiments presented in this publication were carried out using the Grid'5000 experimental testbed, being developed under the INRIA ALADDIN development action with support from CNRS, RENATER and several Universities as well as other funding bodies (see https://www.grid5000.fr).

References

1. Message passing interface forum. http://www.mpi-forum.org/
2. Rabenseifner, R.: Automatic MPI counter proling of all users: first results on a CRAY T3E 900–512. Proceedings of the Message Passing Interface Developers and Users Conference **1999**(MPIDC99), 77–85 (1999)
3. Hasanov, K., Quintin, J.N., Lastovetsky, A.: Hierarchical approach to optimization of parallel matrix multiplication on large-scale platforms. J. Supercomputing., 24p. March 2014 (Springer). doi:10.1007/s11227-014-1133-x
4. van de Geijn, R.A., Watts, J.: SUMMA: scalable universal matrix multiplication algorithm. Concurrency: Practice and Experience **9**(4), 255–274 (1997)
5. Hasanov, K., Quintin, J.-N., Lastovetsky, A.: High-level topology-oblivious optimization of mpi broadcast algorithms on extreme-scale platforms. In: Lopes, L., Žilinskas, J., Costan, A., Cascella, R.G., Kecskemeti, G., Jeannot, E., Cannataro, M., Ricci, L., Benkner, S., Petit, S., Scarano, V., Gracia, J., Hunold, S., Scott, S.L., Lankes, S., Lengauer, C., Carretero, J., Breitbart, J., Alexander, M. (eds.) Euro-Par 2014, Part II. LNCS, vol. 8806, pp. 412–424. Springer, Heidelberg (2014)
6. Hasanov, K., Quintin, J.N., Lastovetsky, A.: Topology-oblivious optimization of MPI broadcast algorithms on extreme-scale platforms. Simulation Modelling Practice and Theory. 10p. April 2015. doi:10.1016/j.simpat.2015.03.005
7. Gabriel, E., Fagg, G., Bosilca, G., Angskun, T., Dongarra, J., et al.: Open MPI: goals, concept, and design of a next generation MPI implementation. In: Proceedings of the 11th European PVM/MPI Users Group Meeting (2004)
8. Grid'5000. http://www.grid5000.fr
9. Bala, V., Bruck, J., Cypher, R., Elustondo, P., Ho, C.-T., Ho, C.-T., Kipnis, S., Snir, M.: CCL: a portable and tunable collective communication library for scalable parallel computers. IEEE TPDS **6**(2), 154–164 (1995)
10. Kielmann, T., Hofman, R.F.H., Bal, H.E., Plaat, A., Bhoedjang, R.A.F.: MagPIe MPIs collective communication operations for clustered wide area systems. In: Proceedings of PPoPP99, **34**(8): 131–140 (1999)

11. Vadhiyar, S.S., Fagg, G.E., Dongarra, J.: Automatically tuned collective communications. In: Proceedings of ACM/IEEE Conference on Supercomputing (2000)
12. Chan, E.W., Heimlich, M.F., Purkayastha, A., Van de Geijn, R.A.: On optimizing collective communication. In: Proceedings of IEEE International Conference on Cluster Computing (2004)
13. Rabenseifner, R.: Optimization of collective reduction operations. In: Proceddings of International Conference on Computational Science, June 2004
14. Sanders, P., Speck, J., Tráff, J.L.: Two-tree algorithms for full bandwidth broadcast. Reduct. Scan. Parallel Comput. **35**(12), 581–594 (2009)
15. MPICH-A Portable Implementation of MPI. http://www.mpich.org/
16. Thakur, R., Gropp, W.D.: Improving the performance of collective operations in MPICH. In: Dongarra, J., Laforenza, D., Orlando, S. (eds.) EuroPVM/MPI 2003. LNCS, vol. 2840, pp. 257–267. Springer, Heidelberg (2003)
17. Venkata, M.G., Shamis, P., Sampath, R., Graham, R.L.l, Ladd, J.S.: Optimizing blocking and nonblocking reduction operations for multicore systems: hierarchical design and implementation. In: Proceedings of IEEE Cluster, pp. 1–8 (2013)
18. Hockney, R.W.: The communication challenge for MPP: intel paragon and Meiko CS-2. Parallel Comput. **20**(3), 389–398 (1994)
19. Lastovetsky, A., Rychkov, V., O'Flynn, M.: MPIBlib: benchmarking MPI communications for parallel computing on homogeneous and heterogeneous clusters. In: Lastovetsky, A., Kechadi, T., Dongarra, J. (eds.) EuroPVM/MPI 2008. LNCS, vol. 5205, pp. 227–238. Springer, Heidelberg (2008)
20. Sack, P., Gropp, W.: A scalable MPI_comm_split algorithm for exascale computing. In: Keller, R., Gabriel, E., Resch, M., Dongarra, J. (eds.) EuroMPI 2010. LNCS, vol. 6305, pp. 1–10. Springer, Heidelberg (2010)
21. Moody, A., Ahn, D.H., de Supinski, B.R.: Exascale algorithms for generalized MPI_comm_split. In: Proceedings of the 18th European MPI Users' Group conference on Recent advances in the message passing interface (EuroMPI 2011) (2011)

On Parallel Computational Technologies of Augmented Domain Decomposition Methods

Y.L Gurieva[1] and V.P Il'in[1,2] [✉]

[1] Institute of Computational Mathematics and Mathematical Geophysics SB RAS,
Novosibirsk, Russia
yana@lapasrv.sscc.ru
[2] Novosibirsk State University, Novosibirsk, Russia
ilin@sscc.ru

Abstract. The performance of the parallel domain decomposition methods (DDM) for solving very large systems of linear algebraic equations with non-symmetric sparse matrices depends on the convergence of the iterative algorithms as well as on the efficiency of the computational technologies. Usually in DDM approach the number of iterations grows together with a growth of the degree of freedom. We consider the algorithms for increasing the convergence rate based on the preconditioning with using deflation and aggregation techniques which take low rank approximations of the original systems of linear algebraic equations. The efficiency of the proposed approaches is demonstrated on the representative set of model tasks.

1 Introduction

In general, the modern domain decomposition methods to solve very large systems of linear equations (SLAEs), which arise in the discrete approximation on the non-structured meshes of the multi-dimensional boundary value problems (BVPs) by the finite element or by other grid methods, can be presented by three main mathematical approaches: external Krylov's type iterative process "on subdomains" which presents the additive Schwarz (or special block Jacobi) algorithm, simultaneous solving the auxiliary BVPs in the subdomains which can be carried out by a direct or an iterative algorithm, and preconditioning procedures to accelerate the external iterations, see [1–5] and the literature cited there.

The last factor is very important for strongly scalable parallelized tasks, because for a very large number of subdomains and corresponding block degree of freedom, one can observe the considerable stagnation of the iterative process. In recent decades, various versions of aggregation, deflation, and coarse grid correction accelerators have been investigated and applied successfully by many authors. The main goal of our paper consists namely in the numerical analysis of

The work is supported partially by Russian Science Foundation grant N 14-11-00485.
The experimental part of the paper is supported by the RFBR grant N 14-07-00128.

© Springer International Publishing Switzerland 2015
V. Malyshkin (Ed.): PaCT 2015, LNCS 9251, pp. 35–46, 2015.
DOI: 10.1007/978-3-319-21909-7_4

several versions of aggregation accelerating based on low rank matrix approximations in different coarse subspaces. The program implemtation of the algorithms
is realized for the universal compressed sparse matrix format which is necessary
to solve the practical problems.

The conventional parallel technologies of DDMs include two levels: application of MPI processes for corresponding subdomains, including interface communications between them at each outer (external) iteration, and implementation of
the multi-thread computing for the "internal" parallelization on the multi-core
processors.

The problems are specified by three levels of the degrees of freedom: the
number of unknowns of SLAE ($10^8 - 10^{11}$), the quantity of subdomains ($10^2 - 10^5$,
block dimension of the broblem), and coarse grid dimension ($10 - 10^3$) which
determine the scalability of the parallelism of the general computational process.

The bottleneck of DDM approach is in a minimization of a communication time. It can be done by simultaneous data transfer and synchronized computations in the subdomains. In general, DDM performance depends on the
convergence properties of the iterative algorithms and on the efficiency of the
computational technologies whose variants are discussed later on. In Sect. 2, we
describe the algebraic and structured representation of the multi-level preconditioned iterative processes in the Krylov subspaces. Section 3 is devoted to the
mapping of the parallel algorithms under consideration onto the computational
multi-processor system with the distributed and shared memory architecture.
The results of numerical experiments and an analysis of the various approaches
is carried out for different orders of the basic interpolation functions and for
different placement of the coarse grid nodes. The efficiency of the proposed algorithms is demonstrated on the representative set of the model examples.

2   Statement of the Problem and Algorithms

Let us consider a SLAE

$$Au = \sum_{l' \in \omega_l} a_{l,l'} u_{l'} = f, \quad A = \{a_{l,l'}\} \in \mathcal{R}^{N,N}, \quad u = \{u_l\}, \quad f = \{f_l\} \in \mathcal{R}^N, \quad (1)$$

with the sparse matrix of large order with real entries arising from some discrete
approximation of a multi-dimensional BVP by the finite element or the finite
volume or other grid methods; ω_l means a set of the indices of off-diagonal
entries in the l-th row of matrix A.

We can divide the total set of the vector indices $\Omega = \{l\}$ into P non-
intersected subsets, or algebraic subdomains,

$$\Omega = \bigcup_{s=1}^{P} \Omega_s, \quad N = \sum_{s=1}^{N} N_s, \quad (2)$$

each containing approximately equal number of elements N_s. For subdomains
Ω_s, let us denote their boundaries Γ_s^0 and closures as the following:

$$\Gamma_s \equiv \Gamma_s^0 = \{l' \in \omega_l, \ l \in \Omega_s, \ l' \notin \Omega_s\}, \quad \bar{\Omega}_s^0 = \Omega_s \bigcup \Gamma_s^0. \quad (3)$$

Also, we can define the boundary layers of Ω_s:

$$\Gamma_s^t = \left\{ l' \in \omega_l, \; l \in \bar{\Omega}_s^{t-1}, \; \bar{\Omega}_s^t = \bar{\Omega}_s^{t-1} \bigcup \Gamma_s^t, \; t = 1, 2, ..., \Delta_s \right\}. \tag{4}$$

Parameter Δ_s presents the measure of an extension of the subdomain Ω_s. The set of $\bar{\Omega}_s^{\Delta_s}$ forms the algebraic decomposition of the original domain Ω into subdomains with parametrized overlapping. Hystorically, it is known that increasing of the overlapping yields the increasing of the iterative convergence of DDMs and increasing of the cost of each iteration. For the subvectors

$$\bar{u}_s = \{u_l, l \in \bar{\Omega}_s^{\Delta_s}\} \in \mathcal{R}^{\bar{N}_s}, \quad u = \bigcup_{s=1}^{P} \bar{u}_s,$$

the original system can be written in a block form

$$A_{s,s}\bar{u}_s + \sum_{s' \in Q_s} A_{s,s'}\bar{u}_{s'} = f_s, \; s = 1, ..., P, \tag{5}$$

where Q_s is the set of subdomains which are adjacent to the extended subdomain $\bar{\Omega}_s^{\Delta_s}$.

To solve (5), the generalized block Jacobi iterative process is used:

$$\bar{B}_s(\bar{u}_s^{n+1} - \bar{u}_s^n) = \bar{f}_s - (\bar{A}\bar{u}^n)_s \equiv \bar{r}_s^n, \; \bar{u}_s^n \in \mathcal{R}^{\bar{N}_s}. \tag{6}$$

Here $\bar{r}_s^n$ is the residual subvector and $\bar{B}_s$ is some preconditioning matrix which takes into account the permutations of the "boundary" rows $l \in \Gamma_s^{\Delta_s}$, because of using special interface conditions of Steklov-Poincare type between the neighbour subdomains in the Schwarz iterations, see [2–4] for details.

The vector u^n of the sought for solution of original SLAEs (1) is not defined uniquely in (6), because in the intersections of the neighbour subdomain $\bar{\Omega}_s^{\Delta_s}$ we have several values of the vector components for the various s. In order to avoid such an indefiniteness, different approaches are used. We apply the restricted alternating Schwarz (RAS) slgorithm, which is based on using the restricting operators $R_s \in \mathcal{R}^{N_s, \bar{N}_s}$:

$$u_s^n = R_s \bar{u}_s^n = \{u_l^n = (R_s \bar{u}_s^n)_l, \; l \in \Omega_s\} \in \mathcal{R}^{N_s}, \tag{7}$$

where the subdomains $\Omega_s, s = 1, ..., P$, define the domain decomposition without overlapping.

The RAS Jacobi type method can be written in the following form:

$$\begin{aligned} u^{n+1} &= u^n + B_{ras}^{-1} r^n, \\ B_{ras}^{-1} &= R\hat{A}^{-1}W^T, \; \hat{A} = W^T A W = \text{block-diag}\{A_{s,s} \in \mathcal{R}^{\bar{N}_s, \bar{N}_s}\}, \end{aligned} \tag{8}$$

$W = [w_1...w_P] \in \mathcal{R}^{N,P}$ is a rectangular matrix, each its column w_s has the entries equal to one in the nodes from $\bar{\Omega}_s$ and has zero entries otherwise. Let us note that generally even if the original SLAE is symmetric, a preconditioning matrix B_{ras} from (8) is not a symmetric one. In addition, the inversion of the blocks $A_{s,s}$ of the matrix $\hat{A}$ is actually reduced to the simultaneous solution of independent subsystems in the corresponding subdomains.

We suppose that SLAE (1) is obtained from the approximation of a multi-dimensional BVP for partial differential equations by the finite element, finite volume or other method on some non-structured grid. For example, let the Dirichlet problem for the diffusion-convection equation

$$-\frac{\partial^2 u}{\partial x^2} - \frac{\partial^2 u}{\partial y^2} + p\frac{\partial u}{\partial x} + q\frac{\partial u}{\partial y} = f(x,y), \ (x,y) \in \Omega, \tag{9}$$
$$u|_{\Gamma} = g(x,y),$$

be solved in a computational domain $\Omega = (a_x, b_x) \times (a_y, b_y)$, where Γ is a boundary of Ω, and the convection coefficients p, q are, for simplicity, the given values. For the sake of brevity, we will use the symbol Ω to denote either the computational domain or the grid domain according to the context.

The given boundary value problem is approximated on a uniform grid

$$x_i = a_x + ih_x, \ y_j = a_y + jh_y,$$
$$i = 0, 1, ..., N_x + 1; \ j = 0, 1, ..., N_y + 1; \tag{10}$$
$$h_x = (b_x - a_x)/(N_x + 1), \ h_y = (b_y - a_y)/(N_y + 1),$$

by a five-point scheme of the form

$$(Au)_l = u_{l,l}u_l + a_{l,l-1}u_{l-1} + a_{l,l+1}u_{l+1} + a_{l,l-N_x}u_{l-N_x} + a_{l,l+N_x}u_{l+N_x} = f_l, \tag{11}$$

where l is a "global", or natural, number of an inner grid node:

$$l = l(i,j) \equiv i + (j-1)N_x = 1, ..., N = N_x N_y. \tag{12}$$

A particular view of the coefficients $a_{l,l'}$ in (11) can be different, and specific formulae can be found in [6,7]. Eq. (11) are written for the inner grid nodes, moreover, for the nodes near the boundary, whose numbers are from a set of the indices $i = 1, N_x$ or $j = 1, N_y$, the values known from the boundary conditions of the solution are substituted into the corresponding equations and moved to their right-hand sides, so that the corresponding coefficients $a_{l,l'}$ in (11) equal zero (it is the so- called "constraining" procedure).

We can think of the isomorphism between the vector entries in (1)–(5) and the grid nodes: u_l is the value of the grid function u in the l-th node at the grid Ω which is a set of all nodes in the computational domain. The subdomains Ω_s in (2) can be redefined as the grid subdomains, and for the model problem (9) we present a simple decomposition of Ω into a union of an identical non-interesecting rectangle subdomains:

$$\Omega = \bigcup_{s=1}^{P} \Omega_s, \ P = P_x P_y,$$

each containing an equal number of the grid nodes

$$M = m_x m_y, \ N_x = P_x m_x, \ N_y = P_y m_y, \ N = PM.$$

One can find that the subdomains form a two-dimensional macrogrid, where each macrovertex can be numbered by a pair of indices p, q (similarly to the grid node indices i, j), and a "continuous" number of a subdomain is defined as

$$s = s(p,q) \equiv p + (q-1)P_x = 1, ..., P,$$
$$p = 1, ..., P_x; \quad q = 1, ..., P_y. \tag{13}$$

We now turn from continuous numbering of nodes to their subdomain-by-subdomain ordering: at first, we number all the nodes in Ω_1, then in Ω_2, etc. The vector components u, f are ordered correspondingly, so that the SLAE (11) takes the block-matrix form (5), where $\bar{u}_s \in \mathcal{R}^{N_s}$ means a subvector of the vector u, whose components correspond to the nodes from the grid subdomain Ω_s, and Q_s means a set of the numbers of the grid subdomains adjacent to subdomain Ω_s. Hereinafter we assume that a local node ordering in every subdomain is a natural one: local pairs of indices $i' = 1, ..., m_x; \ j' = 1, ..., m_y$ are introduced and a continuous number is determined by the formula $l' = i' + (j' - 1)m_x$ similar to (12).

The rate of convergence of the iterative process (8) depends on the number of the subdomains, or more precisely, on the diameter of a graph representing a macrogrid formed by the decomposition. This can be clearly explained by the fact that on a single iteration the solution perturbation in one subdomain is transmitted only to the neighbouring, or adjacent, subdomains. To speed up the iterative process, it is natural to use not only the nearest but also the remote subdomain couplings at every step. For this purpose, different approaches are used in decomposition algorithms: deflation, coarse grid correction, aggregation, etc., which to some extent are close to the multigrid principle as well as the low-rank approximations of matrices, see numerous publications cited at a special site [8].

We will consider the following approach based on an interpolation principle. Let Ω_c be a coarse grid with the number of nodes $N_c \ll N$ in the computational domain Ω, moreover, the nodes of the original grid and the coarse grid may not match.

Let us denote by $\varphi_1, ..., \varphi_{N_c}$ a set of the basis interpolating polynomials of order M_c on the grid Ω_c which are supposed to have a finite support and without loss of generality form an expansion of the unit, i.e.

$$\sum_{k=1}^{N_c} \varphi_k(x, y) = 1.$$

Then a sought for solution vector of SLAE (1) can be presented in the form of an expansion in terms of the given basis:

$$u = \{u_{i,j} \approx u_{i,j}^c = \sum_{k=1}^{N_c} c_k \varphi_k(x_i, y_j)\} = \Phi \hat{u} + \psi, \tag{14}$$

where $\hat{u} = \{c_k\} \in \mathcal{R}^{N_c}$ is a vector of the coefficients of the expansion in terms of the basis functions, ψ is an approximation error, and $\Phi = [\varphi_1...\varphi_{N_c}] \in \mathcal{R}^{N,N_c}$ is a rectangular matrix with every k-th column consisting of the values of the basis function $\varphi_k(x_i, y_j)$ at N nodes of the original grid Ω (most of the entris of Φ equal zero in virtue of the finiteness of the basis). The columns, or the functions φ_k, can be treated to be the orthonormal ones but not necessarily. If at some k-th node P_k of the coarse grid Ω_c only one basis function is a nonzero one $(\varphi_k(P_{k'}) = \delta_{k,k'})$, then $\hat{u}_k = c_k$ is the exact value of the sought for solution at the node P_k. With a substitution of (14) into the original SLAE, one can obtain the system

$$A\Phi\hat{u} = f - A\psi, \tag{15}$$

and if to multiply it by Φ^T one can obtain

$$\hat{A}\hat{u} \equiv \Phi^T A\Phi\hat{u} = \Phi^T f - \Phi^T A\psi \equiv \hat{f} \in \mathcal{R}^{N_c}. \tag{16}$$

Assuming further that the error ψ in (14) is sufficiently small and omitting it, one can obtain a system for an approximate coarse grid solution $\check{u}$:

$$\hat{A}\check{u} = \Phi^T f \equiv \check{f}. \tag{17}$$

If the matrix A is a non-singular matrix and Φ is the full-rank matrix(the rank is much less than N), we assume these facts to hold further, then from (16) we have

$$u \approx \tilde{u} = \Phi\check{u} = \Phi\hat{A}^{-1}\hat{f} = B_c^{-1}f, \; B_c^{-1} = \Phi(\Phi^T A\Phi)^{-1}\Phi^T.$$

For the error of the approximate solution we have

$$u - \tilde{u} = (A^{-1} - B_c^{-1})f. \tag{18}$$

The error of the approximate solution can also be presented via the error of the approximation ψ. Subtracting Eqs. (16) and (17) term by term we have

$$\hat{A}(\hat{u} - \check{u}) = -\Phi^T A\psi$$

what yields the required equation:

$$u - \tilde{u} = \Phi\hat{u} + \psi - \Phi\check{u} = \psi - B_c^{-1}A\psi.$$

The matrix B_c^{-1} introduced above can be regarded as a low rank approximation of the matrix A^{-1} and used as a preconditioner to build an iterative process. In particular, for an arbitrary vector u^{-1} we can choose an initial guess as

$$u^0 = u^{-1} + B_c^{-1}r^{-1}, \; r^{-1} = f - Au^{-1}. \tag{19}$$

In doing so, the corresponding initial residual $r^0 = f - Au^0$ will be orthogonal to a coarse grid subspace

$$\hat{\Phi} = \text{span} \{\varphi_1, ..., \varphi_{N_c}\} \tag{20}$$

in the sense of fulfilling the condition

$$\Phi^T r^0 = \Phi^T (r^{-1} - A\Phi\hat{A}^{-1}\Phi^T r^{-1}) = 0. \tag{21}$$

The relations given in [10] are the basis for the conjugate gradient method with deflation, wherein an initial direction vector is chosen by the formula

$$p^0 = (I - B_c^{-1}A)r^0, \tag{22}$$

which ensures that the following orthogonality condition holds:

$$\Phi^T Ap^0 = 0. \tag{23}$$

Further iterations are implemented using the following relations:

$$\begin{aligned}
u^{n+1} &= u^n + \alpha_n p^n, \quad r^{n+1} = r^n - \alpha_n Ap^n, \\
p^{n+1} &= r^{n+1} + \beta_n p^n - B_c^{-1} Ar^{n+1}, \\
\alpha_n &= (r^n, r^n)/(p^n, Ap^n), \quad \beta_n = (r^{n+1}, r^{n+1})/(r^n, r^n).
\end{aligned} \tag{24}$$

In this method, which we will refer to as DCG, at every step the following relations hold:

$$\Phi^T r^{n+1} = 0, \quad \Phi^T Ap^{n+1} = 0. \tag{25}$$

If now we turn back to the additive Schwarz method (11), we can try to accelerate it by the coarse grid preconditioner B_c^{-1} (in addition to the preconditioner B_{ras}^{-1}). We will consider this point in a more general formulation assuming that matrix A is a non-symmetric one and that there are several but not only two preconditioning matrices. Moreover, the preconditioners can change from iteration to iteration what corresponds to the so-called dynamic, or flexible, preconditioning.

The SLAE with the non-symmetric matrix A is solved by the well-known BiCGStab algorithm [1].

3   Parallel Technologies of DDM

The objectives of our research consist in the verification, testing, and a comparative analysis of the efficiency of different algorithms and computational technologies of solving big sparse SLAEs aimed at their optimization and including into the KRYLOV library [9] of the parallel algebraic solvers. The main requirements to develop a proper software are high and scalable performance and no formal restrictions on the orders of the SLAEs and on the number of the processors and computational cores used. According to [3], a strong and a weak scalability can be distinguished. The first one describes a decrease in the execution time of one big problem with an increase of the number of computing devices, while the second one stands for approximate preservation of the solution time while increasing the dimension (the number of degrees of freedom) of the problem and the number of processors and/or cores.

The algorithms were coded with taking into account the architecture of the SSCC SB RAS cluster [11] (where KRYLOV library is available) but without GPGPU usage as their effective utilization in the considered domain decomposition methods has its own technological and computational complexity and requires a special study.

Computations are carried out in the following natural way: if a computational domain is divided into P subdomains than the solution is performed on $P + 1$ MPI-processes (one is the root process and other ones correspond to their own subdomains). During the program execution, the root MPI-process is used to accumulate partial dot products from the subdomains thus also keeping the synchronization of the computational work in the subdomains and upon completion it accumulates the whole sought for solution vector.

A scalable parallelization of the algorithms is provided by synchronization of the calculations in subdomains and by a minimization of the time losses during interprocessors communication. The solutions to auxiliary algebraic subsystems in the subdomains are obtained simultaneously on the multicore CPUs with the usage of multithread OpenMP calculations. The reduced system 17 is formed and solved in all the processes.

As algorithms from KRYLOV library are designed to solve large sparse SLAEs arising from an approximation of multidimensional boundary value problems on non-structured grids, the well-known compressed sparse row format of the matrix storage is used to keep the non-zero matrix entries. The global matrix A is formed in the root MPI-process (in the simplest implementation) at the preliminary stage, and then the distributed storage of the block rows $\bar{A}_s = \{A_{s,s'}, s' \in Q_s\}$ from (5) is done for the s-th extended subdomain (i.e., on the corresponding MPI-processes). If the original matrix is very big, it can be stored in a row-blocked form already and then the block rows be distributed among computational nodes (subdomains) thus keeping the global matrix on the root MPI-process is not a bottleneck for the problem under consideration.

An important condition for the high performance computing consists in the matching the arithmetic calculations and data communications between the subdomains by using MPI unblocked send-receive means. Moreover, the volume of the data transfer is very small as only the short vectors corresponding to the number of grid ponits on mutual boundary between the subdomains should be exchasnged.

Let us note that for the examined grid boundary value problems, a two-dimensional balanced domain decomposition into subdomains is considered, when for an approximately equal number of nodes $N_S \approx N/P$ in every subdomain the macrogrid daimeter d (for a macrogrid composed of subdomains) is equal, approximately, to $\sqrt{P}$. As the number of the iterations of the additive Schwarz method even with the usage of the preconditioned Krylov methods is proportional to $d^\gamma, \gamma > 0$, this yields a significant advantage over a one-dimensional decomposition for which $d \approx P$.

A solution to the isolated SLAEs in Ω_s is produced by the direct or iterative method requiring $(N/P)^{\gamma_1}, \gamma_1 > 0$ operations at every step of the two-level process. As it is necessary to exchange the data corresponding to peripheral nodes of the adjacent subdomains only, the volume of such an information is much less and proportional to $(N/P)^{\gamma_1/2}$ (for two-dimensional BVPs) thus allowing one to carry out arithmetic and communication operations simultaneously.

A high performance of the code based on the presented approach is ensured by an active usage of the standard functions and vector-matrix operations from BLAS and Sparse BLAS included into Intel MKL [12].

4   Results of Numerical Experiments

We present the results of methodical experiments on solving five-point SLAEs for 2D Dirichlet problem in the unit square computational domain on the square grids with the number of nodes $N = 128^2$ and 256^2. Calculations were carried out via $P = 2^2, 4^2, 8^2$ MPI-processes each of which corresponded to the subdomains forming the square macrogrid. Iterations over the subdomains were realised with the help of BiCGStab algorithm [1] with the stopping criterion $||r^n||_2 \leq \varepsilon ||f||_2$, $\varepsilon = 10^{-8}$. Solving of the auxiliary subdomain subsystems was carried out by the direct solver PARDISO from Intel MKL. The most time-consuming part of LU matrix decomposition was done only once before the iterations.

In the Table 1, each cell contains the numbers of iterations over the subdomains and the times of SLAEs solving (in seconds) on the grids 128^2 and 256^2. The upper figure in each cell corresponds to the zero convection coefficients while the bottom figure – to the convection coefficients $p = q = 4$). Domain decompositions were made for equal overlapping parameters in subdomains: $\Delta_s = \Delta = 0, 1, 2, 3, 4, 5$. Interface boundary conditions of the Dirichlet type between the adjacent subdomains were used in all experiments.

The results demonstrate that with Δ increasing up to 5, the number of the iterations reduces 3 - 4 fold, but when the overlapping value is big, the time of a subdomain solving begins to increase. So, for almost all the grids and the numbers of MPI-processes (subdomains), the optimal Δ value is approximately 3 – 4 in terms of the total execution time. If the convection coefficients p, q are nonzero ones, the number of the iterations increases by approximately 30–50 %. Let us notice that the figures for 4 and 16 subdomains were obtained in the experiments when each MPI-process was ran on its own cluster node in exclusive mode while the data for 64 subdomains were got in a series of experiments on cluster nodes that were given to the tasks in non-exclusive mode yielding some increase of the execution time. So the last line of the Table does not present "pure" speedup of the algorithm.

In the Tables below, for the sake of bravity, the results for the Poisson equation are presented, i.e. when there are no convection coefficients in equation (1). The experiments shown that with the moderate values of p, q ($|p| + |q| < 50$) the behavior of the iterative process varied slightly.

The numerous results for the different model and practical problems shown that the behavior of iterations varied slightly in the considered algorithms when

Table 1. The numbers of iterations and the solution times (in seconds) on the grids 128^2 and 256^2 for different overlapping parameter Δ

P	q	$N \backslash \Delta$	0	1	2	3	4	5
	0		18 2.17	11 1.74	9 1.64	7 1.53	7 1.48	6 1.42
4	4	128^2	31 2.85	17 2.10	13 1.87	12 1.81	11 1.74	10 1.74
4	0	256^2	27 8.34	16 5.38	12 4.21	10 3.68	9 3.33	8 2.93
	4		61 16.88	25 7.74	19 6.52	17 5.47	15 5.28	13 4.25
	0		32 1.46	18 1.29	14 1.25	12 1.17	11 1.03	9 0.98
16	4	128^2	41 1.60	25 1.40	19 1.31	16 1.18	14 1.17	14 1.10
16	0	256^2	40 3.23	24 2.23	20 1.97	17 1.77	14 1.27	14 1.24
	4		58 4.32	35 2.83	28 2.46	22 1.98	19 1.62	18 1.52
	0		43 1.56	26 1.66	19 1.39	16 1.50	14 1.56	12 0.86
64	4	128^2	57 2.02	34 1.91	26 1.78	21 1.98	20 1.69	18 1.35
64	0	256^2	60 4.75	36 4.16	27 3.35	22 3.11	20 3.00	18 4.66
	4		87 7.04	47 5.61	38 4.89	31 4.13	28 4.02	25 4.48

the initial error varied. The experiments given above were hold for the initial guess $u^0 = 0$ and the exact SLAE solution $u = 1$.

Table 2 shows the effect of applying of two deflation methods when the conjugate gradient algorithm without any additional preconditioning and without additive Schwarz method is used for three square grids with different numbers of nodes N and for different macrogrids with the number of the macronodes N_c. The macronodes are taken in the vicinity of the subdomain corners, i.e. when $P = 2^2, 4^2, 8^2$ the numbers of the macronodes, or the values of N_c, are $3^2, 5^2$ and 9^2 respectively. The basis functions $\phi_k(x, y)$ were the bilinear finite functions. Three right columns have the number of the iterations (the upper figures in every cell) for the single orthogonalization of the form (23) while the iteration number for the orthogonalization (25) on every iteration is the bottom figure. If to compare these data with the algorithm when the deflation is not used at all (the column with $P = 0$, i.e. no macrogrid is used) one can see the acceleration up to three times when P increases. However, it should be taken into account that an implementation of the multiple orthogonalization makes each iteration more expensive, so an additional investigation is required to optimize the algirithms on practice.

The results from Table 3 present the same data but when using the additive Schwarz method with the domain decomposition into P subdomains. The numbers of the coarse grid nodes are taken the same as that of the macronodes for Table 2. The basis functions $\phi_k(x, y)$ as in the previous series of experiments from Table 2 were the bilinear finite ones. Every cell of Table 3 contains the numbers of the iterations carried out without deflation (the upper figures in each cell) and the numbers of the iterations for the single orthogonalization of the initial guess (the bottom figures in each cell). In every cell, the first column gives the

Table 2. The deflation influence in the conjugate gradient method without additive Schwarz

$N \setminus P$	0	2^2	4^2	8^2
64^2	176	167	166	103
		118	87	56
128^2	338	309	255	181
		220	159	104
256^2	609	544	442	276
		376	294	190

Table 3. Aggregation influence in the additive Schwarz method (decomposition with different overlapping parameter Δ)

$N \setminus P$	2^2			4^2			8^2		
64^2	19	11	8	26	15	12	37	20	15
	23	9	7	21	12	9	28	15	11
128^2	29	15	11	35	22	17	51	31	21
	24	14	10	26	16	12	36	21	15
256^2	38	21	17	53	31	23	71	43	32
	31	18	15	35	21	17	40	26	21

data for the zero value of the overlapping parameter Δ, the second column – for $\Delta = 1$, and the third column – for $\Delta = 2$.

The presented results for the considered grids and macrogrids have approximately the same character as in Table 2 when the increasing of the deflation space yields to the decreasing of the iteration number together with the increasing of the amount of computations at each step. In these experiments, the outer iterations were carried out by the BiCGStab method.

Let us note that the experiments for Table 3 were hold for the initial guess $u^0 = 0$ and the exact SLAE solution $u(x_i, y_j) = x_i^2 - y_j^2$. Naturally, the efficiency of the considered "interpolation" deflation depends on the behaviour of the solution sought for. For example, if it is, e.g., $u(x_i, y_j) = x - y$, then the usage of the bilinear basis functions $\varphi_k(x, y)$ for $N_c \geq 4$ yields to the convergence in one iteration, and this fact was confirmed in the experiments.

5   Conclusion

We have studied experimentally the efficiency and the performance of several advanced approaches for domain decomposition methods. The results presented demonstrate a considerable increasing of the convergence rate of the iterative process when the corresponding overlapping parameters and a coarse grid correction are used to accelerate the additive Schwarz algorithm. It should be mentioned that the augmented versions of DDM have been implemented without additional communication losses. Obviously, the further research should be held to obtain some practical recommendations to optimize the performance when a combination of various parametrized approaches are used simultaneously. The efficient code optimization for multi-GPGPU and a multi-core implementation is a challenge technology but it is an open question now. For example, it is interesting to analyse two-level iterative FGMRES procedure with some dynamic stopping criterion in subdomains and various basis functions in low rank matrix approximations.

References

1. Saad, Y.: Iterative Methods for Sparse Linear Systems. PWS Publications, New York (2002)
2. Toselli, A., Widlund, O.: Domain Decomposition Methods - Algorithms and Theory. Springer, Heidelberg (2005)
3. Chapman, A., Saad, Y.: Deflated and augmented krylov subspace technique. Numer. Linear Algebra Applic. **4**(1), 43–66 (1997)
4. Il'in, V.P.: Parallel Methods and Technologies of Domain Decomposition (in Russian). Vestnik YuUrGU. Series Computational mathematics and informatics. 46(305), 31–44 (2012)
5. Dubois, O., Gander, M.J., St-Cyr, A., Loisel, S., Szyld, D.: The optimized schwarz method with a coarse grid correction. SIAM J. Sci. Comput. **34**(1), 421–458 (2012)
6. Il'in, V.P.: Finite Difference and Finite Volume Methods for Elliptic Equations. ICMMG Publisher, Novosibirsk (2001). (in Russian)
7. Il'in, V.P.: Finite Element Methods and Technologies. ICMMG Publisher, Novosibirsk (2007). (in Russian)
8. Official page of Domain Decomposition Methods. http://www.ddm.org
9. Butyugin, D.S., Gurieva, Y.L., Il'in, V.P., Perevozkin, D.V., Petukhov, A.V.: Functionality and Algebraic Solvers Technologies in Krylov Library (in Russian). Vestnik YuUrGU. Series Computational mathematics and informatics. 2(3), 92–105 (2013)
10. Gander, M.J., Halpern, L., Santugini, K.: Domain decomposition methods in science and engineering XXI. In: Erhel, J., Gander, M.J., Halpern, L., Pichot, G., Sassi, T., Widlund, O. (eds.) A New Coarse Grid Correction for RAS/AS. LNCSE. Springer-Verlag, Switzerland (2013)
11. Siberian Supercomputer Centre. http://www2.sscc.ru
12. Intel Math Kernel Library (Intel MKL). http://software.intel.com/en-us/intel-mkl

A Modular-Positional Computation Technique for Multiple-Precision Floating-Point Arithmetic

Konstantin Isupov$^{(\boxtimes)}$ and Vladimir Knyazkov

Department of Electronic Computing Machines,
Vyatka State University, Kirov 610000, Russia
{ks_isupov,knyazkov}@vyatsu.ru

Abstract. Floating-point machine precision is often not sufficient to correctly solve large scientific and engineering problems. Moreover, computation time is a critical parameter here. Therefore, any research aimed at developing high-speed methods for multiple-precision arithmetic is of great immediate interest. This paper deals with a new technique of multiple-precision computations, based on the use of modular-positional floating-point format for representation of numbers. In this format, the significands are represented in residue number system (RNS), thus enabling high-speed processing of the significands with possible parallelization by RNS modules. Number exponents and signs are represented in the binary number system. The interval-positional characteristic method is used to increase the speed of executing complex non-modular operations in RNS. Algorithms for rounding off and aligning the exponents of numbers in modular-positional format are presented. The structure and features of a new multiple-precision library are given. Some results of an experimental study on the efficiency of this library are also presented.

Keywords: Multiple-precision arithmetic · Floating-point · Residue number system · Non-modular operation · Rounding · Performance

1 Introduction

The increasing power of current computers enables one to solve more and more complex problems. This, therefore, requires to perform a high number of floating-point operations, each one leading to a rounding error. As Exascale computing (10^{18} operations per second) is likely to be reached within a decade, getting accurate results in machine-precision floating-point arithmetic on such computers will be a challenge [1]. Many problems whose correct solution requires multiple-precision arithmetic are already emerging [2,3]. The situation is complicated by the fact that even with a small amount of computation in machine precision, a result can be obtained, free of any significant digit [4,5].

A 128-bit IEEE format is currently one of the ways of improving the accuracy of computations [6]. Here, the significand field is extended to 113 bits. However, hardware support for this format requires considerable expenses [3] and, apparently, such is not expected in the near future. A more common variant of high-precision arithmetic implemented as software is known as the "double-double"

© Springer International Publishing Switzerland 2015
V. Malyshkin (Ed.): PaCT 2015, LNCS 9251, pp. 47–61, 2015.
DOI: 10.1007/978-3-319-21909-7_5

format. It provides an accuracy of about twice the standard 64-bit arithmetic
[7]. In this format, the number is expressed as a pair of 64-bit integers x_h and x_l,
where x_h is a floating-point number next to the true value, while x_l is the difference between the true value and x_h. T.J. Dekker's algorithms [8] can be used to
add or multiply numbers in this format. The "quad-double" format is based on
similar principles. Besides, there are some freely available software packages that
support arbitrary precision (when the number of digits is limited only by the
available memory). The most famous of these packages are the Arbitrary Precision Computation Package (ARPREC) [9], the GNU Multiple-Precision Arithmetic Library (GMP) [10], the GNU Multiple-Precision Floating-Point Reliable
Library (MPFR) [11], and the Number Theory Library (NTL) [12].

A common disadvantage of methods and software for positional multiple-
precision arithmetic is that computation speed reduces greatly as precision
increases. It is noted in particular that computation in the "double-double" format
is usually five times slower than in the 64-bit format; in the "quad-double", it is 25
times slower. When using arbitrary precision, the computation time can increase
by hundreds of times [3]. This is unacceptable when dealing with many major problems, particularly those related to real-time control of objects. Therefore, developing accelerated methods and software for multiple-precision arithmetic is a topical
area of research in the field of high-performance computing.

Next, we consider a new computation technique for multiple-precision arithmetic. This technique is based on the use of non-positional number system to
represent significands – the residue number system.

2   Residue Number System

The residue number system (RNS) [13–16] is defined by a set of relatively prime
integers $\{p_1, p_2, \ldots, p_n\}$ called modules. Its dynamic range is equal to the product
$P = \prod_{i=1}^{n} p_i$. According to the Chinese remainder theorem (CRT) [16], every
integer $X \in \{0, 1, 2, \ldots, P-1\}$ has a unique RNS (modular) representation:

$$X = \langle x_1, x_2, \ldots, x_n \rangle, \tag{1}$$

where $x_i = |X|_{p_i} \leftrightarrow x_i \equiv X \bmod p_i$. Generally, operations on various modules
p_i are performed independently:

$$Z = X \, op \, Y = \left\langle |x_1 \, op \, y_1|_{p_1}, |x_2 \, op \, y_2|_{p_2}, \ldots, |x_n \, op \, y_n|_{p_n} \right\rangle.$$

As a consequence, operations on large numbers can be effectively divided into
several modular operations executed in parallel and with reduced wordlength
[17]. This makes it possible to consider RNS as a prospective basis for high-
speed parallel multiple-precision computations.

However, the effective use of RNS is complicated due to the highly complex
nature of non-modular operations (magnitude-comparison, overflow detection,
scaling, etc.), which require much longer time to be executed than modular
operations. The problem becomes aggravated by the fact that when performing
long iterative floating-point computations, non-modular operations make up a
significant part of the total volume of arithmetic operations performed.

3   Interval-Positional Characteristic Method for Non-modular Operations in RNS

To perform non-modular operations on numbers in the form (1), one must have information about their positional value, which is defined by the expression

$$X = \left| \sum_{i=1}^{n} x_i |P_i^{-1}|_{p_i} P_i \right|_P , \qquad (2)$$

where $P_i = P/p_i$ and $|P_i^{-1}|_{p_i}$ is the multiplicative inverse of P_i modulo p_i. However, computing (2) directly is slow because it requires multi-digit addition and reduction by large modulus P.

Mixed Radix Conversion (MRC) [14–16] is another method used to estimate the magnitude of RNS numbers. But to calculate all mixed-radix coefficients, you require $O(n^2)$ arithmetic operations, where n is the number of modules p_i. Parallel algorithms [18, 19] reduce this evaluation to $O(n)$, but are still expensive. The MRC-II algorithm [20] allows to calculate the value of an RNS number for $O(n)$ operations. However, this involves dealing with large numbers.

The interval-positional characteristic method [21–23] is an alternative technique for estimating a magnitude in RNS. We will consider this in more detail. Let's denote X/P as the relative value of RNS number X, i.e. the ratio of its positional value to the product of modules P. Dividing the equality (2) by P, we obtain the following expression for the relative value [16]:

$$\frac{X}{P} = \left| \sum_{i=1}^{n} \frac{|x_i |P_i^{-1}|_{p_i}|_{p_i}}{p_i} \right|_1 , \qquad (3)$$

where $|\ |_1$ is the fractional part of the sum. It is obvious that $X/P \in [0, 1)$. Instead of calculating the exact value of X/P, which is quite costly in this context, it would be more efficient to compute its approximation – *interval-positional characteristic*.

Definition. Interval-positional characteristic (IPC) of RNS number X is closed interval $I(X/P) := \left[\underline{X/P}, \overline{X/P} \right]$ with directed rounded positional bounds that satisfy inclusion condition: $\underline{X/P} \le X/P \le \overline{X/P}$.

IPC projects range $[0, P)$ to the interval $[0, 1)$, associating every RNS number X with a pair of directed rounded positional numbers – bounds that localize its relative value (3), as shown in Fig. 1.

We denote operations of addition, subtraction, multiplication and division as $\triangledown, \triangledown, \triangledown, \triangledown$ correspondingly executed with rounding down (toward $-\infty$). Similarly, $\triangle, \triangle, \triangle, \triangle$ will be used to denote the operations with rounding up (toward $+\infty$). The operations are also supposed to be executed with the directed rounding if $\triangledown$ or $\triangle$ are placed before the group operator. For example, $\triangle \sum_{i=1}^{n} x_i$ means that all of the $n - 1$ additions are rounded up. The issues referring to rounding in floating-point arithmetic are described in papers [7, 24].

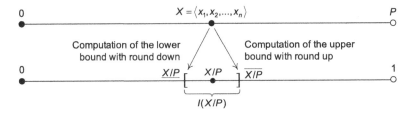

Fig. 1. Interval approximation of the relative value of an RNS number (IPC)

To compute IPC bounds, the following expressions are correct:

$$\underline{X/P} = \left| \triangledown \sum_{i=1}^{n} \left(|x_i| P_i^{-1}|_{p_i}|_{p_i} \triangledown p_i \right) \right|_1, \quad \overline{X/P} = \left| \triangle \sum_{i=1}^{n} \left(|x_i| P_i^{-1}|_{p_i}|_{p_i} \triangle p_i \right) \right|_1.$$

Sequential computation of these expressions requires performing $O(n)$ elementary operations, while parallel computation requires $O(\log n)$ elementary operations, which is much less than MRC. Specialized algorithm ISaC (Iterative Shift and Correction) [22,23] was developed to minimize the relative error when computing the IPCs of small RNS numbers. The algorithm is based on the possibility of error-free power-of-two scaling of normalized binary floating-point numbers. It gives high-speed computation of IPC with a priori given accuracy, regardless of the size of the RNS range, thus allowing to obtain highly accurate information about the magnitude of the RNS number without the need for time-consuming conversion to a positional number system. Studies have shown that at equal high accuracy (the upper limit of the relative error of IPC is less than 1%), ISaC algorithm is faster than its multiple-precision counterpart by more than an order of magnitude [22].

If the IPCs of RNS numbers are computed, performing the main non-modular operations on them becomes easy, as the task is reduced to analyzing positional interval bounds and assessing the validity of the result. For example, for magnitude-comparison of the RNS numbers X and Y, the following steps need to be performed (it is assumed that $X \neq Y$):

1. Compute the IPCs for X and Y: $I(X/P)$ and $I(Y/P)$.
2. Ensure that the formal correctness conditions are met (discussed later).
3. Determine the result, comparing the two IPCs by interval arithmetic rules: if $\underline{X/P} > \overline{Y/P}$, then $X > Y$; if $\overline{X/P} < \underline{Y/P}$, then $X < Y$.

To determine the sign of a number in symmetric RNS, you only need to compare $I(X/P)$ with constant $I\left(\frac{P-1}{2P}\right)$ by the above scheme. To establish whether overflow of arithmetic sum $X+Y$ has occurred, you only need to compare interval sum

$$I(X/P) + I(Y/P) = \left[\underline{X/P} \triangledown \underline{Y/P}, \ \overline{X/P} \triangle \overline{Y/P} \right] \tag{4}$$

with constant $I\left(\frac{P-1}{P}\right)$. If $\overline{X/P} \triangle \overline{Y/P} \leq \frac{P-1}{P}$, then overflow will not occur; if $\underline{X/P} \triangledown \underline{Y/P} > \frac{P-1}{P}$, then overflow will occur. A sign of an overflow when multiplying two RNS numbers $X \times Y$ is defined similarly. In this case, the product of IPCs are calculated using the following formula:

$$I(X/P) \times I(Y/P) = \left[(\underline{X/P} \triangledown \underline{Y/P}) \triangledown \overline{1/P}, \ (\overline{X/P} \triangle \overline{Y/P}) \triangle \underline{1/P} \right], \qquad (5)$$

where $\overline{1/P}$ and $\underline{1/P}$ are constants approximating the ratio $1/P$. Relations (4) and (5) naturally take into account rounding errors, extending the IPC bounds in accordance with the inclusion property of intervals [24].

Formal conditions for verifying the correctness of non-modular operations are defined as follows. The absolute rounding error in computing IPC is characterized by the IPC diameter

$$\operatorname{diam} I(X/P) = \overline{X/P} - \underline{X/P}. \qquad (6)$$

When computing IPC, directed rounding increases the diameter (6), without generally affecting the inclusion property $X/P \in I(X/P)$. However, this is violated in some cases due to lack of accuracy – when the number X is very small in relation to P or vice versa, when X is near P. In the first case, the lower bound of IPC is computed incorrectly, in the second case, it is the upper bound that falls the victim. In any case, $\operatorname{diam} I(X/P) < 0$, and IPC is called "incorrect". Correctness of IPCs for all operands is the first formal condition for correct execution of any non-modular operation. If this condition holds, then final conclusion on whether the result is correct or not is taken based on analysis of the second formal condition (Table 1).

Table 1. Second formal correctness conditions for non-modular operations.

Operation	Condition
Magnitude-comparison	$\operatorname{diam} \left[I(X/P) \cap I(Y/P) \right] < 0$
Sign detection in symmetric RNS	$\operatorname{diam} \left[I(X/P) \cap I \left(\frac{P-1}{2P} \right) \right] < 0 \vee \overline{X/P} = \frac{P-1}{2P}$
Overflow detection in the addition of two unsigned numbers	$\operatorname{diam} \left[I(S/P) \cap I \left(\frac{P-1}{P} \right) \right] < 0 \vee \overline{S/P} = \frac{P-1}{P}$
Overflow detection in the multiplication of two numbers	$\operatorname{diam} \left[I(Z/P) \cap I \left(\frac{P-1}{P} \right) \right] < 0 \vee \overline{Z/P} = \frac{P-1}{P}$

In Table 1, $I(S/P) = \left[\underline{S/P}, \overline{S/P} \right]$ and $I(Z/P) = \left[\underline{Z/P}, \overline{Z/P} \right]$ are the sum of IPCs (4) and product of IPCs (5) respectively. Intersection of IPCs is defined as follows:

$$[I(X/P) \cap I(Y/P)] = \left[\max\{\underline{X/P}, \underline{Y/P}\}, \ \min\{\overline{X/P}, \overline{Y/P}\} \right].$$

If the diameter (6) of this interval is less than zero, then the IPCs do not intersect in a standard set-theoretic sense.

Conjunction of two formal conditions comprises of *sufficient condition* for verifying the correctness of a non-modular operation that is invariant to the number of RNS modules and their bit length. If the sufficient condition doesn't hold, it becomes possible to improve the accuracy of IPCs or use other methods of performing non-modular operations (for example, MRC).

So, IPC is a sufficient universal positional characteristic for estimating the magnitude of RNS numbers. Interval evaluation of rounding error, which does not require considering the specifics of machine computations, makes this characteristic easy for performance of a wide range of non-modular operations. Results of experiments [22] showed significant speedup of the method based on IPC compared with its classical counterparts (MRC and CRT).

4   Format for Representation of Floating-Point Multiple-Precision Numbers

The following *modular-positional floating-point format* (MF-format) is proposed for representation of multiple-precision numbers [25]:

$$x \to \{s, M, \lambda, I(M/P)\}, \tag{7}$$

where $s = \mathrm{sgn}(x)$ is the number sign, $M = \langle m_1, m_2, \ldots, m_n \rangle$ is the modular significand represented in RNS modulo $p_1, p_2, \ldots p_n$, $\lambda \in [\lambda_{\min}, \lambda_{\max}]$ is the integer exponent with a sign, $I(M/P) = \left[\underline{M/P}, \overline{M/P} \right]$ is the IPC of the significand. The value of the number in the form (7) is defined by the expression

$$x = (-1)^s \cdot \left| \sum_{i=1}^{n} m_i \left| P_i^{-1} \right|_{p_i} P_i \right|_P \cdot 2^\lambda.$$

Significand M lies within the range $[0, P)$, where P is the product of all p_i. Thus, variation in the number of modules allows setting arbitrary precision of computations. The MF-format scheme is presented in Fig. 2 with indication of data types.

Sign	Exponent	Modular (RNS) significand				Interval-positional characteristic	
int	int	int	int	...	int	real	real

Fig. 2. MF-format for representation of multiple-precision numbers: int – an integer data type; real – a machine-precision floating-point data type

Inclusion of IPC in number representation is the main difference of MF-format from previously known methods of floating-point representation in RNS [26–29]. In terms of high-precision arithmetic, this offers the following benefits:

- you can use a new arithmetic processing based on fast preliminary estimation of the significand of the result;
- interval arithmetic enables you, while performing modular operations, to reduce IPC computation time to several steps, using an algorithmic specification only when performing non-modular procedures.

For MF-format (7), the following characteristics are defined: machine epsilon $\epsilon = 2^{-\lfloor \log_2(P-1) \rfloor}$, unit in the last place $\mathrm{ulp}(x) = 2^{\lambda - \lfloor \log_2((P-1)/M) \rfloor}$; the absolute and relative rounding errors are limited to the values $\mathrm{ulp}(x)$ and 2ϵ respectively (when rounding the significand by truncation).

Apart from finite numbers, infinite numbers and not-a-number values (NaNs) can be written in MF-format. Their encodings are given in Table 2.

Table 2. Encodings of modular-positional floating-point values. NaNs cannot be the input data of operations, but can appear in the course of performing operations.

Type	Exponent, λ	Sign and modular significand, $\pm M$
Zero	0	$\pm \langle 0, 0, \ldots, 0 \rangle$
Finite numbers	$\lambda_{\min} \leq \lambda \leq \lambda_{\max}$	$\pm \langle m_1, m_2, \ldots, m_n \rangle, \exists m_i \neq 0$
Signed infinities	$\lambda_{\max} + 1$	$\pm \langle 0, 0, \ldots, 0 \rangle$
Not-a-Numbers (NaNs)	$\lambda_{\max} + 1$	$\pm \langle m_1, m_2, \ldots, m_n \rangle, \exists m_i \neq 0$

Algorithms for rounding off, exponent alignment, addition, subtraction, multiplication, division, and comparison of numbers were developed for multiple-precision computations in MF-format. Using IPC for fast estimation of significand allowed to implement a new scheme of rounding, shown in Fig. 3. Under this scheme, the decision to round up is taken based on analysis of the IPCs of operands immediately before arithmetic operation and not after its execution (as in binary floating-point number systems).

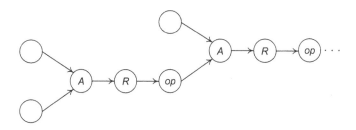

Fig. 3. Rounding scheme in MF-format: A – analysis of operand significands (checking the need for rounding), R – rounding, op – arithmetic operation

This scheme is preferable in terms of precision of computations and number (less) of iterations of modular significand scaling, as it avoids unnecessary rounding in the case if the significand of the result of the subsequent operation will fall within acceptable range $[0, P)$. The greatest effect from the use of the scheme is achieved when additive operations that do not lead to significant increase in the bit length of the result are performed. Effective implementation of the scheme is possible due to availability of IPC in number representation, which allows giving a quick estimate of the significand. In accordance with the presented scheme, the rounding task is divided into two sub-tasks:

1. Checking the need for rounding, the purpose of which is to determine the value of the $\mathrm{rsh}(M)$ function – the smallest scaling factor for significand M to avoid overflow when performing operation op.
2. Direct rounding – a reduction in the value of the modular significand by $2^{\mathrm{rsh}(M)}$ times with an increase in the exponent λ by $\mathrm{rsh}(M)$.

The method of calculating the $\mathrm{rsh}(M)$ function depends on the arithmetic operation to be performed. For example, if the op is multiplication, then the following is calculated for each factor:

$$\mathrm{rsh}(M) = \max\left\{ \lceil \log_2 \underline{M/P} \triangledown \log_2 \overline{\alpha} \rceil, \lceil \log_2 \overline{M/P} \triangle \log_2 \underline{\alpha} \rceil, 0 \right\},$$

where $\alpha = \lfloor \sqrt{P-1} \rfloor / P$ (directed rounding is used when calculating logarithms). If op is binary addition, then $\mathrm{rsh}(M)$ is defined as follows (for each term):

$$\mathrm{rsh}(M) = \begin{cases} 0, & \text{if } \overline{M/P} \leq \frac{P-1}{2P}; \\ 1, & \text{otherwise.} \end{cases}$$

If $I(M/P)$ is known, the complexity of calculating $\mathrm{rsh}(M)$ is invariant to the number of RNS modules. Further, if $\mathrm{rsh}(M)$ is greater than zero, then rounding should be done in accordance with Algorithm 1.

Algorithm 1. Rounding numbers in MF-format.
Input: $x \to \{s, M, \lambda, I(M/P)\}$, $\mathrm{rsh}(M)$.
Output: rounded number $round(x) \to \{s_r, M_r, \lambda_r, I(M_r/P)\}$.
  1: If $\mathrm{rsh}(M) \leq 0$, then accept $round(x) = x$ and end the algorithm, otherwise proceed to the next step.
  2: Calculate the exponent: $\lambda_r = \lambda + \mathrm{rsh}(M)$; set $s_r = s$.
  3: If $\lambda_r > \lambda_{\max}$, then it is taken that the arithmetic overflow criterion holds. In this case, form the corresponding exception code and enter it into the status register, set $\lambda_r = (\lambda_{\max} + 1)$, $M_r = 0$, $I(M_r/P) = [0,0]$ (encoding infinity) and end the algorithm. Otherwise, proceed to the next step.
  4: Scale modular significand M with factor $2^{\mathrm{rsh}(M)}$: $M_r = \lfloor M/2^{\mathrm{rsh}(M)} \rfloor$.
  5: Compute the IPC of rounded significand M_r based on the ISaC algorithm.

The most time-consuming in this algorithm is step 4. Here, a non-modular scaling operation is performed in RNS. A new iterative method was developed for power-of-two RNS scaling [25]. It enables to, in comparison with bisection method, accelerate computations by $q/(\lfloor q/h \rfloor + 1)$ times, where q is the scaling factor logarithm and h is the scaling step logarithm. The maximum value of step 2^h is limited only to the available memory and IPC computation accuracy.

Remark. If $\log_2 P$ is the bit length of the full variation range of the significand (RNS range), then the *effective bit length* (the length of operands involved in operations) is generally defined by the value $\log_2 \sqrt{P-1}$. It is the effective bit length that should be a priori considered as "precision" in the terms of IEEE-754 [6], at least in the first approximation. The actual precision can be higher since the entire significand representation range is used for addition and subtraction.

Exponent alignment operation plays an essential role in addition and subtraction of multi-scale values. The next alignment algorithm allows minimizing loss of accuracy and accelerating computations by compensating the difference in the number exponents through mutual correction of their significands.

Algorithm 2. Alignment of number exponents in MF-format.

Input: $x \to \{s_x, M_x, \lambda_x, I(M_x/P)\}$, $y \to \{s_y, M_y, \lambda_y, I(M_y/P)\}$.
Output: numbers x and y with aligned exponents $\lambda_x = \lambda_y$.
1: For a given exponent difference $\Delta\lambda = \lambda_x - \lambda_y$ (it is assumed that $\Delta\lambda < 0$ otherwise x and y are interchanged), check the condition $I(M_y/P) < 2^{\Delta\lambda}$. If it holds, then set $r = 0$ (where r is the number of rounding digits) and proceed to step 3, otherwise proceed to the next step.
2: Recalculate interval $I(M_y/P)$ using the ISaC algorithm. The number $r = \lceil \log_2 \overline{M_y/P} + |\Delta\lambda| \rceil$ is then calculated, where $\overline{M_y/P}$ is the upper bound of $I(M_y/P)$. If $\log_2 \underline{M_x/P} > (r - \log_2 P)$, where $\underline{M_x/P}$ is the lower bound of $I(M_x/P)$, then proceed to step 3, otherwise reset the number x to zero and end the alignment algorithm (the difference between the exponents is too large to be compensated by mutual alignment of operand significands).
3: Multiply modular significand M_y and both bounds of $I(M_y/P)$ by $2^{|\Delta\lambda|-r}$; deduct $(|\Delta\lambda| - r)$ from λ_y.
4: Round the number x by scaling M_x with factor 2^r; add r to λ_x.

After executing Algorithm 2, nonzero numbers x and y with the same exponents will be obtained or one of them (the one whose exponent is less) will be zero, while the second will not change.

A detailed description of the algorithms of addition, multiplication and division of numbers in MF-format, as well as the algorithm for accelerated power-of-two scaling of modular significands can be found in [25]. The developed algorithms support arithmetic of infinities and exception handling (overflow, invalid operation, division by zero, etc.). All the algorithms, except division, generally have linear complexity in terms of number of RNS modules when calculating the significand digits consecutively. In many cases, parallel processing reduces the estimate to a logarithmic function.

5   High-Precision Arithmetic Library

5.1   Structure and Features

A program library (MF-Library) for multiple-precision modular-positional computations is being developed based on the approach presented. The library is written in C. It uses basic data type (mf_t), which provides an arbitrary length of floating-point numbers. This type corresponds to the MF-format (7) and consists of five fields shown in Table 3.

Table 3. Data type used in MF-Library (mf_t)

Field	Basic C type	Description
sign	int	Sign of number
residue	long[n]	Significand in RNS
exp	int	Binary exponent
ic_bot	double	Lower bound of IPC
ic_top	double	Upper bound of IPC

At the structural level, the library consists of three units: kernel unit, auxiliary routine unit and test unit. Each unit contains same-type subunits. Besides, an auxiliary positional multiple-precision unit is used for automatic configuration and switching to another RNS base (when changing the precision). The centerpiece is the kernel unit whose structure is presented in Fig. 4. The functionality of the library is classified as follows: initialization routines, input/output routines, arithmetic functions, rounding functions, tests of basic operations (arithmetic, rounding, IPC computation, etc.), performance tests. At the moment, the basic feature of the library has been developed and partially covered by tests. For example, in computing the following polynomial proposed by S.M. Rump [5],

$$f(a, b) = 333.75b^6 + a^2(11a^2b^2 - b^6 - 121b^4 - 2) + 5.5b^8 + a/(2b),$$

with $a = 77617.0$ and $b = 33096.0$, correct answer $-0.8273960599\ldots$ was obtained with a 140-digit accuracy.

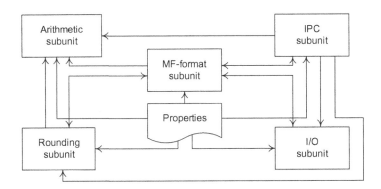

Fig. 4. Structure of the MF-Library kernel: an arrow pointing from one subunit to another indicates that the functions (or data) defined in the first subunit are used by the second subunit

5.2 Efficiency Evaluations

During the experiments, the performance of the MF-Library was compared to its counterparts: MPFR 3.0.1, NTL 6.1.0 and Wolfram Mathematica 10.0. In all experiments, computation precision was 239 bits. The time taken to transform data into the target format was not taken into account. Test system: Intel Core i5-3570 K / 8 Gb RAM / Intel C++ Compiler 13.0. The MF-Library operated in vectorization of cycles of computation of modular significands and IPCs.

First experiment. The time used to execute multiple-precision operations was measured: addition (add), subtraction (sub), multiplication (mult), division (div), comparison (cmp), addition-accumulate (aac, $x = x + y$), subtraction-accumulate (sac, $x = x - y$) and multiply-accumulate (mac, $z = z + xy$). 10^7 iterations were performed. For operations with accumulation, this means that by the end of the cycle, the accumulator contained the results of 10^7 summations or subtractions. The arithmetic average was chosen as the final time. The results obtained are shown in Table 4.

Table 4. Timings in microseconds for some operations. Initial data – pseudo-random 239-bit floating-point numbers. The best results are marked bold. Higher speedup of MF-Library is expected with an increase in precision and length of vector registers.

	MF-library	MPFR	NTL	Wolfram math.
add	**0.041**	0.057	0.154	0.480
sub	**0.043**	0.071	0.153	0.656
mult	**0.028**	0.211	0.549	0.416
div	6.360	**0.357**	0.649	0.946
cmp	0.034	**0.004**	0.099	0.246
aac	**0.027**	0.059	0.193	0.661
sac	**0.037**	0.058	0.190	0.946
mac	**0.230**	0.279	0.730	0.632

Multiplication is the most efficient operation in the library. It is performed 7.54 times faster than in MPFR. A more effective algorithm can be used to accelerate division. The higher speed of the MF-Library as against the results obtained previously [25] is due to code optimization that does not affect operation execution algorithms substantially. Wolfram Mathematica turned out to be the slowest. It was not considered further.

Second experiment. The time used for high-precision multiplication of dense real m-by-m matrices was studied. The order of matrix m varied in the range of 100 to 800. The number of partial products that need to be summed to provide the final element increases linearly with an increase in m. This allows to evaluate the efficiency of exponent alignment and rounding algorithms. The initial matrices were tightly filled with pseudo-random 239-bit numbers. The experiment results are shown in Fig. 5.

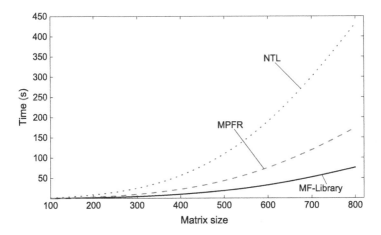

Fig. 5. Time used for high-precision multiplication of dense real matrices. On average, performance of the MF-Library is 2.29 times higher than that of MPFR and 5.75 times higher than that of NTL.

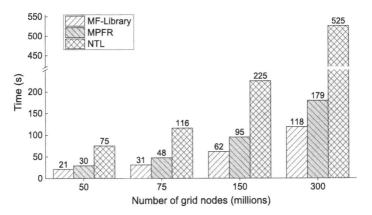

Fig. 6. Time used for high-precision solution of heat equation. On average, performance of the MF-Library is 1.51 times higher than that of MPFR and 4.05 times higher than that of NTL.

Third experiment. The time taken for high-precision solution of the boundary value problem for a one-dimensional heat equation with homogeneous boundary conditions of the first kind was investigated:

$$\frac{\partial u}{\partial t} = D\frac{\partial^2 u}{\partial x^2}, \qquad u(t, 0) = u(t, 1) = 0. \tag{8}$$

Equation (8) describes the dynamics of temperature $u(t, x)$ in a unit-length rod made of an isotropic material. 0.75 s of the physical process was modeled. A four-point explicit difference scheme was used for solution. The computations were performed at a constant ratio $D = 0.03$ on rectangular grids of different dimensions: $10^3 \times 50 \cdot 10^3$ ($h = 10^{-3}, \tau = 1.5 \cdot 10^{-5}$), $10^3 \times 75 \cdot 10^3$ ($h = 10^{-3}, \tau = 10^{-5}$), $10^3 \times 150 \cdot 10^3$ ($h = 10^{-3}, \tau = 0.5 \cdot 10^{-5}$), $10^3 \times 300 \cdot 10^3$ ($h = 10^{-3}, \tau = 0.25 \cdot 10^{-5}$), where h and τ are steps on spatial and temporal coordinates, respectively. The experiment results are presented in Fig. 6.

6   Conclusion

The technique for multiple-precision computations, which was considered, combines the advantages of residue number systems and positional floating-point arithmetic. Representation of significands in RNS allows you to process them faster by separating slow operations on large numbers into several independent modular operations that can be performed in parallel. On the other hand, the presence of a binary exponent ensures a wide dynamic range of values, which is of high importance in scientific computations that often require processing of multi-scale quantities.

By supplementing the number format with interval-positional characteristic, which allows for rapid assessment of the magnitude of the modular significand without conversion to a positional number system, we were able to significantly weaken, and in some cases completely overcome the main disadvantages of RNS that are related to the very complex nature of non-modular operations (magnitude-comparison, overflow detection, scaling, etc.).

Consequently, the algorithms for modular-positional floating-point arithmetic, implemented in new multiple-precision library MF-Library, have higher performance (except division algorithm) than their counterparts that are based on positional multiple-precision arithmetic (MPFR, NTL) and symbolic computation (Wolfram Mathematica), both when performing individual operations and long iterative computations that require multiple exponent alignments and rounding.

Despite the fact that the main algorithms and library routines have been developed, there are a number of important problems that remain to be solved:

1. Defining strict rounding error bounds when performing operations.
2. Precomputing the main mathematical constants.
3. Calculating mathematical functions (sqrt, exp, log, sin, cos, acos, atan etc.).
4. Accelerating division operation: the current division algorithm involves converting the significand from RNS to a positional number system and this is, therefore, slow.
5. Offering cross-platform service with hardware optimization features.

We also plan to investigate the possibility of efficiently implementing MF-Library in parallel on graphics processors and specialized many-core accelerators (Intel Xeon Phi). There is reason to believe that this will significantly speed up computations involving numbers of extra high precision (4096 bits or above).

Acknowledgement. The reported study was supported by RFBR, research project No. 14-07-31075 mol_a.

References

1. Collange, S., Defour, D., Graillat, S., Iakymchuk, R.: Reproducible and accurate matrix multiplication for high-performance computing. In: Nehmeier, M. (ed.) Book of Abstracts of the 16th GAMM-IMACS International Symposium on Scientific Computing, Computer Arithmetic and Validated Numerics (SCAN 2014), pp. 42–43. Würzburg, Germany (2014)
2. Bailey, D.H., Barrio, R., Borwein, J.M.: High-precision computation: mathematical physics and dynamics. Appl. Math. Comput. **218**(20), 10106–10121 (2012)
3. Bailey, D.H., Borwein, J.M.: High-Precision Arithmetic: Progress and Challenges. http://www.davidhbailey.com/dhbpapers/hp-arith.pdf
4. Ghazi, K.R., Lefèvre, V., Théveny, P., Zimmermann, P.: Why and how to use arbitrary precision. Comput. Sci. Eng. **12**(3), 62–65 (2010)
5. Rump, S.M.: Algorithms for verified inclusions - theory and practice. In: Moore, R.E. (ed.) Reliability in Computing, pp. 109–126. Academic Press, New York (1988)
6. IEEE Standard for Floating-Point Arithmetic, IEEE Std. 754–2008, pp. 1–58. IEEE Computer Society, New York (2008)
7. Muller, J.-M., Brisebarre, N., de Dinechin, F., Jeannerod, C.-P., Lefèvre, V., Melquiond, G., Revol, N., Stehlé, D., Torres, S.: Handbook of Floating-Point Arithmetic. Birkhäuser, Boston (2010)
8. Dekker, T.J.: A floating-point technique for extending the available precision. Numerische Mathematik **18**(3), 224–242 (1971)
9. Bailey, D.H., Hida, Y., Li, X.S., Thompson, B.: ARPREC: An Arbitrary Precision Computation Package. Lawrence Berkeley National Laboratory, Technical report (2002). http://www.davidhbailey.com/dhbpapers/arprec.pdf
10. The GNU Multiple Precision Arithmetic Library. https://gmplib.org
11. Fousse, L., Hanrot, G., Lefèvre, V., Pélissier, P., Zimmermann, P.: MPFR: A multiple-precision binary floating-point library with correct rounding. ACM Trans. Math. Softw. **33**(2) (2007). Article No. 13
12. NTL: A Library for doing Number Theory. http://www.shoup.net/ntl
13. Garner, H.L.: The residue number system. IRE Trans. Electron. Comput. **8**(2), 140–147 (1959)
14. Szabo, N.S., Tanaka, R.I.: Residue Arithmetic and its Application to Computer Technology. McGraw-Hill, New York (1967)
15. Omondi, A., Premkumar, B.: Residue Number Systems: Theory and Implementation. Imperial College Press, London (2007)
16. Parhami, B.: Computer Arithmetic: Algorithms and Hardware Designs. Oxford University Press, Oxford (2000)
17. Cardarilli, G.C., Del Re, A., Nannarelli, A., Re, M.: Programmable power-of-two RNS scaler and its application to a QRNS polyphase filter. In: 2005 IEEE International Symposium on Circuits and Systems (ISCAS 2005), pp. 1002–1005. Kobe, Japan (2005)
18. Huang, C.H.: A fully parallel mixed-radix conversion algorithm for residue number applications. IEEE Trans. Comput. **32**(4), 398–402 (1983)
19. Gbolagade, K.A., Cotofana, S.D.: An $O(n)$ residue number system to mixed radix conversion technique. In: IEEE International Symposium on Circuits and Systems (ISCAS 2009), pp. 521–524. IEEE Press, New York (2009)
20. Akkal, M., Siy, P.: A new mixed radix conversion algorithm MRC-II. J. Syst. Archit. **53**, 577–586 (2007)

21. Isupov, K.S.: The method for implementation non-modular operations in modular arithmetic with use of interval positional characteristics. University proceedings. Volga region. Eng. Sci. **3**, 26–39 (2013)

22. Isupov, K.S.: Calculation interval-positional characteristic algorithm for implementation non-modular operations in residue number systems. Bulletin of the South Ural State University. Comput. Technol., Autom. Control, Radio Electron. **14**(1), 89–97 (2014)

23. Isupov, K.S.: On an algorithm for number comparison in the residue number system. Vestnik of Astrakhan State Technical University. Manage., Comput. Sci. Inf. (3), 40–49 (2014)

24. Kulisch, U.: Implementation and Applications Computer Arithmetic and Validity - Theory, implementation, and applications. de Gruyter, Berlin (2008). http://www.degruyter.com/view/product/178972

25. Isupov, K.S., Maltsev, A.N.: A parallel-processing-oriented method for the representation of multi-digit floating-point numbers. Numer. Methods Program. **15**(4), 631–643 (2014)

26. Sasaki, A.: The basis for implementation of additive operations in the residue number system. IEEE Trans. Comput. **17**(11), 1066–1073 (1968)

27. Kinoshita, E., Kosako, H., Kojima, Y.: Floating-point arithmetic algorithms in the symmetric residue number system. IEEE Trans. Comput. **23**(1), 9–20 (1974)

28. Kinoshita, E., Lee, K.-J.: A residue arithmetic extension for reliable scientific computation. IEEE Trans. Comput. **46**(2), 129–138 (1997)

29. Chiang, J.-S., Lu, M.: Floating-point numbers in residue number systems. Comput. Math. Appl. **22**(5), 127–140 (1991)

Creation of Data Mining Algorithms as Functional Expression for Parallel and Distributed Execution

Ivan Kholod[(⊠)] and Ilya Petukhov

Saint Petersburg Electrotechnical University "LETI",
ul. Prof. Popova 5, Saint Petersburg, Russia
iiholod@mail.ru, ioprst@gmail.com

Abstract. The article describes extension of λ-calculation for creation of parallel data mining algorithms. The proposed approach uses presentation of the algorithm **as a consequence of pure functions with unified interfaces.** For parallel execution we use special function that allows to change a structure of the algorithm and to implement various strategies for processing of data set and model.

Keywords: Parallel algorithms · Data mining · Parallel data mining · Distributed data mining · Data mining algorithms

1 Introduction

At present time, modern computing systems allow the accumulation of big data storages. For analyzing of big data need super quickest algorithms. Unfortunately, the most data analyze algorithms use complex mathematical or heuristic approaches and cannot work quickly. Therefore, it is very important to increase their performance by means of algorithm parallelization. However, the creation of parallel algorithms and achieving their high efficiency on parallel or distributed systems is no common task. The key issues are the need for synchronization of access to the data, minimization of interaction between the parallel components, load balancing and so on.

We suggest an approach which allows to transform a sequential algorithm into a parallel one with various structures and implementation of various execution strategies for various conditions. It is possible because we use principles of functional programming and λ-calculus theory [1]. A peculiar feature of functional languages is the absence of program status and therefore the need for its change. The functions process only the variables being the arguments of these functions and not use global variables. Such functions are "pure" functions.

2 Related Work

A lot of research is fulfilled currently in the field of parallel and distributed data mining algorithm developing. As a matter of fact, separate focus areas can be distinguished within the data mining field [2]: parallel data mining (PDM) and distributed data mining (DDM).

© Springer International Publishing Switzerland 2015
V. Malyshkin (Ed.): PaCT 2015, LNCS 9251, pp. 62–67, 2015.
DOI: 10.1007/978-3-319-21909-7_6

There are several problems in developing parallel algorithms for a distributed environment with association discovery data mining which is being considered in research work [3]: data distribution, minimizing communication, maximizing locality and other, etc. Achieving all of the these goals in one algorithm is nearly impossible, as there are tradeoffs between several of the above points.

As for the approach towards the elaboration of parallel data mining algorithms, two main approaches can be singled out:

- individual parallelizing of algorithms: individual approach to each data mining algorithm and choosing most efficient parallel structure for the given conditions;
- universal algorithm parallelizing: generalized approach to the data mining algorithms, suggesting its decomposition into parts which can be run concurrently.

The main work in the sphere of parallel data mining is aimed at individual parallelization of algorithms. Examples of data mining algorithms for specific types of computing systems are reviewed in the paper [3]. With this approach the complexity and effort for developing of parallel algorithms is very high. At that this effort is aimed at adapting the algorithms to execution strictly in the required conditions. The changes to the conditions lead to the necessity of conversion of the algorithm which is in fact a creation of a new algorithm.

For example of the universal algorithm parallelizing approach we can refer to two works: NIMBLE system [4] and method for data mining algorithms corresponding to Statistical Query Model [5]. Both methods are MapReduce paradigm oriented. However not all data mining algorithms can be implemented on the basis of this paradigm without substantial processing.

We offer universal approach to constructing parallel data mining algorithms with the use of functional programming paradigm. This approach allows to easily convert sequence algorithms into its parallel form for various conditions.

3   Data Mining Algorithm as Functional Expression

The main idea using in proposed approach is an algorithm must be decomposed into thread-safe blocks. We modified approach proposed in the paper [6] by using the functional language principles. They based on the λ-calculus theory [1] have this feature, because classic functions in the functional languages are pure functions. According to Church-Rosser theorem [1], reduction of functional expression of pure function can be fulfilled in any order, also concurrently.

A data mining algorithm can be presented as a sequence of function calls:

$$dma = fb_n \circ fb_{n-1} \circ \ldots \circ fb_i \circ \ldots \circ fb_1 = fb_n(d, \ fb_{n-1}(d, \ldots fb_i(d, \ldots fb_1(d, nil) \ldots) \ldots)),$$

where fb_i : is function (function block) of the type $FB:: D \rightarrow M \rightarrow M$, where

- D: is input data set that is analyzed by the function fb_i;
- M: is mining model that is built by the function fb_i.

So according to Church-Rosser theorem reduction (execution) of such functional expression (algorithm) can be doing concurrently.

As example of a presentation of data mining algorithms as functional expression, we consider classification algorithm 1R [7]. It can be present as expression from two functions of the FB type:

$$1R = rulesCycle \circ vectorsCycle, \text{where} \tag{1}$$

- *vectorsCycle*: is function of cycle for vectors which builds the rule and computes its error for each attribute of each vector;
- *rulesCycle*: is function of cycle for rules which selects of rule with minimal error for the each value of the target.

One of the main advantages of elaborating algorithms from the function blocks is the possibility of their concurrent execution. At that of practical value is the parallel execution of the function blocks computing the arguments at applicative reduction order. Thus, for the parallel execution of the data mining algorithm blocks they must be invoked for computing the arguments of one function. For example, in an expression:

$$fb_i\big(d, \ fb_j(d, \ m), \ fb_p(d, \ m), \ fb_q(d, \ m)\big)$$

the blocks fb_j, fb_p and fb_q can be computed concurrently. Such concurrent computations correspond to the task parallelism. The data mining algorithms containing such blocks in their structure have inner parallelism and can be parallelized.

However, the data mining algorithms are mostly characterized with data parallelism. In this case explicit conversion of the functional expression is needed with adding of data partitioning function and subsequent aggregation of results. For the function blocks the input parameters are input data set D and mining model M. Thus the parallelizing can be fulfilled both for the input data D, and for the model M.

To make the concurrent execution of a data functional expression, it must be converted into the form in which the function blocks will be invoked as arguments. For this we added a function which will allow data-parallelizing in the algorithms:

$$<parallelization\_function\_name> \ = join \circ fb \circ (splitD, \ splitM), \text{where}$$

- *join:* the function joining the mining models from the list *[M]* and returning the merged mining model *M*: $join :: [M] \rightarrow M$.
- *fb:* a function block is executed concurrently;
- *splitD*: the function fulfilling the splitting of the data set D and returning list from the n split data sets *[D]*: $splitD :: D \rightarrow M \rightarrow [D]$;
- *splitM*: the function fulfilling the splitting of the mining model M and returning lists from the n split mining models *[M]*: $splitM :: D \rightarrow M \rightarrow [M]$.

A combination of split functions allows to implement the following strategies for concurrent execution of a data mining algorithms:

- in part of data processing:
  - single data set: in this case each parallel block processes a copy of one data set;
  - separated data sets: in this case each parallel block processes a part of the source data set.
- in part of model processing:
  - same model: in this case each parallel block receives a model copy;
  - separated model: in this case each parallel block receives a part of the general model.

Implementations of the parallelization function for various strategies are presented in the Table 1. In this function the fb_i block is invoked to compute the arguments of the *join* function, therefore can be executed concurrently (Church-Rosser theorem).

Table 1. Implementations of the parallelization function for various strategies

Strategy	Single data set	Separated data set
Same mining mode	$join \ (([fb_i(d, m), ..., fb_i(d, m))$	$join \ ([fb_i \ (m, \ splitD(d, m)[0]), ..., fb_i \ (m, \ splitD(d, m)[n])])$
Separated mining model	$join \ (([fb_i \ (splitM(d, m)[0], d), ..., fb_i(splitM(d, m)[n], d)])$	$join \ ([fb_i(splitM(d,m)[0], \ splitD(d,m)[0]), ..., fb_i(splitM(d,m)[n], \ splitD(d, m)[n])])$

Thus, to a data mining algorithm presented as functional expressions can be execute (reductive) parallel need to add the parallelization function to expression. It can be added for any function block in the functional expression thus converting the expression to the parallel form.

For example, we can execute parallel function *vectorsCycle* of the 1R algorithm and implement the "separated data sets" strategy. For this, need to wrap function *vectorsCycle* in parallelization function for the expression (1):

$$vectorsCycleParall = join \circ vectorsCycle \circ splitD$$
$$1RVectorsCycleParallel = rulesCycle \circ vectorsCycleParall. \tag{2}$$

Another variant is parallel execution of *rulesCycle* function of the 1R algorithm and implementation of the "separated model" strategy:

$$rulesCycleParall = join \circ rulesCycle \circ splitM.$$
$$1RRulesCycleParallel = rulesCycleParall \circ vectorsCycle \tag{3}$$

Last variant of parallel form of 1R algorithm is parallel execution of both functions and implementation two strategies at same time:

$$1RWholeParall = join \circ (rulesCycle \circ vectorsCycle)^\circ (splitD, \ splitM) \tag{4}$$

Similarly we can parallel execute any function of a data mining algorithm presented as functional expression.

4 Experimental Results

We modified the framework for multi threads execution of data mining algorithms [6, 8] for the proposed approach and implemented for the 1R algorithm parallel variations (2), (3) and (4). We have performed several experiments for the implemented algorithms. The experiments have been performed with various input data sets (Table 1). These data sets contain various numbers of vectors and attributes. The experiments have been done on a multicore computer the following configuration: CPU: Intel i7 3.4 GHz, RAM: 4 Gb, OS: Windows 7, JDK 1.7. The parallel algorithms have been executed for the numbers of threads equal to 2, 4 and 8, respectively. The experimental results are provided in Table 2. Correctness of algorithms has been verified by comparison of built mining models. All algorithms have built same mining models for same data sets.

Table 2. Data sets

Data set	W1	W3	W5	W10	A1	A3	A5	A10
Number of vectors	10 000	30 000	50 000	100 000	1 000	1 000	1 000	1 000
Number of attributes	10	10	10	10	100	300	500	1 000
Avg. number of classes	5	5	5	5	5	5	5	5

Table 3. Experement results

Algorithms	Threads	W1	W3	W5	W10	A1	A3	A5	A10
1R	1	125	376	614	1 205	360	2 317	5 731	31 776
1RVectorsCycle Parallel	2	**109**	**266**	**390**	**815**	203	1 330	3 236	15 910
	4	**47**	**250**	**371**	**601**	153	953	2 328	11 202
	8	**36**	**130**	**297**	**455**	101	892	2 000	10 339
1RRulesCycle Parallel	2	141	390	673	1 375	344	1 197	**2 442**	**13 357**
	4	156	434	725	1 370	376	883	**1 690**	**10 515**
	8	113	322	620	1 102	432	585	**1 260**	**8 295**
1RWhole Parallel	2	109	251	401	875	223	1 457	3 607	21 956
	4	93	206	329	630	181	1 126	3 092	16 382
	8	78	119	309	550	98	820	2 836	14 515

The experiments show that parallel execution of the 1R algorithms for data sets with various parameters is different (Table 3). The parallel form of algorithm (2) is more efficiently for data sets with a large number of vectors (W*). It is possible because large number of iteration for all vectors is splitted between few threads. Unlike it the parallel form of algorithm (3) is more efficiently for data sets with a large number of attributes (A5 and A10). This is because 1R algorithm executes much iteration for rules in case of these data sets and them processing are divided at the few threads. The parallel form of algorithm (4) is efficiently for both types of data sets, but is less efficiently than the form (2) for data sets W*, because it needs to execute the *splitM* function, additional.

5   Conclusion

Presentation of a data mining algorithm as functional expression makes it possible to divide the algorithm into functions of FB type (functional blocks). Such a splitting of data mining algorithms into blocks allows us to easily create parallel algorithms from the sequential algorithm by adding special structural elements for parallel execution.

The proposed approach in the present article makes it possible to construct different variants of parallel data mining algorithms and to implement different efficiently execution strategies for various data sets. As future work we plan to extend distributed environments for Actor model, MapReduce and other.

Acknowledgments. The work has been performed in Saint Petersburg Electrotechnical University "LETI" within the scope of the contract Board of Education of Russia and science of the Russian Federation under the contract № 02.G25.31.0058 from 12.02.2013. This paper is also supported by the federal project "Organization of scientific research" of the main part of the state plan of the Board of Education of Russia and project part of the state plan of the Board of Education of Russia (task # 2.136.2014/K).

References

1. Church, A., Barkley Rosser, J.: Some properties of conversion. Trans. AMS **39**, 472–482 (1936)
2. Paul, S.: Parallel and distributed data mining. In: Funatsu, K. (ed.) New Fundamental Technologies in Data Mining, pp. 43–54. INTECH Open Access Publisher (2011). http://www.intechopen.com/books/new-fundamental-technologies-in-data-mining/parallel-and-distributed-data-mining
3. Zaki, M.J., Ho, C.-T. (eds.): Large-Scale Parallel Data Mining. LNCS, vol. 1759, pp. 1–23. Springer, Heidelberg (2000)
4. Amol, G., Prabhanjan, K., Edwin, P., Ramakrishnan, K.: NIMBLE: a toolkit for the implementation of parallel data mining and machine learning algorithms on MapReduce. In: Proceedings of the 17th ACM SIGKDD International Conference on Knowledge Discovery and Data Mining (KDD 2011), pp.334–342. San Diego, California, USA (2011)
5. Ng, A.Y., Bradski, G., Chu, C-T., Olukotun, K., Kim, S.K., Lin, Y-A., Yu, Y.Y.: Map-Reduce for machine learning on multicore. In: Proceedings of the Twentieth Annual Conference on Neural Information Processing Systems, pp. 281–288. Vancouver, Canada. (2006)
6. Kholod, I., Karshiyev, Z., Shorov, A.: The formal model of data mining algorithms for parallelize algorithms. In: Wiliński, A., Fray, I.E., Pejaś, J. (eds.) Soft Computing in Computer and Information Science. AISC, vol. 342, pp. 385–394. Springer, Heidelberg (2015)
7. Holte, R.C.: Very simple classification rules perform well on most commonly used datasets. Mach. Learn. **11**, 63–90 (1993)
8. Kholod, I.: Framework for multi threads execution of data mining algorithms. In: Proceeding of 2015 IEEE North West Russia Section Young Researchers in Electrical and Electronic Engineering Conference. (2015 ElConRusW), pp. 74–80. IEEE Xplore (2015)

Dynamic Parallelization Strategies
for Multifrontal Sparse Cholesky Factorization

Sergey Lebedev, Dmitry Akhmedzhanov, Evgeniy Kozinov,
Iosif Meyerov[(✉)], Anna Pirova, and Alexander Sysoyev

Lobachevsky State University of Nizhni Novgorod, Nizhni Novgorod, Russia
meerov@vmk.unn.ru

Abstract. This paper discusses parallelization of the computationally intensive numerical factorization phase of sparse Cholesky factorization on shared memory systems. We propose and compare two parallel algorithms based on the multifrontal method. Both algorithms are implemented in a task-based fashion employing dynamic load balance. The first algorithm associates OpenMP tasks with the nodes of an elimination tree and relies on the OpenMP scheduler. The second algorithm employs a concurrent priority queue to implement balancing. Experimental results on symmetric positive definite matrices from the University of Florida Sparse Matrix Collection show that our implementation is comparable to MUMPS and Intel MKL PARDISO in terms of performance and scaling efficiency on shared memory systems.

Keywords: Sparse algebra · Cholesky factorization · Numerical phase · Multifrontal method · High performance computing · Dynamic parallelization · Task-based parallelism

1 Introduction

Systems of linear algebraic equations (SLAE) Ax = b with large sparse symmetric positive definite matrix A arise in a wide range of scientific and engineering problems from different domains. The need for solution of such SLAE springs up when modeling various physical processes, e.g. in large scale finite element analysis in mechanical engineering. For solution of sparse SLAE direct and iterative methods are applied. Both approaches have advantages and disadvantages flowing out of the inward nature of the methods.

Direct methods are based on matrix factorization with the following triangular solutions which are much simpler. These methods are distinguished by their reliability and numerical stability, but worse computational complexity. They also require additional extensive memory resources for keeping intermediate matrices. Iterative methods in turn are based on consecutive approaching to the solution and are deprived to a significant extent of the disadvantages mentioned, but in a number of instances may demonstrate slow convergence.

In this paper we discuss the issues of the effective implementation of direct methods of SLAE-solving with symmetric, positive definite matrix, oriented towards modern multi-core architectures.

© Springer International Publishing Switzerland 2015
V. Malyshkin (Ed.): PaCT 2015, LNCS 9251, pp. 68–79, 2015.
DOI: 10.1007/978-3-319-21909-7_7

Prominent progress in the area of sparse direct SLAE-solving methods parallelization has been achieved over the last decade. Thus, multilevel hierarchical memory structure usage (supernode ideas), vector computing, parallelism for shared-memory multiprocessors and for clusters and other approaches have been worked out. Also, implementations for GPU-based computing are being developed [1]. The described efforts have resulted in a variety of software solutions for direct solving – MKL PARDISO, Cluster PARDISO, MUMPS, SuperLU, CHOLMOD, etc. Contemporary software packages allow to solve systems of millions of variables, which opens wide possibilities of numerical modeling in physics, chemistry, biology, medicine, and many other fields.

It should be noted, that even though significant results in solving methods and their implementation have been reached, the research of the issue cannot be considered as finished. Thus, permanent developing of computing architectures and programming paradigms not only provides an opportunity to solve more and more complicated problems, but also demands considerable efforts to increase SLAE-solving algorithms and programs efficiency.

The multifrontal factorization concept is one of the most effective among the direct methods. In this paper we propose a new way of parallelization of the multifrontal method for shared memory systems. As opposed to the static parallelization method our solver uses a dynamic scheme. We do load balancing via task based parallelism. Two ways of parallelization are compared. The first scheme uses OpenMP tasks and the second one operates with a task pool, organized as a priority queue. We show that both approaches have advantages. Series of tests were done to compare our implementation to the two state-of-the-art solvers – Intel MKL PARDISO and MUMPS. The results show comparability of our implementation to the other solvers. The rest of the paper is organized as follows: we summarize related work, give a short overview of the multifrontal method, describe our approach to parallelization for shared memory systems, give numerical results on the matrices from the University of Florida Sparse Matrix Collection, compare performance and scaling efficiency, and discuss the results.

2 Related Work

2.1 Direct Methods for Sparse SLAE

Consider SLAE Ax = b, where A is known square, sparse, symmetric, positive definite matrix, x and b are dense vectors. On the assumption given, there exists a unique Cholesky factorization of the form (1), where L is a lower triangular matrix.

$$A = LL^T \tag{1}$$

Fill-in of the matrix L is one of the central problems arising out of matrix factorization. This problem cannot be solved completely, but matrix A rows and columns reordering before the factorization allows to reduce the fill-in and decrease both memory and processing time expenses. Further, during the direct solving procedure, matrix A is being factorized, two triangular systems are being solved and components of a solution x are being back transposed.

As a general rule, matrix factorization is the most computationally intensive step of the direct solution. Calculations at that are usually divided into two phases: Symbolic phase and Numerical Phase. In the course of the symbolic part a non-zero elements pattern is performed, in the course of the numerical phase – filling the pattern with values. It should be noted that the symbolic phase is carried out much faster than the numerical one; therefore the combined effort of researchers is directed mostly at optimization and parallelization of the numerical phase.

The main methods of numerical Cholesky factorization may be divided into three groups – right-looking Cholesky [2], left-looking Cholesky [2], and Multifrontal method [3]. The first two methods differ in the order of pivotal entry elimination. In the column-oriented right-looking method a yet another column j of the factor is computed. After that, the pivotal entry is eliminated in all following columns with a non-zero element in a row j. The column-oriented left-looking is much the same, except the second step, which treads to the right. The method mentioned demonstrates similar efficiency. This statement is confirmed by the fact that a variety of modern solvers employ various methods. Thus, in both CHOLMOD and MKL PARDISO left-looking Cholesky is implemented, whilst MUMPS uses the multifrontal method [4, 5]. Even as part of the same solver, various approaches may be used, for instance, left-looking method in SuperLU_MT (parallel version for shared memory systems) and right-looking method in SuperLU_Dist (parallel version for distributed memory systems) [6]. In this work the multifrontal method was chosen as primary.

2.2   Multifrontal Method Overview

The multifrontal method is one of the most effective and scalable numerical factorization schemes. It has made the first appearance in Duff and Reid's paper [7, 8] in 1983, has been developed by Liu [3, 9], by Amestoy et al. [10], by Amestoy et al. [11]. The approach lies in the fact that the factorization process is being divided into factorization of small dense matrices that are called frontal. The multifrontal method can effectively employ all of the cache-memory levels; also, if data structures are properly implemented, BLAS 3 operations become dominant. Supernodal multifrontal concept has appeared for the first time in the paper of Ashcraft et al. [12] in 1987, thereafter it has been studied by Ng and Peyton [13, 14]. The key idea of the concept is in allocation and further usage of supernodes [15] – groups of columns with similar or exactly the same structure beneath the upper triangle. The approach given allows to use optimized BLAS 3 dense matrix procedures for factorization in a blocked manner, which significantly accelerates calculations. In general, high parallelization potential can be regarded as an advantage of the multifrontal method. This assertion may be proven by successful applying of the method for distributed memory systems in MUMPS solver. Should be noted, though, that the necessity of intermediate results presentation leads to high memory losses. This, and large number of floating point operations are among limitations of the multifrontal method. The effect mentioned often appears for problems originate from three-dimensional space discretization [16].

Calculations in the multifrontal method are being conducted in accordance with a task graph, which in the context of symmetric, positive definite matrix has the form of tree and is called elimination tree. Each node of the tree corresponds to a matrix column. Thus, in the multifrontal method leaves-to-root tree traversal takes place and, when a succeeding node is attended, original matrix column is being processed with following factor column obtaining. High-level multifrontal method description is presented below.

Algorithm 1. High-level multifrontal method description

1	**foreach** node i **of** elimination tree **in** topological order
2	init_frontal_matrix(F_i)
3	**foreach** son j **of** i **do**
4	$U \leftarrow U \oplus U_j$
5	**end for**
6	assembly_frontal_matrix(F,U)
7	factorize(F)
8	form_update_matrix(U_i)
9	$L_i \leftarrow F_{(1,*)}$
10	**end for**

Herein procedures init_frontal_matrix, assemble_frontal_matrix and form_update_matrix are accomplished through the use of corresponding BLAS functions. More detailed description may be found in [3, 16].

There are two parallelization schemes that can be employed in terms of the multifrontal method: parallel BLAS functions usage and parallel solving of independent tasks in correspondence with elimination tree structure. Let us give considerations to these methods' prospects.

2.3   Parallel BLAS Usage

The majority of computations in the multifrontal method fall within BLAS procedures, such as matrix multiplication and the Gaussian elimination. Therefore, usage of existing high-performance computation libraries, such as, for instance, Intel MKL, is one of the most natural ways to parallel the numerical phase of Cholesky factorization. Unfortunately, as experiments show, applying the approach mentioned more often than not leads to disappointing results. It can be explained by the fact, that most of intermediate matrices, arising during the calculations, have small dimension, so overhead charges, associated with organization of the parallelism are not compensated with following parallel processing benefits. Hence, desirable efficiency can be achieved only if parallelizing on the basis of the elimination tree is selected as primary.

2.4   Parallelization in Terms of Elimination Tree

Paralleling of the multifrontal method can be accomplished drawing on the elimination tree, which contains the knowledge about all the data dependencies that can occur during the calculations. The fact, that the structure of the matrix L row is a subset of the subtree of the elimination tree with a root at corresponding to that row node is a key point here. Therefore, any node from the set of non-zero elements of the column j is a descendant of the node j in the elimination tree. Thus, the column j cannot be processed until the columns with numbers from the mentioned set have been processed. Continuing this reasoning, it can be shown, that for calculating a consecutive column, processing all the columns from the subtree of the conforming node is essential. Nevertheless, other nodes are not employed in calculations, so they can be considered independently. Hence, a couple of columns, i and j, can be processed in a parallel way if and only if subtrees T[i] and T[j] do not intersect, in other words, do not have common nodes. Both static and dynamic strategies can be used to parallel the computations on the basis of the elimination tree.

2.5   Static Parallelization Strategies

There is a plurality of methods that use a static parallelization scheme, the main part of them are based on the Geist-Ng algorithm [17]. The idea of the algorithm is in detection a layer in the elimination tree, a set of nodes that are not compulsive on the same level, but they do not have common descendants. The located layer must be balanced, which means that the number of operations required to process subtrees' nodes with roots among the mentioned layer nodes complies with the preset threshold and is also much the same.

A comparison of static parallelization methods may be found in papers [18, 23].

3   Dynamic Parallelization Strategies

One of the main disadvantages of static schemes is impossibility of accurate evaluation of workload demandable to process each node of the elimination tree. This is the reason why we suggest one more way of balancing the assignment, based on the dynamic scheme.

In terms of dynamic scheme a task pool is being built. At any step of the algorithm a thread takes task from the queue and proceeds to its accomplishment.

Each task conforms to computing the corresponding column of the factor and consists of four subtasks: calculating a node matrix, calculating a frontal matrix, forming the factor column out of the frontal matrix, calculating a renew matrix.

3.1   OpenMP Tasks

One of the ways to implement the dynamic scheme is to use OpenMP tasks to parallelize computations. In this regard it is sufficient to organize the elimination tree traversal in topological order.

Algorithm 2. Parallel multifrontal method using OpenMP tasks (**alg_task**)
1   **procedure** process_node(*node* of *elimination_tree*)
2   **foreach** *son* of *node* in elimination_tree **do**
3   **#pragma omp task**
4   process_node(*son*)
5   **end for**
6   **#pragma omp taskwait**
7   multifrontal_step(*node*)
8   **end procedure**
9
10   **procedure** multifrontal_step(*node* of elimination tree)
11        $i \leftarrow$ number of *node* in elimination tree
12   init_frontal_matrix(F_i)
13   **foreach** son j of i **do**
14            $U \leftarrow U \oplus U_j$
15   **end for**
16   assembly_frontal_matrix(F,U)
17   factorize(F)
18   form_update_matrix(U_i)
19        $L_i \leftarrow F_{(1,*)}$
20   **end procedure**

3.2 Priority Queue

Speaking of the second algorithm, the task pool is organized as a priority queue (Fig. 1).

To form the task pool mentioned we use the algorithm that takes into account varying tree node characteristics to achieve better balancing. The algorithm traverses all the nodes of the tree in accordance with topological order that has been formed already.

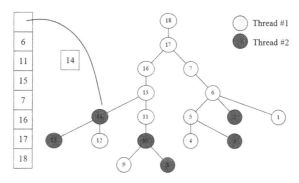

Fig. 1. The dynamic parallelization strategy on the ground of the priority queue.

It also adds the concerned node to the queue. The priority of the node is a combination of the primary and the secondary priorities. The first one corresponds to the proper column traversal in the parallel multifrontal method, the second one corresponds to an improvement of the balancing. The primary priority equals to the quantity of the node children in the elimination tree, the secondary can be calculated as an estimation of corresponding subtasks solving complexity. This complexity may be estimated as the number of floating-point operations.

Algorithm 3. Parallel multifrontal method using priority queue (**alg_queue**)

1	**#pragma omp parallel for**
2	**for** $k = 0$ **to** n **do**
3	**#pragma omp critical (queue)**
4	$i \leftarrow$ task_queue.get_task_with_highest_priority();
5	init_frontal_matrix(F_i)
6	**foreach** son j **of** i **do**
7	$U \leftarrow U \oplus U_j$
8	**end for**
9	assembly_frontal_matrix(F,U)
10	factorize(F)
11	form_update_matrix(U_i)
12	$L_i \leftarrow F_{(1,*)}$
13	**#pragma omp critical (queue)**
14	task_queue.increase_task_primary_priority(parent(i))
15	**end for**

---

4 Numerical Results and Discussion

Experimental results have been measured using a cluster node that contains two eight-core processors Intel Sandy Bridge E5-2660 2.2 GHz, 64 GB RAM under Linux CentOS 6.4. Intel C++ Compiler and Intel MKL BLAS library from Intel Parallel Studio XE 2013 SP1 have been used. For conducting experiments matrices from the University of Florida matrix collection [19] have been chosen. Properties of the test matrices are given below (Table 1). All of them are positive definite and symmetric.

4.1   Dependency of alg_queue Algorithm Performance on the Parameters

There are some downsides of paralleling the multifrontal method using the elimination tree. Thus, unnecessary synchronizations happen during the processing of the lower part of the tree. This effect appears because an access to a shared resource – the queue – should be synchronized, even though it has a lot of independent tasks. As a solution to the problem, the dynamic strategy algorithm has been changed – now not tasks are to

Table 1. Test matrices properties

Matrix name	Dimension	Non-zeros in A	Non-zeros in L
Pwtk	217 918	5 926 171	49 025 872
Msdoor	415 863	10 328 399	51 882 257
parabolic_fem	525 825	2 100 225	25 571 376
tmt_sym	726 713	2 903 835	28 657 615
boneS10	914 898	28 191 660	266 173 272
Emilia_923	923 136	20 964 171	1 633 654 176
audkiw_1	943 695	39 297 171	1 225 571 121
bone010_M	986 703	12 437 739	363 650 592
bone010	986 703	36 326 514	1 076 191 560
ecology2	999 999	2 997 995	35 606 934
thermal2	1 228 045	4 904 179	50 293 930
StocF-1465	1 465 137	11 235 263	1 039 392 123
Hook_1498	1 498 023	31 207 734	1 507 528 290
Flan_1565	1 564 794	59 485 419	1 451 334 747
G3_circuit	1 585 478	4 623 152	90 397 858

be extracted, but blocks of tasks. This approach allows reducing the number of synchronizations. A sequence of experiments has been conducted with matrices pointed above and the block size as a parameter (Fig. 2). It could be seem, that this parameter significantly influences the runtime, particularly with greater number of threads. Thus, for one thread the execution time ranges within 6 %, for 8 threads it ranges within up to 16 %, and for 16 threads – up to 48 %. The optimal value has to be assorted independently for every matrix. However, an appropriate value may be selected beforehand on an assumption of matrix properties.

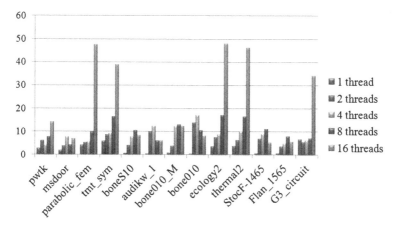

Fig. 2. A percentage difference between the minimal and maximal execution time depending on the task block size.

4.2   A Comparison with State-of-the-Art Solvers

A series of experiments has been conducted with test matrices using two prominent and widely applied solvers:

- MKL PARDISO: Intel Math Kernel Library (as part of Intel Parallel Studio XE 2014)
- MUMPS (ver. 4.10.0)

For all the packages the exact same METIS-generated [20] orderings have been used, but others reordering tools may be utilized as well [21, 22]. Also BLAS and ScaLAPACK functions from Intel MKL library have been employed.

The results obtained are shown on the diagrams below (Fig. 3). The following conclusions could be drawn from the results. On a single core our implementation demonstrates performance similar to MUMPS.

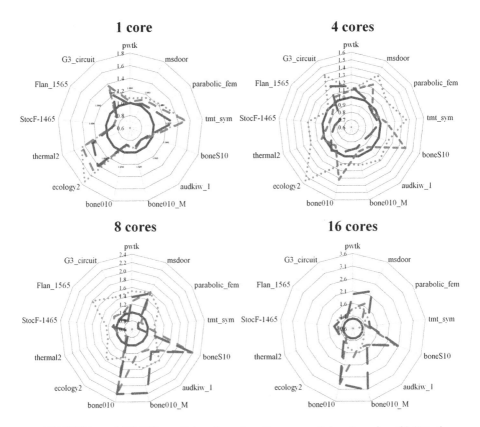

―PARDISO   ··MUMPS   ---Solver based on Queue ――Solver based on OMP tasks

Fig. 3. A comparison of numerical phase of solvers. Axes represent time. MKL PARDISO execution time is taken for 1.

In comparison with MKL PARDISO our implementation is behind on 5 matrices, and is ahead on 5 matrices. The implementation shows the greatest acceleration of 13 % on audikw_1 matrix. Comparing alg_task and alg_queue it could be noted, that in most cases the first algorithm demonstrates better performance, even though calculations are identical. The explanation of this is in the fact that for running one thread OpenMP tasks use stubs, whereas alg_queue builds the task queue, computes priorities and conducts operations with the queue during the calculations. All of those processes bring in additional expenses. For running 4 threads MUMPS performance is remarkably lower in comparison with other solvers. The both implemented algorithms show decent acceleration, leaving behind MUMPS and PARDISO with 7 and 6 matrices (alg_queue) and 12 and 9 matrices (alg_task).

For running 8 threads, the execution time of the implemented algorithms and MUMPS runtime resemble. Alg_queue is ahead of MUMPS with 9 matrices, alg_task – with 8. Significant runtime leaps may be observed on the diagram for experiments with matrices boneS10 and bone010, also with four matrices (pwtk, msdoor, StocF-1465, Flan_1565) results are very similar. Need to note, that greater consistence and larger sizes of supernodes of these matrices are their peculiarities. It means that BLAS functions performance becomes major. In both of the implemented algorithms a consequent version of BLAS is used whereas PARDISO and MUMPS employ a parallel variant [23], which allows them to achieve greater acceleration in experiments with matrices mentioned. This observation may explain the execution time of the solvers for running 16 threads. The lower part of the elimination tree consists for the most part of small tasks, so it can be quickly processed by a larger number of threads, whilst the top part of the tree contains less possibilities for paralleling. Herein processing tasks in parallel is not so important, but effective accomplishment of linear algebra operations is essential. These operations are, for instance, matrix multiplication and Gaussian elimination.

Conducting an insulated comparison of the algorithms proposed (Fig. 4) the following learnings can be made. Alg_task demonstrates better execution time for running

	1 core	2 cores	4 cores	8 cores	16 cores
pwtk	0.96	1.19	1.09	1.08	1.42
msdoor	0.93	0.92	0.96	0.98	1.32
parabolic_fem	0.92	0.87	0.93	0.90	1.34
tmt_sym	0.86	0.82	0.85	0.77	1.08
boneS10	0.95	0.89	0.93	0.99	1.05
audikw_1	1.00	1.01	0.99	1.05	0.99
bone010_M	0.96	1.00	1.42	1.83	2.20
bone010	0.99	1.06	0.93	1.01	1.00
ecology2	0.93	0.91	0.96	0.92	1.17
thermal2	0.87	0.79	0.78	0.74	0.99
StocF-1465	0.99	0.96	0.96	1.02	1.14
Flan_1565	0.99	0.95	0.93	0.91	0.97
G3_circuit	0.88	1.07	0.86	0.78	0.93

Fig. 4. The ratio of alg_task to alg_queue. The runs with alg_task superior to alg_queue are shown in light grey, the opposite case is shown in dark grey, the runs with comparable performance of both implementations are shown in white.

one thread with every matrix. Further, the OpenMP task scheduler deals with his job quite well, but notable runtime leaps can appear, for example, with bone010_M matrix for running 8 threads. Nevertheless, the information about specifics of the tasks becomes a determinative factor for greater numbers of threads. This knowledge enables the alg-queue algorithm to show better results for running 16 threads.

5 Conclusion and Future Work

We have presented a new parallel implementation of the multifrontal method for the Numerical phase of Cholesky factorization. The key idea of our approach is to employ load balancing using task-based parallelism to improve performance and scaling efficiency on shared memory systems. We discuss and compare two ways of implementation. The first algorithm uses OpenMP tasks and works better on 1 to 8 cores on the majority of tests. The second one employs concurrent priority queue to operate with the task pool and effectively converts the information about the matrix into an additional performance gain on 16 cores. Computational experiments on the symmetric, positive definite matrices from the University of Florida Sparse Matrix Collection show that both algorithms demonstrate reasonable performance and scaling efficiency compared to MUMPS. Both are comparable to Intel MKL PARDISO on up to 8 cores, but PARDISO outperforms our implementation on 16 cores thanks to better utilization of computational resources and multilevel memory hierarchy.

The main directions of future work are improving performance and scaling efficiency of our implementation. We plan to employ nested parallelism to use sequential BLAS for small tasks at the bottom levels of an elimination tree and parallel BLAS for large tasks at the top levels of the tree. Hybrid MPI + OpenMP code and improved concurrent priority queue implementation could improve performance as well.

Acknowledgments. The study was partially supported by the RFBR, research project No. 14-01-3145514 and by the grant 02.B.49.21.0003 of The Ministry of education and science of the Russian Federation.

References

1. Rennich, S.C., Stosic, D., Davis, T.A.: Accelerating sparse Cholesky factorization on GPUs. In: Proceedings of the Fourth Workshop on Irregular Applications: Architectures and Algorithms. pp. 9–16. IEEE Press (2014)
2. Davis, T.A.: Direct Methods for Sparse Linear Systems. Fundamental of Algorithms, vol. 2. SIAM, Philadelphia (2006)
3. Liu, J.W.: The multifrontal method for sparse matrix solution: theory and practice. SIAM Rev. **34**(1), 82–109 (1992)
4. Duff, I.S., et al.: Direct Methods for Sparse Matrices. Clarendon Press, Oxford (1986)
5. Davis, T.A.: User Guide For Cholmod: A Sparse Cholesky Factorization and Modification Package. Department of Computer and Information Science and Engineering, University of Florida, Gainesville (2008)

6. Li, X.S.: An overview of SuperLU: algorithms, implementation, and user interface. ACM Trans. Math. Softw. (TOMS) **31**(3), 302–325 (2005)
7. Duff, I.S., Reid, J.K.: The multifrontal solution of indefinite sparse symmetric linear. ACM Trans. Math. Softw. (TOMS) **9**(3), 302–325 (1983)
8. Duff, I.S., Reid, J.K.: The multifrontal solution of unsymmetric sets of linear equations. SIAM J. Sci. Stat. Comput. **5**(3), 633–641 (1984)
9. Liu, J.W.: The multifrontal method and paging in sparse Cholesky factorization. ACM Trans. Math. Softw. (TOMS) **15**(4), 310–325 (1989)
10. Amestoy, P.R., et al.: Vectorization of a multiprocessor multifrontal code. Int. J. High Perform. Comput. Appl. **3**(3), 41–59 (1989)
11. Amestoy, P.R., et al.: A fully asynchronous multifrontal solver using distributed dynamic scheduling. SIAM J. Matrix Anal. Appl. **23**(1), 15–41 (2001)
12. Ashcraft, C.C., Grimes, R.G., Lewis, J.G., Peyton, B.W., Simon, H.D., Bjorstad, P.E.: Progress in sparse matrix methods for large linear systems on vector supercomputers. Int. J. High Perform. Comput. Appl. **1**(4), 10–30 (1987)
13. Ng, E.G., Peyton, B.W.: Block sparse Cholesky algorithms on advanced uniprocessor computers. SIAM J. Sci. Comput. **14**(5), 1034–1056 (1993)
14. Ng, E., Peyton, B.W.: A supernodal Cholesky factorization algorithm for shared-memory multiprocessors. SIAM J. Sci. Comput. **14**(4), 761–769 (1993)
15. Demmel, J.W., Eisenstat, S.C., et al.: A supernodal approach to sparse partial pivoting. SIAM J. Matrix Anal. Appl. **20**(3), 720–755 (1999)
16. L'Excellent, J.Y.: Multifrontal Methods: Parallelism, Memory Usage and Numerical Aspects. Ph.D. thesis, Ecole normale superieure de lyon-ENS LYON (2012)
17. Geist, G., Ng, E.: Task scheduling for parallel sparse Cholesky factorization. Int. J. Parallel Prog. **18**(4), 291–314 (1989)
18. Ashcraft, C., Eisenstat, S.C., Liu, J.W., Sherman, A.H.: A comparison of three column-based distributed sparse factorization schemes. Technical report, DTIC Document (1990)
19. Davis, T.A., Hu, Y.: The university of Florida sparse matrix collection. ACM Trans. Math. Softw. (TOMS) **38**(1), 1 (2011)
20. Karypis, G., et al.: A fast and highly quality multilevel scheme for partitioning irregular graphs. SIAM J. Sci. Comput. **20**(1), 359–392 (1999)
21. Pellegrini, F.: Scotch and libScotch 6.0 User's Guide. Technical report, LaBRI (2012)
22. Pirova, A.Yu., Meyerov, I.B.: MORSy – a new tool for sparse matrix reordering. In: Proceedings of an International Conference on Engineering and Applied Sciences Optimization, pp. 1952–1963 (2014)
23. L'Excellent, J.Y., Sid-Lakhdar, M.W.: Introduction of shared-memory parallelism in a distributed-memory multifrontal solver (2013)

Distributed Algorithm of Data Allocation in the Fragmented Programming System LuNA

Victor E. Malyshkin[1,2,3], Vladislav A. Perepelkin[1,2(✉)], and Georgy A. Schukin[1,3]

[1] Institute of Computational Mathematics and Mathematical Geophysics, Siberian Branch of Russian Academy of Sciences, Novosibirsk, Russia
{malysh,perepelkin,schukin}@ssd.sscc.ru
[2] Novosibirsk State National Research University, Novosibirsk, Russia
[3] Novosibirsk State Technical University, Novosibirsk, Russia

Abstract. The paper presents distributed algorithm with local communications Rope-of-Beads for static and dynamic data allocation in the LuNA fragmented programming system. LuNA is intended for implementation of large-scale numerical models on multicomputers with large number of processors and various network topologies. The algorithm takes into account the structure of a numerical model, provides static and dynamic load balancing and can be used in various network topologies.

Keywords: Dynamic data allocation · Distributed algorithms with local interactions · Fragmented programming technology · Fragmented programming system LuNA

1 Introduction

Implementation of large-scale numerical models on supercomputers is a challenging problem in high-performance parallel computing. In the light of growing size of supercomputers (in terms of memory capacity, number of cores, etc.) new system algorithms are to be developed for data processing and computations' organization, because to achieve good efficiency and scalability of parallel programs one has to provide its dynamic properties, such as dynamic load balancing, effective resources allocation strategy, etc. So, the complexity of application parallel programming becomes comparable to the one of system parallel programming. To simplify creation of parallel programs, enabled to achieve good performance, LuNA system was developed, that provides automatic parallel program generation [1–4].

In LuNA a program is assembled out of fragments of data and computations on these data. Each fragment of computation (CF) can be viewed as independent process, computing output data fragments (DF) from its inputs. Each DF is single assigned and each CF is executed only once. Fragmented structure is preserved during execution, which allows fragments to migrate between processing elements (PEs) of multicomputer and be executed in parallel.

© Springer International Publishing Switzerland 2015
V. Malyshkin (Ed.): PaCT 2015, LNCS 9251, pp. 80–85, 2015.
DOI: 10.1007/978-3-319-21909-7_8

The quality of parallel programs, generated by LuNA, heavily depends on the quality of resources distribution. In the paper the authors propose a distributed local algorithm of dynamic resources allocation, employed in LuNA system.

2   Related Works

The problem of efficient and scalable data allocation is actively researched.

Worth mentioning are scalable diffusion-like algorithms [6–8], since they do not require global interactions, but they lack concerning data structures, balancing speed and allow global imbalance with low load gradient.

Data allocation problems are common in distributed databases [12–15] and cloud services [16–18]. Due to relative small count of objects to distribute and low migration rates, these systems use centralized algorithms, which, because of potentially unlimited number of DFs and high migration rates, are not suitable for large distributed multicomputers.

Good efficiency can be achieved for allocating data of particular structures, such as meshes [19,20]. However, these algorithms have very limited application domain.

Worth mentioning are static analysis algorithms, employed in compilers [9–11]. Their limitation is static decision making at compile time.

Algorithms in [21,22] do not take data structure into account and do not solve the problem of data search. In [15] a relatively scalable algorithm is presented, but it employs global communications, which should be avoided.

3   Requirements for Data Allocation Algorithm

In order to provide high efficiency and scalability, data allocation algorithm should meet the following requirements:

- To provide equal load of available PEs (static and dynamic load balancing)
- To reduce communications length by taking into account the structure of data
- To tune to behavior of phenomena being modeled
- To be decentralized and use mostly local communications

4   Distributed Algorithms of Data Allocation

Two distributed algorithms of data allocation were developed: Hash-and-Track (HaT) and Rope-of-Beads (RoB), RoB being an improvement over HaT. Next sections present description and comparison of these algorithms.

4.1   Hash-and-Track Algorithm

The basic idea of the algorithm is that each PE is responsible for tracking actual location of a subset of DFs. Each DF is tracked by exactly one PE (called *tracker PE*), defined by static hash-function on DF identifier and thus known to all other PEs.

Whenever a DF is created or transferred to another PE, the tracker PE is notified about its new location. If a DF is required on some PE, the tracker PE is queried first and then forwards the request to the actual location of the DF. Thus fixed (three in the worst case) number of interactions is required to obtain any DF from any PE.

The main advantage of the HaT algorithm is its scalability in terms of computational load and extra storage usage. Given the hash function provides uniform distribution of values among PEs, the number of DFs to track will be nearly equal for all PEs.

Drawback of the algorithm is extra non-local communications with tracker PEs, which impede scalability (this is confirmed in "Experiments" section).

To overcome the shortcomings of HaT algorithm, RoB algorithm was developed.

4.2   Rope-of-Beads Algorithm

In the RoB algorithm each DF is statically mapped onto a line segment $[0,1]$ (like beads on a rope – hence the name of the algorithm) and is assigned a real number called *coordinate*. The segment $[0,1]$ is divided into sub-segments, one for each PE, adjacent sub-segments are mapped to neighbouring (connected by physical link) PEs, thus creating a line of PEs. In such a way, each DF is mapped to a PE. For each DF each PE can compute its coordinate and, if it belongs to its sub-segment, access required DF, or, depending on the value of the coordinate, forward query to a next or previous PE in the line.

In many problems spatially close data elements in domain are related by data dependencies. One good way to construct mapping from multi-dimensional data domain to $[0, 1]$ segment, which preserves locality of DFs, is to use space-filling curves (SFC), for example, Hilbert curve [23,24].

Distribution of DFs with Hilbert curve on 4 PEs is shown on Fig. 1. Figure 1a shows distribution of equally sized DFs among similar PEs, whereas Fig. 1b shows distribution in a case of non-equally sized DFs and/or PEs having different processing capabilities. In both cases uniform load of PEs is achieved.

The RoB algorithm has the following advantages:

- All communications are local, memory consumption is constant and computational overhead is constant.
- By usage of SFC spatially close DFs will be allocated on the same or neighbor PEs.
- Load balancing is done by dynamic shift of the boundaries of sub-segments and migrating DFs between adjacent PEs. Diffusion-like algorithms may be used for load balancing.

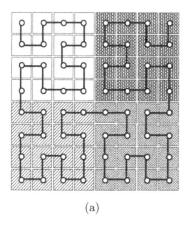 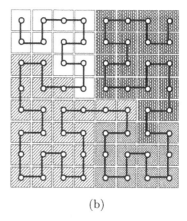

(a) (b)

Fig. 1. Usage of a Hilbert curve for domain decomposition

Drawbacks of the algorithm are:

- DFs structure is reduced to one dimension, therefore some potential locality of the problem may not be utilized.
- The count of hops to find a DF is linear function of a number of PEs in the worst case. However, the search distance is expected to be mostly equal to 0 or 1 for applications with good locality.
- For problems with statically unknown number of DFs, irregular domain structure or dynamically changing data locality it isn't always possible to compute good DFs mapping to $[0, 1]$ segment statically. Therefore, communication overhead may increase.

5   Experiments

To compare performance of the algorithms, LuNA implementation of a solver of Poisson equation with an explicit finite-difference scheme on a regular 3D mesh [5] was chosen.

Experiments were conducted on the cluster of Siberian Supercomputing Center with quad core Intel Xeon 5540 processors. The test contained 100 iterations of the solver on $400 \times 400 \times 400$ mesh, subdivided into 16 fragments per each of two decomposition dimensions.

For both algorithms network traffic and total execution time were monitored. Hilbert curve and generic hash function were used for DF and CF distribution.

5.1   Experiment Results

Table 1 shows execution times for the both algorithms on different number of PEs. HaT algorithm with hash function for distribution shows the worst results. Usage of Hilbert curve greatly improved execution times. RoB algorithm showed the best results in the majority of cases.

Table 1. Execution times (sec.) for the poisson solver

Number of PEs	1	2	4	8	16
HaT (hash)	214.5	1297.2	2360.4	2400.7	2568.7
HaT (Hilbert)	187.9	101.9	54.5	32.7	17.2
RoB (Hilbert)	175.6	95.6	44.8	26.4	15.2

6   Conclusion

The problematics of data distribution automation for implementation of large-scale numerical models for supercomputers is considered. The RoB algorithm for dynamic data allocation for LuNA fragmented programming system is proposed. Performance tests of the RoB algorithm are presented.

Acknowledgments. This work was supported by Russian Foundation for Basic Research (grants 14-07-00381 a and 14-01-31328 mol_a).

References

1. Malyshkin, V.E., Perepelkin, V.A.: LuNA fragmented programming system, main functions and peculiarities of run-time subsystem. In: Malyshkin, V. (ed.) PaCT 2011. LNCS, vol. 6873, pp. 53–61. Springer, Heidelberg (2011)
2. Malyshkin, V.E., Perepelkin, V.A.: Optimization Methods of parallel execution of numerical programs in the LuNA fragmented programming system. J. Supercomputing **61**(1), 235–248 (2012)
3. Malyshkin, V.E., Perepelkin, V.A.: The PIC implementation in LuNA system of fragmented programming. J. Supercomputing **69**(1), 89–97 (2014)
4. Kraeva, M.A., Malyshkin, V.E.: Assembly technology for parallel realization of numerical models on MIMD-multicomputers. J. Future Gener. Comput. Syst. **17**(6), 755–765 (2001)
5. Kireev, S.E., Malyshkin, V.E.: Fragmentation of numerical algorithms for parallel subroutines library. J. Supercomputing **57**(2), 161–171 (2011)
6. Kraeva, M.A., Malyshkin, V.E.: Dynamic load balancing algorithms for implementation of pic method on MIMD multicomputers. J. Programmirovanie, no. 1, pp. 47–53 (1999) (In Russian)
7. Hu, Y.F., Blake, R.J.: An improved diffusion algorithm for dynamic load balancing. J. Parallel Comput. **25**(4), 417–444 (1999)
8. Corradi, A., Leonardi, L., Zambonelli, F.: Performance comparison of load balancing policies based on a diffusion scheme. In: Lengauer, C., Griebl, M., Gorlatch, S. (eds.) Euro-Par 1997. LNCS, vol. 1300, pp. 882–886. Springer, Heidelberg (1997)
9. Anderson, J.M., Lam, M.S.: Global optimizations for parallelism and locality on scalable parallel machines. In: ACM-SIGPLAN PLDI 1993, pp. 112–125. ACM, New York (1993)
10. Li, J., Chen, M.: The data alignment phase in compiling programs for distributed-memory machines. J. Parallel Distrib. Comput. **13**(2), 213–221 (1991)

11. Lee, P.: Efficient algorithms for data distribution on distributed memory parallel computers. J. IEEE Trans. Parallel Distrib. Syst. **8**(8), 825–839 (1997)
12. Kwok, Y.-K., Ahmad, I.: Design and evaluation of data allocation algorithms for distributed multimedia database systems. IEEE J. Sel. Areas Commun. **14**(7), 1332–1348 (1997)
13. Iacob, N.M.: Fragmentation and data allocation in the distributed environments. Annals of the University of Craiova - Mathematics and Computer Science Series 38(3), 76–83 (2011)
14. Jagannatha, S., Geetha, D.E., Suresh Kumar, T.V., Rajani Kanth, K.: Load balancing in distributed database system using resource allocation approach. J. Adv. Res. Comput. Commun. Eng. **2**(7), 2529–2535 (2013)
15. Honicky, R.J., Miller E.L.: Replication under scalable hashing: a family of algorithms for scalable decentralized data distribution. In: 18th International Parallel and Distributed Processing Symposium (2004)
16. Alicherry, M., Lakshman, T.V.: Network aware resource allocation in distributed clouds. In: INFOCOM 2012, pp. 963–971. IEEE (2012)
17. AuYoung, A., Chun, B.N., Snoeren, A.C., Vahdat, A.: Resource allocation in federated distributed computing infrastructures. In: First Workshop on Operating System and Architectural Support for the On-demand IT InfraStructure (2004)
18. Raman, R., Livny, M., Solomon, M.: Matchmaking: distributed resource management for high throughput computing. J. Cluster Comput. **2**(1), 129–138 (1999)
19. Reddy, C., Bondfhugula, U.: Effective automatic computation placement and data allocation for parallelization of regular programs. In: 28th ACM International Conference on Supercomputing, pp. 13–22. ACM, New York (2014)
20. Baden, S.B., Shalit, D.: Performance tradeoffs in multi-tier formulation of a finite difference method. In: Alexandrov, V.N., Dongarra, J., Juliano, B.A., Renner, R.S., Tan, C.J.K. (eds.) ICCS-ComputSci 2001. LNCS, vol. 2073, pp. 785–794. Springer, Heidelberg (2001)
21. Ishikawa, K.-I.: ASURA: Scalable and Uniform Data Distribution Algorithm for Storage Clusters. Computing Research Repository, abs/1309.7720 (2013)
22. Chawla, A., Reed B., Juhnke, K., Syed, G.: Semantics of Caching with SPOCA: A Stateless, Proportional, Optimally-Consistent Addressing Algorithm. In: USENIX Annual Technical Conference 2011, pp. 33–33. USENIX Association (2011)
23. Lawder, J.K., King, P.J.H.: Using space-filling curves for multi-dimensional indexing. In: Jeffery, K., Lings, B. (eds.) BNCOD 2000. LNCS, vol. 1832, pp. 20–35. Springer, Heidelberg (2000)
24. Moon, B., Jagadish, H.V., Faloutsos, C., Saltz, J.H.: Analysis of the Clustering Properties of the Hilbert Space-Filling Curve. J IEEE Trans. Knowl. Data Eng. **13**(1), 124–141 (2001)

Control Flow Usage to Improve Performance of Fragmented Programs Execution

V.E. Malyshkin[1,2,3], V.A. Perepelkin[1,2(✉)], and A.A. Tkacheva[1,2]

[1] Institute of Computational Mathematics and Mathematical Geophysics
of the Siberian Branch of Russian Academy of Sciences,
Novosibirsk, Russia
{malysh,perepelkin,tkacheva}@ssd.sscc.ru
[2] National Research University of Novosibirsk, Novosibirsk, Russia
[3] Novosibirsk Technical State University, Novosibirsk, Russia

Abstract. Dataflow-based systems of parallel programming, such as LuNA fragmented programming system, often lack efficiency in high performance computations due to a high degree of non-determinism of a parallel program execution and execution overhead it causes. The authors concern defining control flow in LuNA programs in order to optimize their execution performance. The basic idea is to aggregate several fragments of a program and to execute them under control flow, thus reducing both surplus parallelism and system overhead. Tests presented show effectiveness of the proposed approach.

Keywords: High performance computing · Fragmented programming technology · Luna fragmented programming system · Control flow

1 Introduction

Implementation of large-scale numerical models on supercomputers is often challenging, because a programmer has to develop a program, possessing a number of dynamic properties, such as dynamic load balancing and tuning to available resources in order to make the program efficient and scalable. To increase the accessibility of supercomputers, a variety of parallel programming systems and tools were developed. To hide low-level programming details they automate the provision of the dynamic properties of programs. Examples of such systems are: LuNA [1], Charm++ [2], SMP Superscalar [3], DPLASMA [4].

High level of a program brings forth problems of its efficient parallel execution on a supercomputer. Since a programming system is able to execute the program in many ways (non-determinism of a program), appropriate to different hardware configurations or input data, the system faces the problem of dynamic choice of an efficient way of the program's execution. Often being NP-complete, this problem has to be overcome with heuristics or particular solutions.

One of the common ways to improve parallel program execution performance for such dataflow-based systems as LuNA is control flow usage. Control flow can presume for a number of subtasks of a program the order of their execution statically, eliminating the need to track data dependencies in run-time. This reduces the run-time system

© Springer International Publishing Switzerland 2015
V. Malyshkin (Ed.): PaCT 2015, LNCS 9251, pp. 86–90, 2015.
DOI: 10.1007/978-3-319-21909-7_9

overhead, thus improving the efficiency of the program execution. Although the parallelism degree of the program also decreases, this is often an advantageous trade-off.

The aim of the current work is the development of control flow support for LuNA programming system.

2 Related Works

Control flow means are widely used to improve performance of declarative parallel programming languages and systems.

In the systems SMP Superscalar [3], Cell Superscalar [5], ProActive Parallel Suite [6], one can define control flow with priorities. This mechanism, while being flexible, is not capable of defining complex program behavior. It also lacks readability for large programs, which results in control errors probability increase. In the functional languages of parallel programming Haskell [7], SISAL [8] in order to improve performance of algorithm's execution there are dedicated language means to mark a loop as parallel or sequential.

The library DPLASMA for linear algebra subroutines for dense matrices is built over PaRSEC [9] system. This system analyses informational dependencies at compile time, and the algorithm is represented in the Directed Acyclic Graph (DAG) form.

In such a way, while the idea is widely exploited, no general to construct control flow has been developed, thus further research of the problem is necessary.

3 LuNA Fragmented Programming System

LuNA (Language for Numerical Algorithms) is a language and a parallel programming system [1] intended for implementation of large-scale numerical models on supercomputers. It is being developed in the Institute of Computational Mathematics and Mathematical Geophysics of the Siberian Branch of Russian Academy of Sciences.

In LuNA an application algorithm is represented in a single-assignment coarse-grained explicitly parallel language LuNA as a bipartite graph of *data fragments* (DF) and *computational fragments* (CF). DFs are basically blocks of data (submatrixes, array slices, etc.). CFs are applications of pure functions on DFs. A CF has a number of input DFs and a number of output DFs. Values of output DFs are computed by the CF from the values of input DFs. Such representation is called *fragmented algorithm* (FA).

LuNA program consists of the FA description in LuNA language and a dynamic load library with a set of conventional sequential procedures. CFs are implemented as calls to these procedures with input and output DFs. Execution of all the CFs is done in accordance with partial order, that is imposed on the set of CFs by the information dependencies, forms the FA execution.

A FA is executed by the LuNA run-time system. Fragmented structure of the FA is kept in run-time, allowing the run-time system to dynamically assign CFs and DFs to different computing nodes, execute CFs in parallel (if possible), balance computational workload by redistributing CFs and DFs and so on.

Since the run-time system makes most decisions on FA execution dynamically, many checks for CFs being ready take place. This leads to significant overhead.

4 Suggested Approach

The basic idea of the suggested approach is to combine a subset of CFs into a single CF. For example, a loop or a subroutine may be collapsed into a single CF. All DFs, necessary to compute the CF, become input DFs, and all DFs produced become its outputs. All the CFs combined are executed sequentially in a fixed order under the control flow. Also, the number of CFs reduces, causing reduced system overhead.

Such FA transformation is possible for a set of CFs if the following condition is true: once all input DFs are provided, the whole chain of CFs may be sequentially executed without having to wait for other CFs to execute.

The transformation may be applied to both dependent and independent CFs. If the latter is the case, then the overall degree of parallelism of the FA is reduced, therefore control flow usage should not be overused.

To determine permissible order of CFs execution information dependencies analysis must be made. Creation of new aggregated CF means generation of new procedure, which invokes other procedures in established order. Both dependencies analysis and procedure generation were implemented as a part of LuNA project for a particular case, i.e. when the part of FA being transformed contains statically defined set of CFs.

It is worth mentioning, that it is algorithmically hard to determine effectiveness of control flow optimization in a given case. This problem is out of the scope of this work. Instead, the LuNA language was extended with annotations, which explicitly define, whether control flow transformation is required or not for given part of FA. Currently the annotations are provided by a user, but annotations generation may be automated.

5 Performance Tests

To investigate the effectiveness of the suggested approach a number of performance tests were conducted on a shared memory multiprocessor with 6 cores (Intel Xeon CPU X5600 2.8GHz) and on the distributed memory cluster of National Research University of Novosibirsk [10].

An application tested is a typical reduction problem on an array of numbers. The array to reduce is split into a number of slices. Each slice is reduced sequentially, while different slices may be processed independently. Two tests were performed: in the first one the slices were reduced sequentially, in the second one the slices were reduced in parallel. Both sequential and parallel tests were performed with and without control flow optimization. Control flow was applied to all the slices. Slice size was a parameter of the tests.

In all the tests "LuNA" stands for non-optimized execution and "static" stands for execution with control-flow optimization.

From Fig. 1 it is seen that control flow significantly reduces the FA execution time on the multiprocessor. The greater the slice size is, the more the benefit is. One can also notice, that for small sizes parallel version executes slower. This is due to extra overhead, which arises from extra CFs in parallel version of the test, and is not relevant to the subject of the work.

Figure 2 shows the result of the similar testing on the distributed memory cluster. It is seen, that usage of direct control benefits in performance by reducing system overhead.

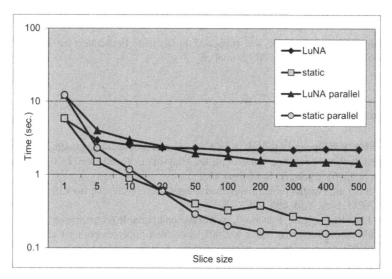

Fig. 1. Multiprocessor test. Dependency of the FA execution time on the slice size is shown. Array size is 3×10^4.

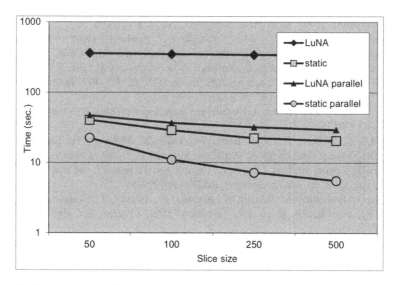

Fig. 2. Multicomputer test. Dependency of the FA execution time on the slice size is shown. Array size is 5×10^5. Number of computer nodes is 8.

6   Conclusion

Profitability of direct control usage in LuNA fragmented programming system was concerned. An approach to employ direct control in LuNA system was suggested and implemented as a part of LuNA software. Performance tests demonstrate that control flow usage significantly increases performance of program execution on the example of the reduction problem.

Acknowledgements. This work was supported by Russian Foundation for Basic Research (grants 14-07-00381 a and 14-01-31328 mol_a).

References

1. Malyshkin, V.E., Perepelkin, V.A.: LuNA fragmented programming system, main functions and peculiarities of run-time subsystem. In: Malyshkin, V. (ed.) PaCT 2011. LNCS, vol. 6873, pp. 53–61. Springer, Heidelberg (2011)
2. Acun, B., et al.: Parallel programming with migratable objects: charm++ in practice. In: SC 2014. ACM, New York, NY, USA (2014)
3. Perez, J.M., Badia, R.M., Labarta, J.: A flexible and portable programming model for SMP and multi-cores. Technical report 03/2007, Barcelona Supercomputing Center, Barcelona, Spain (2007)
4. Bosilca, G., et al.: Flexible development of dense linear algebra algorithms on massively parallel architectures with DPLASMA. In: Proceedings of the Workshops of the 25th IEEE International Symposium on Parallel and Distributed Processing (IPDPS 2011 Workshops), pp. 1432–1441. IEEE, Anchorage, Alaska, USA (2011)
5. Bellens, P., Perez, J.M., Badia, R.M., Labarta, J.: CellSs: a programming model for the cell BE architecture. In: SC 2006: Proceedings of the 2006 ACM/IEEE Conference on Supercomputing, p. 86. ACM Press, New York, NY, USA (2006)
6. Caromel, D., Leyton, M.: ProActive parallel suite: from active objects-skeletons-components to environment and deployment. In: César, E., Alexander, M., Streit, A., Träff, J.L., Cérin, C., Knüpfer, A., Kranzlmüller, D., Jha, S. (eds.) Euro-Par 2008 Workshops - Parallel Processing. LNCS, vol. 5415, pp. 423–437. Springer, Heidelberg (2009)
7. Coutts, D., Loeh, A.: Deterministic parallel programming with haskell. Comput. Sci. Eng. **14** (6), 36–43 (2012)
8. Gaudiot, J.-C., DeBoni, T., Feo, J., Böhm, W., Najjar, W., Miller, P.: The Sisal project: real world functional programming. In: Pande, S., Agrawal, D.P. (eds.) Compiler Optimizations for Scalable Parallel Systems: Languages, Compilation Techniques, and Run Time Systems. LNCS, vol. 1808, pp. 45–72. Springer, Heidelberg (2001)
9. Bosilca, G., Bouteiller, A., Danalis, A., Faverge, M., Herault, T., Dongarra, J.: PaRSEC: exploiting heterogeneity to enhance scalability. IEEE Comput. Sci. Eng. **15**(6), 36–45 (2013)
10. National Research University of Novosibirsk cluster. http://www.nusc.ru/

Towards Application Energy Measurement and Modelling Tool Support

Kenneth O'Brien[1(✉)], Alexey Lastovetsky[1], Ilia Pietri[2],
and Rizos Sakellariou[2]

[1] Heterogeneous Computing Laboratory,
School of Computer Science and Informatics, University College Dublin,
Dublin, Ireland
kenneth.obrien@ucdconnect.ie, alexey.lastovetsky@ucd.ie
http://hcl.ucd.ie
[2] The University of Manchester, Manchester, UK

Abstract. We present a prototype toolkit for researchers to accurately measure and model their application's power and energy usage. We provide an analysis of a matrix multiplication application using our api *libhclenergy*.

Keywords: Energy efficiency · Energy measurement · Software tools · High performance computing

1 Introduction

Energy has emerged as a new finite resource that must be considered by application developers. Currently, developers optimise their applications for performance by making the most efficient use of processor clock cycles, memory hierarchies and network bandwidth, in order to reduce execution time.

In recent years Dennard scaling has ended. Dennard scaling was a law that stated as transistor sizes decreased, the power a processor constructed of these transistors requires, remained proportional to the area of that chip.

The practicalities of this breakdown are that processor manufacturers cannot develop processors consisting of smaller transistors without drawing more power and producing more heat. These factors have resulted in the stagnation of processor clock frequencies and the rise of multicore and accelerator computing, as performance improvements can be achieved by adding more cores without shrinking transistors. This is not a complete solution as increasing the number of cores in a processor increases the overall power consumption, more so than what could be achieved if Dennard scaling had held.

This research is supported by the Structured PhD in Simulation Science which is funded by the Programme for Research in Third Level Institutions (PRTLI) Cycle 5 and co-funded by the European Regional Development Fund. This work is partially supported by EU under the COST Program Action IC1305: Network for Sustainable Ultrascale Computing (NESUS).

© Springer International Publishing Switzerland 2015
V. Malyshkin (Ed.): PaCT 2015, LNCS 9251, pp. 91–101, 2015.
DOI: 10.1007/978-3-319-21909-7_10

Our background is in application performance optimisation, therefore we focus our efforts to application power monitoring and analysis. Efforts at the hardware and data center configuration levels are beyond the scope of our study.

In order to optimise for this new resource we require tool support similar to that which exists for performance optimisation. We first need the ability to measure an application's energy usage, as well as the application's power usage, which is the rate of consumption over time.

Advanced mathematical modelling methods and tool support exists for applications with respect to performance [9]. For application energy modelling to develop further, we require equally advanced tool support.

In Sect. 2 we introduce our api for application power and energy monitoring. In Sect. 3 we introduce a tool for power and energy model construction. In Sect. 4, we demonstrate the capabilities of our tools. In Sect. 5 we describe related efforts and in Sect. 6 we conclude and describe our future works.

2 *Libhclenergy*

2.1 Measurement Infrastructure

There are two application metrics we wish to study, power and energy. Power is a rate of consumption of electricity, measured in units of Watts. Energy is a measure of work done. It is measured in Watt Hours (Wh) or Watt Seconds, commonly known as Joules (J). Power and energy are related by Eq. 1, where E is energy in Joules, P is power in Watts and t is time in seconds.

$$E = P * t \tag{1}$$

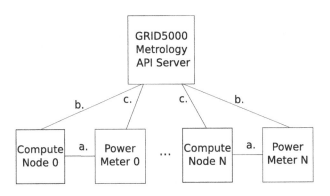

Fig. 1. Power measurement infrastructure

For our measurements we assume the existence of an external measurement device that records the power consumption of our node at regular intervals without affecting the nodes consumption. Figure 1 shows the mechanism by which we obtain our data.

Our *Compute Node 0* is instrumented by an external power measurement device (*Power Meter 0*) which records the power draw through the mains electricity socket (shown as *a*). This data is reported from all compute nodes to a centralised server(*c*), which can then be requested via HTTP, by any web capable device (*b*).

Given the hostname of node and a time frame, the API returns a series of measurements and the resolution at which the measurements were made. We use this infrastructure at GRID5000 as the basis for our api *libhclenergy*.

2.2 Experimental Platform

We chose GRID5000 [4] as our prototyping platform. It is large scale testbed for parallel computing. It provides a highly heterogeneous platform that can be configured by the researcher for their experiments. Many of the nodes contain accelerators and high speed interconnects representative of the current landscape in supercomputing. These include Nvidia GPU, Intel Xeon Phi and Infiniband. We believe our work here to be easily portable to other infrastructure. Recently the HPC job scheduling software Torque [1] added support for power measurement and management providing a suitable basis for production implementations of our *libhclenergy*.

2.3 Measurement of Distributed Applications

In the case of a single process, our API will measure energy and power between start and end time events as specified by the programmer. There can be multiple events and they may overlap, providing the programmer with the granularity of measurement that they require.

For the case of an application with multiple processes, we expose the total value for the node only at present.

In the case of an MPI application, we produce an energy reading for each participating node executing the application.

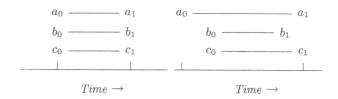

Fig. 2. Multiple processes on single node, both cases

There are two cases to be considered when measuring multiprocess or distributed applications. Firstly we can measure the power of all processes from the launch of the application until the final process exits. This is useful in the context of a grid, where resources are reserved and the idle machines must be taken into account. Secondly we can measure each node only when it's executing our application. This is the actual power of our application. We do not want to measure a node with a high power draw if we're only using it for a small unit of time. Both cases are shown graphically in Fig. 2 for processes a, b and c. In the first diagram, the energy calculated will be the same for either method of calculation. For the second diagram however, the energy could be calculated for all processes a, b and c for the duration of process a's execution. Our alternative measurement method only accounts for energy of processes while they're executing.

2.4 API Features

Power consumption of a node can be characterised by the static(or idle) power of the components powered on, and the dynamic power, which is the power consumed by devices performing work. Our idle power consumption function allows the user to measure both the static and dynamic power of their application by simple subtraction.

We provide a utility function to calculate the idle power consumption of a node. This function puts the calling process to sleep, after which it requests the consumption for that period. The idle power is the baseline from which deviations are considered characteristic of the running application. In addition, we provide the energy cost in euros of running the application for a given price per kWh, provided by the utility company.

Table 1 demonstrates our API's key functions. Functions 1–3 provide a way to initialise, start and stop a measurement. Functions 4–6 provide measurements. The Raw variant of the PowerSeries function captures measurements for all participating processes at all times, as opposed to only when executing. Functions 7 and 8 are only available when instrumenting MPI applications. Function 7 gathers all measurements from all compute processes to the root process. Functions 9–11 are utilities, providing idle power of a compute node, average power of a series of power measurements and cost of electricity for a given event.

As researchers we want to be confident in the accuracy of our measurements. As such the developer may specify a confidence level and a tolerance for both power and energy measurements. For a segment of instrumented code, the measurements are repeated until the confidence level falls within tolerance or a set maximum iterations is reached.

Figures 3 and 4 show examples of our API in use. Each time an application is measured, the raw data is written to a file. In order to attain our required accuracy we script a repetitive execution until the confidence interval is below tolerance.

3   Greenman

We have developed a tool to profile applications, measure their energy consumption and fit existing state of the art statistical models. To the authors' knowledge,

```
int main() {
hclenergy_t *event = hcl_init();

hcl_start(event);
// Execute code for instrumentation
hcl_stop(event);

double energy = energy_consumed(event);
}
```

Fig. 3. Libhclenergy example of energy measurement

```
int main() {
hclenergy_t *event = hcl_init();

hcl_start(event);
// Execute code for instrumentation
hcl_stop(event);

struct host_power_series *power = getPowerSeries(event);
}
```

Fig. 4. Libhclenergy example of power measurement

these are representative of the current models found in the literature. This tool builds on our hclenergy API to gather power and energy measurements.

Presented in Table 2 are some of the existing models we have implemented in the tool. They are all statistical regression models. U components of the models denote utilisation as a percentage of clock cycles for CPU, and total bytes written and read in the cases of memory, disk and network.

The models parameterise the power consumed by a compute node. Energy can be derived from Eq. 1 when execution time is known.

All models are fitted using a variety of standard and robust methods (bisquare, cauchy, fair, huber, welsch and ordinary) from GSL [14]. Robust methods are used to counteract the effect of outliers in the data. As a system is composed of many processes, another scheduled process may interfere with data collection. Robust methods dampen their effect on our model fitting.

We collect statistics at a per core granularity for c-state and p-state occupancy, as well as percentage of clock cycles spent in our application. Modern processor cores operate in various states of alertness known as c-states. In the highest c-state $C0$, all features of the processor including clocks, caches and voltages are at maximum capacity. In the lowest c-state $C6$ in the case of the Core 2 Duo, a processor can reduce the voltage of it's cores as low as 0 volts, with all internal clocks and caches disabled. There are gradual steps between these two extremes. Processors cores also have p-states representing each of the discrete frequencies a processor core can execute. Lower frequencies mean lower power consumption, but also lower performance. An application running at a low

Table 1. Key hclenergy API calls

No	Function
1	hclenergy_t *hcl_init();
2	void hcl_start(hclenergy_t *event);
3	void hcl_stop(hclenergy_t *event);
4	double energy_consumed(hclenergy_t *event);
5	struct host_power_series *getRawPowerSeries(hclenergy_t *event)
6	struct host_power_series *getPowerSeries(hclenergy_t *event);
7	struct host_power_series *gatherHostSeries(struct host_power_series *local);
8	double *energy_per_host(hclenergy_t *event);
9	double idlePower();
10	double avg_series(struct timeseries *series);
11	double cost(hclenergy_t *event, double price);

Table 2. Selection of models currently implemented

Model
$P = C_{base} + C_1 * U_{cpu}$, [13]
$P = C_{base} + \sum_{core_i=1}^{n} P_{core_i}$, [6]
$P = C_{base} + C_1 * U_{cpu} + C_2 * U_{cpu}^r$, [20]
$P = C_{base} + C_1 * U_{cpu} + C_2 * U_{I/O}$, [21]
$P = C_{base} + C_1 * U_{cpu} + C_2 * U_{disk} + C_3 * U_{net}$, [17]
$P = C_{base} + C_1 * U_{cpu} + C_2 * U_{disk} + C_3 * U_{net} + C_4 * U_{mem}$, [12]
$P = C_{base} + \sum_{core_i=1}^{n} C_i * f^3$, [22]

frequency taking a longer time may use more energy. We also record network packets per second per interface, memory footprint of the application, and bytes read and written to disk drives.

The sample rate of our power measurements is relatively low compared to our sample rate of application statistics. We interpolate our power readings in order to approximate the correct measurement for the given point in time. We provide approximation by akima, linear, cspline and polynomial splines.

The tool is executed on the Linux commandline as:

```
greenman <greenmanArguments> <resultsFolder> <Application> <Arguments>
```

As such the application under analysis does not require alteration. Any executable can be non intrusively instrumented. The source code of the application is not required for analysis.

If the user wishes to instrument segments of code in an application, the user must alter their code to tell greenman where to start and stop measuring. Models within this segment are calculated only using measurements collected inside these segments.

For each model we provide the researcher R^2, $R^2 Adj$, F statistic, and p value for each model parameter, χ^2, covariance matrix and correlation.

The tool is available as opensource software and is extensible as newer models arise. New models can be added by implementing a standard interface we provide. All models implemented so far use this interface, providing many examples on which to base new ones. We foresee new measurements to be required in the future and so we implement our measurement code in a similar extensible fashion.

As our tool is built in part on top of the PAPI library [24], any counters exposed now or in the future by it are supported.

When greenman is executed, it calls the fork() and exec() system calls to begin executing the researcher's application. We use the ptrace API which is primarily designed for implementing system call tracing and breakpoint debugging of applications. Ptrace allows us to control the application under analysis, frequently sampling it's application data from the/proc filesystem to build a time-series profile of the applications performance.

4   Applying Our API

Here we demonstrate the use of our tools to analyse a matrix multiplication application. We measure only during the kernel's execution. Allocation and initialisation of memory are not considered. We wish to understand the effect of number of the number of threads used in the computation. Using a non-distributed multi-threaded implementation we vary the number of threads and measure the energy consumed separately on two compute nodes (Sagittaire 30 and 72). Both nodes are of dualcore x86_64 architecture(AMD Opteron 250) and are identical, with the exception of Sagittaire 72 having an additional hard disk drive and 16 GB instead of Sagittaire 30's 2 GB of ram.

The results of our experiment are shown in Table 3. We report confidence intervals at the 95 % level. We note that Power while executing the application does not vary with the number of threads used, but that it is heavily influenced by the idle power of the machine. As the CPU is the most power demanding component of most servers [13], the similarity between idle and active power led us to investigate if the CPU was not using frequency scaling features. We confirmed this to be true in our environmental setup. Enabling these features would cause a reduction in power consumption.

We also observe that the execution time for both machines is similar for the given number of threads. Energy however varies dramatically. For the same computation, *Sagittaire-72*, uses 727.17 J and 369.12 J more than *Sagittaire-30* for 1 and 2 threads respectively. This analysis tells us that we should use *Sagittaire-30* for this computation as we will use less energy and not suffer any performance degradation.

The cost of energy is shown in Table 4. Though the costs are low, we must consider how they scale for longer running applications on a greater number of compute nodes.

Table 3. Power and energy measurements

Machine	#Threads	Power (W)	Energy (J)	Idle	Time(s)
Sagittaire-30	1	175.14 ± 0.40	2900.37 ± 6.90	174.58 ± 0.38	16.56
Sagittaire-30	2	175.28 ± 0.32	1560.72 ± 3.33	174.586 ± 0.38	8.90
Sagittaire-72	1	218.53 ± 4.26	3627.53 ± 70.81	215.975 ± 0.69	16.60
Sagittaire-72	2	217.29 ± 0.91	1929.84 ± 8.21	215.975 ± 0.69	8.88

Table 4. Energy costs

Machine	#Threads	Cost(4.125 c/KWh)
Sagittaire-30	1	€ 0.0033
Sagittaire-30	2	€ 0.0017
Sagittaire-72	1	€ 0.0041
Sagittaire-72	2	€ 0.0022

5   Related Works

Related tools for model prediction include JouleTrack [26], a web based tool for application profiling and energy estimation for StrongARM and Hitachi SH-4 processors. Dunkels [11], provides an energy estimation framework for small sensor web devices based on work by [29] which assumes that a larger infrastructure would be able to measure it's own energy via ACPI [2]. While power measurement via ACPI is part of the standard since version 4, we do not have access to machines that support it. Neither of these tools target the architectures and infrastructure that we do.

Barrachina [5] presents pmlib, a software package for measuring energy states on CPUs. This library provides whole node level measurements accurately, but does not capture finer grained measurements such as that of components and accelerators. However, it has the advantageous ability to interface with high frequency external measurement devices.

Cabrera provides EML [8] which are similar contributions to our *libhclenergy*, but lacks the ability to report statistical confidence and also to transparently calculate per node power in MPI applications.

In addition to these software methods, there are such as PowerPack [15] and PowerMon2 [7] that intercept power rails of components to give component level. These methods are difficult to deploy in real systems and are disadvantaged by the complexity of measuring devices with multiple power rails [18].

To the authors' knowledge there is no existing tool for energy profiling and model testing.

5.1 Existing Tools

We limit ourselves to a node level granularity, but should the reader be interested in finer grained measurement at the device level, we advise you to consult the following tools which are performance counter based.

Intel provide the RAPL interface [10] which is a software power model and similarly AMD provides APM [3]. Though easily accessible, both have disadvantages. Intel RAPL fails to provide power measurements, only providing energy with no timestamp data, hindering indepth analysis [16] and AMD APM has been shown to be inaccurate due to assumptions during sleep modes [16].

Likwid [28], a lightweight performance tool offers RAPL measurements from Intel SandyBridge and IvyBridge x86 processors.

Nvidia, through their management library (NVML [25]) provides access to milliwatt power consumption metrics, accurate to 5 %, as well as current Pstate of each graphics card in a system. NVML also provides related metrics such a fan speed and temperature.

6 Conclusion

We have provided a prototype implementation of a power and energy monitoring API for a modern parallel infrastructure as well as a tool for model fitting. These tools allow us to test a variety of schemes for power approximation when the origin sample frequency is low. Both tools will be released under an opensource license in the coming weeks.

Future hardware will likely by necessity include higher precision energy measurement capability. Current accelerator devices are capable of updating their power consumption data as frequently as 10 Hz.

7 Future Works

A current limitation of greenman is an inability to profile MPI applications due to the design patterns used to construct the tool. We aim to resolve this in subsequent releases.

Our api provides us with the ability to instrument sections of code. We plan to use this api to instrument different tasks of workflow applications to best allocate tasks for energy efficiency.

Current generations of Nvidia GPGPU and Intel Xeon Phi accelerators have the facility to report their own board power consumption through their vendor APIs [19,25]. As these are essentially performance counters, we will be adding them to the metrics that greenman can record. From there, we will implement existing accelerator power models [23,27] and provide an extensible interface for researchers to add their own models.

We will be exploring functional models of applications on heterogeneous platforms. The Heterogeneous Computing Lab has produced Fupermod [9] for producing optimal data partitioning in heterogeneous environments. We will be augmenting this software with energy measurements.

Acknowledgment. Experiments presented in this paper were carried out using the Grid'5000 experimental testbed, being developed under the INRIA ALADDIN development action with support from CNRS, RENATER and several Universities as well as other funding bodies (see https://www.grid5000.fr).

References

1. Adaptive Computing, I: Torque resource manager (2015). http://www. adaptivecomputing.com/products/open-source/torque/
2. H.P.C., et al.: Acpi v4.0a (2010). http://www.acpi.info/DOWNLOADS/ACPIspec40a.pdf
3. AMD: Bios and kernel developers' guide(bkdg) for amd family 15h models 00h–0fh processors (2013). http://amd-dev.wpengine.netdna-cdn.com/wordpress/media/2012/10/42301_15h_Mod_00h0Fh_BKDG1.pdf
4. Balouek, D., et al.: Adding virtualization capabilities to the Grid'5000 testbed. In: Ivanov, Ivan I., van Sinderen, Marten, Leymann, Frank, Shan, Tony (eds.) CLOSER 2012. CCIS, vol. 367, pp. 3–20. Springer, Heidelberg (2013)
5. Barrachina, S., Barreda, M., Catalán, S., Dolz, M.F., Fabregat, G., Mayo, R., Quintana-Ortí, E.S.: An integrated framework for power-performance analysis of parallel scientific workloads. In: ENERGY 2013, The Third International Conference on Smart Grids, Green Communications and IT Energy-Aware Technologies, pp. 114–119 (2013)
6. Basmadjian, R., Ali, N., Niedermeier, F., de Meer, H., Giuliani, G.: A methodology to predict the power consumption of servers in data centres. In: Proceedings of the 2nd International Conference on Energy-Efficient Computing and Networking, pp. 1–10. ACM (2011)
7. Bedard, D., Lim, M.Y., Fowler, R., Porterfield, A.: Powermon: fine-grained and integrated power monitoring for commodity computer systems. In: Proceedings of the IEEE SoutheastCon 2010 (SoutheastCon), pp. 479–484, March 2010
8. Cabrera, A., Almeida, F., Arteaga, J., Blanco, V.: Measuring energy consumption using EML (energy measurement library). Comput. Sci. Res. Dev. **30**(2), 135–143 (2015). http://dx.doi.org/10.1007/s00450-014-0269-5
9. Clarke, D., Zhong, Z., Rychkov, V., Lastovetsky, A.: Fupermod: a software tool for the optimization of data-parallel applications on heterogeneous platforms. J. Supercomput. **69**(1), 61–69 (2014)
10. David, H., Gorbatov, E., Hanebutte, U.R., Khanna, R., Le, C.: Rapl: memory power estimation and capping. In: 2010 ACM/IEEE International Symposium on Low-Power Electronics and Design (ISLPED), pp. 189–194, August 2010
11. Dunkels, A., Osterlind, F., Tsiftes, N., He, Z.: Software-based on-line energy estimation for sensor nodes. In: Proceedings of the 4th Workshop on Embedded Networked Sensors, pp. 28–32. ACM (2007)
12. Economou, D., Rivoire, S., Kozyrakis, C., Ranganathan, P.: Full-system power analysis and modeling for server environments. In: Proceedings of Workshop on Modeling, Benchmarking, and Simulation, pp. 70–77 (2006)
13. Fan, X., Weber, W.D., Barroso, L.A.: Power provisioning for a warehouse-sized computer. ACM SIGARCH Comput. Archit. News **35**(2), 13–23 (2007)
14. Galassi, M., et al.: Gnu Scientific Library Reference Manual, 3rd edn. Network Theory Ltd., Bristol (2009). http://www.gnu.org/software/gsl/manual/gsl-ref.ps.gz

15. Ge, R., Feng, X., Song, S., Chang, H.C., Li, D., Cameron, K.: Powerpack: energy profiling and analysis of high-performance systems and applications. IEEE Trans. Parallel Distrib. Syst. **21**(5), 658–671 (2010)
16. Hackenberg, D., Ilsche, T., Schone, R., Molka, D., Schmidt, M., Nagel, W.: Power measurement techniques on standard compute nodes: A quantitative comparison. In: 2013 IEEE International Symposium on Performance Analysis of Systems and Software (ISPASS), pp. 194–204, April 2013
17. Heath, T., Diniz, B., Horizonte, B., Carrera, E.V., Bianchini, R.: Energy conservation in heterogeneous server clusters, pp. 186–195 (2005)
18. Hsu, C.H., Poole, S.: Power measurement for high performance computing: state of the art. In: 2011 International Green Computing Conference and Workshops (IGCC), pp. 1–6, July 2011
19. Intel Corporation: Intel manycore platform software stack (2015). https://software.intel.com/en-us/articles/intel-manycore-platform-software-stack-mpss
20. Jung, G., Hiltunen, M.A., Joshi, K.R., Schlichting, R.D., Pu, C.: Mistral: dynamically managing power, performance, and adaptation cost in cloud infrastructures. In: 2010 IEEE 30th International Conference on Distributed Computing Systems (ICDCS), pp. 62–73. IEEE (2010)
21. Kansal, A., Zhao, F.: Fine-grained energy profiling for power-aware application design. ACM SIGMETRICS Perform. Eval. Rev. **36**(2), 26 (2008). http://portal.acm.org/citation.cfm?doid=1453175.1453180
22. Kim, K.H., Beloglazov, A., Buyya, R.: Power-aware provisioning of virtual machines for real-time cloud services. Concurrency Comput. Pract. Exp. **23**(13), 1491–1505 (2011)
23. Lai, Z., Lam, K.T., Wang, C.L., Su, J.: A power modelling approach for many-core architectures. In: 2014 10th International Conference on Semantics, Knowledge and Grids (SKG), pp. 128–132, August 2014
24. Mucci, P.J., Browne, S., Deane, C., Ho, G.: Papi: a portable interface to hardware performance counters. In: Proceedings of the Department of Defense HPCMP Users Group Conference, pp. 7–10 (1999)
25. Nvidia Corporation: Nvidia management library (2015). https://developer.nvidia.com/nvidia-management-library-nvml
26. Sinha, A., Chandrakasan, A.P.: Jouletrack: a Web based tool for software energy profiling. In: Proceedings of the 38th Annual Design Automation Conference, pp. 220–225. ACM (2001)
27. Song, S., Su, C., Rountree, B., Cameron, K.W.: A simplified and accurate model of power-performance efficiency on emergent gpu architectures (2013)
28. Treibig, J., Hager, G., Wellein, G.: Likwid: a lightweight performance-oriented tool suite for x86 multicore environments. In: 2010 39th International Conference on Parallel Processing Workshops (ICPPW), pp. 207–216. IEEE (2010)
29. Zhao, Y.J., Govindan, R., Estrin, D.: Residual energy scan for monitoring sensor networks. In: 2002 IEEE Wireless Communications and Networking Conference, WCNC 2002, vol. 1, pp. 356–362. IEEE (2002)

The Mathematical Model and the Problem of Optimal Partitioning of Shared Memory for Work-Stealing Deques

Andrew Sokolov$^{(\boxtimes)}$ and Eugene Barkovsky

Institute of Applied Mathematical Research of the Karelian Research
Centre of the Russian Academy of Sciences, Pushkinskaya Str. 11,
185910 Petrozavodsk, Russia
avs@krc.karelia.ru, barkevgen@gmail.com

Abstract. In this paper we propose the mathematical model and solve the problem of optimal partitioning of shared memory for work-stealing deques. Operations have probabilistic characterisation and along with sequential execution it is possible to execute operations on deques (with given probabilities) in parallel.

Keywords: Work-stealing · Deques · Data structures · Markov chains · Random walks

1 Introduction

Currently, there are two ways of dynamic (regardless of the specific tasks) scheduling of multi-thread parallel calculations on multi-core architecture [1,2]. Work-sharing is a centralized method of redistribution of tasks — scheduler moves tasks from the most loaded processors to the less loaded ones. In the work-stealing method the approach is different. Processors, which became free, try to "steal" a portion of work from other processors. This method is implemented in many systems like Cilk, Cilk++; TBB; TPL; X10 and others.

In this method of load balancing each processor (thread) has a pool of tasks. Information about these tasks is stored inside a processor's deque. We suppose that elements, which are stored inside deque, have fixed size. When a thread (processor) creates a new task, it adds an element to its deque; when it needs a task, it takes an element from the top of its deque. If a thread detects that the deque is empty, it "steals" a task from its "victim" — another thread. The first two operations are similar to stack, while thefts occur from the base of stack — like FIFO-queue. That is, we work with (in terminology of D.E. Knuth) an input-restricted deque [3]. In [1] it was proposed to steal one element from a random deque, in [4] — half of the elements.

There are different methods of representation of input-restricted deques in memory. It is possible to implement "the linked representation" method. The model of this method will be similar to the already constructed models of the

V. Malyshkin (Ed.): PaCT 2015, LNCS 9251, pp. 102–106, 2015.
DOI: 10.1007/978-3-319-21909-7_11

linked representation of stacks and queues [5]. In [6] the method of representation of stacks and queues as a double-linked list of arrays (the paged implementation) is proposed and analysed. In [7] this method is proposed for deques. In the linked and paged implementations a portion of memory is allocated for pointers. In the serial methods of work with several data structures, losses are occur when one of the data structures is in overflow while there are still pieces of free memory.

In [8] the model of a work-stealing load balancer was proposed. This model was constructed on the basis of apparatus of the queuing theory using Poisson law of occurrences of events. In this model a specific method of representation of deques in memory was not considered.

Here, we will construct the model for the serial circular method of representation of work-stealing deques (similar to FIFO-queues), where each deque is located in a separate part of memory [9]. In the model on each step of discrete time some operations on deques are performed with given probabilities. Previously, such models were constructed by our team to represent some dynamic data structures: stacks, queues, priority queues [10–12] and others.

2   The Mathematical Model

Here, we describe case of two deques. This particular case is important not only as a stepping stone in construction of the model, but it also has an independent significance. Note that among multi-core architectures there are ones without cache memory. For example, in the architecture AsAP-II each core has two FIFO buffers and in the architecture SEAforth — two stacks (for storing data and return addresses) [13]. In these architectures, queues and stacks are implemented circularly and separate from each other with the possibility to lose items due to overflow. But in our works we research situations, where to store several data structures we use shared memory. In some cases this can minimize a number of lost items.

Work-stealing deques may be implemented in the hardware, on the basis of these architectures. In this case it is important to research an optimal organization of two deques and then, in case of an arbitrary number of cores, obtain the desired chips by composing them from a "two-deques" ones.

Suppose that in the memory size of m we work with two serial circular deques, where elements are of fixed size in one conventional unit. For serial representation of dequeues we will allocate each a number of memory units from the total volume of m units. Suppose that s is a number of units allocated to the first deque, then $(m - s)$ is a number of units allocated to the second deque.

Some probabilistic characteristics of operations performed on the deques are known in advance:

- insertion of an element in the first deque with the probability p_1;
- insertion of an element in the second deque with the probability p_2;
- insertion of elements in parallel in both deques — p_{12};
- deletion of an element from the first deque with the probability q_1;

- deletion of an element from the second deque with the probability q_2;
- deletion of elements in parallel from both deques — q_{12};
- insertion in the first deque and deletion from the second one — pq_{12};
- insertion in the second deque and deletion from the first one — pq_{21};
- deques do not change their length with the probability r (reading or no operation).

$p_1 + p_2 + p_{12} + q_1 + q_2 + q_{12} + pq_{12} + pq_{21} + r = 1$. If you exclude an element from an empty deque shutdown does not occur — empty deque begins to "steal" elements from another deque (provided that victim deque has any elements).

Assume that x and y are current lengths of the first and the second deques, respectively. We consider random walks in two-dimensional space on an integer lattice in the area $-1 \leq x \leq s+1$, $-1 \leq y \leq m-s+1$ (Fig. 1) as a mathematical model.

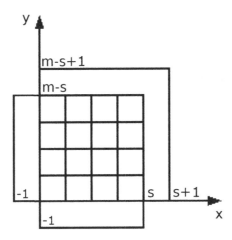

Fig. 1. Area of walk for two work-stealing deques.

Walk starts in the origin of coordinates. Lines $x = s+1$ and $y = m-s+1$ form two absorbing screens — when we get to these lines the memory becomes full and the program crashes (or memory is redistributed according to some principle); lines $x = -1$, $y = -1$ form two reflective screens, which are designated for cases of deletion of an element from the empty deque. Introduction of this screen takes into account "theft" of an element. Formally, deque goes on the screen, but in fact into the area $1 \leq x < s$, $1 \leq y < m - s$. Specific transition point depends on the strategy of theft. For the case of stealing of one element: if we are located on points $(x, -1)$ or $(-1, y)$ we will move to points $(x - 1, 1)$ or $(1, y - 1)$. For the case of stealing half of the elements: we will move to points $(x/2, x/2)$ or $(y/2, y/2)$ respectively. In case of uneven x or y (or if there is not enough memory to allocate stolen elements) we round to the desired integer.

As the criterion of optimality we consider the maximum mean time to the memory overflow (i.e. before hitting lines $x = s + 1$ and $y = m - s + 1$). Based on the model described above we have solved following problems:

1. Problem of optimal partition of memory if we steal an element from one deque when the other becomes empty;
2. Problem of optimal partition of memory if we steal half of the deque;
3. Comparison of these two strategies of thefts and issuing recommendations.

Table 1. The numerical results

The input data	Maximum mean time to the memory overflow $(m = 10)$			
	Stealing of one element		Stealing half of the elements	
	The optimal partition	Splitting in half $(s = 5)$	The optimal partition	Splitting in half $(s = 5)$
$p_1 = 0.15$, $p_2 = 0.05$, $p_{12} = 0.1$ $q_1 = 0.05$, $q_2 = 0.15$, $q_{12} = 0.1$ $pq_{12} = 0.1$, $pq_{21} = 0.1$, $r = 0.2$	48 $(s = 6)$	44	49 $(s = 6)$	47
$p_1 = 0.19$, $p_2 = 0.01$, $p_{12} = 0.1$ $q_1 = 0.01$, $q_2 = 0.19$, $q_{12} = 0.1$ $pq_{12} = 0.1$, $pq_{21} = 0.1$, $r = 0.2$	43 $(s = 6)$	37	47 $(s = 6)$	40

Table 1 shows some results of calculations. Here, probabilities are theoretical — for greater visibility of results. In practice these probabilities should be obtained by the preliminary statistical research.

Judging by results we can conclude following. When we portion memory optimally the system works longer. For example, when we steal one element, and the probabilities of insertions and deletions are unequal, the differences of mean time to the memory overflow between the optimal partition and splitting in half are 4 in the first case and 6 in the second one. That is, system works by 4 (or 6) operations longer if we partition memory optimally.

If we steal half of the elements (with the same probabilities as above), the differences of mean time are 2 and 7. We can recommend this particular strategy for systems with very asymmetrical probabilities — in this situation stealing half of the elements is more beneficial that stealing of only one element.

3 Conclusion

The mathematical model describing the process of working with two parallel cyclic work-stealing deques in shared memory was created. We have considered two strategies of stealing: theft of one element; and theft of half of the elements. The algorithm and the program for finding the optimal memory partitioning between deques, depending on their probabilistic characteristics, were created for both strategies. To solve the problems we use apparatus of controlled random walks, Markov chains.

Acknowledgments. This research work was supported by the Russian Foundation for Basic Research, grant 15-01-03404-a.

References

1. Blumofe, R.D., Leiserson, C.E.: Scheduling multithreaded computations by work stealing. J. ACM. **46**, 720–748 (1999)
2. Herlihy, M., Shavit, N.: The Art of Multiprocessor Programming. Elsevier, Amsterdam (2008)
3. Knuth, D.: The Art of Computer Programming, vol. 1. Addison-Wesley, Redwood City (2001)
4. Hendler, D., Shavit, N.: Non-blocking steal-half work queues. In: The Twenty-first Annual ACM Symposium on Principles of Distributed Computing, pp. 280–289 (2002)
5. Sokolov, A.V., Drac, A.V.: The linked list representation of n LIFO-stacks and/or FIFO-queues in the single-level memory. Inf. Process. Lett. **13**(19–21), 832–835 (2013)
6. Aksenova, E.A., Lazytina, A.A., Sokolov, A.V.: About optimal methods of representation of dynamic data structures. Rev. Appl. Ind. Math. **10**(2), 375–376 (2003) (in Russian)
7. Hendler, D., Lev, Y., Moir, M., Shavit, N.: A dynamic-sized nonblocking work stealing deque. Distrib. Comput. - Spec. Issue: DISC 2004 **18**(3), 189–207 (2006)
8. Mitzenmacher, M.: Analyses of load stealing models based on differential equations. In: The ACM Symposium on Parallel Algorithms and Architectures, pp. 212–221 (1998)
9. Chase, D., Lev, Y.: Dynamic circular work-stealing deque. In: The Seventeenth Annual ACM Symposium on Parallelism in Algorithms and Architectures, pp. 21–28 (2005)
10. Aksenova, E.A., Lazutina, A.A., Sokolov, A.V.: Study of a non-markovian stack management model in a two-level memory. Program. Comput. Softw. **30**(1), 25–33 (2004)
11. Aksenova, E., Sokolov, A.: The optimal implementation of two FIFO-queues in single-level memory. Appl. Math. **2**(10), 1297–1302 (2011)
12. Sokolov, A.V., Drac, A.V.: Simulation of some methods of representation of n FIFO-queues in the single-level memory. Heuristic Algorithms Distrib. Comput. **1**(1), 40–52 (2014) (In Russian)
13. Kalachev, A.V.: Multi-core processors. Binom (2010) (In Russian)

Dynamic Load Balancing Based on Rectilinear Partitioning in Particle-in-Cell Plasma Simulation

Igor Surmin[1], Alexei Bashinov[1,2], Sergey Bastrakov[1], Evgeny Efimenko[1,2], Arkady Gonoskov[1,2,3], and Iosif Meyerov[1(✉)]

[1] Lobachevsky State University of Nizhni Novgorod, Nizhny Novgorod, Russia
{bastrakov,meerov}@vmk.unn.ru,
{i.surmin,evgeny.efimenko,arkady.gonoskov}@gmail.com
[2] Institute of Applied Physics of the Russian Academy of Sciences,
Nizhny Novgorod, Russia
avbk@mail.ru
[3] Chalmers University of Technology, Gothenburg, Sweden

Abstract. This paper considers load balancing in Particle-in-Cell plasma simulation on cluster systems. We propose a dynamic load balancing scheme based on rectilinear partitioning and discuss implementation of efficient imbalance estimation and rebalancing. We analyze the impact of load balancing on performance and accuracy. On a test plasma heating problem dynamic load balancing yields nearly 2 times speedup and better scaling. On the real-world plasma target irradiation simulation load balancing allows to mitigate particle resampling and thus improve accuracy of the simulation without increasing the runtime. Balancing-related overhead in both cases are under 1.5 % of total run time.

Keywords: Load balancing · High performance computing · Plasma simulation · Particle-in-cell

1 Introduction

Simulation of plasma dynamics is actively used in solving many fundamental and applied physical problems, such as laser-driven particle acceleration using various targets. Numerical plasma simulation is often based on the Particle-in-Cell method [1], often abbreviated as PIC, with proper extensions [2]. Solving up-to-date problems can require simulation of up to $\sim 10^9$ particles and $\sim 10^8$ grid nodes, which makes it computationally challenging. Therefore, there is a need for high performance computing.

There are several widely used Particle-in-Cell plasma simulation codes, such as VLPL [3], OSIRIS [4], and PIConGPU [5]. Since 2010 we develop the Particle-in-Cell code PICADOR [6,7] oriented at heterogeneous cluster systems with GPUs and Xeon Phi coprocessors.

© Springer International Publishing Switzerland 2015
V. Malyshkin (Ed.): PaCT 2015, LNCS 9251, pp. 107–119, 2015.
DOI: 10.1007/978-3-319-21909-7_12

A key problem in developing a high performance Particle-in-Cell code is efficient data and workload distribution between nodes of a cluster system [8–11]. The Particle-in-Cell method operates with two main data sets: an ensemble of particles and grid values of the electro-magnetic field. Independent distribution of particle and field data between computational nodes is impossible because Particle–Grid interactions are spatially local. Thus, each node stores a subset of particles and all spatially close grid values of the field. The particle distribution may significantly change due to complex particle dynamics, potentially leading to high workload imbalance between the computational nodes. All nodes are synchronized after each time step, this raises the load balancing problem of achieving uniform distribution of workload and memory consumption between nodes of a cluster system and minimization of data transfers.

The most widely used approach to load balancing in Particle-in-Cell plasma simulation is spatial decomposition: the simulation area is subdivided into domains, each handled by a computational node that stores all particle and field data of the domain. Due to spatial locality of the Particle-in-Cell method, each domain needs data exchanges only with adjacent domains. Spatial decomposition methods mostly differ in domain geometry and control of the maximum number of adjacent domains. The orthogonal bisection [12] and Octree [13] methods are based on recursive subdivision and provide good particle balancing but result in a big amount of adjacent domains and, as consequently, intricate data exchange pattern. The Quicksilver method [14] is based on dynamically floating domain boundaries and intermediate window area. Its modification One-handed help [15] provides ideal particle balancing and close to ideal grid balancing but at a price of intensive exchanges of the grid values.

We propose a dynamic load balancing strategy based on rectilinear partitioning scheme [16]. Rectilinear partitioning is topologically equivalent to uniform parallelepiped partitioning, therefore data exchange patterns are the same and each domain communicates only with 26 adjacent domains. We discuss application of rectilinear partitioning scheme to dynamic load balancing in Particle-in-Cell plasma simulation and implementation of efficient imbalance estimation and rebalancing. Performance and scaling of our implementation is demonstrated on a benchmark plasma heating problem and the real-world simulation of irradiation of a plasma target by two counter-propagating laser pulses.

The paper is organized as follows. The Particle-in-Cell method and our implementation PICADOR are described in Sect. 2. Load balancing problem, rectilinear partitioning, the proposed dynamic load balancing strategy and implementation details are presented in Sect. 3. Experimental results and discussion on test and real problems are given in Sects. 4 and 5, respectively.

2   PICADOR Particle-in-Cell Code

PICADOR [6,7] is a tool for three-dimensional plasma simulation using the Particle-in-Cell method. The Particle-in-Cell method is based on a self-consistent mathematical model representing plasma as an ensemble of negatively charged electrons and positively charged ions, additional modules may include positively charged positrons and chargless photons. The entire particle ensemble is represented by a smaller amount of macroparticles (super-particles) with constant mass m and charge q, that have the same charge-to-mass ratio as the real particles, resulting in equivalent plasma dynamics. For brevity, macroparticles are hereafter referred as particles. The Particle-in-Cell method operates on two main sets of data: each particle of a particular type is characterized by its position r and momentum p; electric field E, magnetic field B and current density j are set on a uniform space grid.

The method implies numerical solving the equations of particle motion together with the Maxwell's equations on a discrete grid. Each time step is divided into four stages: field update, field interpolation, particle push, and current deposition. We use the finite-difference time-domain method [18] to update the grid values of the electric and magnetic field according to the Maxwell's equations. During the particle push stage the field values are interpolated to compute the corresponding force; for each particle only values in several nearest grid nodes are used. We use Boris method [1] to solve particles' equations of motion. The current deposition stage implies computing the current density based on the positions and velocities of the particles; each particle contributes to the current density values in only several nearest grid nodes. The final current density values are determined by summarizing contributions of all particles and are used during the next field update, thus closing the loop.

PICADOR computational scheme for cluster systems is organized as follows. We employ spatial decomposition of the simulation area: it is subdivided into sub-domains, each processed by a node of a cluster using MPI. Each node stores particle and field data corresponding to the sub-domain being processed. MPI exchanges are used to keep near-boundary values up-to-date, at the boundaries thin layers of ghost cells are stored in both nodes. Due to constraints on time and spatial steps a particle can not move across multiple sub-domains in a single time step, so all MPI exchanges concern only adjacent sub-domains. Field and current density exchanges are done using 6 adjacent sub-domains, particle exchanges require 26 adjacent sub-domains.

Features of PICADOR include FDTD and NDF field solvers, periodic boundary conditions and absorbing boundary layer PML [17], Boris particle pusher, CIC and TSC particle form factors, Esirkepov current deposition, ionization, and moving frame. The code is capable of running on heterogeneous cluster systems with CPUs, GPUs and Xeon Phi coprocessors.

3   Load Balancing Based on Rectilinear Partitioning

3.1   Rectilinear Partitioning

A scalable implementation of the Particle-in-Cell method requires a balanced distribution of particles and grid values of the field. Decomposition of the simulation area into spatially equal domains does not provide a balanced workload distribution as particle distribution is in many cases significantly non-uniform.

Another extreme approach is uniform distribution of particles between computational nodes and storing the whole set of grid values in each node, resulting in massive exchanges of grid values of the field. We need a trade-off between overheads caused by imbalance and communications.

We consider the following load balancing problem statement. Grid cells are enumerated with 3D indices $(i, j, k), 0 \leq i < N_x, 0 \leq j < N_y, 0 \leq k < N_z$, where N_x, N_y, N_z are the numbers of cells along the corresponding axes. The numbers of domains along the x, y and z axes are fixed and equal to m, n and l, respectively. A rectilinear partitioning is defined by a triple of integer vectors (P, Q, R). We denote the set of all possible ways of rectilinear partitioning as

$$S = \{(P, Q, R) : P = (0, p_1, p_2, \ldots, p_{m-1}, N_x),$$
$$Q = (0, q_1, q_2, \ldots, q_{n-1}, N_y), R = (0, r_1, r_2, \ldots, r_{l-1}, N_z)\}. \quad (1)$$

Let $(P, Q, R) \in S$ be a rectilinear partitioning in form (1). A domain (I, J, K) contains cells with the following indices defined by the partitioning:

$$\{(i, j, k) : p_I \leq i < p_{I+1}, q_J \leq j < q_{J+1}, r_K \leq k < r_{K+1}\}.$$

For each cell (i, j, k) we define a workload value $L_{i,j,k} = nParticles_{i,j,k} + 1$, where $nParticles_{i,j,k}$ is the number of particles in the cell . This workload function is additive, the workload of a domain is a sum of workloads of its cells, namely, the total number of particles and grid cells. We introduce a workload of a domain (I, J, K) with partitioning (P, Q, R):

$$Work(P, Q, R, I, J, K) = \sum_{i=p_I}^{p_{I+1}-1} \sum_{j=q_J}^{q_{J+1}-1} \sum_{k=r_K}^{r_{K+1}-1} L_{i,j,k}.$$

The choice of this particular workload function is determined by two factors. First, theoretical complexity of the Particle-in-Cell method is linear in number of particles and grid values. Second, the sum of number of particles and grid values is nearly proportional to memory consumption. However, the real performance is not linear in number of particles, especially for the low number of particles per cell due to hardware-specific features. A more sophisticated performance and workload models such as [19,20] could be used instead.

Imbalance of a partitioning is the ratio of the maximum workload to the average:

$$Imbalance(P, Q, R) = \frac{\max\limits_{I=\overline{0,m-1}, J=\overline{0,n-1}, K=\overline{0,l-1}} Work(P, Q, R, I, J, K)}{\frac{1}{mnl} \sum_{I=0}^{m-1} \sum_{J=0}^{n-1} \sum_{K=0}^{l-1} Work(P, Q, R, I, J, K)}. \quad (2)$$

The average workload in the denominator of (2) does not depend on a partitioning, but is a convenient normalization.

We need to solve the optimal rectilinear decomposition problem:

$$(P^{opt}, Q^{opt}, R^{opt}) = \arg \min_{(P,Q,R) \in S} Imbalance(P, Q, R). \quad (3)$$

The problem (3) is NP-complete [16]. Therefore we use a heuristic algorithm to quickly find a reasonably good solution. The algorithm iteratively solves 1D optimal rectilinear decomposition problem with fixed decomposition along the other two axes, while the next decomposition decreases the imbalance.

3.2 Implementation Overview

Implementation of load balancing based on rectilinear partitioning requires efficient computation of the imbalance on cluster systems. Each MPI process handles a sub-domain and stores the corresponding particles and grid values. Gathering information about workload in all cells in one process is inefficient because of both high data transfer overhead and memory consumption.

We use a distributed scheme of workload and imbalance computation based on prefix sums. Each process computes a 3D prefix sum of cell workloads and exchanges it with adjacent domains. After all prefix sums are known, computing workload in each domain takes constant time.

Load balancing is implemented in two forms: static, performed once at initialization of the simulation, and dynamic that periodically estimates the imbalance and performs rebalancing if the imbalance exceeds a threshold. The dynamic scheme has two adjustable parameters: frequency of imbalance estimation and imbalance threshold. These values are currently fixed throughout the whole simulation. During rebalancing the simulation area is divided into blocks with respect to union of boundaries of old and new partitioning. Data transfers is performed in two stages to avoid deadlocks. First, blocks are transferred from processes with lesser ranks to processes with greater ranks. Then senders and receivers are interchanged. This scheme allows to only transfer field and particle data that actually changes a domain and do not touch the rest of the data.

4 Evaluation of Load Balancing Efficiency on a Test Plasma Heating Problem

As a test problem we consider simulation of heating of a small ball of plasma with 42 M particles and $128 \times 128 \times 128$ grid. During the heating particles drift from the center in random directions. The experiments were done using 256 MPI processes on cluster MVS-100K of Joint Supercomputing Center of RAS with 2 Intel Xeon E5450 CPUs and 8 GB RAM on each node, and Infiniband DDR.

Dynamic partitioning starts from the same domain geometry as the static partitioning, but domain boundaries change during the simulation if the imbalance exceeds the threshold. Dependence of the imbalance on time iteration is presented at Fig. 1. In spatially uniform partitioning all particles are initially located in central domains and gradually drift from the center reducing imbalance. This effect is inverse for static rectilinear partitioning: the imbalance is low initially, but increases over time. In dynamic load balancing scheme the imbalance drops after each rebalancing and gradually grows in between.

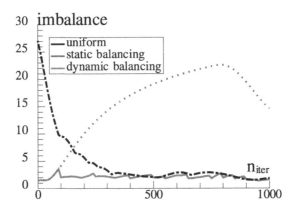

Fig. 1. Imbalance of uniform, static and dynamic domain decomposition on the test plasma heating simulation with 256 MPI processes

As the result static rectilinear partitioning is expectedly not beneficial compared to uniform partitioning, but dynamic load balancing is superior by factor of 2, see Fig. 2. In order to estimate how close is our balancing to the optimum we consider ideally balanced problem with the same number of particles and grid cells. Throughout the whole simulation particles are uniformly distributed among the simulation area. Simulation of plasma heating with dynamic load balancing takes $1.5 \times$ longer compared to simulation of the ideally balanced problem of the same size, see Fig. 2.

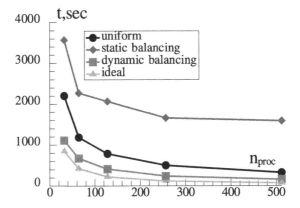

Fig. 2. Performance of uniform, static and dynamic rectilinear domain decomposition on the test plasma heating problem. As ideally balanced problem we use a problem with the same number of particles and grid cells that are equally distributed among all processes

Figure 3 presents computational time distribution between processes, colors correspond to different stages of the Particle-in-Cell method. In uniform partitioning scheme most particles are located in central domains shown as pikes in the center of the plot. In static rectilinear partitioning scheme most time is spent on corner domains. In dynamic rectilinear partitioning computational time of all processes is much closer.

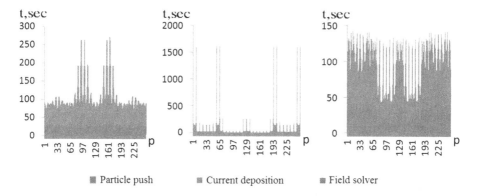

Fig. 3. Computational time of processes on the test plasma heating simulation with 256 MPI processes. Left: uniform decomposition. Middle: static rectilinear decomposition. Right: dynamic rectilinear decomposition

In all experiments for the dynamic scheme we checked imbalance each 50 time steps, imbalance threshold was 1.2. Overhead on estimation of imbalance and rebalancing is under 1 % of total simulation time, thus our scheme could be also applied for fairly well-balanced problems.

5   Load Balancing in Simulation of Plasma Target Irradiation by Two Laser Pulses

5.1   Problem Statement

In order to reveal the benefits of dynamic load balancing scheme to the state-of-the-art physical problems we performed numerical simulations of the micron-sized plasma target irradiation by two counter-propagating relativistically strong laser pulses with circular polarization. This problem is interesting from the point of view of efficient conversion of laser energy into the energy of gamma radiation [21]. Complex dynamics of laser-plasma interaction, including target compression, layer rotation and expansion stages, results in rapid change of characteristic scales by more than order of magnitude, thus leading to a strongly non-uniform particles distribution.

The important distinctive feature of this problem in comparison with the test problem from the previous section is the increasing number of particles during simulation due to quantum electrodynamics (QED) effects of photon generation and electron-positron pair production [22]. The resampling algorithm used in our simulations [2] limits the maximum number of particles in a domain by the chosen value to prevent memory shortage and control computational speed. The resampling inevitably reduces accuracy of simulation, increasing level of computational noise. At the same time, the dynamic load balancing may split the region of quick growth of particle number between several domains thus preventing additional resampling. These considerations allow us to expect that dynamic load balancing can speed up simulations and/or decrease level of computational noise even in case of the rapidly growing number of particles.

The simulation area of $1.6\,\mu\text{m} \times 4\,\mu\text{m} \times 4\,\mu\text{m}$ was covered with $256 \times 1024 \times 1024$ uniform grid. This subdivision was dictated by specific features of the initial plasma distribution and laser-plasma interaction. The plasma target with density $8 \cdot 10^{23}\,\text{cm}^{-3}$, corresponding to the electron density of the fully ionized aluminum foil, had dimensions $0.36\,\mu\text{m} \times 2.4\,\mu\text{m} \times 2.4\,\mu\text{m}$ and was located in the center of the simulation area. Laser pulses with wavelength of $0.8\,\mu\text{m}$ had a super-Gaussian transverse profile with diameter of $2.4\,\mu\text{m}$ at FWHM level and the intensity at the maximum $10^{24}\,\text{W/cm}^2$. Laser pulses propagated along x axis. Time envelope of laser pulses $f(t)$ is defined as follows:

$$f(t) = \begin{cases} \sin^2(\omega t/8), & t < \frac{4\pi}{\omega} \\ 1, & \text{otherwise,} \end{cases} \tag{4}$$

where $\omega = 2.36 \cdot 10^{-15}\,\text{s}^{-1}$ is laser frequency. The interaction of plasma with laser pulses was considered during 12 periods of the laser field. PML [17] was used to prevent the reflection of the electromagnetic field from the boundaries of the simulation area, along with absorbing boundary conditions for particles. With these parameters, the conversion efficiency of the laser energy into the gamma radiation is about 35 %.

5.2   Load Balancing Efficiency

To estimate the efficiency of dynamic load balancing we performed two runs with the parameters given in the previous subsection with and without dynamic balancing. The numerical experiments were done on the MVS-100K cluster of the Joint Supercomputing Center of RAS. The simulation area was subdivided into $4 \times 16 \times 8$ domains using 512 MPI processes. We measured the total run time from the start of the simulation and the total number of particles as a function of time iteration number, shown in Fig. 4.

The interaction dynamics is very complex and can be divided into several stages. The initial stage corresponds to the compression of plasma target by incident laser pulses in the direction of field propagation. During this stage the imbalance of the uniform partitioning does not change, because particles are

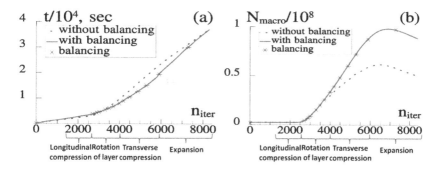

Fig. 4. (a) The total run time from the start of the simulation and (b) the total number of particles N_{macro} as a function of time iteration number n_{iter}

compressed inside the same central domains they were initially and redistribution between domains does not occur. The dependence of the imbalance on time iteration is depicted in Fig. 5. With dynamic load balancing enabled, the initial decomposition is non-ideal, so the repartition is performed once to reduce the imbalance and the resulting decomposition is also unchanged until the end of the stage, when a thin electron layer with thickness much smaller than the wavelength is formed. This size is also 30 times smaller than the initial longitudinal size of the plasma target. At the next stage the compressed plasma layer rotates at the frequency of the incident wave and emits gamma photons. The number of particles starts to grow approximately five times over a laser period mainly due to photon emission, and to a lesser degree due to photon decay into electron-positron pairs in a strong electromagnetic field [23,24]. Electrons and positrons remain mainly in the center of the simulation area, while photons propagate to the periphery. The imbalance starts to grow quickly which is clearly seen for the uniform decomposition in Fig. 5, but dynamic load balancing immediately

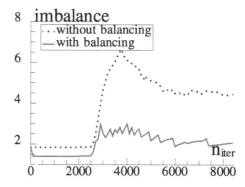

Fig. 5. Imbalance with and without dynamic load balancing

starts repartitioning of the simulation area, reducing imbalance approximately by a factor of 2.5 and speeding up the computation at this stage.

During the next stage, rotating electron layer in the field of two counter-propagating laser pulses produces electromagnetic field, that not only holds plasma, but compresses it in the transversal direction. The transverse dimension of the plasma is reduced by the order of magnitude. The dynamic load balancing tracks changes in the geometric dimensions of the plasma and perform repartitions, so that the spatial region of the rapid particle production is shared between larger number of domains than in the calculation without rebalancing. It leads to approximately 1.5 times larger number of particles and reduces the number of resampled particles. On the one hand, larger number of particles increases execution time of one iteration, however, on the other hand, it decreases computational noise in the field, current and particle distribution. In the considered case the computational noise with dynamic load balancing is approximately 1.5 times less at the end of the simulation in comparison to no balancing. The corresponding electron distributions along y axis are presented at Fig. 6. Quadratic averaging over 5 points of particle distribution was used to assess the level of noise, then the standard deviation is obtained from comparison of exact and smoothed distributions.

The final stage is the expansion of the plasma. The process of birth of new particles extends toward the incident laser pulses as the laser filed is strongly screened, so the particle generation takes place in a thin layer of plasma, where the laser field can penetrate and particle growth rate slows down. At this stage, most of the photons reach the boundaries of simulation area, which leads to decrease in the number of particles. The imbalance reaches the minimum value of about two and slowly increases. As a result durations of calculations are approximately equal with and without dynamic load balancing, although this method shows speedup of approximately 1.2 times, after that increasing number

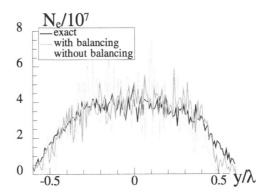

Fig. 6. Final electron distributions along y axis for simulations with load balancing, shown in green, and no load balancing, shown in yellow. The black line corresponds to a much less noisy simulation with an order of magnitude larger amount of particles (Colour figure online)

of particles slows down the calculation. Dynamic load balancing maintains the imbalance at the constant level, as it is supposed to do. However in case of growth of particle number developed method equalizes and increases load at the same time. But an advantage of it is better accuracy because of lesser number of resampled particles. Load balancing overhead was $\sim 1.34\%$ of total run time.

6 Summary

This paper considers load balancing for three-dimensional Particle-in-Cell plasma simulation on cluster systems. We propose a dynamic load balancing scheme based on rectilinear partitioning and discuss implementation of efficient imbalance estimation and rebalancing. We analyze the impact of load balancing on performance and accuracy of simulation. First, we consider the significantly imbalanced synthetic plasma heating problem and show 2 times speedup compared to uniform partitioning. We found that the overhead on dynamic load balancing is under 1% of total run time thus the scheme could also be applied for fairly balanced problems.

Then, we assess the applicability of the implemented load balancing scheme to the up-to-date physical problems. For that, we simulate a micron-sized plasma target irradiated by two counter-propagating relativistically strong laser pulses with circular polarization. The increasing number of particles during simulations due to gamma photon generation and electron-positron pair production results in the need of using a particle resampling algorithm to prevent memory shortage and control the run time by means of merging of particles with close properties. The computational noise expansion is the obvious disadvantage of the resampling procedure. Our experimental results show that dynamic load balancing significantly improves accuracy of the simulation by increasing the overall number of particles and reducing the number of merging of particles. This effect is the result of more uniform distribution of the particles between computational nodes. Balancing-related overhead is under 1.5% of total simulation time.

Our future work includes employing a more sophisticated workload model and auto-tuning of parameters of the dynamic load balancing scheme: frequency of imbalance estimation and imbalance threshold.

This study was partially supported by the RFBR, research projects No. 14-07-31211 and No. 14-02-31495, and by the grant (the agreement of August 27, 2013 No. 02.B.49.21.0003 between The Ministry of education and science of the Russian Federation and Lobachevsky State University of Nizhni Novgorod). A. Bashinov acknowledges the Dynasty Foundation support.

References

1. Birdsal, C., Langdon, A.: Plasma Physics via Computer Simulation. Taylor & Francis Group, New York (2005)
2. Gonoskov, A., Bastrakov, S., Efimenko, E., Ilderton, A., Marklund, M., Meyerov, I., Muraviev, A., Surmin, I., Wallin, E.: Extending PIC Schemes for The Study of Physics in Ultra-Strong Laser Fields. arXiv:1412:6426 (2014)

3. Pukhov, A.: Three-dimensional electromagnetic relativistic particle-in-cell code VLPL. J. Plasma Phys. **61**, 425–433 (1999)
4. Fonseca, R.A., Silva, L.O., Tsung, F.S., Decyk, V.K., Lu, W., Ren, C., Mori, W.B., Deng, S., Lee, S., Katsouleas, T., Adam, J.C.: OSIRIS: a three-dimensional, fully relativistic particle in cell code for modeling plasma based accelerators. In: Sloot, P.M.A., Hoekstra, A.G., Tan Kenneth, C.J., Dongarra, J.J. (eds.) ICCS-ComputSci 2002, Part III. LNCS, vol. 2331, p. 342. Springer, Heidelberg (2002)
5. Burau, H., Widera, R., Honig, W., et al.: PIConGPU: a fully relativistic particle-in-cell code for a GPU cluster. IEEE Trans. Plasma Sci. **33**, 2831–2839 (2010)
6. Bastrakov, S., Donchenko, R., Gonoskov, A., Efimenko, E., Malyshev, A., Meyerov, I., Surmin, I.: Particle-in-cell plasma simulation on heterogeneous cluster systems. J. Comput. Sci. **3**, 474–479 (2012)
7. Bastrakov, S., Meyerov, I., Surmin, I., Efimenko, E., Gonoskov, A., Malyshev, A., Shiryaev, M.: Particle-in-cell plasma simulation on CPUs, GPUs and Xeon Phi coprocessors. In: Kunkel, J.M., Ludwig T., Meuer, H.W. (eds.) ISC 2014. LNCS, vol. 8488, pp. 513–514. Springer (2014)
8. Liewer, P.C., Decyk, V.K.: A general concurrent algorithm for plasma particle-in-cell codes. J. Comput. Phs. **85**, 302–322 (1989)
9. Walker, D.W.: Characterising the parallel performance of a large-scale, particle-in-cell plasma simulation code. Concurr. Pract. Experience **2**, 257–288 (1990)
10. Kraeva, M.A., Malyshkin, V.E.: Implementation of PIC method on MIMD multicomputers with assembly technology. In: Hertzberger, B., Sloot, P. (eds.) High-Performance Computing and Networking. LNCS, vol. 1225, pp. 541–549. Springer, Heidelberg (1997)
11. Kraeva, M.A., Malyshkin, V.E.: Assembly technology for parallel realization of numerical models on MIMD-multicomputers. Int. J. future Gener. Comput. Syst. **17**, 755–765 (2001)
12. Fox, G.C.: A review of automatic load balancing and decomposition methods for the hypercube. Numer. Algorithms Mod. Parallel Comput. Architect. **13**, 63–76 (1988)
13. Barnes, J., Hutt, P.: A hierarchical O(N logN) force calculation algorithm. Nature. **324**, 446–449 (1986)
14. Plimpton, S.J., Seidel, D.B., Pasik, M.F., Coats, R.S., Montry, G.R.: A load-balancing algorithm for a parallel electromagnetic particle-in-cell code. Comput. Phys. Commun. **152**, 227–241 (2003)
15. Nakashima, H., Miyake, Y., Usui, H., Omura, Y.: OhHelp: a scalable domain-decomposing dynamic load balancing for particle-in-cell simulations. In: 23rd International Conference on Supercomputing, pp. 90–99. ACM New York (2009)
16. Nicol, D.N.: Rectilinear partitioning of irregular data parallel computations. J. Parallel Distrib. Comput. **23**, 119–134 (1994)
17. Berenger, J.P.: A perfectly matched layer for the absorption of electromagnetic waves. J. Comput. Phys. **114**, 185–200 (1994)
18. Taflove, A.: Computational Electrodynamics: The Finite-Difference Time-Domain Method. Artech House, London (1995)
19. Corradi, A., Leonardi, L., Zambonelli, F.: Performance comparison of load balancing policies based on a diffusion scheme. In: Lengauer, C., Griebl, M., Gorlatch, S. (eds.) Euro-Par'97 Parallel Processing. LNCS, vol. 1300, pp. 882–886. Springer, Heidelberg (1997)
20. Zhong, Z., Rychkov, V., Lastovetsky, A.: Data partitioning on multicore and multi-GPU platforms using functional performance models. IEEE Trans. Comput. **12**, 14 (2014)

21. Bashinov, A.V., Kim, A.V.: On the electrodynamic model of ultra-relativistic laser-plasma interactions caused by radiation reaction effects. Phys. Plasmas **20**, 113111 (2013)
22. Bell, A.R., Kirk, J.G.: Phys. Rev. Lett. **101**, 200403 (2008)
23. Ritus, V.: Quantum effects of the interaction of elementary particles with an intense electromagnetic field. J. Sov. Laser Res. **6**, 497–617 (1985)
24. Nikishov, A.: Problems of intense external-field intensity in quantum electrodynamics. J. Sov. Laser Res. **6**, 619–717 (1985)

Unconventional Computing - Cellular Automata

A Behavioral Analysis of Cellular Automata

Jan M. Baetens$^{(\boxtimes)}$ and Bernard De Baets

KERMIT, Department of Mathematical Modelling, Statistics and Bioinformatics,
Ghent University, Coupure Links 653, 9000 Ghent, Belgium
{jan.baetens,bernard.debaets}@ugent.be

Abstract. Although gaining full insight into the dynamics of cellular automata still poses a challenge, significant advances have been made through establishing an appropriate dynamical systems theory, similar in spirit to the one that is in place for analyzing the dynamics of their continuous counterparts such as (partial) differential equations. In this work we will show not only how it can be relied on to characterize the dynamics of cellular automata, but also how to quantify the effect of the involved design parameters on the evolved dynamics. Furthermore, we will illustrate that its scope is not limited to two-state CAs, as it is also applicable to cellular automata that are based upon multiple states.

1 Introduction

Catalyzed by the emergence of modern computers, cellular automata (CAs) became a full-fledged research domain in the eighties of the previous century. Essentially, the relevant literature is of a dichotomous nature in the sense that studies either focus on the spatio-temporal dynamics that is evolved by CAs [20,37,38,40], while others merely use the CA paradigm to build a model for a given biological, natural or physical process [10,16,32]. It goes without saying that a profound understanding of CA dynamics is a prerequisite for building realistic, identifiable CA-based models, though this is not straightforward due the fact that a CA is discrete in all its senses (state, time, space). In an attempt to quantify CA behavior in a meaningful and reproducible way, several so-called behavioral measures have been proposed during the last two decades.

Roughly speaking, these measures can be divided into three categories. A first one encloses measures that are directly based on the space-time diagrams evolved by CAs, such as the Hamming distance and the (sequence) density [14,36,37], and give insight into the local properties of CAs. The second class of measures groups those that may be referred to as true complexity measures, such as the Kolmogorov [9,19,33], Shannon [30] and Rényi entropies [27] (see [22] for a comprehensive review) and the Lempel-Ziv complexity [41]. Finally, the third class of measures encloses those that have been proposed from a dynamical systems point of view and includes measures like the Langton parameter [20], Boolean derivatives [35], Derrida coefficients [11] and Lyapunov exponents [5,6]. Especially the latter two enable a deeper understanding of CAs as dynamical systems in the sense that they reflect how CAs react to small perturbations in their

© Springer International Publishing Switzerland 2015
V. Malyshkin (Ed.): PaCT 2015, LNCS 9251, pp. 123–134, 2015.
DOI: 10.1007/978-3-319-21909-7_13

configurations, i.e. they quantify the sensitive dependence on initial conditions, one of the prerequisites to the emergence of chaos in dynamical systems [12].

Here, we will show how Lyapunov exponents and Boolean derivatives can be used to get a complete picture of CA dynamics in the sense that they not only make it possible to unravel the nature of a given CA, but also allow for assessing the effect of changing model design parameters on the CA behavior [2–4], an understanding that is a prerequisite for CA-based models to become appreciated as a full-fledged modeling paradigm. Besides, we will also investigate whether there is a link between the complexity of the configuration evolved by a CA and its dynamics. Finally, we will illustrate how the scope of Lyapunov exponents and Boolean derivatives is not limited to two-state CAs, as they can easily be extended to CAs that are based upon multiple states.

This paper is organized as follows. In Sect. 2 we present the CA formalism and we introduce the measures that will be used throughout this paper. The main results of this paper are presented in Sect. 3, together with a description of the experimental set-up.

2 Cellular Automata and Their Behavior

2.1 Cellular Automata

In this paper, we stick to the following definition of a CA.

Definition 1. *A cellular automaton $\mathcal{C}$ is a fivetuple*

$$\mathcal{C} = \langle \mathcal{T}, S, s, N, \Phi \rangle,$$

where

(i) *$\mathcal{T}$ is a countably infinite tessellation of an n-dimensional Euclidean space $\mathbb{R}^n$, consisting of cells c_i, $i \in \mathbb{N}$.*

(ii) *S is a finite set of k states.*

(iii) *The output function $s : \mathcal{T} \times \mathbb{N} \to S$ yields the state value of cell c_i at the t-th discrete time step, i.e. $s(c_i, t)$.*

(iv) *The neighborhood function $N : \mathcal{T} \to \cup_{p=1}^{\infty} \mathcal{T}^p$ maps every cell c_i to a finite sequence $N(c_i) = (c_i)_{j=1}^{|N(c_i)|}$, consisting of $|N(c_i)|$ distinct cells c_{i_j}.*

(v) *Φ is a transition function governing the dynamics of cell c_i, i.e.,*

$$s(c_i, t+1) = \Phi(\tilde{s}(N(c_i), t))$$

where $\tilde{s}(N(c_i), t) = (s(c_{i_j}, t))_{j=1}^{|N(c_i)|}$.

Throughout this paper two CA families will be considered, being the family of elementary CAs (ECAs) and the family of 2D $(2, 7)$ irregular totalistic CAs. The former can be recovered from Definition 1 upon restricting to one dimension, i.e. $n = 1$, and choosing $S = \{0, 1\}$ and $N = (c_{i-1}, c_i, c_{i+1})$. For the second family, it holds that $n = 2$, $S = \{0, 1\}$, the underlying tessellation is irregular, the neighborhood of a cell c_i consists of those cells that share at least a vertex with c_i (Moore neighborhood) and the next state of a cell solely depends on the sum of the states in c_i's neighborhood. In order to account for the fact that this sum is theoretically unbounded in the case of irregular CAs, all inputs equal or larger than 7 are mapped to the same state [5]. In practice, CAs have to be evolved on finite lattices $\mathcal{T}^*$ of size $|\mathcal{T}^*|$. For the sake of brevity, the * will be dropped in the remainder of this paper. Besides, from the context it will be clear that we refer to a finite tessellation.

2.2 Measures of Cellular Automaton Behavior

Although a detailed discussion of most measures that will be used throughout this paper is beyond its scope and it might be redundant for some readers as they are well established, we briefly present their definitions in the remainder of this section. Special attention will be given to the Lempel-Ziv complexity in Sect. 2.3 because the use of this complexity measure in the field of CAs has been limited to a few recent exploratory papers [24–26].

Density. The density of a CA configuration at a given time step, denoted $\rho(t)$, is defined as the proportion of cells in state one, i.e.

$$\rho(t) = \frac{1}{|\mathcal{T}|} \sum_{i=1}^{|\mathcal{T}|} s(c_i, t).$$

Sensitivity. A measure quantifying the sensitivity of a CA can be formalized on the basis of the Boolean counterpart of a Jacobian matrix for continuous state dynamical systems. More specifically, the entries of the Boolean Jacobian matrix J_{ij}^t at a given time step t are given by Boolean derivatives of $s(c_i, t+1)$ with respect to $s(c_j, t)$ [35], and the sensitivity measure $\bar{\mu}$ is given by the geometric mean of

$$\mu(t) = \frac{1}{|\mathcal{T}|} \sum_{i=1}^{|\mathcal{T}|} \frac{1}{|N(c_i)|} \sum_{j:c_j \in N(c_i)} J_{ij}^t \tag{1}$$

for a large number of time steps T [6]. Informally, $\bar{\mu}$ may be understood as the average proportion of cells in the neighborhood that affects the state of a given cell c_i at the $t+1$-th time step if their state was perturbed, i.e. it reflects the sensitivity of a CA. Given Eq. (1), a CA will evolve in exactly the same way irrespective of whether and how many cells are perturbed at a given time step if $\bar{\mu} = 0$ (e.g. Rule 0), whereas introducing the smallest possible perturbation at a given time step will always have an effect on the further evolution if the sensitivity is equal to one (e.g. Rule 150).

Lyapunov Exponents. Following Wolfram's suggestion to quantify the rate of divergence between two initially close configurations [37], both directional and non-directional Lyapunov exponents were established [6,31], the latter overcoming the fact that the directionality of damage spreading gets blurred in higher-dimensional CAs. In [6] the maximum Lyapunov exponent (MLE) of a CA is defined as

$$\lambda = \lim_{t \to \infty} \frac{1}{t} \log \left(\frac{\epsilon_t}{\epsilon_0} \right), \tag{2}$$

where ϵ_t denotes the total number of defects that are accumulated as the CA evolved until the t time step starting from an initial configuration containing ϵ_0 defects, where, typically, $\epsilon_0 = 1$. The important finesse that should be mentioned with regard to this approach is that it involves tracking the defects in tangent space rather than in configuration space as is the case in [31,37]. The reader who is not yet familiar with this approach is referred to [1] for a more comprehensive explanation.

Among ECAs the MLE varies between $\{-\infty\} \cup [0, \log(3)]$, where $\log(3)$ is the maximum rate of exponential divergence that can be reached by any of the ECAs, while it varies between $\{-\infty\} \cup [0, \log(\bar{V})]$ for $(2, 7)$ irregular totalistic CAs, where $\bar{V}$ denotes the average neighborhood size. Typically, the MLE is used to classify CAs on the basis of their robustness to perturbations. More specifically, when evaluating the MLE across an ensemble of initial configurations, there will be CAs for which the MLE will always be $-\infty$. Such CAs may be referred to as unconditionally superstable, while the ones for which the contrary is true, i.e. their MLE is positive irrespective of the initial configuration, may be referred to as unconditionally unstable CAs. In addition to these unconditionally superstable and unstable CAs, the discrete nature of CAs makes that there are also CAs for which the MLE is positive only for some initial configurations, while it is $-\infty$ for others. Such rules are referred to as conditionally unstable or superstable depending on the prevailing behavior across the ensemble.

2.3 Lempel-Ziv Complexity

Although the Lempel-Ziv complexity was originally conceived to quantify the randomness of binary sequences for constructing random sequences [21], after which it became established for data compression applications [41,42], it seems to have been making its entrance in several disciplines during the last decade, such as computer vision [8], medicine [18] and biochemistry [17]. Only very recently it has been introduced in the field of CAs to quantify the randomness of the consecutive CA configurations [24–26]. In order to define the Lempel-Ziv complexity it is easier to first define a so-called Lempel-Ziv partition of a binary sequence. For that purpose, let $v = v_1 v_2 ... v_k$ and $w = w_1 w_2 ... w_n$ be two binary sequences and let us refer to a prefix v of w if it holds that $v_i = w_i$ for $1 \leq i \leq k$, while v is said to be a proper prefix of w if $k < n$. Then the partition $y_1|...|y_r$ is called the Lempel-Ziv partition of w if (1) $y_i \neq y_j$ for all $i = 1, ..., r - 1$ and

$j = 1, \ldots, i - 1$; (2) $w = y_1 y_2 \ldots y_r$; and (3) every proper prefix of y_i is equal to some y_j for all $i = 1, \ldots, r$ and some $j = 1, \ldots, i - 1$.

If $y_r = y_i$ for some $i = 1, \ldots, r - 1$, w is referred to as an open sequence. Otherwise, w is called a closed sequence. Given the Lempel-Ziv partition of a binary sequence w, the Lempel-Ziv complexity of w, denoted $\ell(w)$, is defined as the number of distinct patterns in the Lempel-Ziv partition of w. In the case of an open sequence, the Lempel-Ziv complexity will be equal to $r - 1$, rather than r, thus agreeing with the maximal number of unique substrings in a given binary sequence w. For example, the Lempel-Ziv partition of 111000111110100 is given by $1|11|0|00|111|110|10|0$, such that we have $\ell(111000111110100) = 7$ because the last substring in the Lempel-Ziv partition is not unique.

Since the Lempel-Ziv complexity obviously depends on the length of the considered binary sequence, it makes sense to normalize it in such a way that it becomes possible to compare it across binary strings of different lengths. Here, the normalized Lempel-Ziv complexity is defined as

$$\ell_n = \frac{\ell - \ell_{min}}{\ell_{max} - \ell_{min}}, \tag{3}$$

where ℓ_{min} and ℓ_{max} represent the minimum and maximum Lempel-Ziv complexity for a sequence of a given length, respectively. The normalized Lempel-Ziv complexity obviously lies between zero and one. Given the definition above, the normalized Lempel-Ziv complexity of a sequence ℓ_n increases as its degree of randomness grows. As such, the normalized Lempel-Ziv complexity is zero for a uniform configuration.

3 Behavioral Analysis

3.1 Experimental Setup

For all ECA simulations, a tessellation consisting of 500 cells with periodic boundary conditions was used, whereas a tessellation of 675 irregular cells was used to evolve the $(2, 7)$ irregular totalistic CAs. Since a CA might not be able to evolve to its most stable state as a consequence of its discrete nature [6], all values reported in the remainder of this paper represent averages over an ensemble $\mathcal{E}$ of 30 and 8 random initial configurations, respectively for the ECAs and $(2, 7)$ irregular totalistic CAs. Every representative CA was evolved for 500 time steps for each of the configurations in the ensemble. All simulations were performed in Mathematica (Wolfram Research Inc., Champaign, USA) using the high performance computing infrastructure of Ghent University.

Figure 1 depicts the behavioral signatures of both CA families. More specifically, it visualizes the relationship between the sensitivity $\bar{\mu}$ and the MLE λ for the members of both CA families. Generally speaking, this figure indicates that there exists a positive relationship between the sensitivity and the MLE, especially in the case of the $(2, 7)$ irregular totalistic CAs. Besides, it is remarkable that many CAs for which $\lambda \neq -\infty$ for all members of E give rise to MLEs that are close to the respective theoretical upper bound.

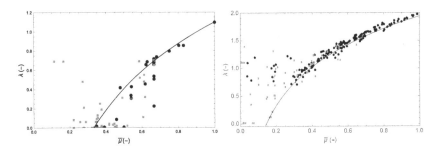

Fig. 1. Maximum Lyapunov exponent (λ) for the ECAs and $(2,7)$ irregular totalistic CA versus sensitivity $\bar{\mu}$ after 500 time steps together with the theoretical upper bounds on the MLE. Results are averages calculated over an ensemble of different initial configurations E, and are only shown for those CAs for which $\lambda \neq -\infty$ for all members of E (circles), and for rules giving rise to $\lambda = -\infty$ for at most all but one member of E (squares) [5].

3.2   Complexity versus stability

Whether or not there exists a link between the Lempel-Ziv complexity of an ECA and its corresponding MLE, can be investigated by examining Fig. 2(a), which depicts the normalized Lempel-Ziv complexity of the final ECA configuration versus the corresponding normalized MLE for the 88 representative ECAs. Recall that results represent averages calculated over an ensemble of different initial configurations E, and note that they are only shown for those ECAs leading to $\lambda \neq -\infty$ for all members of E (circles), and those giving rise to $\lambda = -\infty$ for at most all but one member of E (squares). First of all, it is clear that the rules giving rise to the highest possible MLE (i.e. Rules 105 and 150) also evolve configurations that reach the highest possible degree of randomness, but apparently there exist significantly more stable rules that evolve similarly random configurations, such as Rules 15, 30, 45, 51, 60, 90, 106, 154, 170 and 204. This already hints that complexity in configuration space does not necessarily imply pronouncedly unstable behavior from a dynamical systems point of view, and vice versa. Secondly, Fig. 2(a) also points to the fact that the entire range of Lempel-Ziv complexities is covered by rules of which the nMLE is only marginally positive, which is a further confirmation of the decoupling between complexity in configuration space and the nature of the dynamics underlying the ECA evolution, which corroborates the recent findings of [26]. Moreover, it is obvious that ECAs can evolve complex configurations, irrespective of whether or not their dynamics is unstable, i.e. whether or not there is sensitive dependence on initial conditions. So, contrary to what one could expect, perturbing the state of a cell in the evolution of an ECA that produces relatively random configurations might not cause a cascade of defects that makes the subsequent ECA configurations diverge exponentially from the ones that would be evolved in absence of such a perturbation. Of course, in configuration space the evolved configurations might still differ significantly, but this does not hold for the tangent space where the defects are tracked. In Fig. 2(a) it can also be observed that there are a few peculiar rules (57, 72, 104) with nMLEs between 0.6 and 0.65 that give rise to comparatively low complexity values. Essentially, these three rules are the only ones in an otherwise unfrequented zone below the 1:1 line in this scatterplot and their location in the (λ_n, ℓ_n)-plane can be understood by acknowledging that they give rise to (diagonally) striped patterns.

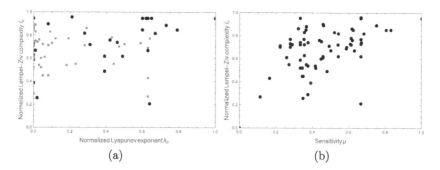

Fig. 2. Normalized Lempel-Ziv complexity ℓ_n of the final ECA configuration (500 time steps) versus the corresponding normalized Lyapunov exponent λ_n (a) and sensitivity $\bar{\mu}$ (b) for the 88 representative ECAs. Results are averages calculated over an ensemble of different initial configurations E, and are only shown for those ECAs leading to $\lambda \neq -\infty$ for all members of E (circles), and those giving rise to $\lambda = -\infty$ for at most all but one member of E (squares).

To get an even better insight in how the complexity of ECA configurations and the ECA dynamics are related, Fig. 2(b) depicts the normalized Lempel-Ziv complexity ℓ_n of the final ECA configuration versus the corresponding sensitivity $\bar{\mu}$ for the 88 representative ECAs. It shows that a critical sensitivity of approximately 0.1 must be exceeded in order for an ECA to be able to evolve configurations that have a positive normalized Lempel-Ziv complexity, and moreover, that a sensitivity of at least 0.25 is a prerequisite for evolving truly random configurations ($\ell_n > 0.6$). Beyond this threshold there does not seem to be a link between the randomness of the evolved ECA configurations, on the one hand, and the sensitivity of the governing transition function $\bar{\mu}$, on the other hand, because the normalized Lempel-Ziv complexity apparently can take any value once this threshold is exceeded, irrespective of $\bar{\mu}$. Besides, Fig. 2(b) uncovers the existence of zones in the $(\bar{\mu}, \ell_n)$-plane that are (almost) unreachable by evolving ECAs.

3.3   Interference Between CA Design and Behavior

Many studies have addressed how the choice of CA design parameters, such as the kind of tessellation, the neighborhood structure, or the update mechanism affects the CA behavior [13–15, 23, 28, 29, 34], though the majority of these focuses on the qualitative differences between the results, obtained with different design parameters. On the other hand, by resorting to the measures introduced in Sect. 2, it becomes possible to investigate these interferences in a quantitative way as their values can be compared across a series of designs. As an illustration, we will demonstrate how the effect of the update method and underlying tessellation can be assessed in this way.

Figure 3 depicts the MLEs obtained by relying on the random order update and exponential clocked update methods [7] versus the ones found by synchronously updating the $(2, 7)$ irregular totalistic CA. It should be mentioned that the presented MLEs are normalized with respect to the governing upper bounds so that they can be compared across different designs. Clearly, these upper bounds are much lower in the case of asynchronously updated CAs because much fewer cells get updated at every consecutive time step than in a synchronous setting. As can be inferred from Fig. 3, the MLE of a synchronously updated CA is typically higher than the one of its asynchronous counterpart, though there are a few rules for which the contrary is true. Similar discrepancies between synchronously and asynchronously updated CAs have been observed for other asynchronous update methods [4], and reflect that these discrete dynamical systems may exhibit completely distinct behavior if their evolution is based upon a different update method. Consequently, when it comes to the development of a CA-based model for describing a given process, particular care should be given to choosing the appropriate update method as it constitutes an inseparable part of the CA-based model as a whole.

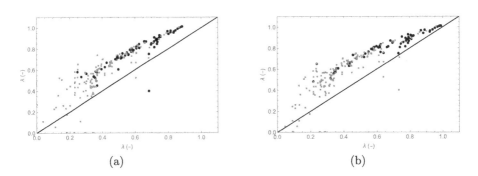

(a)   (b)

Fig. 3. Normalized MLE obtained by relying on the random order update method (a) and exponential clocked update method (b) versus the normalized MLE found by synchronous updating. The results are only shown for those rules within the family of $(2, 7)$ irregular totalistic CA that, for both concerned update methods, give rise to $\lambda^* \neq -\infty$ for at least one member of the ensemble E.

To illustrate the interference between a CA's topology and its dynamics, Fig. 4 illustrates how the normalized MLE of four different $(2, 7)$ irregular totalistic CAs changes as a function of the average neighborhood size $\bar{V}$. This figure demonstrates that there can be a significant impact of a CA's topology on its dynamics. There are rules like numbers 20 and 235 for which the MLE stays positive across the entire range of neighborhood sizes, though its magnitude varies substantially, which indicates that the underlying topology affects how fast close phase trajectories are diverging. In addition, there are rules for which not only the rate of divergence is affected, but their stability properties as a whole in the sense that they give rise to diverging trajectories for some neighborhood sizes, whereas there exists sensitive dependence on initial conditions for others (e.g. rules 244 and 246). Apparently, for the latter type of rules there seem to exist critical neighborhood sizes where the dynamics of the underlying rule changes dramatically. Consequently, such neighborhood sizes may be referred to as topological bifurcation points. All together this indicates that particular care should be given to the definition of the topology underlying a CA-based model because it is an indivisible part of the model.

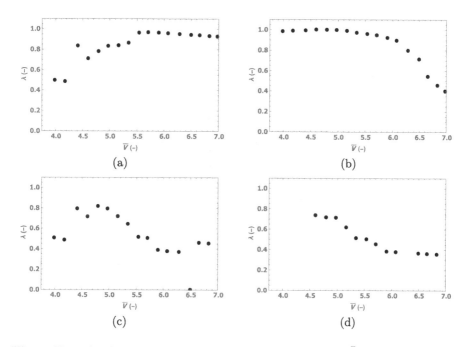

Fig. 4. Normalized MLE λ versus average neighborhood size $\bar{V}$ for $(2,7)$ irregular totalistic CAs 20 (a), 235 (b), 244 (c) and 246 (d).

3.4   The Nature of Multi-state CAs

Although a detailed elaboration on how the Lyapunov exponents defined by Eq. (2) should be computed in the case of multi-state CAs is considered to be beyond the cope of this paper, it is important to realize that different types of

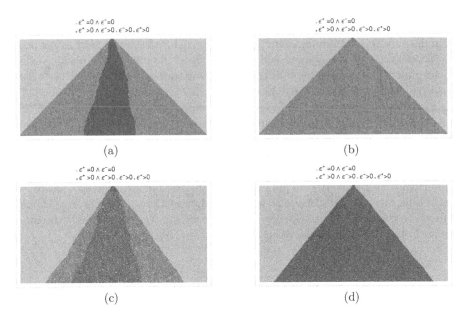

$. \epsilon^+ =0 \wedge \epsilon^- =0$
$. \epsilon^+ >0 \wedge \epsilon^- >0 . \epsilon^- >0 . \epsilon^+ >0$

$. \epsilon^+ =0 \wedge \epsilon^- =0$
$. \epsilon^+ >0 \wedge \epsilon^- >0 . \epsilon^- >0 . \epsilon^+ >0$

(a)   (b)

$. \epsilon^+ =0 \wedge \epsilon^- =0$
$. \epsilon^+ >0 \wedge \epsilon^- >0 . \epsilon^- >0 . \epsilon^+ >0$

$. \epsilon^+ =0 \wedge \epsilon^- =0$
$. \epsilon^+ >0 \wedge \epsilon^- >0 . \epsilon^- >0 . \epsilon^+ >0$

(c)   (d)

Fig. 5. Defect cone after 500 time steps for rules 1776 (a), 1893 (b), 1994 (c) and 2014 (d) starting from a random initial condition with defects in the 11 most centrally located cell of a system consisting of 1001 cells.

defects can emerge in such a setting. More specifically, at most $\frac{k(k-1)}{2}$ different types of defects can emerge during the course of the evolution of a k-state CA, one for every possible combination of states, namely $(0,1)$, $(0,2)$, $\ldots$, $(0,k)$, $(1,2)$, $\ldots$, $((k-1),k)$. Obviously, in the case of two-state CAs, there exists only one type of defect, being the one that involves flipping state zero to one, or vice versa, and sensitive dependence on initial conditions traces back unambiguously to that particular type of defect. In contrast, in the case of three-state CAs for which the state set is endowed with a cyclic ordering, there exist two types of defects, namely so-called L-defects that invoke a shift to a left shift of a cell's state and R-defects that invoke an R-shift. For each of them, a distinct Lyapunov exponent can be defined that quantifies how fast either L- or R-defects are accumulating.

In order to exemplify the nonuniform nature of defect propagation in multistate CAs, Fig. 5 visualizes the defect cone for four three-state one-dimensional totalistic CAs with rule numbers, according to [39], 1776, 1893, 1994 and 2014 which are evolved for 500 time steps, starting from a random initial condition on a system consisting of 1001 elements, upon introduction of both L- and R-defects in the 11 most centrally located cells at $t = 0$. The defect cones depicted in this figure demonstrate that the propagation of defects in multi-state CAs can be of a very complicated nature in the sense that the different types of defects do not necessarily have to coexist, while the way the different types of defects spread is often spatially very heterogeneous.

References

1. Baetens, J.M., De Baets, B.: A lyapunov view on the stability of two-state cellular automata. In: Zenil, H. (ed.) Irreducibility and Computational Equivalence. ECC, vol. 2, pp. 31–40. Springer, Heidelberg (2013)
2. Baetens, J.M., De Baets, B.: Topology-induced phase transitions in cellular automata. Phys. D: Nonlinear Phenom. **249**, 16–24 (2013)
3. Baetens, J.M., De Loof, K., De Baets, B.: Influence of the topology of a cellular automaton on its dynamical properties. Commun. Nonlinear Sci. Numer. Simul. **18**, 651–668 (2013)
4. Baetens, J.M., Van der Weeën, P., De Baets, B.: Effect of asynchronous updating on the stability of cellular automata. Chaos, Solitons Fractals **45**, 383–394 (2012)
5. Baetens, J.M., De Baets, B.: Phenomenological study of irregular cellular automata based on Lyapunov exponents and Jacobians. Chaos **20**, 033112 (2010)
6. Bagnoli, F., Rechtman, R., Ruffo, S.: Damage spreading and Lyapunov exponents in cellular automata. Phys. Lett. A **172**, 34–38 (1992)
7. Bandini, S., Bonomi, A., Vizzari, G.: What do we mean by asynchronous CA? a reflection on types and effects of asynchronicity. In: Bandini, S., Manzoni, S., Umeo, H., Vizzari, G. (eds.) ACRI 2010. LNCS, vol. 6350, pp. 385–394. Springer, Heidelberg (2010)
8. Batista, L., Meira, M., Cavalcanti Jr., N.: Texture classification using local and global histogram equalization and the Lempel-Ziv-Welch algorithm. In: Proceedings Fifth International Conference on Hybrid Intelligent Systems (2005)
9. Chaitin, G.: On the length of programs for computing finite binary sequences. J. Assoc. Comput. Mach. **5**, 547–569 (1966)
10. Chopard, B., Droz, M.: Cellular automata model for the diffusion equation. J. Stat. Phys. **64**, 859–892 (1991)
11. Derrida, B., Stauffer, D.: Phase transitions in two-dimensional Kauffman cellular automata. Europhys. Lett. **2**, 739–745 (1986)
12. Devaney, R.L. (ed.): An Introduction to Chaotic Dynamical Systems. Addison-Wesley, Redwood City (1987)
13. Fatès, N., Morvan, M.: Perturbing the topology of the game of life increases its robustness to asynchrony. In: Sloot, P.M.A., Chopard, B., Hoekstra, A.G. (eds.) ACRI 2004. LNCS, vol. 3305, pp. 111–120. Springer, Heidelberg (2004)
14. Fatès, N.A., Morvan, M.: An experimental study of robustness to asynchronism for elementary cellular automata. Complex Syst. **16**, 1–27 (2005)
15. Fats, N.A.: Critical phenomena in cellular automata: perturbing the update, the transitions, the topology. Acta Phys. Pol. B Proc. Suppl. **3**, 315–325 (2010)
16. Greenberg, J.M., Hastings, S.P.: Spatial patterns for discrete models of diffusion in excitable media. SIAM J. Appl. Math. **34**, 515–523 (1978)
17. Han, G., Yu, Z.G., Anh, V., Krishnajith, A.P., Tian, Y.C.: An ensemble method for predicting subnuclear localizations from primary protein structures. PLoS ONE **8**, 1 (2013)
18. Jouny, C., Lempel, A.: Characterization of early partial seizure onset: frequency, complexity and entropy. Clin. Neurophysiol. **123**, 658–669 (2012)
19. Kolmogorov, A.: Three approaches to the quantitative definition of information. Probl. Peredachii Informatsii **1**, 3–11 (1965)
20. Langton, C.: Computation at the edge of chaos. Phys. D **42**, 12–37 (1990)
21. Lempel, A., Ziv, J.: On the complexity of finite sequences. IEEE Trans. Inf. Theory **22**, 75–81 (1976)

22. Lindgren, K., Nordahl, M.: Complexity measures and cellular automata. Complex Syst. **2**, 409–440 (1988)
23. Manzoni, L.: Some formal properties of asynchronous cellular automata. In: Bandini, S., Manzoni, S., Umeo, H., Vizzari, G. (eds.) ACRI 2010. LNCS, vol. 6350, pp. 419–428. Springer, Heidelberg (2010)
24. Ninagawa, S., Adamatzky, A., Alonso-Sanz, R.: Phase transition in elementary cellular automata with memory. Int. J. Bifurcat. Chaos **24**, 23–35 (2014)
25. Ninagawa, S., Martinez, G.J.: Compression-based analysis of cyclic tag system emulated by rule 110. J. Cell. Automata **9**, 23–35 (2014)
26. Ninagawa, S., Adamatzky, A.: Classifying elementary cellular automata using compressibility, diversity and sensitivity measures. Int. J. Mod. Phys. C **25**, 1350098 (2014)
27. Rényi, A.: On measures of entropy and information. In: Neyman, J. (ed.) Proceedings of the Fourth Berkeley Symposium on Mathematical Statistics and Probability. pp. 547–561. University California Press, Los Angeles (1961)
28. Rouquier, J.-B., Morvan, M.: Combined effect of topology and synchronism perturbation on cellular automata: preliminary results. In: Umeo, H., Morishita, S., Nishinari, K., Komatsuzaki, T., Bandini, S. (eds.) ACRI 2008. LNCS, vol. 5191, pp. 220–227. Springer, Heidelberg (2008)
29. Schnfisch, B., de Roos, A.: Synchronous and asynchronous updating in cellular automata. Biosystems **51**, 123–143 (1999)
30. Shannon, C.: A mathematical theory of communication. Bell Syst. Tech. J. **27**, 623–656 (1948)
31. Shereshevsky, M.: Lyapunov exponents for one-dimensional cellular automata. J. Nonlinear Sci. **2**, 1–8 (1992)
32. Sirakoulis, G.C., Karafyllidis, I., Thanailakis, A.: A cellular automaton model for the effects of population movement and vaccination on epidemic propagation. Ecol. Model. **133**, 209–223 (2000)
33. Solomonoff, R.: A formal theory of inductive inference part I. Inf. Control **7**, 1–22 (1964)
34. Valsecchi, A., Vanneschi, L., Mauri, G.: A study on the automatic generation of asynchronous cellular automata rules by means of genetic algorithms. In: Bandini, S., Manzoni, S., Umeo, H., Vizzari, G. (eds.) ACRI 2010. LNCS, vol. 6350, pp. 429–438. Springer, Heidelberg (2010)
35. Vichniac, G.: Boolean derivatives on cellular automata. Phys. D **45**, 63–74 (1990)
36. Wolfram, S.: Cellular automata. Los Alamos Sci. **9**, 2–21 (1983)
37. Wolfram, S.: Universality and complexity in cellular automata. Phys. D **10**, 1–35 (1984)
38. Wolfram, S.: Two-dimensional cellular automata. J. Stat. Phys. **38**, 901–946 (1985)
39. Wolfram, S.: Cellular Automata and Complexity: Collected Papers. Addison-Wesley, Reading, United States (1994)
40. Wuensche, A., Lesser, M.: The Global Dynamics of Cellular Automata, vol. 1. Addison-Wesley, London, United Kingdom (1992)
41. Ziv, J., Lempel, A.: A universal algorithm for sequential data compression. IEEE Trans. Inf. Theory **23**, 337–343 (1977)
42. Ziv, J., Lempel, A.: Compression of individual sequences via variable-rate coding. IEEE Trans. Inf. Theory **24**, 530–536 (1978)

Contradiction Between Parallelization Efficiency and Stochasticity in Cellular Automata Models of Reaction-Diffusion Phenomena

Olga Bandman$^{(\boxtimes)}$

Supercomputer Software Department ICM and MG, Siberian Branch,
Russian Academy of Sciences, Pr. Lavrentieva, 6, Novosibirsk 630090, Russia
bandman@ssd.sscc.ru

Abstract. Simulation of reaction-diffusion phenomena are usually done using cellular automata with asynchronous mode of operation, which is in accordance with the stochastic nature of such processes, though does not provide acceptable parallelization efficiency, when the model is implemented on a supercomputer. The contradiction is resolved by endowing the asynchronous mode with some synchronization under the condition that the result is not distorted. How much of synchronization is admissible is not known. Moreover, it is not known what is the impact of synchronization on the parallelization efficiency. In the paper an attempt is made to answer these questions basing on the analysis of parallel experimental simulations of typical subclasses of reaction diffusion processes on a supercomputer.

Keywords: Reaction-diffusion processes · Cellular automata models · Asynchronous cellular automata · Composition of cellular automata · Parallel computations · Parallelization efficiency

1 Introduction

Mathematical modeling and computer simulation is now primarily focused on nonlinear dissipative phenomena in chemistry and biology [1], rather than on conventional physics. Most commonly, such phenomena are classified as reaction-diffusion processes being described in terms of elementary displacements and transformations of real or abstract particles. The displacements of particles are represented as diffusion or convection, obeying conservation laws, while transformations simulate phase transitions, chemical reactions, or some biological transmutation, that are dissipative by nature. All these elementary actions are performed in random with probabilities being in accordance with the rates of their intensity. Such kind of kinetic probabilistic models are under intensive development in chemistry and biology [2,3]. Moreover, reaction-diffusion phenomena, expressed in terms of CA [4,5], are also studied and used in different

Supported by Presidium of Russian Academy of Sciences, Basic Research Program N 15-9 (2015).

V. Malyshkin (Ed.): PaCT 2015, LNCS 9251, pp. 135–148, 2015.
DOI: 10.1007/978-3-319-21909-7_14

fields of science. The most advanced CA models are used in microelecronics [8,9], in pattern formation processes investigation [6], and in surface catalytic reactions simulation [7]. Sometimes, they are referred to as "Kinetic Monte Carlo"method, although most of them are versions and extensions of asynchronous CA (ACA), differing from the Monte Carlo method in presentation manner and transition functions composition.

Whereas there exists some experience in ACA simulation [10,11], there is no systematic methodology for synthesize a model of a complex phenomenon given by a description of involved elementary actions and their interactions. The main problem is in constructing a global operator out of given elementary local functions, which provides the best adequacy to the process under simulation on one hand, and satisfies computational correctness and efficiency conditions on the other hand. Moreover, parallel implementation conditions should be taken into account [12]. It means that asynchronous behavior should be partially synchronized, but in such a manner that does not cause distortion of the process. The synchronization is induced [13] by transforming asynchronous mode into an equivalent block-synchronous one. In [14] it is shown on a real life example of a heterogeneous catalytic reaction simulation, that such a transformation induces no errors in asynchronous CA behavior. This fact provokes the following question: how much of synchronization may be admissible in reaction-diffusion CA. The question was partially answered in [15]. But, the results of serial algorithms study cannot be transferred directly to a parallel case, because parallel implementation conditions concerning correctness and performance of coarse grain parallelism impose additional requirements onto the mode of operation. Orientation to parallel implementation needs to allow for these requirements at the earlier stages of model development. Hence, it should be clear how simulation results depend on the chosen mode of elementary operations interaction. Just this dependence is a subject of the investigation reported in the paper. Based on the available experience, the main modes of operations are distinguished. For them, the amount of additional computations needed for parallel computing and the amount of data exchanges are assessed, and computer simulation of typical reaction-diffusion processes are performed in 2D and 3D case. Computations are executed on the cluster NKS-30 of Siberian Supercomputer Center.

Apart from Introduction and Conclusion the paper contains 3 sections. In the second section necessary formal definition of reaction diffusion CA are given. In the third section the peculiarities of parallel implementation are enlightened and analytical assessment of parallel implementation efficiency is obtained. The rest of the article is devoted for presenting the results of computational experiments.

2   Formal Representation of CA-Models

CA is defined by a triplet $\aleph = \langle A, X, \Theta(X) \rangle$. A is a set of symbols, called *alphabet*, which may be interpreted as designations of substances involved in the process under simulation. $X = \{\mathbf{x}\}$ is a *set of cell names*, usually given by a set of coordinates of a finite discrete space, $\mathbf{x} = (i, j, k), i =$

$0, \ldots, I, j = 0, \ldots, J, k = 0, \ldots, K$. A pair $(u, \mathbf{x})$, $u \in A$, $\mathbf{x} \in X$, is called a *cell*. The set of cells $\Omega = \{(u_k, \mathbf{x}_k) | \mathbf{x}_k \neq \mathbf{x}_l\}$, forms a *cellular array*, $\Omega_A = (u_1(\mathbf{x}_1), u_2(\mathbf{x}_2), \ldots, u_{|X|}(\mathbf{x}_{|X|}))$, or, simply, $\Omega_A = (u_1, \ldots, u_{|X|})$, being its *global state*. $\Theta(X)$ is a *global operator* which defines the transition of the CA to the next global state, which constitutes an iterative step or *iteration*. The sequence $\Omega_A(0), \ldots \Omega_A(t) \ldots \Omega_A(\hat{t})$, is referred to as a *CA evolution*, $\hat{t}$ being the number of global operator executions. The global operator $\Theta(X)$ is the result of application of elementary actions to all cells in the array. Each elementary action is expressed as a *substitution* of the form

$$\theta(\mathbf{x}) : S(\mathbf{x}) \xrightarrow{p} S'(\mathbf{x}), \tag{1}$$

where $S(\mathbf{x_0}) = (u_0, \ldots, u_n,)$ and $S'(\mathbf{x}) = (v_0, \ldots, v_m)$, $u, v \in A$, $m \leq n$, are local configurations, expressed as vectors of cell states in the vicinity of $\mathbf{x}$, defined by its *underlying template*

$$T(\mathbf{x}) = \{\mathbf{x}, \mathbf{x} + \mathbf{a}_1, \ldots, \mathbf{x} + \mathbf{a}_{n-1}\}, \tag{2}$$

where $\mathbf{a}_j$ is a shift vector such that the cell $\mathbf{x} + \mathbf{a_j}$ has the state u_j, the maximum component of all $\mathbf{a}_j$ is referred to as the *template radius* $R(T)$.

In (1) local configurations $S(\mathbf{x})$ and $S'(\mathbf{x})$ are vectors whose components are states of cells named from their underlying templates $T(\mathbf{x})$ and $T'(\mathbf{x})$, which are in the ratio

$$T'(\mathbf{x}) \subseteq T(\mathbf{x}). \tag{3}$$

Application of $\theta(\mathbf{x})$ to a certain $\mathbf{x_k} \in X$ replaces the states $\{u_0, \ldots, u_i, \ldots, u_m\} \in S(\mathbf{x_k})$ by the states

$$v_i = f_i(u_0, \ldots, u_i \ldots, u_n), \quad i = 0, \ldots, m, \quad n = |S(\mathbf{x_k})|, \tag{4}$$

which are values of *transition function* $f_i(u_1, \ldots, u_n)$. Application of $\theta(\mathbf{x})$ is performed with probability p, that is determined by physical properties of the corresponding action. Application of $\theta(\mathbf{x_k})$ is considered to be successful, if the subset of cells named from $T(\mathbf{x_k})$ is included into Ω, i.e.

$$\{(u_0, \mathbf{x_k}), \ldots, (u_n, \mathbf{x}_k + \mathbf{a}_n)\} \subseteq \Omega. \tag{5}$$

Otherwise the application of $\theta(\mathbf{x})$ fails, and nothing is changed in $\Omega(t)$.

If the global operator $\Theta(X)$ contains only one substitution, then it is called a *simple global operator*. Accordingly, a CA with simple $\Theta(X)$ is a simple CA. Application of a simple $\Theta(X)$ consists of application $\theta(\mathbf{x})$ to all $\mathbf{x} \in X$, the order of cells to be chosen being defined by the *mode of operation* ρ. There are two basic modes of simple CA operation.

Synchronous mode, $\rho = \sigma$, prescribes the following algorithm of application $\Theta(X)$ to $\Omega(t)$:

(1) a cell $(u, \mathbf{x})$ is chosen from $\Omega(t)$, any order of choice being admissible,

(2) if condition (5) is satisfied, the next state of $\mathbf{x}$ is computed according to (4),

(3) when next states are obtained for all $\mathbf{x} \in X$, they are adjusted all at once transforming $\Omega(t)$ into $\Omega(t+1)$.

Asynchronous mode, $\rho = \alpha$, suggests the following algorithm of a global operator application:

(1) a cell $(u, \mathbf{x})$ is chosen from $\Omega(t)$ with probability $p = 1/|X|$,

(2) if condition (5) is satisfied, $\theta(\mathbf{x})$ is applied to $\mathbf{x}$, adjusting the states of cells from $T'(\mathbf{x})$ immediately,

(3) the global operator is considered to be executed, when (1) and (2) are executed $|X|$ times, completing the t-th iteration.

In reality, asynchronous computation process is not divided into iterations, so the concept of iteration is accepted conditionally for making the comparison with synchronous case more conceptual.

In reaction-diffusion CA $\Theta(\mathbf{x})$ is a *complex* global operator, being composed of several (at least two) local operators

$$\Theta(X) = \Phi(\theta_1(\mathbf{x}), \dots, \theta_q(\mathbf{x})), \tag{6}$$

which determines the CA functioning. For providing functioning correctness $\Theta(X)$ must satisfy the following condition: no pair of substitution application should aim at adjusting the same cell at the same time, i.e.

$$T'_k(\mathbf{x}) \bigcap T'_l(\mathbf{y}) = \emptyset \quad \forall (\mathbf{x}, \mathbf{y}) \subset X, \tag{7}$$

where $T'_k(\mathbf{x})$ and $T'_l(\mathbf{y})$ are underlying templates for the right hand sides of (1) in $\theta_k(\mathbf{x})$ and $\theta_l(\mathbf{x})$, including the case $l = k$.

In simple synchronous CA the correctness condition (7) is satisfied, when $|T'_k(\mathbf{x})| \leq 1$. It yields a strong constraint in reaction-diffusion process synthesis, but simplifies parallel implementation of a large scale CA on a cluster. Unlike the synchronous case, asynchronous computation on one processor is always correct due to its serial character, but its implementation on parallel processor is absolutely inefficient.

Global operator $\Theta(X)$ (6) of a complex CA defines common behavior of included in it substitutions, further referred to as *complex mode of operation*. To be more concise let us further focus on reaction-diffusion CA, whose global operator deals with two substitutions: $\theta_d(\mathbf{x})$ and $\theta_r(\mathbf{x})$, simulating elementary acts of diffusion and reaction, respectively. The following complex modes deserve detailed consideration.

Stochastic complex mode $\Phi_S(\theta_d(\mathbf{x}), \theta_r(\mathbf{x}))$, when a randomly chosen substitution $\theta_d(\mathbf{x})$ or $\theta_r(\mathbf{x})$ is applied to a randomly chosen cell $\mathbf{x} \in X$, and, if the condition (6) is satisfied, the cells of corresponding underlying template $T'(\mathbf{x})$ are adjusted immediately. Coordination of the component reactions rates is done by setting the probability of each θ_l execution equals to

$$p_l = k_l/(k_d + k_r), \quad l \in \{d, r\}, \tag{8}$$

where k_l is the rate constant of the l-th action [7].

Local complex mode $\Phi_L(\theta_d(\mathbf{x}), \theta_r(\mathbf{x}))$, when $\Theta(X)$ implies asynchronous application of the superposition of substitutions, i.e. $\theta_d(\mathbf{x})$ is applied to the result of $\theta_r(\mathbf{x})$ application to a randomly chosen cell $\mathbf{x} \in X$,

$$\Theta(X) = \Phi_L(\theta_d(\mathbf{x}), \theta_r(\mathbf{x})) = \Phi(\theta_d(\theta_r(\mathbf{x}))), \tag{9}$$

each substitution is applied with its own probability.

Global complex mode $\Phi_G(\theta_d(\mathbf{x}), \theta_r(\mathbf{x}))$, when $\Theta(X)$ is a superposition of simple global operators

$$\Theta(X) = \Phi(\Theta_d(\Theta_r(X))), \tag{10}$$

where each $\Theta_i(X)$ is a global operator of $\theta_l(\mathbf{x})$, $l = d, r$, that may operate in its own mode and with its own probability. It is worth to be mentioned, that synchronous $\Theta_\sigma(X)$ may be included only in global superposition.

In [15] the comparative study of the above three composition modes applied to three typical asynchronous CA, showed that their evolutions differ insignificantly. The question is whether this thesis is valid for large scale CA parallel implementation and how operation ordering in reaction-diffusion CA influences parallel implementation performance.

3   Modes of CA Operation in Multiprocessor Environment

The homogeneous structure of CA prescribes using domain decomposition method for distributing the computations among processors. There is an ingrained opinion that coarse grained CA parallelization causes no problem: the single thing that should be done is to cut the cellular array into parts of suitable size. But, this is true only for synchronous CA. Parallelization of CA that contains asynchronous operators requires inducing some synchronization for making parallel implementation performance acceptable. This statement enters in contradiction with the preposition about the stochastic character of reaction-diffusion processes. To achieve a compromise, or at least come up close to it, modes of CA operation in multiprocessors are further considered in detail.

3.1   Parallelization Costs for Simple CA

Let us at first assess time and space costs of parallelization for simple synchronous $\aleph_\sigma$ and asynchronous $\aleph_\alpha$ CA. For that, let us assume, that CA is allocated on N processors of a cluster. Accordingly, CA's naming set X is divided in N equal parts called domains, $X = X_1 \ldots \cup X_l \ldots \cup X_N$, each being of the size $(I_l + 2R(T)) \times (J_l + 2R(T)) \times (K_l + 2R(T))$, where $R(T)$ is the substitution template radius (2).

In case of a simple *synchronous CA* the exchange of data is done once per iteration yielding in the total amount of data to be transmitted by the lth domain per iteration

$$V_{syn} = 2R(T)(I_l + J_l + K_l), \quad l = 1, \ldots, N, \tag{11}$$

Synchronous diffusion CA $\aleph_{\sigma,Diff} = \langle A, X, \Theta(X) \rangle$ may be used only in global complex modes. In 2D case $\aleph_{\sigma,Diff}$ is known as "CA with Margolus neighborhood", proposed in [16], and studied in [17]. It is not really a simple CA. Its global operator

$$\Theta_{Diff}(x) = \Phi(\Theta_\sigma(X_0), \Theta_\sigma(X_1)),$$
$$X_0 = \{(i,j) : (i+j)_{mod2} = 0\}, \quad X_1 = \{(i,j) : (i+j)_{mod2} = 1\}, \tag{12}$$

is performed in two stages: (1) $\theta(i,j)$ ia applied to all $(i,j) \in X_0$ (even stage), and (2) the same $\theta(i,j)$ ia applied to all $(i,j) \in X_1$ (odd stage). The substitution $\theta(i,j)$ has a 4-cell's underlying template $T(i,j) = \{(i,j), (i,j+1), (i+1,j+1), (i+1,j)\}$ and a probabilistic transition function

$$(v_0, v_1, v_2, v_3) = \begin{cases} (u_1, u_2, u_3, u_0) & \text{if} \quad r < p, \\ (u_3, u_2, u_1, u_0) & \text{if} \quad r \geq (1-p), \end{cases} \tag{13}$$

where p is a probability of substitution application, and r is a random number, $0 < r < 1$. When implemented in a cluster the exchange of data should be done twice per iteration, the amount of transferred data being equal to V_{syn}, since each exchange is done only in one direction: "west $\rightarrow$ east" and "south $\rightarrow$ north" at even stages, and "east $\rightarrow$ west" and "north $\rightarrow$ south " at odd stages.

The 3D case is more complicated. It is executed as a superposition of three synchronous 2D global operators, applied over the three sets of planes: (1) (i,j)-planes for all $k = 0, \ldots, K$, (2) (i,k)-planes for all $j = 0, \ldots, J$, and (3) (j,k)-planes for all $i = 0, \ldots, I$. So, the amount of exchanges per iteration is six, and the amount of exchanged data is $V(3D) = 3V_{syn}$.

Synchronous reaction CA may represent quite different elementary transformations: chemical reaction, phase transition, dissociation, emerging, disappearing, etc. To satisfy correctness conditions (7) simple reaction CA contain a substitution of a single-cell form:

$$\theta(\mathbf{x}) : (a, \mathbf{x}) \xrightarrow{p} (b, \mathbf{x}), \quad a, b \in A. \tag{14}$$

Such single cell substitutions mimic elementary actions like adsorption and desorption on catalytic surface, phase transition (evaporation), etc. In parallel implementation no exchange is needed.

In *asynchronous case* the application of $\theta(\mathbf{x})$ requires immediate adjustment of several cells from $T'(\mathbf{x})$. Hence, any state change in a border domain cell should be sent to the adjacent domain at once, requiring $P \times R(T)$ acts of data exchange per iteration, where P is the perimeter of the domain, $R(T)$ — the template (2) radius. This yields unacceptable parallelization efficiency and cannot be used. The problem is solved by introducing some synchronization [13], that transforms $\aleph_\alpha$ into a block-synchronous $\aleph_\beta$ according to the following algorithm.

(1) A template $B(\mathbf{x})$, further referred to as a *block*, is defined in X such that

$$B(\mathbf{x}) \supseteq T(\mathbf{x}), \quad B(\mathbf{x}) = \{\mathbf{x}, \mathbf{x} + \mathbf{a_1}, \ldots, \mathbf{x} + \mathbf{a_n}\}, \quad B(\mathbf{x_i}) \cap B(\mathbf{x_j}) = \emptyset,$$
$$\bigcup_{i=1}^{N} B(\mathbf{x_i}) = X, \quad i, j = 1, \ldots N, \quad N = |X|/n, \tag{15}$$

where $T(\mathbf{x})$ is the underlying template of the local configuration $S(\mathbf{x})$ of $\theta(\mathbf{x})$. Most efficient parallelization is achieved with $T_B(\mathbf{x}) = T(\mathbf{x})$.

(2) A partition $\Pi = \{\Pi_1, \Pi_2, \ldots, \Pi_n\}$ is generated by the set of blocks $B = \{B(\mathbf{x_i}) : i = 1, \ldots, N\}$, in such a way that

$$\Pi_k = \{\mathbf{x} : \mathbf{x} = \mathbf{x_i} + \mathbf{a_k} \in B(\mathbf{x_i})\}, \quad i = 1, \ldots, N. \tag{16}$$

(3) The iteration is divided into n stages, on each kth stage $\theta(\mathbf{x})$ is applied synchronously to all $\mathbf{x} \in \Pi_k$.

So, the implementation of a simple asynchronous CA requires n exchanges of data per iteration, the total amount of data to be transmitted by a domain being the same as in synchronous case.

Asynchronous 2D diffusion CA-model, called "naive diffusion" [16] is usually represented by a substitution $\theta(i, j)$, based on the underlying template $T_5 = \{(i, j), (i-1, j), (i, j+1), (i+1, j), (i, j-1), \}$, that identifies a local configuration $S(i, j) = (u_k : k = 0, 1, 2, 3, 4)$, adjusted by the substitution

$$\theta(i, j) : (u_0, \ldots, u_4) \xrightarrow{p_d} (u_k, u_0), \quad (k-1)/4 < r \le k/4, \tag{17}$$

where p_d is a probability of substitution application, r a random number between 0 and 1. To implement a "naive diffusion" in a cluster, $\Theta_d(X)$ should be transformed into a block-synchronous mode using the above algorithm. Thereby it may be done in two ways: (1) with the minimal block equal to the template T_5, and with the template T_9, such that $T_9 \supset T_5$. The first way yields 5 exchange stages per iteration. While the second — needs 9 stages, but due to the square template form it is easy to deal with.

Asynchronous reaction simulate chemical transformations, where several particles interact when occur allocated in neighboring cells. For example, reaction of oxidation, oxygen dissociation, production water molecules from hydrogen and oxygen, etc. They are described by substitutions of the following form:

$$\theta(\mathbf{x}) : \{(a_0, \mathbf{x}), \ldots, (a_q, \mathbf{x} + \mathbf{a_q}\} \xrightarrow{p} \{(b_0, \mathbf{x}), \ldots, (b_q, \mathbf{x} + \mathbf{a_q})\}, \quad a_l, b_l \in A. \tag{18}$$

Naturally, when implemented on multiprocessor system, they should be transformed into block-synchronous mode with q synchronous stages.

3.2   Parallelization of Reaction-Diffusion Complex CA

When simulating reaction-diffusion phenomena on a supercomputer with distributed memory, the following complex modes of operation are of interest: (1) parallel stochastic mode, (2) parallel local superposition mode, and (3) parallel global superposition mode). In all cases the cellular array $\Omega(0)$ is partitioned into N domains, each domain being enlarged by $R(T_\Phi)$ on each side, i.e.

$$T_\Phi = T'_d \cup T'_r, \tag{19}$$

T'_d and T'_r being the underlying templates of $\theta_d(\mathbf{x})$ and $\theta_r(\mathbf{x})$, respectively.

Parallel stochastic mode Ψ_S is intended to simulate an asynchronous process dealing with two local operators involved in it: $\theta_d(\mathbf{x})$ and $\theta_r(\mathbf{x})$. Interaction between the domains requires transformation into block-synchronous mode. So, the procedure of simulation is as follows.

(1) Determine the block $B = T_\Phi$ and the subsets $\Pi_1, \ldots, \Pi_n$, $n = |B|$ according to (15),(16).

(2) For each domain X_l, $l = 1, \ldots, N$, for each iteration t, and for each subset Π, $i = 1, \ldots, n$:

- choose a random cell $\mathbf{x} \in X$ with probability $p = 1/|X_l|$,
- calculate the probabilities p_d and p_r according to (8),
- generate a random number r,
- if $r < p_d$, then θ_d is applied to $\mathbf{x}$, otherwise θ_r is applied to the same cell,
- send border cells states to the adjacent domains and receive border states from them.

Operation in this mode requires $n = |B|$ acts of data exchange between adjacent domain borders per iteration, and $2 \times N \times |X|$ calls to random numbers generator.

Parallel local superposition mode Ψ_L is also a process with asynchronous $\theta_d(\mathbf{x})$ and $\theta_r(\mathbf{x})$. Hence for parallel implementation it should be transformed into a block synchronous mode just in the same way, that it is done for stochastic mode. The difference is in the interaction of the substitutions (point 2) which is here sequential instead of random in Ψ_S:

- choose a random cell $\mathbf{x} \in X$ with probability $p = 1/|X_l|$,
- generate a random number r,
- if $r < p_d$ apply θ_d to $\mathbf{x}$,
- generate a random number q,
- if $q < p_r$ apply θ_r to $\mathbf{x}$,
- send border cells states to the adjacent domains and receive border states from them.

Operation in this mode requires $n = |B|$ acts of data exchange between adjacent domains per iteration and $N \times |X|$ calls to random numbers generator.

Parallel global superposition mode Ψ_G is a superposition of two simple global operators (Subsect. 3.1), each operating in its own mode. In all cases complexity of parallel implementation is the sum of the corresponding values of each global operator.

Experimental investigation of these three parallel modes aims to obtain and compare the following computation characteristics:

- Time in **sec** and in iteration numbers ($\hat{t}$).
- Weak parallelization efficiency

$$\eta = t_{p=1}/t_{p=N}, \tag{20}$$

where $t_{p=1}$ and $t_{p=N}$ are "times" (in seconds) needed for executing an iteration under identical computing conditions.

– Difference between the evolutions. In comparison procedures stochastic composition is taken as a reference, because it represents the most studied case, being similar to Kinetic Monte-Carlo methods [7].

Two typical reaction-diffusion phenomena are simulated on multi processor cluster with variation of parallel composition modes.

4   Simulation Results

4.1   Wave Front Propagation CA Models

The first mathematical model of a wave front propagation was studied in [18,19]. In [18] the process had been called "diffusion, combined with increase of substance". The results turned out to be extremely fruitful not only from the point of view of nonlinear mathematical analysis, but also for modeling a wide class of processes, such as fire propagation, epidemics and weeds spreading, chemical reactions expansion in active media. For contemporary computer simulation the process of *front propagation* may be considered as a typical component of more complex phenomena [20,21]. So, its CA model deserves to be investigated in detail.

The propagation front kinetics is regarded as a discrete space, a part of which is occupied by cloud of particles. The particles diffuse from the place of high density towards the free space, the displacement being accompanied by a reaction of the medium, which is described by a nth order polynomial function of the substance density $v(\mathbf{x})$. In continuous models [18,19] $n = 2$, and the function looks like

$$F(v) = \alpha v(1 - v), \quad v \in [0,1]. \tag{21}$$

In a CA $\aleph = \langle A, X, \Theta(X) \rangle$ with Boolean $A = \{0,1\}$, $u \in \{0.1\}$, the substance density is simulated by a real number $\langle u(\mathbf{x}) \rangle$, equal to averaged value over an area $Av(\mathbf{x})$ around the cell $\mathbf{x}$,

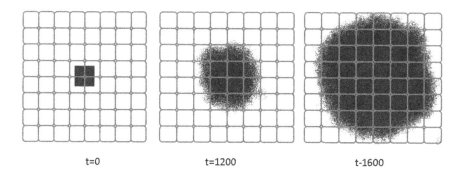

t=0   t=1200   t-1600

Fig. 1. Evolution of wave propagation process implemented in 64 processors.

A CA, simulating a 2D wave front propagation process, $\aleph = \langle A, X, \Theta(X) \rangle$ has $A = \{0, 1\}$, $X = \{(i.j) : i, j = 0, \ldots, N\}$, and $\Theta(X) = \Psi_z(\theta_d(i, j), \theta_r(i, j))$, where Ψ_z is any of parallel mode of operation from $\{\Psi_S, \Psi_L, \Psi_G\}$, $\theta_d(i, j)$ being a naive diffusion substitution (17), and $\theta_r(i, j)$ simulating reaction,

$$\theta_r(i, j) : u \xrightarrow{p_r} u', \quad u' = \begin{cases} 1, & \text{if } r < \alpha \langle u \rangle (1 - \langle u \rangle), \\ 0 & \text{otherwise,} \end{cases} \qquad (22)$$

where r is a random number, $0 < r \leq 1$, α is a constant, further taken as $0{,}5$, probabilities of diffusion and reaction being $p_d = 0.9, p_r = 0.1$

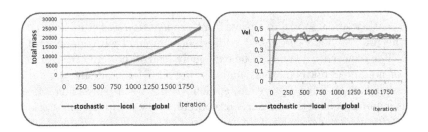

Fig. 2. Simulation of wave front propagation process: (a) dependence of total mass on time, (b) dependence of propagation velocity on time

The simulation task was to implement front-propagation CA with different modes of application to a cellular array Ω of size $N = 2400 \times 2400 = 5.76 \times 10^9$ cells using 64 processors in parallel. The cellular array was divided into equal sized domains, size of $n = 300 \times 300$. In the center of the array there was a dense square cloud 100×100. During CA operation the dense cloud enlarges, gradually filling the whole space (Fig. 1). Simulation was done for three parallel modes of operation $\{\Psi_S, \Psi_L, \Psi_G\}$. In each case 100 runs have been done using different random number sequences, statistical expectation $\bar{u}(i, j)$ being taken as a result. The following characteristic values were output at each iteration for each parallel mode of operation: the total mass $\Gamma_z(t)$, and the velocity of front propagation $Vel_z(t)$,

$$\Gamma_z(t) = \sum_{(i,j) \in X} \bar{u}(i, j), \quad Vel_z(t) = \frac{1}{\sqrt{\pi}} (\sqrt{\Gamma_z(t)} - \sqrt{\Gamma_z(t-1)}), \qquad (23)$$

The results of simulation for three parallel modes of operation: stochastic, local and global asynchronous are shown in Fig. 2 from which it follows, that all three operation modes are qualitatively identical. Computational time and weak efficiency are given in Table 1. The remarkable fact is that the weak parallelization efficiency is larger that 1. It follows from two reasons:

1. block-synchronous mode requires a smaller amount of random generator calls,
2. periodic border conditions used in serial mode require calculation of mod_K neighboring cells names, that takes additional time.

Table 1. Computation time (sec) for sequential ($n = 90 \times 10^6$ cells) and parallel (64n) versions. Propagation velocity (Vel), and weak parallelization efficiency (η) for three parallel modes of operation

Mode	Stochastic		Local		Global	
Size	n	$64n$	n	$64n$	n	$64n$
Time	89,8	62.7	92,3	78.3	115,0	89.5
Vel	0.438	0.435	0.427	0.4434	0.43	0.44
η	1.432		1.17		1.28	

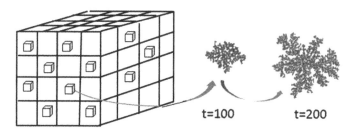

t=100   t=200

Fig. 3. A schematic picture of simulation a process of a 3D diffusion-limited aggregation initialized by four nuclei, the whole task being allocated in 64 processors

4.2 Diffusion Limited Aggregation

A phenomenon, identified as *diffusion limited aggregation* hereinafter called *aggregation* cannot be represented by a continuous function or by differential equations. Its first mathematical model was created for computer implementation and described in terms of interactions and displacement of particles [22]. The model is a representative of a wide range of phase transition processes such as electro galvanization [23], formation of crystal structure [24], growing of settlements [25] etc. . Most popular are CA models, where diffusion is given as a random walk, and reaction transforms a walking particle into an immobile one, if it occurs close enough to another immobile (sticking process) (Fig. 3).

A CA model of aggregation in 3D case is $\aleph = \langle A, X, \Theta(X) \rangle$, where $A = \{0, 1, b\}$, $X = \{(i, j, k)\}$, $\Theta(i, j) = \Psi_z(\theta_d(i, j, k), \theta_r(i, j, k))$, Ψ_z representing one of the parallel modes of operation: Ψ_S, Ψ_L or Ψ_G, where $\theta_d(i, j, k)$ is given by (12) (for synchronous mode) or by (17) (for asynchronous mode). Sticking is simulated by

$$\theta_r(i, j, k)) : \{(1, (i, j, k)), (b, (i', j', k')), \xrightarrow{p_r} \{(b, (i, j, k))\}, \quad (i', j', k') \in T(i, j, k), \tag{24}$$

where $T(i, j, k)$ contains six adjacent to (i, j, k) cells.

Computer simulation was performed on the cellular array size of $396 \times 396 \times 396$. The initial global state $\Omega(0)$ contained uniformly distributed "ones" and "zeros", and 32 nuclei cubes $5 \times 5 \times 5$ cells with $u = b$, allocated inside the array. The whole array was put on 64 cores arranged in a cube $4 \times 4 \times 4$, each core processing a domain of Ω size of $99 \times 99 \times 99$ cells. To provide data exchange between processes, each domain was enlarges up to $101 \times 101 \times 101$ cells, the added border cells ($i = 0, j = 0, i = 100, j = 100$) serving for sending-receiving data. Two characteristics of the process were obtained by simulation: dependence of the mass of the constructed structure on time ($\Gamma_z(t)$), and its fractal dimension δ_z at the end of process at $\hat{t} = 500$ for three parallel modes of operation (Ψ_s, Ψ_L, Ψ_G).

$$\Gamma_z = \sum_X (b, (i, j, k,)), \quad \delta_z = \log \left(\sum_{Sph} (b, (i, j, k,)/\log(R))) \right), \qquad (25)$$

where Sph is the number of cells in the sphere of radius R around the nucleus. To assess possibility of using synchronous global mode $\Psi_G(\Theta_{r,\sigma}(X), \Theta_{d,\sigma}(X))$. was also tested.

The results of simulation for three parallel composition modes: stochastic, local and global synchronous are shown in Fig. 4. From Fig. 4 it follows, that all three asynchronous modes are qualitatively identical, differing in computational time, synchronous being faster both in sequential and in parallel case. In parallel case it is more faster, because requires minimal number of interprocessor exchanges and no random generator calls.

Computational time and weak efficiency are given in Table 2.

It is worth to emphasize the fact that in all asynchronous modes independently of the composition used, the weak efficiency is larger than 1, like in case of front propagation. But when synchronous mode is used, which is initially free

Table 2. Computation time per 500 iterations (**sec**) for sequential ($n = 970, 299$cells) and parallel ($64n$) versions, fractal dimension (δ), and weak parallelization efficiency (η) for three modes of operation

Mode	Stochastic		Local		Global		Synchronous	
Size	n	$64n$	n	$64n$	n	$64n$	n	$64n$
Time	2521	3408	2068	1849	2906	2656	198	955
δ	2.68	2.643	2.678	2.649	2.67	2.65	2.649	2.655
η	0.77		1.12		1.09		0.22	

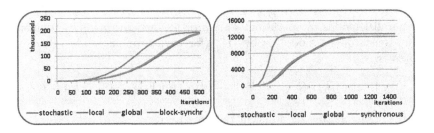

Fig. 4. Simulation of 3D diffusion limited aggregation process. Dependence of total mass of immobile particles on time when implementing a) $480 \times 480 \times 400$ cells on 64 cores, and b) sequentially $60 \times 60 \times 400$ on one core.

of additional computation of random numbers, performance is affected only by data exchanges, which decrease the efficiency, resulting nonetheless in less time consuming.

5 Conclusion

The results of comparative study of parallel simulation of reaction-diffusion processes by CA differing by parallel modes of operation are presented. The investigation aims to find a compromise between parallelization efficiency of synchronous CA and inherent natural stochasticity of simulated phenomena. Two typical reaction-diffusion processes (front propagation and diffusion limited aggregation) have been simulated by CA, operating in stochastic mode, local superposition mode and global superposition. Simulation results have been compared by the obtained evolutions and basic characteristics of the processes. The comparison revealed two points. The first is that all investigated modes of parallel operation are qualitatively equivalent, differing, however, in computational time. The second is that weak parallelization efficiency is larger than 1 when asynchronous modes are used. This information may be essentially helpful on the stage of CA-model development, when the elementary operations interaction should be determined allowing for parallel implementation conditions.

References

1. Hoekstra, A.G., Kroc, J.K., Sloot, P.M.A. (eds.): Simulating Complex Systems by Cellular Automata. Springer, Heidelberg (2010)
2. Desai, R.C., Kapral, R.: Dynamics of Self-organized and Self-assembled Structures. Cambridge University Press, Cambridge (2009)
3. Echieverra, C., Kapral, R.: Molecular crowding and protein enzymatic dynamics. Phys. Chem. **14**, 6755–6763 (2012)
4. Bandini, S., Bonomi, A., Vizzari, G.: What do we mean by asynchronous CA? a reflection on types and effects of asynchronicity. In: Bandini, S., Manzoni, S., Umeo, H., Vizzari, G. (eds.) ACRI 2010. LNCS, vol. 6350, pp. 385–394. Springer, Heidelberg (2010)
5. Bouré, O., Fatès, N., Chevrier, V.: First steps on asynchronous lattice-gas models with an application to a swarming rule. In: Sirakoulis, G.C., Bandini, S. (eds.) ACRI 2012. LNCS, vol. 7495, pp. 633–642. Springer, Heidelberg (2012)
6. Kireeva, A.: Parallel implementation of totalistic cellular automata model of stable patterns formation. In: Malyshkin, V. (ed.) PaCT 2013. LNCS, vol. 7979, pp. 330–343. Springer, Heidelberg (2013)
7. Matveev, A.V., Latkin, E.I., Elokhin, V.I., Gorodetskii, V.V.: Turbulent and stripes wave patterns caused by limited CO_{ads} diffusion during CO oxidation over Pd(110) surface: kinetic Monte Carlo studies. Chem. Eng. J. **107**, 181–189 (2005)
8. Nurminen, L., Kuonen, A., Kaski, K.: Kinetic Monte-Carlo simulation on patterned substrates. Phys. Rev. B **63**, 03540:17–03540:7 (2000)
9. Chatterjee, A., Vlaches, D.: G.: An overview of spatial microscopic and accelerated kinetic Monte-Carlo methods. J. Comput. Aided Mater. Des. **14**, 253–308 (2007)

10. Bandman, O.: Parallel composition of asynchronous cellular automata simulating reaction diffusion processes. In: Bandini, S., Manzoni, S., Umeo, H., Vizzari, G. (eds.) ACRI 2010. LNCS, vol. 6350, pp. 395–398. Springer, Heidelberg (2010)

11. Kalgin, K.: Comparative study of parallel algorithms for asynchronous cellular automata simulation on different computer architectures. In: Bandini, S., Manzoni, S., Umeo, H., Vizzari, G. (eds.) ACRI 2010. LNCS, vol. 6350, pp. 399–408. Springer, Heidelberg (2010)

12. Bandman, O.: Implementation of large-scale cellular automata models on multi-core computers and clusters. In: 2013 International Conference on IEEE Conference Publications High Performance Computing and Simulation (HPCS), pp. 304–310 (2013)

13. Bandman, O.: Parallel simulation of asynchronous cellular automata evolution. In: El Yacoubi, S., Chopard, B., Bandini, S. (eds.) ACRI 2006. LNCS, vol. 4173, pp. 41–47. Springer, Heidelberg (2006)

14. Bandman, O., Kireeva, A.: Stochastic cellular automata simulation of oscillations and autowaves in reaction-diffusion systems. Numerical Analysis and Applications, vol. 2 (2015)

15. Bandman, O.: Functioning modes of asynchronous cellular automata simulating nonlinear spatial dynamics. Appl. Discrete Math. 1, 110–124 (2015). (in Russian)

16. Toffolli, T., Margolus, N.: Cellular Automata Machines: A New Environment for Modeling. MIT Press, USA (1987)

17. Bandman, O.: Cellular automata diffusion models for multicomputer implementation. Bull. Novosibirsk Comput. Cent. Ser. Comput. Sci. 36, 21–31 (2014)

18. Kolmogorov, A.N., Petrovski, I.G., Piskunov, I.S.: Investigation of the equation of diffusion, combined with the increase of substance and its application to a biological problem. Bull. Moscow State Univ. A Issue 6, 1–25 (1937). (in Russian)

19. Fisher, R.A.: The genetical Theory of Natural Selection. Oxford University Press, New York (1930)

20. Szakály, T., Lagzi, I., Izsák, F., Roszol, L., Volford, A.: Stochastic cellular automata modeling excitable systems. Central Eur. J. Phys. 5(4), 471–486 (2007)

21. van Saarloos, W.: Front propagation into unstable states. Phys. Rep. 386, 209–222 (2003)

22. Witten Jr., T.A., Sander, I.M.: Diffusion-limited aggregation, a kinetic critical phenomenon. Phys. Rev. Lett. 47(19), 1400–1403 (1981)

23. Ackland, G.J., Tweedie, E.S.: Microscopic model of diffusion limited aggregation and electro deposition in the presence of leveling molecules. Phys. Rev. E 73, 011606 (2006)

24. Bogoyavlenskiy, A., Chernova, N.A.: Diffusion-limited aggregation: a relationship between surface thermodynamics and crystal morphology. Phys. Rev. E. N 2, 1629–1633 (2000)

25. Batty, M., Longley, P.: Urban growth and form: scaling, fractal geometry, and diffusion-limited aggregation. Environ. Plann. A 21, 1447–1472 (1989)

A Parallel Genetic Algorithm to Adjust a Cardiac Model Based on Cellular Automaton and Mass-Spring Systems

Ricardo Silva Campos, Bernardo Martins Rocha, Luis Paulo da Silva Barra, Marcelo Lobosco, and Rodrigo Weber dos Santos[✉]

Programa de Pós-Graduação em Modelagem Computacional, Universidade Federal de Juiz de Fora, Juiz de Fora, MG, Brazil
{ricardo.campos,luis.barra,rodrigo.weber}@ufjf.edu.br,
{bernardomartinsrocha,marcelo.lobosco}@ice.ufjf.br

Abstract. This work presents an electro-mechanical model of the cardiac tissue and an automatic way to tune its parameters. A cellular automaton was used to simulate the action potential propagation, and a mass-spring system to model the tissue contraction. A parallel genetic algorithm was implemented in order to automatically adjust a simple and fast discrete model, to reproduce simulations of another synthetic well known model based on differential equations (DEs). Our results suggest that the discrete model was able to qualitatively reproduce the results obtained by DEs with much less computational effort.

1 Introduction

Cardiac diseases are the major cause of death worldwide. The mechanical contraction of the cardiac tissue that ejects blood is preceded and triggered by a fast electrical wave, i.e. the propagation of the so called action potential (AP). Abnormal changes in the electrical properties of cardiac cells as well as in the structure of the heart tissue can lead to life-threatening arrhythmia and fibrillation.

A widely-used technique to observe the heart behavior is *in silico* experiments. It comprises of mathematical models that can reproduce the heart's tissue function through computational tools. Generally these models are described by differential equations, representing the cell's electrical and mechanical activity by ordinary differential equations (ODEs), and the electrical wave propagation on the tissue and cardiac contraction via partial differential equations (PDEs). Cardiac cells are connected to each other by gap junctions creating a channel between neighboring cells and allowing the flux of electrical current in the form of ions. An electrically stimulated cell transmits an electrical signal to the neighboring cells allowing the propagation of an electrical wave to the whole heart which triggers contraction.

Although PDEs are able to perform realistic tissue simulations, they involve the simulation of thousands of cells, which make its numerical solution quite

© Springer International Publishing Switzerland 2015
V. Malyshkin (Ed.): PaCT 2015, LNCS 9251, pp. 149–163, 2015.
DOI: 10.1007/978-3-319-21909-7_15

challenging. This is an issue for clinical softwares, that may demand accurate results and real-time simulations. In this manner, some effort has been done in speeding up the PDEs solvers by parallel computing, as well as by different techniques to emulate PDEs simulations with less computational cost.

In this manner, this work proposes a new model by joining two previously proposed models: The first one is based on cellular automaton (CA), it was proposed by [1] and has been extensively used [2,3]. It represents the electrical excitation of cells propagating according to simple rules in a regular, discrete and finite network. It uses precomputed profiles of cell AP and force-development that mimics those obtained by complex models based on ODEs. Although it is less accurate than the models based on ODEs, it is much faster than PDE based-simulators, making possible real time simulations of heart behavior [4]. The second model was proposed by [5,6] and it consists in a 3D mass-spring system (MSS) adapted to control anisotropy. This is an important issue on simulating the heart behavior, since the cardiac muscle is composed by fibers. The mechanical model is governed by Hooke's Law and Newton's Second Law.

Once we have the electro-mechanical model, we need to determine its parameters. A genetic algorithm (GA) was developed for automatically tune parameters to the mechanical model based in mass-spring systems and cellular automaton, in order to reproduce the accurate simulations from partial differential equations (PDEs) simulators. The GA generates sets of parameters to the discrete model and compare the resulting simulation with PDE simulations, which are taken as the GA target. The GA was able to find suitable parameters and therefore the discrete model could reproduce simulations similar to those obtained by PDEs. We also added OpenMP directives to GA code in order to decrease its execution time.

The electro-mechanical model and the GA are parts of a software named FisioPacer, that aims to perform low computational cost simulations of the cardiac tissue with the forthcoming purpose of simulating the pacemaker implantation.

2 Methods Part I: Discrete Model

This section presents the electro-mechanical model used by FisioPacer. The electrical model uses cell cellular automate (CA) to represent the action potential (AP) propagation over the tissue. The mechanical contraction is modeled by a mass-spring system (MSS) coupled with the electrical model.

2.1 Modeling Action Potential Propagation with Cellular Automaton

A cellular automaton is the model of a spatially distributed process that can be used to simulate various real-world processes. A 2-dimensional cellular automaton consists of a two-dimensional lattice of cells where all cells are updated synchronously for each discrete time step. The cells are updated according to some updating rules that will depend on the local neighborhood of a particular cell.

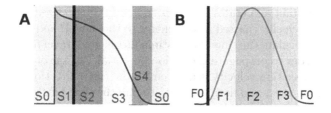

Fig. 1. Part (A) Action potential of a cardiac cell computed by ordinary differential equations and its representation via CA with five states (S0, S1, S2, S3, S4). Part (B) Force development states (F0, F1, F2, F3). The black line represents the moment when the change from S1 to S2 triggers the change in the force state from F0 to F1. (Adapted from [4])

The idea of macroscopically simulating the excitation-propagation in the cardiac tissue with CA was proposed in [1] and extensively used in the literature [2–4]. The CA is build on the idea that a single cell gets excited if the electrical potential exceeds a determined threshold. Once it is excited, it can trigger the excitation of neighboring cells. In this manner, an electrical stimulus will propagate through the CA grid as the time steps are computed. In this work, the CA states are related to the action potential (AP) and force development in a cell. To make CAs more efficient they are usually parametrized using simulated data from accurate models. This means that the states related to the AP in the cell will be related to a specific portion of the cardiac cell potential. Figure 1 Part A presents the AP computed by ODEs, the AP divided into five different states that represents the different physiological stages of the AP.

In state S0 the cell is in its resting potential where it can be stimulated, in S1 the cell was stimulated and can stimulate the neighbors. In S2 the cell is still able to stimulate the neighbors. In S3 the cell is in its absolute-refractory state where it cannot be stimulated and does not stimulate its neighbors. In S4 the cell is in its relative refractory state where it can be stimulated by more than one neighbor but it does not stimulate its neighbors. As described, the states of a cell generate rules for when a cell can stimulate a neighbor and when it can be stimulated. These rules are an important aspect which will allow the stimulus to propagate.

Another important point is how the cells change their states. The AP has a predetermined period so that the states will be spontaneously changed after the AP starts, where the time of each state is a parameter of the system. Our CA is adapted to work with irregular meshes of tetrahedrons. In that case, the cells of the system are the tetrahedrons and a cell is considered a neighbor of other cell if they share at least one vertex. The distance between two neighbors cells is computed as the distance between their barycenters, given by:

$$X_b^i = \frac{1}{4} \sum_{a=1}^{4} x_a^i, \tag{1}$$

where x_a are the coordinates of the vertices from tetrahedron x^i.

Equally important, CA states are updated at every discretized time, dt. Based on the distance, velocity (passed as parameter to the model) and activation time it is possible to calculate the time that a stimulus takes to travel from one CA cell to another, in order to propagate the action potential. An anisotropic tissue was used, so the propagation velocity is different in the three directions of interest in heart tissue: fiber, sheet and normal-sheet. To find the time t for a stimulus travel from one cell to another, first direction between the barycenters is computed and then the distance d:

$$\boldsymbol{D}_{ij} = \boldsymbol{X}_b^j - \boldsymbol{X}_b^i \tag{2}$$

$$d = \|\boldsymbol{D}_{ij}\|, \tag{3}$$

where $\boldsymbol{X}_b^i$ and $\boldsymbol{X}_b^j$ are the positions of the barycenter of elements i and its neighbor j. Next, the total velocity of the AP is calculated, based on the velocities in each one of the directions: v_f, v_s and v_n:

$$\boldsymbol{V} = v_f \boldsymbol{F} + v_s \boldsymbol{S} + v_n \boldsymbol{N}, \tag{4}$$

$$v = |\boldsymbol{V} \cdot \hat{\boldsymbol{D}}_{ij}|, \tag{5}$$

where $\boldsymbol{F}$, $\boldsymbol{S}$ and $\boldsymbol{N}$ are respectively fiber, sheet and normal normalized directions. $\hat{\boldsymbol{D}}_{ij}$ is the normalized direction between element i and j. And then the traveling time t of the propagation between them is:

$$t = \frac{d}{v}, \tag{6}$$

So it verifies if there is enough time to propagate the stimulus by comparing the time since the neighbor has been stimulated and time t.

Finally, electrical potential is coupled with the active force, which is responsible for starting the contraction of the cardiac tissue. When the cell is stimulated, there is an increase in the concentration of calcium ions inside the cell, which triggers the cell contraction. The force development has a delay after the cell is stimulated because of its dependence on the calcium ions. The force development of a cell can be represented in states that change over time like the electrical potential states. Figure 1 Part B presents the force development states and its relation with the electrical states. The force-development states will only pass from state F0 (no contraction force) to state F1 when the electrical state of the cell goes from state S1 to S2. After this change, force development will be time dependent but will not depend on the electrical state of the cell.

2.2   Modeling Mechanical Contraction with Anisotropic Mass-Spring Systems

The active force is responsible for the starting the contraction of the cardiac tissue. This force is coupled with the electrical potential of the cell. When the cell is stimulated, there is an increase in the concentration of calcium ions inside the cell, which triggers the cell contraction. The force development has a delay after the cell is stimulated because of its dependence on the calcium ions. The passive force on our simulator can be modeled as a mass-spring systems, where masses are connected with the neighboring masses by springs. The active force is applied to the system deforming its spatial distribution and then the springs will try to bring the system back to its initial configuration. The cardiac tissue does not have a linear stress-strain relation. However the linear model of the Hooke's law can be used as a simplification

The mechanical model was originally presented in [6]. In each tetrahedron, there are six springs. Three springs are placed in the anisotropic axes and are named as axial springs. The other three springs are named angular and they are placed between each pair of axial springs. The points where the axes intercept the tetrahedron faces are called interception points.

It follows that the active force must be applied on each tetrahedron, only on the fiber direction:

$$a_{2l}^t = +\frac{1}{2}\mu f_a^t \widehat{\zeta}_l^t S_{2l}^t, \tag{7}$$

$$a_{2l+1}^t = -\frac{1}{2}\mu f_a^t \widehat{\zeta}_l^t S_{2l+1}^t, \tag{8}$$

where μ is a parameter of active pressure given in N/mm^2, $f_a^t \in [0,1]$ comes from the force automaton, $\widehat{\zeta}_l$ is the normalized fiber direction and S_{2l}^t and S_{2l+1}^t are the area of the faces intercepted by the fiber. At last, a_{2l}^t and a_{2l+1}^t are the active forces on interception points.

Turning to the passive force, we present the equations that react to the active force and bring the system to its initial geometry. The forces that each axial spring apply on an interception point are based on Hooke's Law:

$$\delta_s^t = \frac{|\zeta_l^t| - |\zeta_l^0|}{|\zeta_l^0|}, \tag{9}$$

$$h_{2l}^t = -k_l \delta_s^t \widehat{\zeta}_l^t S_{2l}^t |\widehat{\zeta}_l^t \cdot n_{2l}^t|, \tag{10}$$

$$h_{2l+1}^t = +k_l \delta_s^t \widehat{\zeta}_l^t S_{2l+1}^t |\widehat{\zeta}_l^t \cdot n_{2l+1}^t|, \tag{11}$$

where δ_s^t represents the displacement of the spring and $|\zeta_l^t|$ and $|\zeta_l^0|$ are the current and initial spring axial length. n is the normal of the face and k_l is the spring stiffness associate to axis l. It is another system parameter given in N/mm^2.

Next, it is presented the equations that control the angular forces (torsion springs) between a pair of axial springs, defined by the interception points $(2m, 2m + 1)$ and $(2l, 2l + 1)$:

$$\delta_a^t = \alpha_{lm}^t - \alpha_{lm}^0, \tag{12}$$

$$g_{2l}^t = -k_a \delta_a^t \widehat{\zeta}_m^t S_{2l}^t |\widehat{\zeta}_l^t \cdot n_{2l}^t|, \tag{13}$$

$$g_{2m}^t = -k_a \delta_a^t \widehat{\zeta}_l^t S_{2l+1}^t |\widehat{\zeta}_l^t \cdot n_{2l+1}^t|, \tag{14}$$

$$g_{2l+1}^t = -g_{2l}^t, \tag{15}$$

$$g_{2m+1}^t = -g_{2m}^t, \tag{16}$$

where δ_a^t is the variation of angle α between axis l and m, k_a is the parameter of angular stiffness of torsion springs. To make computations easier, the angles α can be replaced by dot products, so:

$$\delta_a^t = \widehat{\zeta}_l^t \cdot \widehat{\zeta}_m^t - \widehat{\zeta}_l^0 \cdot \widehat{\zeta}_m^0. \tag{17}$$

Now follows the volume preserving force, that tries to keep the tetrahedrons with the same volume during the mesh contraction. This is a very important feature of this model, whereas that the cardiac tissue is mostly composed by water and therefore its volume does not have big variations. This force is computed differently from the model proposed in [6], since we prefer to apply it in the interception points, considering the axis directions. In the original model, this force is applied from the barycenter to the tetrahedron vertices. So the volume preserving force was changed for the purpose of avoiding instabilities on the system:

$$\delta_v^t = \frac{v^t - v^0}{v^0}, \tag{18}$$

$$l_{2l}^t = -k_v \delta_v^t \widehat{\zeta}_l^t S_{2l}^t |\widehat{\zeta}_l^t \cdot n_{2l}^t|, \tag{19}$$

$$l_{2l+1}^t = +k_v \delta_v^t \widehat{\zeta}_l^t S_{2l+1}^t |\widehat{\zeta}_l^t \cdot n_{2l+1}^t|, \tag{20}$$

where v^t and v^0 are current and initial volume of the tetrahedron and k_v is a parameter associated with the volume preservation on axis l.

Finally it is added a damping force that prevents the system to oscillate forever:

$$\dot{\zeta}_l^t = v_{2l}^t - v_{2l+1}^t, \tag{21}$$

$$d_{2l}^t = -k_d(\dot{\zeta}_l^t \cdot \widehat{\zeta}_l^t) \widehat{\zeta}_l^t S_{2l}^t |\widehat{\zeta}_l^t \cdot n_{2l}^t|, \tag{22}$$

$$d_{2l+1}^t = +k_d(\dot{\zeta}_l^t \cdot \widehat{\zeta}_l^t) \widehat{\zeta}_l^t S_{2l+1}^t |\widehat{\zeta}_l^t \cdot n_{2l+1}^t|, \tag{23}$$

where v_{2l}^t and v_{2l+1}^t are the velocities of interception points and k_d is the constant parameter of damping.

To conclude, the algorithm must compute the total force on every interception force on a tetrahedron:

$$f_l^t = a_l^t + h_l^t + (g_{lm}^t + g_{ln}^t) + l_l^t + d_l^t. \tag{24}$$

Next, the forces on interception points are distributed to vertices of the tetrahedron's face, via the interpolation function. Further details can be found in the original text [6]. With the total force on each vertex in hands, it is possible to find their acceleration by Newton's second law $F = ma$:

$$a_i^t = \frac{f_i^t}{m_i}, \tag{25}$$

where the mass of a vertex is given by:

$$m_i = \frac{1}{4}\rho \sum_{k=1}^{n} v_k, \tag{26}$$

where ρ is the mass density, v_k is the tetrahedron volume, and n is the number of tetrahedrons that vertex i belongs to. Several studies point that the mass density of the cardiac tissue is around $1.055\,g/ml$ [7]. For simplification we convert it to $0.001\,g/mm^3$.

The final step consists in integrating the system of equations for each vertex in the CA to simulate the mechanical deformation of the tissue. Using the Forward Euler's method:

$$v_i^{t+dt} = v_i^t + a_i^t dt \tag{27}$$
$$x_i^{t+dt} = x_i^t + v_i^t dt, \tag{28}$$

where v_i^t and x_i^t are the velocity and position of a mass at time t.

3   Methods Part II: Continuum Model

The electrophysiology of cardiac tissue, considering the effects of deformation, can be described by the monodomain model, which in this case is given by the following equation

$$\frac{\partial(Jv)}{\partial t} + JI_{ion} = \mathrm{Div}(J\boldsymbol{F}^{-1}\boldsymbol{D}\boldsymbol{F}^{-T}\mathrm{Grad}v), \tag{29}$$

where v is the transmembrane potential, I_{ion} is the ion current of the cell model, $\mathbf{F}$ is the deformation gradient tensor and $J = \det(\mathbf{F})$. Here, the spatial derivates are taken with respect to the original (underformed) configuration, as in [8].

The continuum model for cardiac biomechanics is computed by solving the quasi-static equilibrium equations

$$\mathrm{div}(\mathbf{FS}) = 0, \tag{30}$$

where $\mathbf{S}$ is the second Piola-Kirchhoff stress tensor. The second Piola-Kirchhoff stress tensor is computed by differentiating the strain energy function Ψ with respect to $\mathbf{C} = \mathbf{F}^T\mathbf{F}$, the left Cauchy-Green strain tensor. An orthotropic model based on the microstructure of the cardiac tissue, proposed by Holzapfel-Ogden [9], was used in this work. Its strain energy function Ψ is given by

$$\Psi(I_1, I_{4f}, I_{4s}, I_{8fs}) = \frac{a}{2b}\{\exp\left[b(I_1 - 3)\right] - 1\} + \sum_{i=f,s} \frac{a_i}{2b_i}\{\exp\left[b_i(I_{4i} - 1)^2\right] - 1\}$$
$$+ \frac{a_{fs}}{2b_{fs}}\left[\exp\left(b_{fs}\,I_{8fs}^2\right) - 1\right], \tag{31}$$

where a, a_f, a_s, a_{fs}, b, b_f, b_s and b_{fs} are material parameters. The fiber and sheet directions in the undeformed configuration are denoted here by $\mathbf{f}_0$ and $\mathbf{s}_0$, respectively. This model has only 8 parameters and is defined in terms of the $\mathbf{C}$ tensor and the following invariants

$$I_1(\mathbf{C}) = \mathrm{tr}(\mathbf{C}), \quad I_{4k}(\mathbf{C}) = \mathbf{k}\cdot\mathbf{C}\mathbf{k}, \quad I_{8fs} = \mathbf{f}_0\cdot\mathbf{C}\mathbf{s}_0, \quad \text{with } \mathbf{k} = \{\mathbf{f}_0, \mathbf{s}_0\}.$$

Although this model is orthotropic, it can be simplified to the transversely isotropic case by neglecting the terms with $i = s$ and the last term.

The finite element method (FEM) was used for the discretization of the continuum models. For the mechanics, the resulting system of non-linear equations was solved using Newton's method. More details about the numerical methods and parameters used for the simulations can be found in [10].

We used the active stress approach that splits the second Piola-Kirchhoff stress in a passive and an active stress parts. The passive part is given by the Holzapfel-Ogden model, described by equation (31), whereas the active stress contribution is given by

$$\mathbf{S}^a = T_a^{max} T_a \mathbf{f}_0 \otimes \mathbf{f}_0, \tag{32}$$

where T_a is the normalized active force from the Rice myofilament model and T_a^{max} is a scaling factor to achieve the active stress found in cardiac myocytes [10].

4   Methods Part III: Automatic Tuning Parameter with Genetic Algorithm

Genetic algorithms (GA) are stochastic optimization methods inspired on Evolution Theory and Natural Selection. In summary, these techniques starts with random set of candidate solutions. Then the environmental pressure causes natural selection, where the fittest individual has more chances to survive and generates more children. The individual is evaluated and so a value is attributed to it, the so-called fitness. So, the algorithm intend to maximize (or minimize) the population fitness by applying genetic operators (selection, recombination and mutation). In other words, at every generation, a new population is created to replace the population of the previous generation. The operators are applied to generate new individuals with better fitnesses. This is repeated for a limited number of generations or until appears an individual with fitness small (or big) enough.

The goal of our GA is to find the six parameters of the mechanical model presented in Sect. 2. So each individual contains a candidate set of parameters, a floating-point vector named chromosome $p_{i=1}^6$, while a single parameter p_i is called a gene. The fitness comes from a comparison between the discrete model simulation and the continuum model, and then it applies the genetic operators to minimize the difference between the simulations. There are many different ways to select, mutate and recombine individuals, however we will describe only the relevant operators to this work.

4.1   Computing Fitness

For each individual in one generation performed two simulations of $0.01s$ of tissue activity in a coarse mesh. In the first, the fiber direction is parallel to Y-axis and in the other the fiber direction changes from epicardium to endocardium (-70 to 70, respectively). It was applied a constant active force during all time steps of the simulation. At the end of the simulations, we compared the final configuration of the continuum and discrete simulations. Our GA does not need to compute the continuum simulation, since it was previously computed. Two meshes were used with the same dimensions ($10 \times 10 \times 10\,mm$) but with different discretizations, for different purposes. The DE simulation, taken by GA as the correct simulation, is used in the refined mesh. It takes a long time for finishing the simulations but produces accurate results. However, the DE simulations were only run once before the AG starts, therefore this was not an issue to the GA performance. On the other hand, the FisioPacer simulation is demanded two times for each GA individual and so its computational cost must be lower. To achieve this we used a coarse mesh, that resulted in less accurate outcomes but it is very fast to simulate. The meshes can be found in Fig. 2.

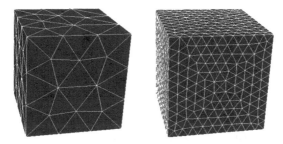

Fig. 2. Meshes: Coarse: 96 points and 295 elements; Refined: 2162 points and 10774 elements.

As for the fitness computation, it is computed the absolute error between the eight vertices of the meshes on their final configurations:

$$\phi_m = \sum_{i=1}^{8}(|\bar{X}_x^i - X_x^i| + |\bar{X}_y^i - X_y^i| + |\bar{X}_z^i - X_z^i|) \tag{33}$$

where i is the index of the vertex and X is its coordinates. $\bar{X}$ represents the vertices computed by DE and X is computed by FisioPacer. Finally, the fitness is computed by the sum of the fitness ϕ of each one of the two mesh configurations (fibers directions parallel to y axis and fibers direction rotating).

4.2   GA Operators

Selection consists in choosing individuals of a spring to be parents of the new individuals to be inserted on the next offspring. It must choose rather the individuals with best fitness, however the other individuals may also have a chance. We have used the rank selection, that consists in choosing individuals by its position (or rank) on a sorted population. The worst individual will be assigned to fitness 1, and the second worst to fitness 2, and so on until the best individual has fitness n, where n is the population size. The selection probability of an individual is proportional to its rank.

Once the parents were selected, they must generate new children. This is done with **crossover** and in this work the blend crossover $BLX - \alpha$ technique was used. It consists in randomly choosing a new gene o^i from a interval that depends on the parent genes p_1^i and p_2^i and a user-defined parameter α, where $i \in [1, 6]$ is the index of the gene in the chromosome. Figure 3 shows the $BLX - \alpha$ algorithm. This is done for each gene of the new individual until its chromosome is complete.

---

1 Input: Parent genes p_1^i, p_2^i, parameter α
2 Output: Children gene o^i
3 $max_i = max(p_1^i, p_2^i)$
4 $min_i = min(p_1^i, p_2^i)$
5 $r_i = max_i - min_i$
6 **return** a random $o^i \in [min_i - r_i\alpha, max_i + r_i\alpha]$

Fig. 3. Blended crossover algorithm.

In order to ensure genetic diversity on the population **mutation** is applied, that consists in a perturbation caused on the genes of a new individual just after the crossover. This works uses non-uniform mutation, that finds a mutated gene $\bar{o}_i$:

$$\bar{o}_i = \begin{cases} c_i + \Delta(t, b_i - c_i) \ if \ \tau = 0 \\ c_i + \Delta(t, c_i - a_i) \ if \ \tau = 0 \end{cases}, \tag{34}$$

$$\Delta(t, y) = y(1 - r^\theta), \tag{35}$$

$$\theta = \left(1 - \frac{g_t}{g_{max}}\right)^b, \tag{36}$$

where $r \in [0, 1]$ is a random number, τ is random number, it can be only 0 or 1. g_t is the current generation and g_{max} is last generation, b is an user-defined parameter. Function $\Delta(t, y)$ generates a value in $[0, y]$, that tends to 0 in the later generations.

When a new offspring takes place of previous one, it is not ensured that the best individual until that generation will remain in the population. However loosing it can be catastrophic to GA success. To avoid this problem it is used the **elitism** technique, that is keeping a certain number λ of the best individuals on the next generation, where λ is a parameter of the algorithm.

4.3 Parallel Code

We add OpenMP directives to the GA code to parallelize it. The genetic operators are not computational expensive and are not worth parallelizing. However each individual must run the discrete model two times, what clearly is a time-consuming task. Moreover it is an embarrassingly parallel problem. So, OMP directives were added to the loop that computes the fitness, like showed on Fig. 4.

```
1  #pragma omp parallel for schedule(static)
2  for (i=0;i<n - λ;i++){
3      φ_i^0 = runSimulation(0);
4      φ_i^1 = runSimulation(1);
5      Φ_i = φ_i^0+φ_i^1;
6  }
```

Fig. 4. OMP parallel genetic algorithm.

5 Results

5.1 Automatic Tuning Parameter

GA was run three times, for 100 generations each. It was used an AMD Opteron Processor 6272 with 64 processors running Red Hat 4.4.7-4 with gcc version 4.4.7. All the AG parameters can be find on Table 1.

Table 1. GA parameters

Threads	64	Generations	100	Prob. Crossover	85 %	α Crossover	0,2
Population	128	Elite	4	Prob. Mutation	2%	b Mutation	5

The initial data interval of all parameters is $[0; 1e6]$. Anyway, only the initial value of the parameters are limited to this interval, since the BLX-alpha crossover is able to generate values out of the limits. Table 2 presents the best set of parameters for each one of the three GA executions. It also presents the total error of the FisioPacer simulation used as fitness, given in mm.

Table 2. Best parameters

-	μ	K_l^f	K_l^t	K_v	K_d	K_a	Fitness
Exec. 1	386831	922663	332752	1.7e6	866	626519	6.1
Exec. 2	300561	687896	293176	1.1e6	515	569372	6.5
Exec. 3	236721	659103	195760	1.2e6	440	313015	7.1

Figure 5 presents the fitness of all individuals during the GA generations. This is the AG execution that resulted in the best set of parameters (Exec. 1). The black line highlights the best fitness of each generation and the gray dots are the other fitnesses. It is possible to observe that in early generations the range of fitness is large and over the generations, the individuals fitnesses gradatively get closer to each other. This happens due the non-uniform mutation, once it causes more genetic diversity in the beginning of the GA execution to allow a good coverage of the search space. In later generations, the search space decreases to allow a more refined tuning. The b *Mutation* is the GA parameter that determines this fitnesses approaching.

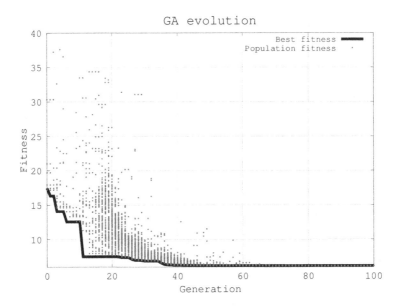

Fig. 5. Fitness of all individuals (Exec. 1).

The final mesh configuration of each simulation is on Fig. 6. Gray volumes are simulated with parameters found by GA execution 1. They are slightly different from DE simulations, however this is expected, since there is an intrinsic lack of accuracy by using a coarse mesh, and furthermore the methods used in this discrete model are expected to be less accurate.

Finally we used the best set of parameter with the electrical model with action potential. The value and time of both AP and active force for each state of the automaton were manually tuned in order to reproduce the simulations. This adjusting is a minor issue and therefore we decided to keep it manual instead of using the GA. FisioPacer simulations propagated the AP through the tissue in a similar manner to the DE simulations. However once again the mesh discretization plays an important role. The AP propagation is not smooth in the coarse mesh as it is in the refined one, since it contains bigger elements. The

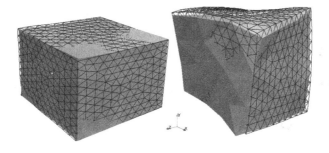

Fig. 6. Final configurations of the meshes: Gray volume is the FisioPacer output and the black wireframe is the DE output

same problem happens to the mechanical model, however the tissue contraction is qualitatively similar to the DE model. Moreover, the execution time of FisioPacer simulation is around $3\,min$, in contrast to the DE simulation that took $4\,h\,30\,min$ to compute, which means the discrete model is 90 times faster. The simulations comparison can be found in Figs. 7 and 8, respectively y-parallel fiber and fibers rotating direction from endo to epicardium.

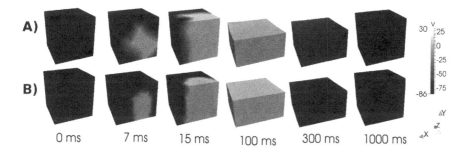

Fig. 7. Simulations with Y-parallel fibers: (A) FisioPacer and (B) Continuum simulator.

5.2   GA Parallel Performance

We perform three executions for each number of threads, and the average execution time, speedup and efficiency are shown in Table 3. The standard deviation was less than 1 % in all cases. The parallel implementation of the code resulted in a 44.5-fold in execution time, with 64 threads. The speedup was almost linear until 16 threads, so in these cases the efficiency was close to 100 %. On the other hand, with 32 and 64 threads there is a good decrease in the overall execution time, however less than expected. This can be explained by the fact that this processor has 64 cores but they share 32 floating-point units. Furthermore IO

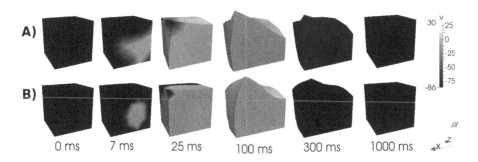

Fig. 8. Simulations with fibers rotating from endo to epicardium: (A) FisioPacer and (B) DE simulator.

operations are a bottleneck in our code that may also have impaired parallel performances: one thread writes the simulation output and the GA reads it to compare with DE solution. When there is a lot of threads trying to access disk simultaneously, each thread must wait some time while another is accessing the hard disk.

Table 3. Performance evaluation

Threads	1	2	4	8	16	32	64
Time (min)	1085	544	272	137	71	42	24
Speedup	-	2.0	3.9	7.9	15.4	25.7	44.5
Efficiency (%)	-	100	100	99	96	80	70

6   Conclusions

This work presented a 3D simulator of the electro-mechanical activity of cardiac tissue via the coupling of CA and mass-spring models and a genetic algorithm to estimate parameters to the model. Although models based on PDEs are more accurate and detailed, they are computationally expensive. CA has shown to be an alternative for real-time simulation because of its fast performance. Our model run 90 times faster than PDE and the pattern of propagation obtained with CA was visually similar to the patterns obtained with PDE-based models. Anyway, a more detailed comparison is still necessary. The tissue deformation obtained with the mass-spring system has shown to be very responsive to the force-development providing a qualitative demonstration of cardiac contraction.

The GA was able to find sets of parameters to the model that qualitatively resulted in similar simulations to a DE model. The parallel implementation performed a 44-fold improvement on computational time, that was very important due the necessity of tuning the model as fast as possible.

As future work, we want to perform a lot of different simulations with both simulators and then do a detailed quantitative analysis on the outputs in order to validate the model. Finally, in near future, we intend using the GA to find parameters that reproduce real data obtained by clinical exams.

Acknowledgments. The authors thank CAPES, CNPq, FAPEMIG and UFJF for supporting this work.

References

1. Gharpure, P.B.: A cellular automaton model of electrical wave propagation in cardiac muscle. Ph.D. thesis, Department of Bioengineering, The University of Utah (1996)
2. Bora, C., Serinagaoglu, Y., Tonuk, E.: Electromechanical heart tissue model using cellular automaton. In: Biomedical Engineering Meeting (BIYOMUT) 1–4 (2010)
3. Gharpure, P.B., Johnson Christopher, R. Harrison, N.E.: A cellular automaton model of electrical activation in canine ventricles: A validation study. Annal. Biomed. Eng, n. pag (1995)
4. Campos, R.S., Lobosco, M., dos Santos, R.W.: A GPU-based heart simulator with mass-spring systems and cellular automaton. J. Supercomputing **69**, 1–8 (2014)
5. Bourguignon, D., Cani, M.P.: Controlling Anisotropy in Mass-Spring Systems. In: Magnenat-Thalmann, N., Thalmann, D., Arnaldi, B. (eds.) Computer Animation and Simulation 2000, pp. 113–123. Springer, Vienna (2000)
6. Oussama Jarrouse: Modified Mass-Spring System for Physically Based Deformation Modeling. Ph.D. thesis, Karlsruher Instituts fur Technologie (2011)
7. Vinnakota, K.C., Bassingthwaighte, J.B.: Myocardial density and composition: a basis for calculating intracellular metabolite concentrations. Am. J. Physiol. - Heart Circulatory Physiol. **286**, H1742–H1749 (2004)
8. Nobile, F., Quarteroni, A., Ruiz-Baier, R.: An active strain electromechanical model for cardiac tissue. Int. J. Numer. Methods Biomed. Eng. **28**, 52–71 (2012)
9. Holzapfel, G.A., Ogden, R.W.: Constitutive modelling of passive myocardium: a structurally based framework for material characterization. Philos. Trans. Roy. Soc. A **367**, 3445–3475 (2009)
10. de Oliveira, B.L., Rocha, B.M., Toledo, E.M., Barra, L.P.S., Sundnes, J., dos Santos, R.W.: Effects of deformation on transmural dispersion of repolarization using in silico models of human left ventricular wedge. Int. J. Numer. Methods Biomed. Eng. **29**, 1323–1337 (2013)

Hexagonal Bravais–Miller Routing by Cellular Automata Agents

Dominique Désérable[1](✉) and Rolf Hoffmann[2]

[1] Institut National des Sciences Appliquées, 20 Avenue des Buttes de Coësmes,
35043 Rennes, France
domidese@gmail.com

[2] Technische Universität Darmstadt, FB Informatik, FG Rechnerarchitektur,
Hochschulstraße 10, 64289 Darmstadt, Germany
hoffmann@ra.informatik.tu-darmstadt.de

Abstract. This paper describes an efficient novel router in which the messages are transported by cellular automata (CA) agents. In order to implement agents more efficiently, the CA-w model (with *write* access) is used. The router is based upon a "Bravais–Miller" algorithm with hexagonal coordinates that explores the symmetries in the triangular lattice to provide a simple, deterministic, minimal routing scheme. As in a previous work, it uses henceforth six channels per node with at most one agent per channel so that one cell can host up to six agents. Each agent in a channel has a computed minimal direction defining the new channel in the adjacent node. In order to increase the throughput an adaptive routing protocol is defined, preferring the direction to an unoccupied channel. A strategy of deadlock avoidance is also investigated, from which the agent's direction can dynamically be randomized.

Keywords: Cellular automata agents · Multi-agent system · Hexagonal Bravais–Miller Routing · Triangular torus

1   Introduction

Problem solving with agents has become more and more attractive [1–6]. Generally speaking, agents are intelligent and their capabilities can be tailored to the problem in order to solve it effectively, and often in an unconventional way. Owing to their intelligence, agents can be employed to design, model, analyze, simulate, and solve problems in the areas of complex systems, real and artificial worlds, games, distributed algorithms and mathematical questions. Simply speaking, a CA agent is an agent that can be modeled within the CA paradigm. Usually an agent performs *actions*. *Internal* actions change the state of an agent, either a visible or a non-visible state, whereas *external* actions change the state of the environment. In this context, the "CA-w model" allows to *write* information onto a neighbor [7,8]. This method has the advantage that a neighbor can

---

D. Désérable—Until 2013.

© Springer International Publishing Switzerland 2015
V. Malyshkin (Ed.): PaCT 2015, LNCS 9251, pp. 164–178, 2015.
DOI: 10.1007/978-3-319-21909-7_16

directly be activated or deactivated, or data can be sent actively to it. Thus the movement of agents can be described more easily [9].

Target searching has been studied in many variations [10, 11]. We consider only stationary targets, and multiple agents having only a local view [12–14]. This contribution continues our preceding work on routing in the cyclic triangular grid with cyclic connections [9, 15–17] and derived from the *arrowhead* family [18, 19]. A relevant network of this family is the *diamond*, isomorphic to the arrowhead and homeomorphic to an orthogonal representation denoted "T_n" therein and used again herein. It is interesting to observe that the k–ary 2–cube [20], denoted "S_n" therein with $k = 2^n$, can be embedded into by eliminating one direction of link, namely the "diagonal" direction in the orthogonal diamond. Note that another family of "augmented" k–ary 2–cubes was investigated elsewhere [21] for any k but which coincide with T_n only when $k = 2^n$.

In a recent work [17], tori S and T were compared; evolved agents, with a maximum of *one* agent per node, were used in both cases. It turned out that routing in T is performed significantly better than in S. Another difference is that in [16] the agent's behavior was controlled by a finite state machine (FSM) evolved by a genetic algorithm [14], whereas the behavior is now handcrafted. Moreover, in [9, 15] each node is provided with six channels so that up to *six* agents per node are henceforth used, with one agent per channel.

The novelty in this paper is the use of hexagonal "Bravais–Miller" coordinates which fit the symmetries in the triangular lattice [22, 23]. The Cartesian coordinates in the *–orthogonal–* T–grid turn into hexagonal coordinates in the *diamond* and this simple transformation is embedded into the router. In short, the "diagonal" direction is no more considered as a specific direction and the three axes of the generating set fit a rotational, hexagonal symmetry. A first use of this coordinate system was investigated in [24] to find a minimal route in the infinite grid and in [25] the route was proven to be normalized by a simple reduction scheme. The effect of this computation is to simplify the router code as far as possible. This work finalizes a previous one investigating this novel router with six channels [9, 15] but using now our Bravais–Miller routing scheme [25].

2 Minimal Routing in the T Cyclic Grid

2.1 Distance and Bravais–Miller Indices

The triangular grid, often denoted "hexavalent grid" in the $2d$ space, is endowed with a hexagonal coordinate system and spanned by the generating set

$$\Sigma_H = \{\varepsilon_1, \varepsilon_2, \varepsilon_3\} = \{(0, 1, -1), (-1, 0, 1), (1, -1, 0)\}. \tag{1}$$

Geometrically, the grid is represented by the graph $H = (V_H, E_H)$ where

$$V_H = \{(x_1, x_2, x_3) \in \mathbb{Z}^3 : x_1 + x_2 + x_3 = 0\} \tag{2}$$

and E_H is such that any $\boldsymbol{x} \in V_H$ is connected to $\boldsymbol{x} \pm \varepsilon_1$, $\boldsymbol{x} \pm \varepsilon_2$, $\boldsymbol{x} \pm \varepsilon_3$. In particular, the origin $\boldsymbol{0} = (0, 0, 0)$ is surrounded by the neighbors given by the

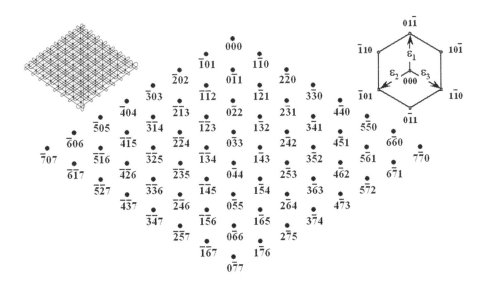

Fig. 1. The diamond $\mathcal{DT}_3$: interconnection pattern and HBM labeling.

six permutations of $(0, 1, -1)$ (negative integers are contracted for convenience). These permutations are known in crystallography as Bravais-Miller indices for the hexagonal arrangement: they are exactly the coordinates of the six points $i_k - i'_k$ ($k \neq k'$) in an orthonormal $3d$ space and where $\{i_1, i_2, i_3\}$ is the set of unit vectors [22,23]. From now on, we denote by "HBM" the hexagonal Bravais-Miller coordinate system defined by (1).

Definition 1. *Let* $d_H : V_H \times V_H \to \mathbb{N}$ *such that, for any* $(\boldsymbol{x}, \boldsymbol{y}) \in V_H \times V_H$:

$$d_H(\boldsymbol{x}, \boldsymbol{y}) = \|\boldsymbol{x} - \boldsymbol{y}\| = \max_{i \in I^*} |x_i - y_i| \quad I^* = (1, 2, 3). \tag{3}$$

Function d_H *defines a distance of shortest path on the triangular grid.*   □

2.2   Shortest Path Routing in the Diamond

Definition 2. *The* diamond *is the graph* $\mathcal{DT}_n = (V_{H,n}, E_{H,n})$ *where*

$$V_{H,n} = \{\boldsymbol{x} \in V_H \mid \boldsymbol{x} = m_2 \varepsilon_2 + m_3 \varepsilon_3 \ (m_2, m_3) \in \mathbb{Z}_{2^n} \times \mathbb{Z}_{2^n}\} \tag{4}$$

and any $\boldsymbol{x} \in V_{H,n}$ *is connected to* $\boldsymbol{x} \pm \varepsilon_1, \boldsymbol{x} \pm \varepsilon_2, \boldsymbol{x} \pm \varepsilon_3$.   □

The diamond is displayed in Fig. 1 for $n = 3$. It has $N = 4^n$ vertices and $3N$ edges. The diameter, given by

$$D_n = \frac{2\sqrt{N} - 1}{3} \quad \text{or} \quad D_n = \frac{2(\sqrt{N} - 1)}{3} \tag{5}$$

depending on the odd-even parity of n, defines an upper bound for the length of any shortest path [26].

Definition 3. *Let $\Sigma_H = \{\varepsilon_1, \varepsilon_2, \varepsilon_3\} = (01\bar{1}, \bar{1}01, 1\bar{1}0)$ and $I^* = (1, 2, 3)$.*
Let x and y be two distinct vertices in $V_{H,n}$ and $u = (u_1\, u_2\, u_3) = y - x$.
Let k, k', k'' in I^* such that $|u_k| \geq |u_{k'}| \geq |u_{k''}|$.
The function below defines a "ρ–reduction" acting on u as:

$$\rho : V_H \longmapsto V_H \quad | \quad \rho(u) = \begin{cases} u - 2^n \lambda \varepsilon_{k''} & u_{kk'} > 2^n \\ u & otherwise \end{cases} \tag{6}$$

where $u_{kk'}$ denotes the sum $|u_k| + |u_{k'}|$ in the reduction condition,

$$\lambda = \frac{u_k}{|u_k|} \cdot \varepsilon_{k''k} = \pm 1 \tag{7}$$

is the sense of reduction of u and wherein $\varepsilon_{k''k}$ is the k–coordinate of $\varepsilon_{k''}$.   □

The role of function ρ is to act on u by a translation of length 2^n while keeping the minimum of the $\{|u_i|\}$ invariant. If the reduction condition is fulfilled by u, then its image is such that $v = \rho(u)$ and $\| v \| < \| u \|$.

Lemma 1. *Let $u = y - x$ and $u^* = \rho^2(u)$. Then u^* is minimal.*   □

In other words, a given $u = y - x$ computed in the infinite grid is normalized in the *diamond* into a minimal u^* after *at most* two ρ–reductions. Practically, for a given pair (x, y) the reduction on u is often not applied, sometimes once, seldom twice and never more. The following proposition provides a normalized shortest path in the diamond.

Proposition 1. *Given a source x and a destination y in $V_{H,n}$.*
Let $u = y - x$ and $u^* = (u_1^*\, u_2^*\, u_3^*)$ be an irreducible representative of u.
Let $\pi : I^* \to I^*$ be the right cyclic permutation (231).
Then the path

$$\sigma_k(x, y) = u_{\pi^{-1}(k)}^* \varepsilon_{\pi(k)} - u_{\pi(k)}^* \varepsilon_{\pi^{-1}(k)} \tag{8}$$

where k stands for the index of a maximum of the $\{|u_i^*|\}$, defines a shortest path from x to y.   □

The application of Lemma 1 and Proposition 1 is illustrated hereafter from a few examples, referring to the diamond $\mathcal{DT}_3$ in Fig. 1. The symbol above '$\equiv$' denotes one of the six possible directions $(-\lambda \varepsilon_{k''})$ acted by the reduction function.

$$x = (2\bar{3}1) \quad y = (\bar{1}45) \quad u = (\bar{3}14) = u^*$$
$$\Rightarrow \sigma_3(x, y) = u_2^* \varepsilon_1 - u_1^* \varepsilon_2 = -\varepsilon_1 + 3\varepsilon_2.$$

$$x = (1\bar{2}1) \quad y = (0\bar{6}6) \quad u = (\bar{1}45) \overset{+\varepsilon_1}{\equiv} v = u + 8(01\bar{1}) = (\bar{1}4\bar{3}) = u^*$$
$$\Rightarrow \sigma_2(x, y) = u_1^* \varepsilon_3 - u_3^* \varepsilon_1 = -\varepsilon_3 + 3\varepsilon_1.$$

$$x = (\bar{5}\bar{1}6) \quad y = (7\bar{7}0) \quad u = ((12)\bar{6}\bar{6}) \overset{-\varepsilon_3}{\equiv} v = u - 8(1\bar{1}0) = (42\bar{6})$$
$$\overset{+\varepsilon_2}{\equiv} w = v + 8(\bar{1}01) = (\bar{4}22) = u^*$$
$$\Rightarrow \sigma_1(x, y) = u_3^* \varepsilon_2 - u_2^* \varepsilon_3 = 2\varepsilon_2 - 2\varepsilon_3.$$

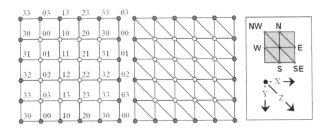

Fig. 2. The orthogonal T_2 of order $N = 16$, labeled in the XY coordinate system; redundant nodes in grey on the boundary. Inset: orientations W–E, N–S, NW–SE according to an XYZ reference frame.

2.3   Bravais–Miller Routing in the Orthogonal T_n

The *diamond* $\mathcal{DT}_n$ is topologically equivalent to the orthogonal T_n displayed in Fig. 2 for $n = 2$. Referring back to the generating system Σ_H in (1) and knowing that $\varepsilon_1 = -(\varepsilon_2 + \varepsilon_3)$, the vertex set is now spanned by the vector basis $\{\varepsilon_2, \varepsilon_3\} = \{\bar{1}01, 1\bar{1}0\}$ on the XY axis system. So let $(e_1, e_2) = (10, 01) = (\varepsilon_2, \varepsilon_3)$ be the usual XY orthonormal reference frame according to the inset of Fig. 2. The orthogonal transformation of the routing scheme is straightforward. Given some pair of vertices that denotes a source (X_s, Y_s) and a destination (X_d, Y_d) in the XY reference frame, it suffices to compute $U_X = X_d - X_s$, $U_Y = Y_d - Y_s$ and to convert (U_X, U_Y) into HBM coordinates by applying the correspondence

$$\boldsymbol{u} = (u_1\, u_2\, u_3) = (U_Y - U_X \quad - U_Y \quad U_X) \tag{9}$$

and to deal with the HBM routing of shortest path in Proposition 1.

2.4   Computing the Minimal Route in T_n

The algorithm below is the core of the router code: from the agent's position (source (X_s, Y_s)) and its target position (destination (X_d, Y_d)), the message header will contain the length of the subpaths of the minimal route and their respective channel identifier. Evidently, if the route is unidirectional, there will only be one subpath.

1. `Coordinate transformation`: Given a source (X_s, Y_s) and a destination (X_d, Y_d) in Cartesian coordinates, compute $\boldsymbol{u}$ from (9).
2. `Normalize u`: Get an irreducible $\boldsymbol{u}^*$ from Lemma 1 and Definition 3.
3. `Length of subpaths`: The lengths are given by $(|\, u^*_{\pi^{-1}(k)}\,|, |\, u^*_{\pi(k)}\,|)$ from (8) in Proposition 1.
4. `Channel identifiers`: The respective directions $(\varepsilon_{\pi(k)}, \varepsilon_{\pi^{-1}(k)})$ of subpaths belong to the set $\{\pm\varepsilon_1, \pm\varepsilon_2, \pm\varepsilon_3\}$ and depend upon the sign of u^*_k as displayed in Table 1.

Note that the apparent subpath order in (8) and Table 1. is in no wise settled (the vector sum is commutative). As in [9], we adopt the cyclic, *deterministic* convention "FIRST dirR THEN dirL" where dirR and dirL define the "right"

Table 1. Length and direction of subpaths as a function of k and $\mathrm{sign}(u_k^*)$

k	$u_k^* > 0$		$u_k^* < 0$	
1	$-\lvert u_3^*\rvert\varepsilon_2$	$+\lvert u_2^*\rvert\varepsilon_3$	$+\lvert u_3^*\rvert\varepsilon_2$	$-\lvert u_2^*\rvert\varepsilon_3$
2	$-\lvert u_1^*\rvert\varepsilon_3$	$+\lvert u_3^*\rvert\varepsilon_1$	$+\lvert u_1^*\rvert\varepsilon_3$	$-\lvert u_3^*\rvert\varepsilon_1$
3	$-\lvert u_2^*\rvert\varepsilon_1$	$+\lvert u_1^*\rvert\varepsilon_2$	$+\lvert u_2^*\rvert\varepsilon_1$	$-\lvert u_1^*\rvert\varepsilon_2$

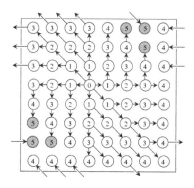

Fig. 3. Deterministic convention"FIRST dirR THEN dirL". This directed graph is a spanning tree of the torus showing the minimal path from a source node "0" to any other node for a 8 x 8 network ($n = 3$, $N = 64$). The number in the nodes represents their distance from the source node "0". The maximal distance is the diameter $D_3 = 5$ for this graph (refer back to Eq. (5)). Six antipodals are highlighted.

minimal subpath and the "left" minimal subpath respectively,[1] viewed from the "observer" agent as depicted in Fig. 3.

2.5   Deterministic, Adaptive and Randomized Routing

From the *deterministic* routing "FIRST dirR THEN dirL" the agent follows always the "right" minimal subpath. The minimal path can be computed only once at the beginning and stored in the agent's state. During the run, the agent updates the remaining path to its target, decrementing its dirR counter until zero, then decrementing its dirL counter if any, until completion. A problem with this protocol is that it may not be optimal with respect to throughput, especially in case of congestion. Formally, the deterministic routing is secure for an agent alone.

The objective for *adaptive* routing is (i) to increase throughput, and (ii) to avoid or reduce the probability of deadlocks. During the run, if the temporary direction (e.g., dirR) points to an occupied channel, then the other channel (e.g., dirL) can be used. A minimal adaptive routing may be roughly denoted as "EITHER dirR OTHERWISE dirL". The path from source to target remains

---

[1] An equivalent symmetric protocol would be "FIRST dirL THEN dirR".

minimal on the condition that it remains inside the boundaries defined by the minimal parallelogram.

It would also be useful to allow the agent to deviate from the minimal route in case of congestion or for deadlock avoidance. The agent could route *out* of its minimal parallelogram and move within an extended area although the minimal route is of course prioritized. As a consequence, *three* possible moving directions, instead of *two*, still remain adaptively possible to move forward. The three other directions backwards are not allowed. But stronger protocols must sometimes be carried out to avoid deadlock, like *randomized* routing where up to the *six* possible directions are allowed. In this case, the agent should recompute its minimal path at each timestep.

Deterministic, adaptive and deadlock-free strategies are investigated in Sect. 4. A general insight on routing protocols can be found in [27].

3 Modeling the Multi-Agent System

This section is a digest of the description of the multi-agent system detailed in [9]. The dynamics of moving agents is recalled and the impact of the *copy–delete* rules in the CA–w model is emphasized.

3.1 Dynamics of the Multi-Agent System

For clarity's sake, the above adaptive protocol where *three* possible directions are allowed to move the agent forward are considered.

Node Structure and Channel State: Each node labeled by its (x, y) coordinates contains the 6–fold set

$$\mathcal{C} = \{C_0, C_1, C_2, C_3, C_4, C_5\} = \{E, SE, S, W, NW, N\} \tag{10}$$

of channels C_i oriented[2] and labeled clockwise. Index i is called *position* or *lane number* in this context. This position defines an implicit direction towards the next *adjacent* node that an agent visits next on its travel. The direct neighbor of channel C_i in the adjacent node is denoted by M_i where

$$\mathcal{M} = \{M_0, M_1, M_2, M_3, M_4, M_5\} \tag{11}$$

and $M_{j \equiv i+3 \pmod 6}$ denotes the i–channel of the adjacent j–neighbor by symmetry. In the cardinal notation, e.g. for $i = 0$, "$W.E$" stands for the E–channel of the W–neighbor (Fig. 4). The i–channel's *state* at time t is defined by

$$c_i(t) = (p, (x', y')) \tag{12}$$

---

[2] Except a homeomorphism.

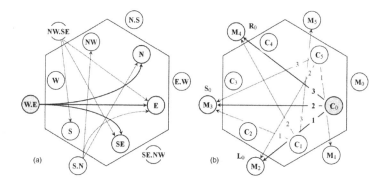

Fig. 4. (a) An agent located in the E–channel of the western node $W.E$ can move to one channel in the "opposite" subset $\{E, N, SE\}$. Two agents in channels $NW.SE$ and $S.N$ are possible competitors for a part of this subset. (b) As a consequence, channel C_0 is the arbiter of three possible concurrent agents in the `right`, `straight`, `left` requesting channels R_0, S_0, L_0 viewed from the "observer" C_0. The same concurrent scheme is valid all around from rotational invariance. A priority order is assigned clockwise by any channel $C_i \in C$ to its own requesting ordered set $\mathcal{M}_i = (R_i, S_i, L_i)$: 1 to `left`, 2 to `straight`, 3 to `right`.

where (x', y') stands for the agent's target coordinates and $p \in \mathcal{P}$ stands for the agent's direction as a pointer to the desired channel in the next node in the set

$$\mathcal{P} = \{-1, 0, 1, 2, 3, 4, 5\} \equiv (\texttt{Empty}, \texttt{toE}, \texttt{toSE}, \texttt{toS}, \texttt{toW}, \texttt{toNW}, \texttt{toN}) \qquad (13)$$

symbolized by $(\rightarrow \searrow \downarrow \leftarrow \nwarrow \uparrow)$ in a graphical representation and including an empty channel encoded by $\omega = -1$.

Agents and Arbiters: An agent can move to a *3–fold* subset of channels at most. For example, coming from channel $W.E$ of W–neighbor at $(x - 1, y)$, going to channel E or N or SE of current node (x, y) as shown in Fig. 4a. In the same way, agents located in outer channels $NW.SE$ and $S.N$ are possible competitors for a part of this subset $\{E, N, SE\}$. The *intersection* of those three requested subsets is the channel E. From this observation, E can be chosen as *arbiter* of three possible concurrent agents. In other words, a priority rule can be locally defined for *this* channel. Arbiter E is C_0 in Fig. 4b and this concurrent scheme is invariant by rotation. This interaction between requesting agents and arbiter channels is formalized hereafter. Let the 3–uple of channels

$$\mathcal{C}_i = (C_{i+1}, C_i, C_{i-1}) \qquad (14)$$

and let us denote by

$$\mathcal{M}_i = (R_i = M_{i+4}, S_i = M_{i+3}, L_i = M_{i+2}) \qquad (15)$$

the ordered 3–uple opposite to C_i and where R_i, S_i, L_i are the `right`, `straight` and `left` outer channels for the three possible incoming agents viewed from the "observer" channel C_i ($i = 0$ assumed herein). Conversely, $\mathcal{C}_i$ is the requested channel subset for S_i, as well as $\mathcal{C}_{i-1}$ for L_i and $\mathcal{C}_{i+1}$ for R_i. Now

$$\mathcal{C}_{i-1} \cap \mathcal{C}_i \cap \mathcal{C}_{i+1} = \{C_i\} \qquad (16)$$

from (14). This simple but important property allows to define a *local* priority rule for channel C_i and invariant by rotation.

Priority Rule: Each channel $C_i \in \mathcal{C}$ computes the three exclusive conditions selecting the incoming agent that will be hosted next, with a priority assigned clockwise (Fig. 4b):

1. Agent wants to move from L_i to C_i, priority 1: $\texttt{LtoC} = (l = i)$
2. Agent wants to move from S_i to C_i, priority 2: $\texttt{StoC} = (s = i) \wedge \neg\texttt{LtoC}$
3. Agent wants to move from R_i to C_i, priority 3: $\texttt{RtoC} = (r = i) \wedge \neg\texttt{StoC}$.

In other words, this rule selects a winner among the three possible concurrent agents requesting channel C_i and the selection is assigned clockwise: first to `left`, second to `straight`, third to `right`, orientation viewed from the observer *channel*.

Moving the Agents: The above priority scheme ensures a *conflict-free* dynamics of moving agents[3] in the whole network. The new target coordinates $(x', y')^*$ in the channel's state are either invariant if the agent stays at rest or are copied from L_i or S_i or R_i exclusively, depending of the selected incoming agent hosted and to be received by the channel. Since the agent's target coordinates are stuck within its state, the agent must clearly carry them with it when moving. The agent's direction is then updated as

$$p^* = \varphi_{xy}\,((x', y')^*) \quad (p^* \in \mathcal{P}) \qquad (17)$$

according to the used protocol which yields the new desired channel, either in the current node (agent not moving) or in the adjacent node (after moving). Function φ_{xy} is computed locally in the current node (x, y). Finally, the i–channel's *state* at time $t + 1$ becomes

$$c_i(t + 1) = (p^*, (x', y')^*) \qquad (18)$$

and the new state is updated synchronously.

3.2   The CA–w *Copy–Delete* Rule

The synchronous transition (18) to the next timestep is governed by the *copy–delete* operating mode of the CA–w model [7,8]. The CA–w model is especially useful if there are no write conflicts by algorithmic design. This is here the case, because an agent is copied by *its* receiving channel, after applying the

---

[3] Except special deadlock or livelock situations pointed out in Sect. 4.

abovementioned priority scheme. Thus only *this* receiving channel is enabled to delete the agent on the sending channel at the same time. A further advantage is that only the short-range *copy*–neighborhood is sufficient to move an agent, the wide-range *delete*–neighborhood (necessary for CA modeling) is not needed. Therefore the 3–fold *copy*–neighborhood $\mathcal{M}_i$ that needs to be checked by C_i in order to receive the hosted agent is given by (15), this agent in $\mathcal{M}_i$ is released by C_i when firing the transition (18) and following the CA–w *delete–copy* operating mode.

4   Simulating HBM Routing Protocols

4.1   Deadlock Situations and Protocols

A trivial deadlock can be produced if all $6N$ channels contain agents (fully packed), thus no moving is possible at all. Another deadlock appears if $\sqrt{N} = 2^n$ agents line up in a loop on all the channels belonging to the same lane, and all of them want to travel in the same direction. Then the lane is completely full and the agents are stuck. To escape from such a deadlock is only possible if the agents could deviate from the shortest path, e.g. by choosing a random direction from time to time. More interesting are the spatial cyclic deadlocks where the agents form a cycle and are blocking each other (no receiving channel is free in the cycle).

Several ways to resolve such deadlocks have been investigated elsewhere, e.g. spatial inhomogeneity, redistribution of the agents on a node, deviation from the minimal route [9], internal control by finite state machine [16] or optimization by genetic algorithms [14]. Randomization is investigated thereafter and four protocols are tested, according to Subsect. 2.5.

1. ProtR: the *deterministic* routing "FIRST dirR THEN dirL".
2. ProtRL: the *minimal* adaptive routing "EITHER dirR OTHERWISE dirL".
3. Rand(ProtR): ProtR is *randomized*. The rightmost i–channel of the minimal route to be used in the next node is modified to $i - 1$ (mod 6) counterclockwise, with the basic probability $p_0 = 1/2$ and with the additional probability $p = 1/Q$ at each simulation time step. Thus the total probability is $p_0 \times p$.
4. Rand(ProtRL): ProtRL is *randomized*. The two possible channels (the rightmost i and the leftmost $j = i - 1$ (or $j = i$ if there is only one) are modified, either turned to the left ($i \rightarrow i - 1$, $j \rightarrow j - 1$) or to the right ($i \rightarrow i + 1$, $j \rightarrow j + 1$) with an equal chance. This modification is done at each timestep with the probability $p = 1/Q$.

4.2   Test Cases

Three test cases are used for evaluation, with k agents, s source nodes and m target nodes:

1. **First Test Case: All-to-One** ($m = 1, k = s$). All agents move to the same target. We set $k = N - 1$, meaning that an agent is initially placed on each site, except on the target. The optimal performance would be reached if the target consumes six messages at every timestep ($t = (N - 1)/6$). In order to check the routing scheme exhaustively, the target location was varying, yielding maximal N test configurations. We recall that the T–grid is vertex-transitive, so the induced routing algorithm must yield the same result for all N cases.

2. **Second Test Case: Random Routes** ($k = s = m$). The sources and target are chosen randomly. The sources are mutually exclusive (each source is used only once in a message set) as well as the targets. Source locations may act as targets for other agents, too. We set $k = N/2$ which was also used in our preceding works for comparison. Note that the minimum number of timesteps t to fulfil the task is the longest distance between source and target contained in the message set. For a high initial density of agents the probability is high that the longest distance is close to the diameter of the network. Thus the best case would be $t = D_n$ (5).

2. **Third Test Case.** This is a special test case defined for the evaluation of the routing protocol `ProtRL`. The sources and targets are chosen randomly as in the second case, but the number of agents is now higher: $k = N$. Differing from the second case, the target locations may now be used more than once.

4.3   Router Efficiency

First Test Case. All possible N initial configurations differing in the target location were tested for N=64, 256, 1024, and 32 for N=4096 (Table 2).

`ProtR`, `ProtRL` (**B**): The time is independent of the position of the target. This means that the router works totally symmetric as expected. An optimal router would consume in every generation six agents at the target, leading to an optimum of $t_{opt} = (N - 1)/6$. But this optimum cannot be reached because the agents need an empty receiver channel in front for moving and they are jammed in a queue. As it can be seen on the two first timesteps of the simulation sequences in Fig. 5, the agents are consumed by the target at *odd* timesteps, generally at the rate of *six* per odd timestep that yields a mean flow rate of *three* per timestep. Thus in Table 2 we observe a transfer time (**B**) close to the optimum $t_{opt} = (N - 1)/3$. As a consequence, the ratio t/D_n (**B/A**) is close to $\sqrt{N}/2$ from (5) and **B/D** ≈ 2 for large N. This phenomenon is easy to understand and has a close relationship with Traffic Rule 184 in a 1–dimensional CA: a car with a car straight ahead cannot move and must wait for the next timestep.

Both deterministic and minimal adaptive protocols need the same number of time steps, although we expected that `ProtRL` could behave better than `ProtR`. The reason is that the channels nearby the target are heavily congested in both cases.

`Rand(ProtR)`, `Rand(ProtRL)` (**C1,C2**): The deterministic and adaptive protocols did not produce any deadlocks. Nevertheless the randomized protocols

Table 2. First test case: $k = N - 1$ messages travel from all disjoint sources to the same common target. Message transfer time (in *timesteps*) in the T–grid, averaged over the number of checked initial configurations.

Nodes N	(A) Diameter	(B) protR protRL	(C1) rand(protR) p=1/128	(C2) rand(protRL) p=1/128	Ratio B/A	(D) N/6	Ratio B/D
64=8 x 8	5	23	23.70	23.44	4.6	11	2.09
256=16 x 16	10	89	91.06	91.59	8.9	43	2.07
1024=32 x 32	21	351	353.48	359.52	16.7	171	2.05
4096=64 x 64	42	1385	1386.61	1400.87	32.95	683	2.03

were tested, in order to check if they work securely. Table 2 shows that the number of time steps is slightly higher compared to the deterministic and minimal adaptive protocols (**B**).

Second Test Case. The number of used initial configurations for averaging were 256 for $N=64$, 256, 1024, and 32 for $N=4096$.

The best performance yields the protocol ProtRL (**B2**) in Table 3 with a time ratio $t_{RL}/t_R \approx 0.95$ (**B2,B1**). The randomized protocols are again slightly slower, if we compare ProtR, Rand(ProtR) (**B1,C1**) and ProtRL, Rand(ProtRL) (**B2,C2**).

The ratio $t_{RL}/D_n \approx 1$ (**B2/A**) is noteworthy and shows that the mean transfer time is close to the diameter. This phenomenon is again easy to understand from Traffic Rule 184 but now in a fluid traffic.

Third Test Case. The field size was $N=4096$ with N agents and random routes and 32 configurations were simulated for averaging the time.

The mean time was $t = 56.59, 50.91, 56.75, 51.41$ respectively for ProtR, ProtRL, Rand(ProtR), Rand(ProtRL) and the randomization probability was

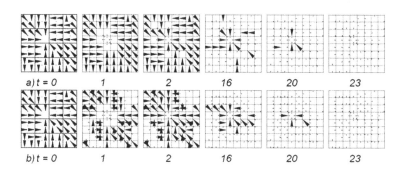

Fig. 5. Simulation snapshots for the first scenario in a 8×8 grid T_3, $N - 1$ agents moving to the same target position. Agents are depicted as large triangles, visited channels as small triangles: directions are symbolized by $(\rightarrow \searrow \downarrow \leftarrow \nwarrow \uparrow)$
(a) Routing protocol ProtR, (b) routing protocol ProtRL.

Table 3. Second test case: $k = N/2$ messages travel from disjoint sources to disjoint targets. Message transfer time (in *timesteps*), averaged over the number of checked initial configurations.

Nodes N	(A) Diameter	(B1) protR	(B2) protRL	(C1) rand(protR) p=1/128	(C2) rand(protRL) p=1/128	Ratio B2/A
64=8 x 8	5	5.926	5.543	6.035	5.609	1.109
256=16 x 16	10	12.222	11.293	12.320	11.387	1.129
1024=32 x 32	21	23.699	23.066	24.219	23.340	1.098
4096=64 x 64	42	46.227	45.355	46.926	45.839	1.080

$p = 1/128$. We get a time ratio $t_{RL}/t_R \approx 0.90$ and a similar ratio $t_{r(RL)}/t_{r(R)}$ with randomization. All results show that ProtRL performs generally better than ProtR.

Testing Deadlock Situations. Two deadlocks $\mathcal{D}_1$, $\mathcal{D}_2$, shown in Fig. 6 at time $t = 0$, were designed in order to demonstrate the behavior of the deterministic and randomized protocols. In $\mathcal{D}_1$ the agents have to travel strictly on their cycle. In $\mathcal{D}_2$ the agents can leave their cycle by choosing another minimal route (direction dirL). The effective resolving of the deadlocks depends upon the used protocol:

– ProtR: Both deadlocks cannot be resolved (not shown).

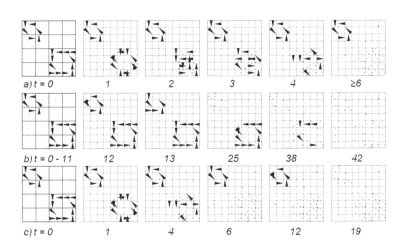

Fig. 6. Two designed deadlock situations $\mathcal{D}_1$ (upper-left) and $\mathcal{D}_2$ (lower-right). Source–target positions $((xy), (x'y'))$ for: $-\mathcal{D}_1$ ((00,12),(01,22),(12,21),(22,10),(21,00),(10,01)), $-\mathcal{D}_2$ ((44,57),(45,60),(46,77),(47,76),(57,05),(67,74),(77,64),(76,53),(75,44),(74,45),(64,36),(54,47)). (a) ProtRL: only $\mathcal{D}_2$ is resolved by using the alternate minimal route dirL. (b) Rand(ProtR): the deadlocks can only be resolved by deviation from the minimal route dirR. (c) Rand(ProtRL): $\mathcal{D}_1$ is resolved by randomization, $\mathcal{D}_2$ mainly by using the alternate minimal route. Probability for deviation is $p = 1/128$.

- (a) `ProtRL`: Only $\mathcal{D}_2$ can be dissolved, because the alternate minimal path `dirL` is used. Deadlock $\mathcal{D}_1$ cannot vanish because `dirR = dirL` and there is no alternative.
- (b) `Rand(protR)`: At $t = 12$ one agent decides to deviate from the minimal route. Thereby $\mathcal{D}_1$ is broken and at $t = 19$ it is dissolved. At $t = 25$ $\mathcal{D}_2$ is broken and dissolved at $t = 42$.
- (c) `Rand(protRL)`: Deadlock $\mathcal{D}_2$ is dissolved in 6 time steps as in (a), the alternate minimal path `dirL` is used. Deadlock $\mathcal{D}_1$ is broken at $t = 12$ and dissolved at $t = 19$.

5 Conclusion

An efficient novel router was presented in which the messages are transported by cellular automata agents, implemented from the CA-w model. The minimal routes are computed by a "Bravais–Miller" algorithm with hexagonal coordinates that explores the symmetries in the triangular lattice. Four protocols were investigated, using a determined minimal route, an adaptive minimal route, as well as their two randomized counterparts. The non-randomized ones were proven to be not deadlock-free, though the deadlock events rarely occur owing to the number of six channels per node. For the first all-to-one scenario, no deadlock appears for all protocols and the performance is close to the optimum $t_{opt} = (N-1)/3$. For the second scenario with a population of $N/2$ agents routing to $N/2$ targets, the best performance $t_{RL} \approx D_n$ is reached by the minimal adaptive protocol, close to the diameter of the network. For the third scenario with random routes, no deadlocks appears for all protocols, and with a performance in timesteps slightly higher than the diameter. Our Bravais–Miller routing scheme yields the same routes than our XY–scheme in [9] but its symmetry now simplifies the implementation of the router code.

References

1. Woolridge, M., Jenning, N.R.: Intelligent agents: theory and practice. Knowl. Eng. Rev. **10**(2), 115–152 (1995)
2. Franklin, S., Graesser, A.: Is it an agent, or just a program?: a taxonomy for autonomous agents. In: Jennings, Nicholas R., Wooldridge, Michael J., Müller, Jörg P. (eds.) ECAI-WS 1996 and ATAL 1996. LNCS, vol. 1193. Springer, Heidelberg (1997)
3. Holland, J.H., Emergence: From chaos to order, Perseus Book (1998)
4. Woolridge, M.: An Introduction to Multiagent Systems. Wiley & Sons, New York (2002)
5. Pais, D.: Emergent collective behavior in multi-agent systems: an evolutionary perspective, PhD Dissertation. Princeton University, Princeton NJ (2012)
6. Schweitzer, F.: Brownian Agents and Active Particles: Collective Dynamics in the Natural and Social Sciences. Springer Series in Synergetics. Springer, Heidelberg (2003)

7. Hoffmann, R.: The GCA-w massively parallel model. In: Malyshkin, V. (ed.) PaCT 2009. LNCS, vol. 5698, pp. 194–206. Springer, Heidelberg (2009)
8. Hoffmann, R.: GCA-w: global cellular automata with write-access. Acta Phys. Polonica B Proc. Suppl. **3**(2), 347–364 (2010)
9. Hoffmann, R., Désérable, D.: Routing by cellular automata agents in the triangular lattice. In: Sirakoulis, G., Adamatzky, A. (eds.) Robots and Lattice Automata. Emergence, Complexity and Computation, pp. 117–147. Springer, Switzerland (2015)
10. Loh, P.K.K., Prakash, E.C.: Performance simulations of moving target search algorithms. Int. J. Comp. Games Tech. **3**, 1–6 (2009)
11. Korf, R.E.: Real-time heuristic search. Artif. Intell. **42**(2–3), 189–211 (1990)
12. Ediger, P., Hoffmann, R.: CA models for target searching agents. São José dos Campos **ENTCS 252**(2009), 41–54 (2009)
13. Ediger, P., Hoffmann, R.: Routing based on evolved agents. In: 23rd PARS Workshop on Parallel System and Algorithms, ARCS, Hannover, Germany, pp. 45–53 (2010)
14. Ediger, P.: Modellierung und Techniken zur Optimierung von Multiagentensystemen in Zellularen Automaten, Dissertation, TU Darmstadt, Darmstadt, Germany (2011)
15. Hoffmann, R., Désérable, D.: Efficient minimal routing in the triangular grid with six channels. In: Malyshkin, V. (ed.) PaCT 2011. LNCS, vol. 6873, pp. 152–165. Springer, Heidelberg (2011)
16. Ediger, P., Hoffmann, R., Désérable, D.: Routing in the triangular grid with evolved agents. J. Cellular Automata **7**(1), 47–65 (2012)
17. Ediger, P., Hoffmann, R., Désérable, D.: Rectangular vs triangular routing with evolved agents. J. Cellular Automata **8**(1–2), 73–89 (2013)
18. Désérable, D.: A family of Cayley graphs on the hexavalent grid. Discrete Appl. Math. **93**(2–3), 169–189 (1999)
19. Désérable, D.: Systolic dissemination in the arrowhead family. In: Wąs, J., Sirakoulis, G.C., Bandini, S. (eds.) ACRI 2014. LNCS, vol. 8751, pp. 75–86. Springer, Heidelberg (2014)
20. Dally, W.J., Seitz, C.L.: The Torus routing chip. Dist. Comp. **1**, 187–196 (1986)
21. Xiang, Y., Stewart, I.A.: Augmented k-ary n-cubes. Inf. Sci. **181**(1), 239–256 (2011)
22. Miller, W.H.: A Treatise on crystallography. J. & J.J. Deighton, London (1839)
23. Buerger, M.J.: Introduction to Crystal Geometry. McGraw-Hill, New York (1971)
24. Désérable, D.: Minimal routing in the triangular grid and in a family of related tori. In: Lengauer, Christian, Griebl, Martin, Gorlatch, Sergei (eds.) Euro-Par 1997. LNCS, vol. 1300. Springer, Heidelberg (1997)
25. Désérable, D.: Hexagonal Bravais-Miller routing of shortest path, IR#CU13220.1, pp. 1–15 (2013) – Désérable, D., Dumont, E.: Routing algorithm in torus T6n, IR#CU13220.2, pp. 1–8 (2013)
26. Désérable, D.: Arrowhead and diamond diameters, (submitted to Discrete Applied Math. & Applications)
27. Duato, J., Yalamanchili, S., Ni, L.: Interconnection Networks, Morgan Kaufmann (2002)

The Influence of Cellular Automaton Topology on the Opinion Formation

Tomasz M. Gwizdałła[✉]

Department of Solid State Physics, University of Łódź,
Pomorska 149/153, 90-236 Łódź, Poland
tomgwizd@uni.lodz.pl

Abstract. We use the Cellular Automata to study the process of opinion formation in the community. The crucial property characterizing the existing models is the method of updating the system. In the paper we choose the randomized Glauber method and concentrate on the influence of topology of the system on the opinion understood as the support for specific real parties. We study also the relation between the topology and the parameters of the Glauber method. We propose to perform the analysis of the results based on the Fourier transform. This form of presentation discloses some interesting properties of both real-world and simulation results.

1 Introduction

The problem of opinion formation has been studied since many years by using many different approaches. It was already in 1973 when Clifford and Sudbury [1] proposed so called voter model. They analysed the competition between two species and used the Glauber dynamics to perform the update of their system in order to find the prevailing one. The approaches which found later a great popularity was the Galam's majority model [2] or the Sznajd model [3] characterized by the outflow dynamics. It can be noticed that as well continuous (e.g. Galam's one) as discrete (e.g. Sznajd one) models were in extensive use. Following the discrete models we can observe that a well-known physical model - the Ising model is widely used in the study of opinion formation process. The Ising-like models characterized by the discrete set of states, the short range of interaction/information exchange provides an interesting instrumentation for the analysis of opinion changes. Such an approach can be also easily described within the frame of Cellular Automata related notions. Some recent Ising-based attempts can be found in [4–9]. On the other hand, the topology describing the community can lead to substantial differences in the simulation results. Usually, when considering the topology one takes into account such network models like: the Erdös-Renyi, the Watts-Strogatz or the Barabasi-Albert ones (see e.g. [10,11]).

In our paper we try to consider both the Ising-like model of opinion formation and the problem of network topology. The crucial question which in our opinion

V. Malyshkin (Ed.): PaCT 2015, LNCS 9251, pp. 179–190, 2015.
DOI: 10.1007/978-3-319-21909-7_17

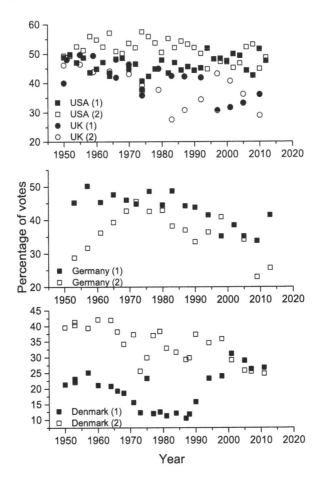

Fig. 1. The percentage support for different parties in several countries. Upper plot: USA (squares), United Kingdom (circles), middle plot: Germany, lower plot: Denmark. The countries are grouped according to their election system. The names of particular parties are not presented, they are distinguished only by numbers.

should be answered when mention "opinion" is the one concerning the real-world equivalent of the number produced in simulations. Quite often scientists deal with the results related to some consumer's opinions expressed in polls. In our paper we take into account the more distinct effect of opinion formation, i.e. the results of elections. We hope that despite of limited data (due to the frequency of elections) it gives us the opportunity to analyse the data which are not charged with the polling procedure and reflect the real data.

In Figs. 1 and 2 we show the results of parliamentary elections for several countries. The plots in Fig. 1 present the percentage results of major parties active in these countries. We choose such countries where the system was relatively stable through last 60 years. The countries are divided in the way that those ones with similar electoral systems are shown in the same plot. That is

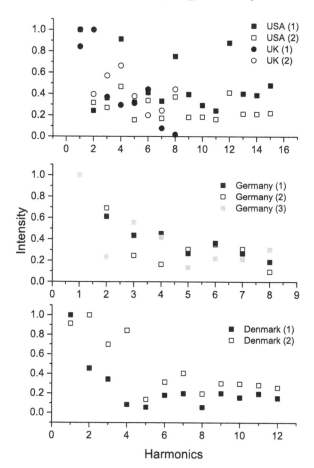

Fig. 2. The amplitudes of Fourier spectra for countries and parties presented in Fig. 1. The description of symbols is the same as in the earlier figure. The spectra are normalized in the way that the highest amplitude corresponds to 1.

why United States and United Kingdom where the First Past The Post system (one round majority system) exists are presented commonly. The second plot contains data for Germany (the mixed system) while the last one shows the results in Denmark (proportional d'Hondt system). Figure 2 contains the Fourier transforms of data shown in the upper one. The data in both figures corresponds one to another so the symbols in the second figure represents the data for the same party as in the first one.

Certainly, some simplifications have to be made. We do not take into account the correct date of elections but only order them in time and these data are used in discrete Fourier transform. The time periods between the elections differ not only between the countries (2 years for USA, 4 years for UK, 3/4 years in Scandinavian countries) but can be different also for the particular country due to e.g. early elections or the change of electoral law. We also have to point out that the result of particular party may not correspond strictly to the notion of opinion. This is due to the fact that parties form coalitions either after the election or sometimes before it. This is e.g. typical for Denmark where, during the last election in 2011, we observed two coalitions. Finally, the better result obtained the alliance of as much as 4 parties (Socialdemokraterne, Det Radikale Venstre, Socialistisk Folkeparti, Enhedslisten De Rod-Gronne) with the opposition formed by 5 organizations.

When considering the spectrum especially two features can be taken into account: the scaling properties and the excess observed for some specific harmonics. The general remark is that it doesn't exist the same scaling for all spectra. When deducing the scaling properties with the exclusion of extremely deviated points we can say that for majority systems we can observe linear dependence (for the US parties it is almost constant) while for the other ones we can assume either linear one or a power-law one. Concerning the distinctive harmonics, it is interesting that often the fourth harmonic exceeds the neighboring ones. This property can be seen for all countries except of Denmark where we should mention rather the 5th harmonic to be lower. For the USA we can notice also that the 8th and 12th harmonics are visibly higher than another ones. We can mention here also the similarity between the spectra for UK and Germany (in both cases the same number of harmonics is determined). When calculating the correlation coefficient between the sets of amplitudes for these countries we obtain value $\rho \approx 0.944$.

2   The Ising-Based CA Opinion Formation Model

Cellular Automata (CA) presents the approach which is often used when studying the processes of opinion formation. Generally CA can be described as a discrete dynamical system with the well-defined function describing the process of system update. The discrete character manifests itself in all aspects of CA simulation starting from the methods of representing the system space up to the definition of particular cell description. The n-dimensional space beeing the arena of simulation is divided into separate cells of shape conforming to the properties of problem under consideration. The most popular is here the seminal (known e.g. from the Conway's Game of Life) idea of building the n-dimensional cubes. These cells can only be empty or filled by the state chosen from some closed set S_i^t where i denotes position and t time. The crucial is however for CA the notion of rule - the function which makes it possible to perform the transformation $S_i^t \rightarrow S_i^{t+1}$. Typically $S_i^{t+1} = f(S_{j1}^t, S_{j2}^t, ...S_{jn}^t)$, where $S_{j1}, S_{j2}, ...S_{jn}$ belong to some neighborhood of cell i.

When applying the general model presented above to the particular case studied in the presented paper we have to point out some features which are in more detail described in the following paragraphs. The structure of the space is the main point of the paper and is addressed as the topology issue. We should mention here that while the typical, most common understanding of space definition is related to the physical, configurational space, here this view is a little changed. Our space should be rather considered as the phase space, where the distance between the cells corresponds to the difference described by some, other than geometric length, value. The set of states for opinion formation problems can be defined in different ways. When looking for some seminal papers, where only two possible opinions were under consideration we can observe that while some figures from Sznajd paper [3] can be understood like the application of $\{0, 1\}$ set, the Stauffer's paper uses rather the physical spin analogy, corresponding to the set $\{-1, 1\}$. In the situation, where the larger number of states is considered, like in our paper, we have the possibility to relate the representation with some context. The good example is here the understanding of different opinions within the frame of Nolan's diagram, presented in the further part of this section. Therefore, we want to discuss the particular representation in detail only in the connection with the transition rule described below.

In the paper we generally follow the approach used in our earlier papers (see [12,13] and references therein). However, since we are going to deal only with the problems of opinion formation and not with the mandate assignment, we present here the details related to the system definition and update.

In our earlier papers we often used different methods of system update such as especially the Glauber's and the Stauffer's [14] rules were of interest. Here, we concentrate on the former one. There are several reasons for this choice. The main one is that we are going to study different topologies and the Stauffer's rules are only of use in the case of two dimensions. When studying the samples with higher dimensionalities we would have to define the new rules and to discuss initially their properties and usefulness. Additionally, the Stauffer's rule is purely deterministic and always leads to stable states, sometimes producing strange patterns when approaching this stability (see e.g. [12]). Typically, such patterns are quantified as the percentage of support for given opinion as well as about secondary characteristics, like e.g. disproportionality indices.

In contrast to the above remarks, the Glauber's rule can be applied without any modification to arbitrary topology although some characteristic values of parameters can be different for different dimensionalities. The basic property of Glauber's rule is the inclusion of a random factor when trying to change the state of cell. The probability is given by the formula 1.

$$P = \frac{1}{1 + \exp(\frac{-\Delta E}{k_B T})} = \frac{1}{1 + \exp(-\beta \Delta E)}. \tag{1}$$

As it can be seen, the probability depends on two values that have specific physical meaning: the temperature and the energy difference. The temperature can be treated as the parameter, and represented in one of two equivalent physical forms shown in Eq. 1 (T or β, $\beta = \frac{1}{k_B T}$). The k_B value, the Boltzmann constant, can be in our calculations neglected ($k_B = 1$).

The value of energy change depends on the set of values describing the state of every cell and on details of the approach, usually reflecting some physically-based model. We will follow the typical Ising-like approach where the energy change can be calculated as:

$$\Delta E = -J * \sum_j \left(s_{i,updated} * s_{j,updated} - s_{i,old} * s_{j,old} \right). \tag{2}$$

In the formula 2 s_i and s_j are the cell states as selected from the set $\{-1,1\}$ and J is the constant characterizing the intensity of interactions, here assumed to equal 1. The index *updated* corresponds to the state after update while *old* is the state before update. The above formula has to be, however, interpreted differently, depending on the number of possible cell states. In our calculations we use systems with two and four states. While the two-opinion system can be well described by the $\{-1,1\}$ set, four-opinion system can be understood e.g. within the frame of the Nolan's diagram (see [12]).

Then, every opinion can be interpreted as containing two opinions concerning two different aspects (originally social and economic). Such an approach can lead also to 2 possible methods of energy change consideration (only for 4 opinions). According to the first one the difference does not depend on the particular value for single aspect. It means the energy difference will be the same independently whether we pass to the state differing in the one or two aspects. As an example we can show the update of the single cell from the state $s_{i,old} = (1,1)$. ΔE would the same for all changed states of cell, those with one aspect different ($s_{i,updated} = (-1,1)$, $s_{i,updated} = (1,-1)$ and for two aspects different $s_{i,updated} = (-1,-1)$. We will further refer to this approach as to the "reduced energy difference". According to the second approach, the difference in the above example is for $s_{i,updated} = (-1,-1)$ twice as high as for $s_{i,updated} = (-1,1)$ since the opinions differs in both aspects. This approach will be referred as "two-mode option".

In the paper we study the dependence of the system on four topologies. Three of them are typical array-like geometries defined by using the n-dimensional rectangular system and we choose $n = 2,3,4$. The difference between them lies in the fact that Their sites have still increasing number of neighbors, such that opinions can possibly change faster. The fourth topology is the well-known Barabasi-Albert (BA) network [15], where the number of neighbors is described by a power-law distribution. The BA network is well recognized as describing the structure of real world as well as virtual social networks, see e.g. [16,17].

All calculations in the paper are performed for the synchronous update. All results are obtained after averaging of harmonics for 200–500 independent runs.

3   Results and Conclusions

All results presented in this section are shown in the form of Fourier spectra since these can be easily compared to the curves in Fig. 1. The spectra are normalized in such a way that the intensity of first harmonics always equals 1.

In the Fig. 3 we show the comparison of spectra for different number of opinions and different topologies. Since the result of simulation can depend also on

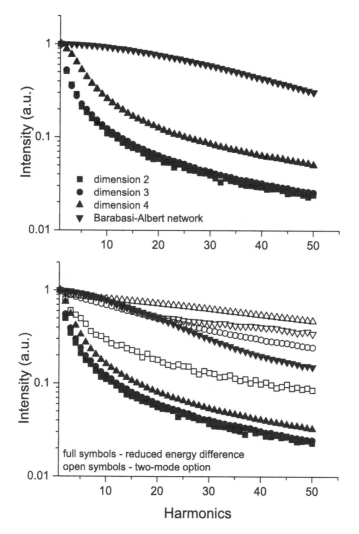

Fig. 3. Typical Fourier spectra obtained for 2-opinion (upper plot) and 4-opinion (lower plot) simulations. Symbols are for both cases same for the same topologies. In the lower plot two methods of energy calculation are distinguished.

the temperature used in Glauber scheme (see Eq. 1) we choose the same value of this parameter for all curves, $T = 1.0$.

The crucial observation is that by using the Glauber update we can produce almost arbitrary spectrum of opinion support. The general observation is that the typical array-like topologies lead to a power-law character of presented curves. It is quite clear, especially in the upper plot of Fig. 3 where the only difference between those curves is the small shift for the $4 - dimensional$ case. Since the dependencies are normalized for the amplitude of first harmonic the breaking of power-law scaling for some small harmonic index ($n \approx$ 5–10) can also be noticed. When considering $4 - opinion$ scheme (Fig. 3, lower plot) we can observe similar effects. The dependencies for array-like topologies can always be described by a power-law functions but there often exist a breaking point, beyond which the power-law scaling does not longer hold. The situation differs for BA topology where there is no clear model to generally describe all three shown cases. While for the upper plot the dependency can be easily scaled by the exponential function, for the lower one there exist either the breaking of exponential scaling or the scaling best described by the form $exp(x^{-3/4})$.

An important observation connected to the Fig. 3 is that all presented dependencies are relatively smooth and cannot reproduce the typical rough character of spectra in Fig. 1. It should however be noticed that a choice of temperature can strongly influence the results. Some characteristics for the $4 - opinion$ simulation and for different temperatures are shown in Figs. 4 and 5.

Due to some similarities between array-like topologies we choose only square array, i.e. $dimension = 2$ for presentation. The results are averaged over 200 simulations. It is interesting that the increase of number of runs (we tested it up to 500 independent runs) leads only to the slight smoothing of presented dependencies and does not influence the visible maxima. We should draw the attention of the reader to the fact that all plots are shown here, contrary to the Fig. 3, in the linear scale.

Plots in Figs. 4 and 5 confirm only partially earlier observations made for $T = 1$ and we observe however some new behavior. The case described earlier ($T = 1.0$) turns out to present the dependence which is typical for low temperatures. These values of temperatures correspond, according to Eq. 1 to the situations where the probabilities of opinion change are relatively high so we can say about the significant volatility of agents composing the sample. We can also observe that for increasing temperatures the spectra for square array present somehow "monotonic" behavior. It means that the curves for successive, increasing temperatures lies one over another what can be understood as pursuing the independence of the amplitude on the number of harmonic. It is however not reached for the parameters range presented. One should also point out the change of scaling of the curves. Being initially, for low temperatures, of power-law character it changes to the exponential or either linear form often with some breaking point.

We can expect that for the Barabasi-Albert topology some changes may occur for lower temperatures, since, due to generally larger number of neighbors, the value of energy difference in the Eq. 1 can be greater. It turns out that the shapes of curves strongly depend on the method how the total energy change is computed as well as it almost arbitrary form of scaling can describe the observed dependencies. It can be most clearly visibly in the Fig. 4 when with increasing temperatures for BA topology we start from linear scaling through power-law-one for $T = 6.0$ and finish once more with the linear one. It can be also observed that a roughness of spectrum is always stronger for square array topology, as

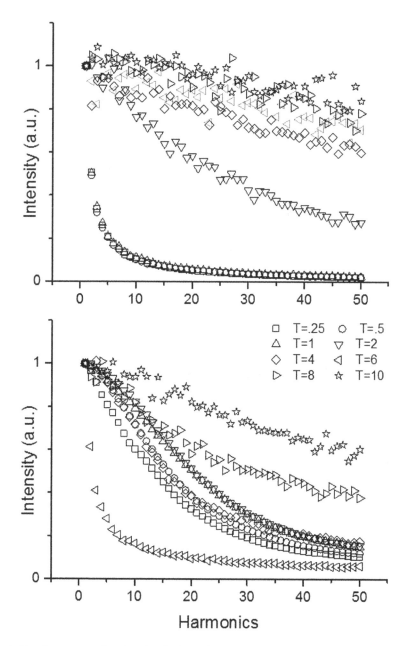

Fig. 4. Fourier spectra for the simulation of 4 opinions with the reduced energy difference. Upper plot corresponds to the dimension 2 while the lower one to the Barabasi-Albert network. The description of temperatures concerns both plots.

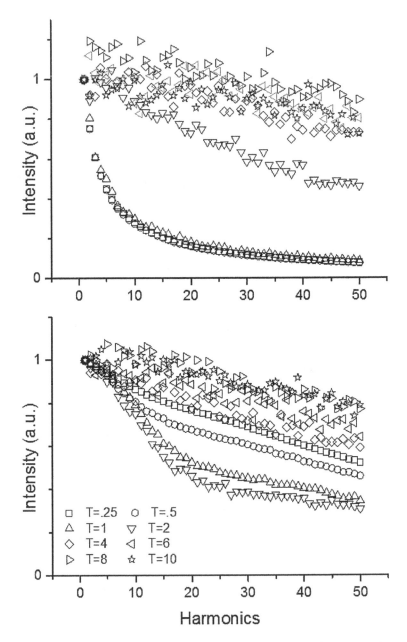

Fig. 5. Fourier spectra for the simulation of 4 opinions with the two-mode option. Upper plot corresponds to the dimension 2 while the lower one to the Barabasi-Albert network. The description of temperatures concerns both plots.

compared with the BA one and for the reduced energy mode as compared to the two-mode one.

Considering the visible maxima, we have to emphasize the significance of high temperatures. It seems that we have to once more take into account especially the BA topology in Fig. 4 where the $6 - th$, $11 - th$ and $19 - th$ harmonics have visibly higher amplitudes that neighboring ones.

The original data concerning the opinion formation in the community may present different behaviors as described by means of some mathematical approaches. We show that the Cellular Automata method based on some physical notions with the stochastic Glauber update can reproduce that different behaviors when considering some simple steering parameters especially for the, in our opinion more realistic, Barabasi-Albert topology. It seems that most interesting areas of future research would be to find the method of more precise description of real results and to find the sociological interpretation of simulation parameters.

References

1. Clifford, P., Sudbury, A.: A model for spatial conflict. Biometrika **60**, 581–588 (1973)
2. Galam, S.: Minority opinion spreading in random geometry. Eur. Phys. J. B **25**, 403–406 (2002)
3. Sznajd-Weron, K., Sznajd, J.: Opinion evolution in closed community. Int. J. Mod. Phys. C **11**, 1157 (2000)
4. Grabowski, A., Kosinski, R.: Ising-based model of opinion formation in a complex network of interpersonal interactions. Physica A **361**, 651–664 (2006)
5. Iniguez, G., Barrio, R.A., Kertesz, J., Kaski, K.K.: Modelling opinion formation driven communities in social networks. Comput. Phys. Commun. **182**, 1866–1869 (2011). Computer Physics Communications Special Edition for Conference on Computational Physics Trondheim, Norway, 23–26 June 2010
6. Bordogna, C.M., Albano, E.V.: Dynamic behavior of a social model for opinion formation. Phys. Rev. E **76**, 061125 (2007)
7. Schmittmann, B., Mukhopadhyay, A.: Opinion formation on adaptive networks with intensive average degree. Phys. Rev. E **82**, 066104 (2010)
8. Krause, S.M., Bornholdt, S.: Opinion formation model for markets with a social temperature and fear. Phys. Rev. E **86**, 056106 (2012)
9. Sobkowicz, P., Kaschesky, M., Bouchard, G.: Opinion formation in the social web: agent-based simulations of opinion convergence and divergence. In: Cao, L., Bazzan, A.L.C., Symeonidis, A.L., Gorodetsky, V.I., Weiss, G., Yu, P.S. (eds.) ADMI 2011. LNCS, vol. 7103, pp. 288–303. Springer, Heidelberg (2012)
10. Prettejohn, B.J., McDonnell, M.D.: Effect of network topology in opinion formation models. In: Guttmann, C., Dignum, F., Georgeff, M. (eds.) CARE 2009/2010. LNCS, vol. 6066, pp. 114–124. Springer, Heidelberg (2011)
11. Hammer, R.J., Moore, T.W., Finley, P.D., Glass, R.J.: The role of community structure in opinion cluster formation. In: Glass, K., Colbaugh, R., Ormerod, P., Tsao, J. (eds.) Complex 2012. LNICST, vol. 126, pp. 127–139. Springer, Heidelberg (2013)

12. Gwizdałła, T.M.: The dynamics of disproportionality index for cellular automata based sociophysical models. In: Sirakoulis, G.C., Bandini, S. (eds.) ACRI 2012. LNCS, vol. 7495, pp. 91–100. Springer, Heidelberg (2012)

13. Sendra, N., Gwizdałła, T.M.: Sznajd model with memory. In: Was, J., Sirakoulis, G.C., Bandini, S. (eds.) ACRI 2014. LNCS, vol. 8751, pp. 349–356. Springer, Heidelberg (2014)

14. Stauffer, D., Sousa, A.O., de Oliveira, S.: Generalization to square lattice of Sznajd sociophysics model. Int. J. Mod. Phys. C **11**, 1239–1245 (2000)

15. Barabasi, A.L., Albert, R.: Emergence of scaling in random networks. Science **286**, 509–512 (1999)

16. Schnettler, S.: A structured overview of 50 years of small-world research. Soc. Netw. **31**, 165–178 (2009)

17. Lattanzi, S., Panconesi, A., Sivakumar, D.: Milgram-routing in social networks. In: Srinivasan, S., Ramamritham, K., Kumar, A., Ravindra, M.P., Bertino, E., Kumar, R. (eds.) Proceedings of the 20th International Conference on WWW, pp. 725–734. ACM (2011)

Cellular Automata Model of Electrons and Holes Annihilation in an Inhomogeneous Semiconductor

A.E. Kireeva$^{(\boxtimes)}$ and K.K. Sabelfeld

Institute of Computational Mathematics and Mathematical Geophysics SB RAS,
Pr. Lavrentjeva, 6, Novosibirsk, Russia
kireeva@ssd.sscc.ru, karl@osmf.sscc.ru

Abstract. An asynchronous CA model of annihilation of electrons and holes in an inhomogeneous semiconductor is presented. The model is based on the Monte Carlo algorithm of electron-hole annihilation. CA model allows us to study the dynamics of electron-hole spatial distribution. The annihilation process is simulated for different values of the modeling parameters. The spatial distributions of particles are analyzed. It is found out that a segregation, i.e., a spatial separation of electron and hole clusters occurs. This happens under certain conditions on the diffusion and tunneling rates. In addition, the parallel implementation of the CA model of the annihilation is performed using OpenMP standard. The parallel implementation makes it possible to perform averaging over a rich ensemble of initial distributions of particles.

Keywords: Cellular automata · Parallel implementation · Electron-hole annihilation · Semiconductor · Recombination centers · Radiative intensity

1 Introduction

Semiconductors are a foundation of modern electronics and optoelectronics industry. Semiconductor devices and integrated circuits are used in computer microprocessors, communications equipment, lighting equipment and other electronic devices which have significant impact on science and economy development. In semiconductors a lot of phenomena arise that can not be observed in metals and dielectrics. Therefore, a great attention of academia and industry is attracted to study semiconductors.

Gallium nitride (GaN) is a direct wide band gap semiconductor, promising applications for developing of high-frequency, thermostable and high-power semiconductor devices. Moreover, this semiconductor and its solid solutions (InGaN) and (AlGaN) are considered as one of the most perspective materials in the fields of short wavelength optoelectronic devices [1]. Gallium nitride is

Supported by Russian Science Foundation under Grant 14-11-00083.

V. Malyshkin (Ed.): PaCT 2015, LNCS 9251, pp. 191–200, 2015.
DOI: 10.1007/978-3-319-21909-7_18

intensively studied by numerous research groups both by experimental methods [2,3] and by means of computer simulation [4,5].

In [4,5], a Monte Carlo method based on spatially inhomogeneous nonlinear Smoluchowski equations with random initial distribution density is used to simulate the annihilation of spatially separate electrons and holes in a disordered semiconductor. The main idea of the Monte Carlo method for solving spatially inhomogeneous Smoluchowski equations lies in the probabilistic interpretation of the evolution of the interacting particles as a Markov chain [6]. Based on the Monte Carlo algorithm [5] an asynchronous cellular automaton model of electrons and holes annihilation in the inhomogeneous semiconductor is developed.

Cellular automaton is a discrete dynamical system whose behavior is defined by local rules [7]. Cellular automaton includes a set of cells corresponding to a space. Cells have states which correspond to the elements of the system under study. Cell states are changed according to the rules imitating the system behaviour. The rules define new cell states depending on the states of their neighbour cells. Local rules allow of describing complex multicomponent systems whose behavior is determined by the local behavior of their constituent elements [8,9]. The main advantage of cellular automaton approach for simulation of the electrons and holes annihilation in the semiconductor is the possibility to model and study in great details the spatial distribution of ensembles of interacting particles progressing in time.

Simulation of electrons and holes annihilation in inhomogeneous semiconductors is a highly challenging problem because the particle kinetics for a very long time (up to about 10^{14} nanoseconds) and for large number of interacting particles is desired. Moreover, since stochastic processes are investigated, the averaging over a large ensemble of initial distributions of particles is required to obtain reliable results. Therefore, the purpose of this paper is developing of a parallel implementation of cellular automaton model of the electrons and holes annihilation and estimation of its efficiency.

The paper consists of Introduction, three sections and Conclusion. In the first section a mechanism of electron-hole annihilation is described and a formal definition of the cellular automaton model of this process is given. The second section presents the parallel implementation of the cellular automaton and its efficiency analysis. In the third section simulation results are discussed.

2   The Model of Annihilation of Electrons and Holes in a Semiconductor

2.1   The Mechanism of Annihilation of Electrons and Holes in a Semiconductor

In [4,5], a mechanism of annihilation of electrons and holes in an inhomogeneous semiconductor is described as follows. On the gallium nitride semiconductor a high frequency pulsed laser induces generation of an instant electron excess. The electrons and holes can recombine with each other radiatively with emission of a light quantum, a photon, or nonradiatively via capturing in recombination centers. A recombination center is an immobile site of the semiconductor usually formed by defects. Initially, a part of recombination centers is free for capturing electrons, and the other part is free to capture holes. When an electron and a hole meet each other in a recombination center, diffusively or via tunneling, they recombine nonradiatively, and the center becomes free. In addition, electrons and holes are generally able to diffuse in the semiconductor.

In [4,5], distributions of electrons, holes and recombination centers in a sample X with densities $\rho_n(\mathbf{r}; t)$, $\rho_p(\mathbf{r}; t)$, $\rho_N(\mathbf{r})$ are considered, here $\mathbf{r}$ is a spatial position, t is a time moment. A part of the recombination centers with a density $\rho_{N_n}(\mathbf{r}; t)$ are in the state waiting for capturing an electron, while the remaining centers with a density $\rho_{N_p}(\mathbf{r}; t) = \rho_N(\mathbf{r}) - \rho_{N_n}(\mathbf{r}; t)$ are waiting for capturing a hole. The total density of all recombination centers $\rho_N(\mathbf{r})$ keeps constant with time, while the other densities are changed in time according to the following differential equations:

$$\frac{\partial \rho_n(\mathbf{r}; t)}{\partial t} = D_n(\mathbf{r}) \Delta \rho_n(\mathbf{r}; t) - \rho_n(\mathbf{r}; t) \int B(|\mathbf{x}|) \rho_p(\mathbf{r}+\mathbf{x}; t) d\mathbf{x}$$

$$- D(\mathbf{r}) \rho_n(\mathbf{r}; t) \rho_p(\mathbf{r}; t) - \rho_n(\mathbf{r}; t) \int b_n(|\mathbf{x}|) \rho_{N_n}(\mathbf{r}+\mathbf{x}; t) d\mathbf{x};$$

$$\frac{\partial \rho_p(\mathbf{r}; t)}{\partial t} = D_p(\mathbf{r}) \Delta \rho_p(\mathbf{r}; t) - \rho_p(\mathbf{r}; t) \int B(|\mathbf{x}|) \rho_n(\mathbf{r}+\mathbf{x}; t) d\mathbf{x} \tag{1}$$

$$- D(\mathbf{r}) \rho_p(\mathbf{r}; t) \rho_n(\mathbf{r}; t) - \rho_p(\mathbf{r}; t) \int b_p(|\mathbf{x}|) \rho_{N_p}(\mathbf{r}+\mathbf{x}; t) d\mathbf{x};$$

$$\frac{\partial \rho_{N_n}(\mathbf{r}; t)}{\partial t} = -\rho_n(\mathbf{r}; t) \int b_n(|\mathbf{x}|) \rho_{N_n}(\mathbf{r}+\mathbf{x}; t) d\mathbf{x} + \rho_p(\mathbf{r}; t) \int b_p(|\mathbf{x}|) \rho_{N_p}(\mathbf{r}+\mathbf{x}; t) d\mathbf{x}.$$

The relevant terms of these equations have the following physical sense. Electron-hole pairs tunnel with the rate $B(|\mathbf{x}|) = B_0 \cdot \exp(-|\mathbf{x}|/a_{np})$, where $|\mathbf{x}|$ is a distance between an electron and a hole, a_{np} is a characteristic distance of the electron-hole interaction. Analogously, $b_n(|\mathbf{x}|) = b_{n0} \cdot \exp(-|\mathbf{x}|/a_{nN_n})$ and $b_p(|\mathbf{x}|) = b_{p0} \cdot \exp(-|\mathbf{x}|/a_{pN_p})$. Further, $D_n(\mathbf{r})$ is an electron diffusion coefficient, $D_p(\mathbf{r})$ is a hole diffusion coefficient, and $D(\mathbf{r}) = D_n(\mathbf{r}) + D_p(\mathbf{r})$. At the initial time, the electrons, holes and recombination centers are randomly and uniformly distributed with a mean number concentrations C_{n0}, C_{p0}, C_{N_n0}, $C_{N0} \in \mathbb{Z}$, which are the numbers of electrons, holes, recombination centers for electrons and all recombination centers, respectively. The total number of all recombination centers is $C_{N0} = C_{N_n0} + C_{N_p0}$, where C_{N_n0} is the number of recombination centers free for electrons, and C_{N_p0} is the number of the recombination centers free for holes.

The photon flux is defined by the following formula:

$$\phi(t) = \left\langle \int \frac{1}{|X|} d\mathbf{r} \int B(|\mathbf{x}|) \rho_n(\mathbf{r}; t) \rho_p(\mathbf{r}+\mathbf{x}; t) d\mathbf{x} \right\rangle, \tag{2}$$

where the angle brackets stand for the mathematical expectation with respect to the initial random distribution of electrons, holes and recombination centers, and $|X|$ is the sample area.

In this paper, based on the [4,5] a cellular automata model of the electron-hole annihilation is developed.

2.2 The Cellular Automata Model of Annihilation of Electrons and Holes in a Semiconductor

A cellular automaton (CA) is a discrete dynamical system consisting of a set of cells [9,10]. A cell is determined as a pair $(u, \mathbf{x})$, where $u \in A$ is a state of a cell from a set of admissible in a model states, which is named *cells state alphabet* A. The second element of the pair $\mathbf{x} \in X^d$ is a coordinate of a cell in a discrete space. All possible coordinates in a model space is named *coordinate set X*. A set of cells with different coordinates is named *cellular array*. Simulation process consists in the calculation of the cells states by special rules depending on own states of cells and states of interacting with them cells. These rules are named *substitutions*. A set of all rules specified in a model is named an *operator* Θ. The interacting cells are specified by a *template T*. The template can be defined either as the set of cells fixed related to a central cell, for example, adjacent cells, or as a set of random cells of cellular array. The cells states are updated according to a *operation mode* μ. There are synchronous σ and asynchronous α modes. In the synchronous mode (σ), the operator is applied to all cells of a cellular array, all being updated simultaneously. The asynchronous (α) mode prescribes the operator to be applied to all cells of a cellular array in random order, all cell states being updated immediately.

According to the definition of cellular automata (CA) given in [9], CA model of the electrons and holes annihilation can be defined by the following notion:

$$\aleph = \langle A, X, \Theta, \alpha \rangle \tag{3}$$

Based on [5], A, X and Θ are defined as follows. The state alphabet is $A = \{n, p, N_n, N_p, \emptyset\}$, where n denotes an electron, p is a hole, N_n denotes an unoccupied recombination center being able to capture an electron, N_p is an occupied by an electron recombination center being able to capture a hole and $\emptyset$ denotes an empty site. The coordinates set $X = \{\mathbf{x} = (i, j),\ i = 1 \ldots Size_i,\ j = 1 \ldots Size_j\}$ is a two-dimensional square discrete space. Size of the cellular array corresponds to the semiconductor surface size measured in nanometers (nm).

The operator is $\Theta = R\{\theta_1, \theta_2, \theta_3, \theta_4\}$. The symbol R denotes a random order of substitutions choice. The substitutions (4) correspond to the following events: θ_1 is a radiative electron-hole recombination, θ_2 is a capturing of an electron by an empty recombination center, θ_3 is a capturing of a hole by a recombination center filled with an electron, θ_4 is an electron diffusion on an empty site or a hole, or an empty recombination center:

$$
\begin{aligned}
\theta_1(\mathbf{x}) &: \{(n, \mathbf{x}), (p, \varphi(\mathbf{x})\} \xrightarrow{p_1 \cdot \omega_1} \{(\emptyset, \mathbf{x}), (\emptyset, \varphi(\mathbf{x})\}, \\
\theta_2(\mathbf{x}) &: \{(n, \mathbf{x}), (N_n, \varphi(\mathbf{x})\} \xrightarrow{p_2 \cdot \omega_2} \{(\emptyset, \mathbf{x}), (N_p, \varphi(\mathbf{x})\}, \\
\theta_3(\mathbf{x}) &: \{(p, \mathbf{x}), (N_p, \varphi(\mathbf{x})\} \xrightarrow{p_3 \cdot \omega_3} \{(\emptyset, \mathbf{x}), (N_n, \varphi(\mathbf{x})\}, \\
\theta_4(\mathbf{x}) &: \{(n, \mathbf{x}), (u, \psi(\mathbf{x})\} \xrightarrow{p_4 \cdot 1} \{(\emptyset, \mathbf{x}), (u', \psi(\mathbf{x})\},
\end{aligned}
\tag{4}
$$

$$u' = \begin{cases} n, & \text{if } u = \emptyset, \\ \emptyset, & \text{if } u = p, \\ N_p, & \text{if } u = N_n, \end{cases}$$

where, $\varphi(\mathbf{x}) \in T_{rnd} = \{\mathbf{y} = rand(X), \mathbf{y} \neq \mathbf{x}\}$ is a cell interactive with the cell $\mathbf{x}$, $\varphi(\mathbf{x})$ being randomly selected from the set X. Analogously, $\psi(\mathbf{x}) \in T_4 = \{(i, j-1), (i+1, j), (i, j+1), (i-1, j)\}$ is a neighbour of the cell $\mathbf{x}$ randomly chosen by the template T_4, being a cross with a center in the cell $\mathbf{x}$. An electron diffusion coefficient D_n is considerably greater than a hole diffusion coefficient D_p, therefore a hole diffusion is neglected in the CA model.

A performance of Θ consists in the following. One of the substitutions θ_l, $l = 1, 2, 3, 4$ is chosen with probability p_l, calculated by the formula (5):

$$\begin{aligned} & p_l = \lambda_l / \lambda, \ l = 1, \ldots, 4, \ \lambda = \sum_{l=1}^{4} \lambda_l, \\ & \lambda_1 = C_n \cdot C_p \cdot B_0 \cdot exp\left\{-r_{np}^{min}/a_{np}\right\}, \\ & \lambda_2 = C_n \cdot C_{N_n} \cdot b_{n0} \cdot exp\left\{-r_{nN_n}^{min}/a_{nN_n}\right\}, \\ & \lambda_3 = C_p \cdot C_{N_p} \cdot b_{p0} \cdot exp\left\{-r_{pN_p}^{min}/a_{pN_p}\right\}, \\ & \lambda_4 = C_n \cdot D, \end{aligned} \tag{5}$$

where C_u, $u \in \{n, p, N_n, N_p\}$ is the number of particles of type u in the cellular array, B_0 is a rate of electron-hole recombination, b_{n0} is a rate of an electron recombination in one of the nonradiative centers, b_{p0} is a rate of a hole recombination in one of the nonradiative centers filled by an electron, D is an electron diffusion coefficient, r_{uv}^{min} is a minimum distance between all particles of type u and v (where $u \in \{n, p\}$, $v \in \{p, N_n, N_p\}$), analogously, a_{uv} is a characteristic distance of interaction between particles of type u and v.

For the chosen substitution two cells ($\mathbf{x}$ and its neighbour φ or ψ) with the relevant particles are randomly selected. A realization of the substitution occurs with a probability $\omega_l = exp\left((r_{uv}^{min} - r_{uv})/a_{uv}\right)$, $l = 1, \ldots, 4$, where u, v correspond to types of particles in the substitution. If a random number $rand < \omega_l$, $rand \in (0, 1)$, then states of chosen cells are replaced by the states in the right part of the corresponding substitution. As mentioned above, an electron and a hole recombine with an emission of photon, therefore, the number of photons ϕ is calculated during the simulation as follows. In the case of substitution θ_1 realization or θ_4 for $u = p$ realization, the number of photons is increased by 1.

To compare simulation results with an experiment two time counters are included in the CA model: a local time counter τ and a global time counter t. The local time counter is increased after each application of the operator Θ by $\Delta\tau = -ln(rand_1)/\lambda$, $rand_1 \in (0, 1)$ being a random number. The global time counter is increased after a number of local time steps by $\Delta t = t_0 \cdot q^k$, where t_0 is an initial time step, q is a parameter responsible for a global step duration, k is a global step number.

The main characteristics experimentally observed are the particle concentrations and a radiative intensity. In the CA model, the particle concentration C_u is calculated as the number of particles of type $u \in \{n, p, N_n, N_p\}$ in the cellular array. The radiative intensity is calculated as the number of photons obtained during a time step t_k divided by a length of the time interval: $I(t_k) = \phi/(t_k - t_{k-1})$.

At the initial time, electrons, holes and recombination centers are randomly and uniformly distributed. So, we deal with stochastic initial conditions, so the annihilation process should be considered as a stochastic process. Therefore, to obtain reliable values of statistical characteristics, an averaging over a sufficiently large ensemble of initial distributions of particles is required. Simulations for a large number of different initial conditions demand a lot of computer time. A parallel implementation of these tasks allows us to essentially decrease a computational time.

3   Parallel Implementation of the CA Model of Electrons and Holes Annihilation

A conventional approach of a parallel implementation of a cellular automata is a decomposition of a cellular array into subdomains. However, this method is inefficient for the CA model of electrons and holes annihilation owing to the possibility of an interaction of particles distributed over the whole cellular array. A simultaneous execution of independent tasks for different initial conditions and an averaging of obtained results is a reasonable approach in our case.

Some optimizations of the CA algorithm and its program implementation have been performed. A calculation of minimum distances r_{uv}^{min} between all particles is computationally expensive operation. Therefore the calculation of minimum distance r_{uv}^{min} between all particles of type u and v is performed only at the initial time or when either a particle of type u or v corresponding to the r_{uv}^{min} has been deleted. When a new particle of type u emerges, distances between the new particle and all particles of type v are calculated. If a minimum of these distances is less than a current value of r_{uv}^{min}, then r_{uv}^{min} is assigned to the new minimum distance. In addition, optimizations of arithmetic operations and memory usage have been done.

Parallel implementation of the simulation process for different initial conditions is performed using OpenMP standard. A set of tasks with different initial particles allocations is distributed between available threads. The threads simultaneously calculate values of particle concentrations and radiative intensity. On each global time step each thread summarizes the obtained values of concentrations and intensity in its own array. After all global steps, master thread summarizes the values calculated by all threads.

To estimate the efficiency of the parallel implementation of the CA algorithm, computing experiments have been performed for the following parameter values: the rates $B_0 = 0.04 \ ns^{-1}$, $b_{n0} = b_{p0} = 0.02 \ ns^{-1}$, the characteristic interaction distances $a_{np} = 4 \ nm$, $a_{nN_n} = a_{pN_p} = 2 \ nm$, the initial particle concentrations $C_n(0) = C_p(0) = 10000$, $C_{N_n}(0) = 5000$, $C_{N_p}(0) = 0$, the diffusion coefficient

$D = 1 \ nm^2 \cdot ns^{-1}$, the surface size $Size_i = Size_j = 1000 \ nm$, the initial time $t_0 = 0.5 \ ns$. The simulations are performed for 1024 different initial conditions for times up to $t_{fin} = 10^8 \ ns$ on cluster "NKS-30T" of the Siberian Supercomputer Center SB RAS.

Table 1 presents a speed-up $S(th) = T(1)/T(th)$, and a strong scaling efficiency $Q(th) = T(1)/(T(th) \cdot th)$ of the parallel implementation of ℵ, where $T(1)$ is the computation time obtained for the sequential version of the CA algorithm, $T(th)$ is computation time obtained for the distribution of the tasks between th threads.

Table 1. Parallel implementation characteristics of the CA algorithm.

th	1	2	4	6	8	10	12
$T(th), hour$	25.8	13.7	7.2	4.7	3.7	2.9	2.5
$S(th)$	1	1.9	3.6	5.4	6.9	8.9	10.5
$Q(th)$	1	0.94	0.89	0.9	0.87	0.89	0.87

The table shows that the parallel implementation efficiency amounts to 0.87 for the distribution of the tasks between 12 threads.

4   Simulation Results

The annihilation of electrons and holes in an inhomogeneous semiconductor is simulated by means of ℵ with different values of the modeling parameters: B_0, b_n, b_p, a_{np}, a_{nN_n}, a_{pN_p}, D, t_0, for various initial particle concentrations $C_n(0)$, $C_p(0)$, $C_{N_n}(0)$, $C_{N_p}(0)$ and different cellular array size $Size_i$, $Size_j$. Periodic boundary conditions are used in the model.

In the course of the simulation, the following characteristics of the electron-hole annihilation process are obtained on each global time step $t \in [t_0; t_{fin}]$: the particle densities per one cell $\rho_u(t) = C_u(t)/(Size_i \cdot Size_j)$, $u \in \{n, p, N_n, N_p\}$ and the radiative intensity $I(t)$, with $C_u(t)$ and $I(t)$ being calculated as described above in Sect. 2.

Computing experiments are carried out in three cases: (1) pure electron-hole annihilation without recombination centers, (2) electron-hole annihilation in the vicinity of recombination centers, (3) electron-hole annihilation in the vicinity of recombination centers and diffusion of electrons. Consider for example annihilation dynamics for the following parameter values: $B_0 = 0.04 \ ns^{-1}$, $b_{n0} = b_{p0} = 0.02 \ ns^{-1}$, $a_{np} = 4 \ nm$, $a_{nN_n} = a_{pN_p} = 2 \ nm$, $D = 1 \ nm^2 \cdot ns^{-1}$, $t_0 = 0.5 \ ns$, $C_n(0) = C_p(0) = 400$, $C_{N_n}(0) = 200$, $C_{N_p}(0) = 0$, $Size_i = Size_j = 200 \ nm$, where in the first case $b_{n0} = b_{p0} = 0 \ ns^{-1}$, $C_{N_n}(0) = 0$, $D = 0 \ nm^2 \cdot ns^{-1}$ and in the second case $D = 0 \ nm^2 \cdot ns^{-1}$.

In the case of pure annihilation, the particle dynamics is as follows. At the beginning, electrons and holes are randomly and uniformly distributed over the surface (Fig. 1a). The near particles interact more likely than distant particles. Therefore, during the simulation, all pairs of electrons and holes, situated on the close distance, disappear, that causes a spatial separation of electrons and holes and cluster formation (Fig. 1b). Further, particles on the cluster boundaries are annihilated. As a result, the clusters are slowly decreasing (Fig. 1c).

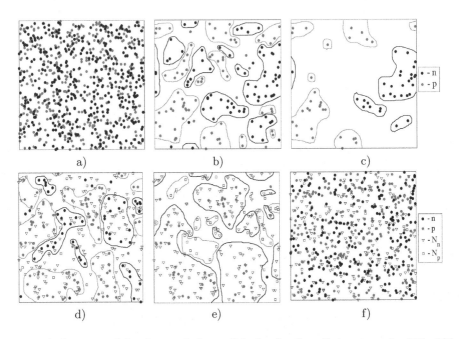

Fig. 1. A character of the electron-hole annihilation for the cellular array size 200×200 in the case of a) pure annihilation for the initial time $t = 0.5$ ns; b) pure annihilation for $t = 374$ ns; c) pure annihilation for $t = 1.06 \cdot 10^5$ ns; d) annihilation with presence of the recombination centers for $t = 374$ ns; e) annihilation in the vicinity of the recombination centers for $t = 6.6 \cdot 10^4$ ns; f) annihilation in the vicinity of the recombination centers and electron diffusion for $t = 35$ ns.

In the case of electron-hole annihilation in the vicinity of the recombination centers, a formation of the electrons and holes clusters occur as well (Fig. 1d). However, clusters are more rarefied than in the case of pure annihilation, because electrons and holes inside the clusters are captured by the recombination centers. Moreover, some separation of unoccupied recombination centers and centers with electrons emerges (Fig. 1e). In the case of diffusion presence for the parameter values mentioned above, particle separation did not occur (Fig. 1f) due to continuous particle mixing. So, the initially distant electrons diffuse to the holes and all electrons and holes rapidly annihilate. However, when the diffusion rate is small compared to the tunneling recombination rate, a formation of the electrons and holes clusters may also occur.

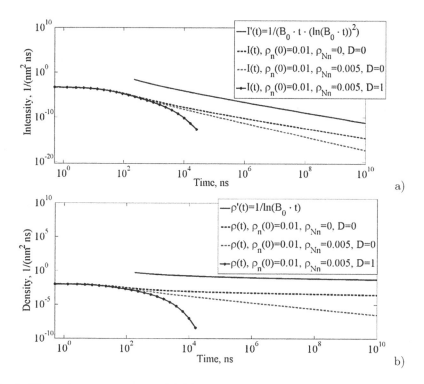

Fig. 2. The radiative intensities $I(t)$ (a) and the electron densities $\rho_n(t)$ (b), obtained by the CA simulation for the cellular array size 200×200 in three cases: 1) the pure annihilation with $\rho_n(0) = \rho_p(0) = 0.01$, 2) the annihilation in the vicinity of the recombination centers $\rho_{N_n}(0) = 0.005$ and 3) the annihilation in the vicinity of the recombination centers and electron diffusion $D = 1 \ nm^2 \cdot ns^{-1}$.

The electron densities $\rho_n(t)$ and the radiative intensities $I(t)$, obtained by the CA simulation in these three cases are presented in Fig. 2. In [4], a long-time asymptotics for the purely radiative annihilation is obtained theoretically by means of correlation analysis. The asymptotics for the electron density is $\rho'_n(t) \sim 1/ln(B_0 \cdot t)$, the asymptotics for the radiative intensity is $I'(t) \sim 1/(B_0 \cdot t \cdot ln(B_0 \cdot t)^2)$. The values of the asymptotics $\rho'_n(t)$ and $I'(t)$ agree well with the values of $\rho_n(t)$ and $I(t)$ obtained by the CA simulation. Both the recombination centers and the electron diffusion accelerate the electron-hole annihilation in comparison with the case of pure annihilation.

5   Conclusion

Based on the Monte Carlo algorithm [4,5], an asynchronous CA model of the annihilation of electrons and holes in inhomogeneous semiconductors has been developed and investigated. The developed CA model has allowed us to study the evolution of electron-hole spatial distribution in great details. Dynamics of the process has been investigated in the case of pure electron-hole annihilation, electron-hole annihilation in the vicinity of recombination centers and electron-hole annihilation in the vicinity of recombination centers and electron diffusion. It has been found out that in the case of low diffusion rates, clusters of electrons and holes are formed on the surface. The cluster sizes are slowly decreasing due to the annihilation of particles on their boundaries. In the case of high diffusion rate, the cluster formation does not occur due to continuous particle mixing.

In addition, the parallel implementation of the CA model of the electron-hole annihilation has been performed using OpenMP standard. The parallel implementation has allowed us to calculate the characteristics of the simulated process for a large ensemble of different initial distributions of the particles. The radiative intensity and electron density, obtained by $\aleph$ for the pure annihilation, agree well with the large time asymptotics for these characteristics, that confirms an accuracy of the CA model of the electron-hole annihilation.

References

1. DenBaars, S.P.: Gallium-nitride-based materials for blue to ultraviolet optoelectronics devices. Proc. IEEE **85**(11), 1740–1749 (1997). doi:10.1109/5.649651
2. Brosseau, C.-N., Perrin, M., Silva, C., Lenonelli, R.: Carrier recombination dynamics in $In_xGa_{1-x}N/GaN$ multiple quantum wells, Phys. Rev. B 88 (2010), Article ID 085305
3. Caro, M., Schulz, S., and O'Reilly, E.: Theory of local electric polarization and its relation to internal strain: Impact on polarization potential and electronic properties of group-III nitrides, Phys. Rev. B, vol. 88 (2013), Article ID 214103
4. Sabelfeld, K.K., Brandt, O., Kaganer, V.M.: Stochastic model for the fluctuation-limited reaction-diffusion kinetics in inhomogeneous media based on the nonlinear Smoluchowski equations. J. Math. Chem **53**(2), 651–669 (2015)
5. Sabelfeld, K.K., Levykin, A.I., Kireeva, A.E.: Stochastic simulation of fluctuation-induced reaction-diffusion kinetics governed by Smoluchowski equations. Monte Carlo Methods Appl. **21**(1), 33–48 (2015)
6. Sabelfeld, K.K., Kolodko, A.A.: Stochastic Lagrangian models and algorithms for spatially inhomogeneous Smoluchowski equation. Math. Comput. Simul. **61**, 115–137 (2003)
7. Toffoli, T., Margolus N.: Cellular Automata Machines: A New Environment for Modeling. MIT Press, USA (1987)
8. Weimar, J.R., Tyson, J.J.W., Layne, T.: Diffusion and wave propagation in cellular automaton model of excitable media. Physica D **55**, 309–327 (1992)
9. Bandman, O.L.: Mapping physical phenomena onto CA-models, AUTOMATA-2008. In: Adamatzky, A., Alonso-Sanz, R., Lawniczak, A., Martinez, G.J., Morita, K., Worsch, T. (eds.) Theory and Applications of Cellular Automata, pp. 381–397. Luniver Press, UK (2008)
10. Bandman, O.L.: Cellular automatic models of spatial dynamics. Syst. Inform. Methods Models Mod. Program. **10**, 59–113 (2006). (In Russian)

Constructions Used in Associative Parallel Algorithms for Directed Graphs

Anna Nepomniaschaya[(✉)]

Institute of Computational Mathematics and Mathematical Geophysics,
Russian Academy of Sciences, Pr. Lavrentieva, 6, Novosibirsk 630090, Russia
anep@ssd.sscc.ru

Abstract. The paper selects constructions that are used to implement a group of algorithms for directed graphs on a model of associative parallel systems with vertical processing (the STAR–machine). Moreover, a new implementation on the STAR–machine of Dijkstra's algorithm for finding the single–source shortest paths is proposed.

Keywords: The single-source shortest path · Spanning tree · Adjacency matrix · Transitive closure of a directed graph · Access data by contents

1 Introduction

Associative (content addressable) parallel processors of the SIMD type with bit–serial (vertical) processing and simple processing elements (PEs) perform the massively parallel search by contents and use 2D tables as the basic data structure. In particular, such an architecture is best suited for natural and efficient implementation of graph algorithms. In [4], we propose an abstract model of the SIMD type (the STAR–machine) that simulates the run of such systems at the micro level. Associative parallel algorithms are represented as corresponding procedures for the STAR–machine. In [5], we present basic associative parallel algorithms that are used to design different associative algorithms for different applications. In [6], we select a group of constructions used to represent on the STAR–machine the classical graph algorithms of Prim–Dijkstra and Kruskal for finding a minimal spanning tree (MST) of an undirected graph, and the Gabow algorithm for finding the smallest spanning tree with a degree constraint of a vertex. In [7], we select constructions being used to implement on the STAR–machine associative parallel algorithms for the dynamic edge update of an MST and for the dynamic reconstruction of an MST after deleting and after inserting a vertex along with its incident edges.

In this paper, we select constructions that are used to implement on the STAR–machine a group of algorithms for a directed graph G having n vertices, m edges, and a function $wt(e)$ that assigns a weight to every edge. This group includes Dijkstra's algorithm for finding the single-source shortest paths, the Italiano algorithms for the dynamic update of the transitive closure after inserting and after deleting an edge, and the Ramalingam algorithms for updating the shortest paths subgraph with a sink after inserting and after deleting an edge.

© Springer International Publishing Switzerland 2015
V. Malyshkin (Ed.): PaCT 2015, LNCS 9251, pp. 201–209, 2015.
DOI: 10.1007/978-3-319-21909-7_19

2 Simultaneous Finding the Single-Source Shortest Paths and Distances

We first recall the main idea of Dijkstra's algorithm [1]. It assigns temporal labels $l(v)$ for each vertex v of the given directed graph G so that $l(v) \geq dist(s, v)$, where $dist(s, v)$ is the weight of the shortest path from the source vertex s to the vertex v. The algorithm constructs a set of vertices F, where the shortest path from s to any vertex of F passes only through vertices in F. Initially, $F = \{s\}$, $l(s) = 0$ and $\forall v \notin F\ l(v) = \infty$. Let F consist of k vertices $(k < n)$ and u be the last vertex added to F. Then the $(k + 1)$-th vertex for F is defined as follows. One first defines all arcs (u, v_i), where $v_i \notin F$. Then for every vertex v_i, one determines the label $l(v_i)$. After that the vertex whose label has the minimum value is included in the set F.

The associative version of Dijkstra's algorithm [8] selects *simultaneously* both the shortest path and the distance for all vertices of G. It is given as procedure DistPath that uses, in particular, the number of bits h required for representing infinity, and the binary representation of infinity inf. It returns both the matrix $Dist$, whose every i-th row saves the distance from s to v_i, and the matrix $Paths$, whose every j-th column saves by bits $'1'$ positions of vertices included in the shortest path from s to v_j.

In this paper, we first propose a new implementation of Dijkstra's algorithm on the STAR–machine. It simplifies the execution of the procedure DistPath due to including the adjacency matrix Adj in the data structure and due to taking into account the fact that only a single arc enters every vertex in a tree. Then we select the main constructions used in this implementation.

The new implementation of Dijkstra's algorithm uses the following data structure: a Boolean matrix Adj; a matrix $Weight$ that consists of n fields having h bits each; a matrix $Dist$ described above.

This implementation of Dijkstra's algorithm uses, in particular, a slice U to save by bits $'1'$ positions of vertices that have not been included yet in the set F and a matrix $R1$ to save a field of the matrix $Weight$.

```
procedure DistPath1(Adj:table; Weight:table; s,h:integer;
    inf:word(Dist); var Dist:table; var Paths:table);
/* Here, s is the source vertex, h is the number of bits for representing
   infinity, inf is the binary representation of infinity. */
var R1,R2:table;
    U,X,Z:slice(Adj); v1:word(Dist); v2:word(Adj);
    i,k:integer;
1. Begin SET(U); U(s):='0'; k:=s;
Here, k saves the last vertex included in the set F. */
2.   WCOPY(inf,U,Dist);
3.   while SOME(U) do
4.     begin TCOPY1(Weight,k,h,R1);
5.       X:=COL(k,Adj);
6.       X:=X and U;
```

```
/* In the slice X, we save positions of vertices that do not
   belong to F, but they are adjacent to the vertex v_k. */
7.       v1:=ROW(k,Dist);
8.       ADDC(R1,X,v1,R2);
/* The result of adding dist(s, v_k) and wt(v_k, v_i) is written
   in every i-th row of R2, which corresponds to X(i) =' 1'. */
9.       SETMIN(R2,Dist,X,Z);
10.      TMERGE(R2,Z,Dist);
/* We decrease the label l(v_i) to l(v_k) + wt(v_k, v_i) in every i-th row
   of the matrix Dist which corresponds to Z(i) =' 1'. */
11.      MIN(Dist,U,X); k:=FND(X);
12.      U(k):='0';
/* A new vertex is included in F. */
13.        v2:=ROW(k,Adj);
14.        X:=CONVERT(v2);
15.        X:=X and (not U);
16.        i:=FND(X);
/* The vertex v_i is the next to the vertex v_k
   in the shortest path from s to v_k. */
17.        X:=COL(i,Paths); X(i):='1';
18.        COL(k,Paths):=X;
19.    end;
20. End.
```

Let us select the main constructions used in the procedure DistPath1.

Construction 1. (*Finding vertices adjacent to the current vertex included in the set F.*)

Let a slice U save the vertices $v_i \notin F$. Let v_k be the current vertex included in the set F. Then the vertices from U adjacent to v_k are defined as intersection of the k-th column of the matrix Adj and the slice U. Let a slice, say X, save these vertices.

Construction 2. (*Finding new labels for vertices adjacent to the current vertex included in F.*)

Let v_k be the current vertex included in F. Let a slice X save the vertices from the slice U adjacent to v_k. Then we first select the k-th field in the matrix $Weight$ and the distance from s to v_k in the matrix $Dist$. After that, we simultaneously define $l(v_i) = dist(s, v_k) + wt(v_k, v_i)$ for vertices v_i that are saved in the slice X. Let a matrix, say $R2$, save the labels $l(v_i)$.

Construction 3. (*Finding a new vertex for including in F.*)

Let v_k be the current vertex included in F. Let a slice X save the vertices adjacent to v_k. Let a matrix $R2$ save the labels of vertices adjacent to v_k. Then we first simultaneously decrease the labels $l(v_i)$ in the matrix $Dist$ to the new labels in the matrix $R2$ for the vertices that are saved in the slice X. After that, among the vertices not included in the set F, we select a vertex having the minimal label in the matrix $Dist$. This vertex is included in the set F.

Construction 4. (*Finding a new arc to include in F.*)

Let a slice U save the vertices $v_r \notin F$. Let v_k be the current vertex included in F. A vertex $v_i \in F$ forming an arc (v_i, v_k) is defined by intersection of the k-th row of the matrix Adj with the slice $not\,U$. Then we include the vertex k in the i-th column of the matrix $Paths$.

It should be noted that the procedure DistPath1 takes the same time $O(hn)$ as the procedure DistPath [8].

Now we compare three implementations of Dijkstra's algorithm on the STAR–machine. The first implementation simultaneously finds the distances from s to all vertices of G and builds a protocol of this computation. This protocol allows one to restore the shortest path from s to a given vertex of G. The next two implementations simultaneously find both the distances and the shortest paths from s to all vertices of G. However, the new implementation of Dijkstra's algorithm simplifies the selection of vertices adjacent to the current vertex included in the set F and the building of the matrix $Paths$.

3   Updating the Shortest-Paths Subgraph

In this section, we select constructions that are used to implement associative versions of the Ramalingam algorithms [13] for updating the shortest-paths subgraph.

Informally, the Ramalingam algorithms define $affective$ vertices for which the new shortest paths to the sink should be defined after deleting and after inserting an arc to the given graph G. Associative versions of these algorithms employ a data structure that includes along with the matrices Adj, $Dist$, and $Weight$ an adjacency matrix SP of the shortest paths subgraph; a matrix $Cost$ that consists of n fields having h bits each, where the weight of an arc (i, j) is written in the i-th row of the j-th field; a slice $AffectedV$ that saves with bits $'1'$ positions of all affected vertices.

We first consider constructions that are used in the associative version of the Ramalingam decremental algorithm [11]. Let us recall the main idea of this algorithm. Let an arc (i, j) be deleted from $SP(G)$ and $outdegree(i) = 0$, that is, the number of arcs $outgoing$ from the vertex i is equal to zero. At first, one determines the set of affected vertices and arcs obtained after deleting the arc (i, j) from $SP(G)$. Then affected arcs are deleted from $SP(G)$. After that for every affected vertex v_i, one computes a new shortest path to the sink s and updates $SP(G)$.

The next two constructions use a slice WS to save the candidates among which affected vertices are selected.

Construction 5. (*Initial update of slices $AffectedV$ and WS.*)

Let an arc (i, j) be deleted from matrices G and SP. Let $outdegree(i) = 0$. Then the slices $AffectedV$ and WS are set to zeros, that is, $AffectedV = \emptyset$ and $WS = \emptyset$. After that the vertex i is included into the slice WS.

Construction 6. (*Updating the matrix SP after selection of an affected vertex.*)

Let the slice $WS \neq \emptyset$. Then the position of the uppermost (first) bit $'1'$ (say k) is deleted from the slice WS and included into the slice $AffectedV$. After that all arcs entering the vertex k are simultaneously deleted from SP. Finally, the tail r of every deleted arc (r, k) is included into the slice WS if $outdegree(r) = 0$.

The following construction uses a similar idea as Construction 2.

Construction 7. (*Finding a new distance from an affected vertex to the sink.*)

Let matrices G, $Weight$, and $Dist$ be given. Let k be an affected vertex. Then, at first, by means of a slice (say Z), one saves non–affected heads of arcs outgoing from the vertex k in the matrix G. After that by means of a matrix (say $W1$), one saves the k-th field of the matrix $Weight$. Knowing the slice Z and the matrices $W1$ and $Dist$, one simultaneously defines weights of different paths from k to the sink. Finally, one selects the new distance from k to the sink and saves it in the k-th row of the matrix $Dist$.

Construction 8. (*Updating the arcs outgoing from an affected vertex.*)

Let the current slice $AffectedV$ and matrices G, SP, $Weight$, and $Dist$ be given. Let k be the current updated affected vertex. By means of the method proposed in Construction 7, one first simultaneously defines the weights of different paths from the vertex k to the sink. Then by means of a slice (say Y), one defines positions of those arcs (k, l) for which $dist_{new}(k) = wt(k, l) + dist_{old}(l)$. Finally, positions of these arcs are saved in the matrix SP.

Construction 9. (*Updating the arcs entering an affected vertex.*)

Let the current matrices G, $Cost$, and $Dist$ be given. Let k be the current updated affected vertex. One first defines positions of arcs entering the vertex k in G. Then knowing $dist_{new}(k)$ and the weights of arcs entering the vertex k in the matrix $Cost$, one determines the weights of different paths to the sink starting with the arc (r, k). After that by means of a slice one saves positions of the tails of arcs (l, k) for which $dist_{new}(l) < dist_{old}(l)$. Finally, one writes $dist_{new}(l)$ in the corresponding rows of the matrix $Dist$.

On the STAR–machine, the associative version of the Ramalingam decremental algorithm is implemented as procedure DeleteArc [11]. It takes $O(hk)$ time, where h is the number of bits for coding the infinity and k is the number of affective vertices obtained after deleting an arc.

Now, we select constructions that are used in the associative version of the Ramalingam incremental algorithm for updating the shortest-paths subgraph [12]. Let us briefly recall the main idea of this algorithm.

Let an arc (i, j) be added to G and its weight be added to matrices $Weight$ and $Cost$. If $wt(i, j) + dist(j) = dist(i)$, then the arc (i, j) is added to the matrix SP. If $wt(i, j) + dist(j) < dist(i)$, then $dist(i) := wt(i, j) + dist(j)$, the vertex i becomes an affected one and it is assigned the maximal priority. Then one updates every arc outgoing from the affected vertex and every arc entering it. It should be noted that updating the arcs outgoing from an affected vertex is performed by analogy with Construction 8.

To select affected vertices, the Ramalingam incremental algorithm uses a data structure called a *heap* or a priority queue with keys. To simulate this data structure on the STAR–machine, one uses a slice $Z1$ to save the selected affected vertices and a matrix $Queue$ whose every l-th row saves the distance from the vertex l to the vertex i.

Construction 10. (*Updating the arcs entering an affected vertex.*)

Let the current matrices G, SP, $Cost$, and $Dist$ be given. Let k be the current updated affected vertex. By analogy with Construction 9, one first simultaneously determines the weights of different paths to the sink each starting with an arc that enters the vertex k. After that by means of a slice, one saves positions of those arcs (l, k) for which $dist_{new}(l) = wt(l, k) + dist_{new}(k)$ and includes positions of these arcs into the matrix SP. Finally, by means of a slice (say Y), one saves the tails of arcs (l, k) for which $dist_{new}(l) < dist_{old}(l)$ and writes $dist_{new}(l)$ in the corresponding rows of the matrix $Dist$.

The following construction uses a slice $Z1$ described above, a slice Y obtained in Construction 10, the matrix $Queue$, a row $v2$ to save the distance from the vertex i to the sink *before* inserting the arc (i, j), and a row $v3$ to save the weight of the shortest path to the sink that starts with the arc (i, j). Initially $Z1 = \emptyset$.

Construction 11. (*The initial update of matrices $Dist$ and $Queue$ after adding a new arc to G.*)

Let an arc (i, j) be added to G and its weight be added to matrices $Weight$ and $Cost$. Let $v3 < v2$. Then the vertex i is included into the slice $Z1$, the weight $v3$ is written in the i-th row of the matrix $Dist$, and the i-th row of the matrix $Queue$ consists of zeros.

Construction 12. (*Finding new affected vertices.*)

Let the matrix $Queue$ and the slices $Z1$ and Y be given. Let $Z1 \neq \emptyset$. Then one selects the vertex (say k) that corresponds to the position of the minimal row in the matrix $Queue$, and deletes this vertex from the slice $Z1$. After that the arcs outgoing from the vertex k and the arcs entering it are updated as described before. Let the slice Y save the tails of arcs (l, k) for which $dist_{new}(l) < dist_{old}(l)$. Then one includes such vertices into the slice $Z1$, simultaneously finds $dist(l) - dist(i)$ and writes the results in the corresponding rows of the matrix $Queue$.

In [12], the associative version of the Ramalingam incremental algorithm is implemented on the STAR–machine as procedure InsertNewArc. It takes the same $O(hk)$ time as procedure DeleteArc.

4   Updating the Transitive Closure of a Digraph

In this section, we select the main constructions that are used in [10] to represent the associative versions of the Italiano algorithms for updating the transitive closure of a digraph.

In [2,3], Italiano proposed the following data structure to support the efficient deletion and insertion of arcs in a digraph. The transitive closure of a graph G is represented by associating to each vertex u a set $Desc[u]$ of all descendants of u in G. Any $Desc[u]$ is organized as a spanning tree rooted at the vertex u. In addition, a matrix of pointers $Index$ is maintained, where $Index[i, j]$ points to the vertex j in the tree $Desc[i]$ if $j \in Desc[i]$ and it is $Null$, otherwise.

In [10], associative versions of the Italiano algorithms for updating the transitive closure use the following data structure: a Boolean matrix Adj; a Boolean matrix $Parent$ that consists of n submatrices ($blocks$) A_1, A_2, ..., A_n, where A_i is the adjacency matrix that corresponds to the spanning tree T_i; a Boolean matrix $Nodes$, whose every i-th column saves with bits '1' all vertices of the spanning tree T_i.

We first select the main constructions that are used in the associative version of Italiano's decremental algorithm. Let us briefly recall the main idea of this algorithm.

Let an arc (i, j) be deleted from G. Then it is deleted from all spanning trees in which it appears. Let (i, j) belong to $Desc[u]$. After deleting the arc (i, j) from $Desc[u]$ it splits into two subtrees. To obtain a new tree, it is necessary to check whether there is such a vertex z in $Desc[u]$ that (z, j) is an arc of G and the corresponding $u - j$ path avoids the vertex i. Such a vertex z is called a $hook$ for j. In this case, the arc (i, j) is replaced by the arc (z, j) and $Desc[u]$ does not change. Otherwise, the vertex j along with its outgoing edges are deleted from $Desc[u]$, and the seach for a hook for each son of j is recursively performed.

Construction 13. ($Finding\ possible\ hooks\ for\ j$.)

Let an arc (i, j) be deleted from the matrix Adj and the spanning tree T_r. Then possible hooks for the vertex j are defined by intersection of the r-th column of the matrix $Nodes$ and the j-th row of the matrix Adj. Let the result of this intersection be saved in a row, say u.

Construction 14. ($A\ case\ when\ there\ is\ a\ hook\ for\ j$.)

Let the assumption of Construction 13 be true and let a row u save the possible hooks for the vertex j. Let $u \neq \emptyset$. Then a hook for j is defined as the leftmost vertex, say p, that corresponds to bit '1' in the row u. The arc (p, j) is included into the spanning tree T_r.

The following construction uses a slice A to save the vertices that have not been updated yet. Initially, $A = \emptyset$.

Construction 15. ($A\ case\ when\ there\ is\ no\ hook\ for\ j$.)

Let the assumption of Construction 13 be true and let a row u save the possible hooks for the vertex j. Let $u = \emptyset$. Then the vertex j is deleted from the r-th column of the matrix $Nodes$. After that all sons of j in the spanning tree T_r are saved in the slice A. While $A \neq \emptyset$, one selects the vertex that corresponds to the uppermost bit '1' and updates it by means of Constructions 13 and 14.

In [10], the associative version of Italiano's decremental algorithm is implemented on the STAR–machine having no less than n PEs as procedure DelArc1. For the considered data structure, it takes $O(n)$ time and $O(n^3)$ space.

Now, we select constructions that are used in the associative version of Italiano's incremental algorithm. We first recall the main idea of Italiano's algorithm.

Let a new arc (i, j) be added to G. Then this arc is added to the trees that include the vertex i and do not include the vertex j. Let an arc (i, j) be added to the spanning tree $Desc[r]$. Then this tree is updated as follows. The common vertices in the trees $Desc[r]$ and $Desc[j]$ are deleted from the copy of $Desc[j]$. Then the pruned copy of $Desc[j]$ is linked to the vertex i in $Desc[r]$.

Construction 16. ($Finding\ new\ vertices\ for\ T_r\ after\ adding\ an\ arc\ (i, j)$.)

Let a new arc (i,j) be added to the spanning tree T_r. Then by means of the matrix $Nodes$, one first defines common vertices in the trees T_r and T_j and deletes these vertices from the copy of T_j. After that one saves the pruned copy of T_j in a slice, say Z. Finally, one includes the vertices from the slice Z into the r-th column of the matrix $Nodes$.

Construction 17. (*Finding new arcs for T_r after adding an arc (i,j).*)

Let a new arc (i,j) be added to the spanning tree T_r. Let a slice Z save the pruned copy of T_j. Then one first includes the arc (i,j) into the spanning tree T_r. Then for every vertex $p \neq j$ saved in the slice Z, one determines the position of an arc entering the vertex p in T_j and includes it in the spanning tree T_r.

In [10], the associative version of Italiano's incremental algorithm is implemented on the STAR–machine as procedure InsertArc1 that takes $O(n)$ time per an insertion and its space complexity is $O(n^3)$.

In [9], the associative version of Italiano's decremental algorithm uses another data structure that includes along wth the matrices Adj and $Nodes$ a matrix $Code$, whose every i-th row saves the binary representation of the vertex i; an association of matrices $Left$ and $Right$ and a global slice X, where positions of arcs belonging to G are marked with bits $'1'$; a matrix $Trans$, whose every i-th column saves positions of arcs belonging to the spanning tree T_i. For the considered data structure, the associative version of Italiano's decremental algorithm is represented on the STAR–machine having no less than m PEs as procedure DeleteArc that takes $O(n \log n)$ time per an deletion and its space complexity is $O(mn)$.

5   Conclusions

We have selected constructions that are used in parallel implementation on the STAR–machine a group of sequential algorithms for directed graphs. The proposed constructions and methods can be used, in particular, to solve in a natural way other graph problems on vertical processing systems.

As shown in our previous papers, the access data by contents, the use of the simple data structure given as 2D tables, and the vertical data processing allowed us, in particular, to implement efficiently on the STAR–machine a lot of important graph algorithms. Unfortunately, now there is no modern and proper hardware to implement our approach. Nevertheless, there are some modern SIMD systems that allow one to update data at the micro level. We are planning to simulate the STAR–machine run by means of the Graphics Processing Unit.

References

1. Dijkstra, E.W.: A note on two problems in connection with graphs. J. Numerische Mathematik. **1**, 269–271 (1959)
2. Italiano, G.F.: Amortized efficiency of a path retrieval data structure. J. Theoret. Comput. Sci. **48**(2–3), 273–281 (1986)

3. Italiano, G.F.: Finding paths and deleting edges in directed acyclic graphs. J. Inf. Process. Lett. **28**, 5–11 (1988)
4. Nepomniaschaya, A.S., Dvoskina, M.A.: A simple implementation of Dijkstra's shortest path algorithm on associative parallel processors. J. Fundamenta Informaticae **43**, 227–243 (2000). IOS Press
5. Nepomniaschaya, A.S.: Basic associative parallel algorithms for vertical processing systems. In: Bulletin of the Novosibirsk Computing Center, IIS Special Issue 29, pp. 63–77. NCC Publisher (2009)
6. Nepomniaschaya, A.S.: Constructions used in associative parallel algorithms for undirected graphs. Part 1. In: Bulletin of the Novosibirsk Computing Center, IIS Special Issue 35, pp. 67–81. NCC Publisher (2013)
7. Nepomniaschaya, A.S.: Constructions used in associative parallel algorithms for undirected graphs. Part 2. In: Bulletin of the Novosibirsk Computing Center, Issue 36, pp. 65–78. NCC Publisher (2014)
8. Nepomniaschaya, A. S.: Concurrent selection of the shortest paths and distances in directed graphs using vertical processing systems. In: Bulletin of the Novosibirsk Computing Center, Issue 19, pp. 61–72. NCC Publisher (2003)
9. Nepomniaschaya, A.S.: Associative version of italiano's decremental algorithm for the transitive closure problem. In: Malyshkin, V.E. (ed.) PaCT 2007. LNCS, vol. 4671, pp. 442–452. Springer, Heidelberg (2007)
10. Nepomniaschaya, A.S.: Efficient Implementation of the Italiano algorithms for updating the transitive closure on associative parallel processors. J. Fundamenta Informaticae. **89**(2–3), 313–329 (2008). IOS Press
11. Nepomniaschaya, A.S.: Efficient parallel implementation of the Ramalingam decremental algorithm for updating the shortest paths subgraph. J. Comput. Inform. **32**, 331–354 (2013)
12. Nepomniaschaya, A.S.: Associative version of the Ramalingam algorithm for the dynamic update of the shortest paths subgraph after inserting a new edge. J. Cybern. Syst. Anal. **3**, 45–57 (2012). Kiev: Naukova Dumka (in Russian) (English translation by Springer)
13. Ramalingam, G.: Bounded Incremental Computation. LNCS, vol. 1089. Springer, Heidelberg (1996)

Oscillatory Network Based on Kuramoto Model for Image Segmentation

Andrei Novikov$^{(\boxtimes)}$ and Elena Benderskaya

St.-Petersburg State Polytechnical University, St.-Petersburg 194064, Russia
spb.andr@yandex.ru, helen.bend@gmail.com

Abstract. Oscillatory networks represent biologically inspired models that implement cognitive functions such as vision, motion and memory. Vision functions are most attractive brain ability. Despite recognition problems have been successfully solved by traditional neural networks, segmentation problems still require close attention. In this paper, we propose oscillatory network based on Kuramoto phase oscillator for image segmentation where each allocated feature is encoded by ensemble of synchronous oscillators like in biologically plausible systems. The proposed model is designed to perform color segmentation and object segmentation using synchronization phenomena. Processes of synchronization between oscillators during image processing and multi-core implementation of the network for simulation are discussed. Experimental results of segmentation by the network using formal and real images have been presented.

Keywords: Oscillatory network · Phase oscillator · Synchronization · Kuramoto model · Image segmentation

1 Introduction

Visual functions such as spatial orientation, segmentation and recognition are trivial operations for vast of majority representatives of the animal world, but at the same time these functions are complex problems for machines. For example, our brain easily remembers new faces, objects and easily recognizes them in different places in the presence of other objects in the field of view and most animals have these abilities. Using conventional algorithms for achieving similar abilities requires many computing and time resources. Therefore, close attention to biologically inspired models is explained by desire to reach the same abilities for solving computer vision problems using similar processing mechanisms [2, 5].

In contrast to the classical concept of neurons, oscillators are closer to the biological plausible models of neurons, which are essentially non-linear dynamical systems. There are many research papers that hypothesize and present experimental confirmation that cognitive functions are implemented by synchronization processes between neurons in the brain [1, 4]. Object features are encoded in-phase activity of neurons in different areas of the cerebral cortex, in other words ensembles of synchronous neurons encode only one feature of the object. Thus, understanding of the principles of the

© Springer International Publishing Switzerland 2015
V. Malyshkin (Ed.): PaCT 2015, LNCS 9251, pp. 210–221, 2015.
DOI: 10.1007/978-3-319-21909-7_20

processes in the mammalian brain gives a wide theoretical base for creation of fundamentally new algorithms which provide a parallel and distributed processing.

Splitting images into disjoint fragments that correspond to visual features such as colors, contours and objects is a relevant issue that is referred to segmentation. In an informal formulation, image segmentation is a process of separation of a digital image into several segments. The aim of segmentation is to change representation of the image to make it simpler and easier to analyze, for instance for further recognition of each segment and scene analysis. From the oscillatory theory point of view, each feature of input image such as color or separate object or contour is encoded by one ensemble of synchronous oscillators [7, 10].

The synchronization is important issue in the oscillatory theory because this process is a key concept to the understanding of self-organization phenomena. Synchronization occurs in pendulum clock, electrical, electromagnetic and quantum generators, and even between organisms in collective, for instance, firefly flickers, poultry and fish in flocks, marching or applauding people. In case of coupled oscillators synchronization should occurs between oscillators that represent the same feature and at the same time de-synchronization should be between ensembles of oscillators that represent different features.

Many studies have been performed in order to build biologically inspired models of oscillatory neural networks of visual cortex based on synchronization principles for the last decades. The model of oscillatory network based on competition between synchronous ensembles of oscillators using global inhibitor has been implemented in LEGION [23, 24] based on the Van Der Pol model and performs object segmentation of isolated objects on images. The idea of central element has been also implemented in oscillatory network model based on Hodgkin-Huxley neuron model [8]. The model is designed for color segmentation and the most interesting feature is that peripheral neurons do not have any connections between themselves, all interaction takes place though central element [4, 11, 12]. There are several papers devoted to contour segmentation based on the Eckhorn model of pulse-coupled neural network [6, 14–16, 22] that has been initially proposed for modeling a cat's visual cortex [5].

In this paper, we propose a model of a double layer oscillatory network for image segmentation where the first layer of the network encodes colors and the second layer encodes objects using output information of the first layer. The network uses another approach of synchronization between oscillators based on the modified Kuramoto model [14]. Unit of the network is phase oscillator and each oscillator is associated with pixel or with area of pixels whose description determines spatial position of each oscillator in the layer. The end of the segmentation process is determined by the evaluation of the synchronization in the network.

2 Preliminaries

Synchronization represents one of the forms of self-organization between objects. There is dependence between coupling strength and convergence rate of synchronization, for example coupling strength can be hard like two interconnected wheels or it can be soft like two pendulums that are fixed on the same beam, nevertheless synchronization can be reached in both cases.

Consider the formal mathematical statement of the synchronization problem. Suppose there is an interconnected system with k dynamic objects where state of each object i is defined by r-dimensional vector $x(s) = [x1(s), x2(s), \ldots, x_r(s)]$, $s = 1, \ldots, k$, components $x_j(s)$ are coordinates of object in phase space of the system. State of the system is determined by the set of vectors $x(s)$ and by the v-dimensional vector $u = [u_1, u_2, \ldots, u_v]$ that describes states of connections between objects. Phase space of the system has $r_1 + r_2 + \ldots + r_k + v$ dimensions. Thus, the system is synchronous if phase coordinates are changed in line with following rule:

$$\begin{cases} x_j^{(s)} = n_j^{(s)}\omega t + y_j^{(s)}\left(m_j^{(s)}\omega t\right), & j = 1, \ldots, r_s, \\ u_p = n_p\omega t + v_p\left(m_p\omega t\right), & p = 1, \ldots, v \end{cases} \tag{1}$$

In these equations, ω –positive constant, $n_j(s)$ and n_p –integer variables, $m_j(s)$ and m_p – positive integer variables, $y_j(s)$ and v_p – periodical functions with periods $2\pi/m_j(s)$ and $2\pi/m_p$ on ωt. If $n_j(s)$ or n_p are equal to zero than corresponding coordinate $x_j(s)$ can be considered as oscillatory. Using averaging operator for both parts of the equation:

$$\begin{cases} <\dot{x}_j^{(s)}> = n_j^{(s)}\omega \\ <\dot{u}_p> = n_p\omega \end{cases} \tag{2}$$

In accordance to Eq. (2), oscillatory or average uniform motion for each phase coordinate corresponds to synchronous movements of the system.

There are three general types of synchronization that can be observed in dynamic systems:

- global synchronization means state of dynamic system where each oscillator has the same phase coordinate in each dimension;
- local synchronization means state of dynamic system that consist of two or more groups of oscillators with different phase coordinates but at the same time with the same phases within groups;
- de-synchronization means state of dynamic system where no oscillators with the same phase coordinates.

Kuramoto has proposed one of the well-known models for the synchronization. Because of simplicity the model is flexible and can be adapted for solving various problems. The model allows studying of synchronization processes in non-linear dynamic systems such as oscillatory networks. The Kuramoto model that consists of a population of N full-interconnected phase oscillators is described by the following differential equation [13]:

$$\dot{\theta}_i = \omega_i + \frac{K}{N}\sum_{j=1}^{N}\sin\left(\theta_j - \theta_i\right) \tag{3}$$

Phase of oscillator θ_i is basic state variable that is distributed in the interval from 0 to 2π. Intrinsic frequency of oscillator ω_i can be considered as bias that is randomly initialized in line with some probability distribution. Couplings between oscillators are defined by strength K is important parameter that affects processes of synchronization.

Synchronization plays important role in oscillatory neural networks, where each ensemble synchronized oscillators corresponds to a single encoded feature. Oscillatory networks that are based on Kuramoto model ensure all general states: global synchronization, local or partial synchronization and de-synchronization. These states can be set by coupling strength K between oscillators. High value of coupling strength $K \geq N$ ensures quick switch to global synchronization state. De-synchronization state can be ensured by coupling strength K that is less than critical value of coupling strength K_c. The critical coupling strength K_c is defined by width of distribution of the intrinsic oscillator frequency: $K_c = 2\gamma$. Thus when oscillatory network contains inhomogeneous connections whose coupling strengths are greater than critical coupling strength then state of local synchronization can be established in the network. The state of local synchronization implies existence of more than one ensemble of synchronized oscillators.

Feature allocation is driven by state of synchronization that should be accurately identified. Evaluation of synchronization degree can be used for state identification and defined by following expression [13]:

$$r = \left| \frac{1}{N \exp(i\varphi)} \sum_{j=1}^{N} \exp(i\theta_j) \right| \tag{4}$$

Where r is distributed from 0 to 1, and φ is average phase of all oscillators in the oscillatory network:

$$\varphi = \frac{1}{N} \sum_{j=1}^{N} \theta_j \tag{5}$$

In case of synchronization degree r tends to 1 the state can be identified as a global synchronization, in case of r tends to 0 then the state corresponds to de-synchronization. It can be assumed that local or partial synchronization are roughly defined by values $0 < r < 1$, but this expression is not always satisfied because of boundary of de-synchronization area that is defined by critical value of coupling strength K_c between oscillators. Condition of local synchronization in line with critical coupling strength is following:

$$r \rightarrow \sqrt{1 - \frac{K_c}{K}} \tag{6}$$

Convergence rate of synchronization process depends on structure of connections between oscillators and also depends on number of oscillators. But convergence rate can be improved by increasing coupling strength between oscillators in case of networks that have differ structures from full-interconnected especially in network with non-uniform punctured connections. Oscillatory network with a "grid" structure where each oscillator has connections with four neighbors (right, left, top and bottom) is characterized by quadratic dependence of convergence rate of synchronization $O(n^2)$. Network with a "list" structure where each oscillator has connection with two neighbors (right and left) is characterized by cubic dependence $O(n^3)$. These dependences affect possibilities to use oscillatory networks for real practical problems [18]. One of the ways for increasing convergence rate is usage of bio-inspired principles that have been proposed in the paper [19] where input data space is encoded by self-organized feature map that saves possibility to ensure parallel execution that is the hallmark of neural networks.

There are studies devoted to adaptation of the Kuramoto model for solving various problems, for example, several algorithms and approaches of cluster analysis [3, 17, 20], image compression method [9], learning patterns [21], also there are papers devoted to graph coloring problem that is solved by oscillatory network based on the considered model with negative connections [25, 26]. In the following section, the oscillatory network based on modified Kuramoto model has been presented for image segmentation problem.

3 The Oscillatory Network for Image Segmentation

3.1 Oscillatory Network Architecture

The oscillatory network consists of two layers of phase oscillators. Total number of oscillators in each layer determined by total number of pixel or area of pixels. Each pixel or area of pixels defines spatial position of each oscillator. Such architecture forms columns of oscillators that correspond to certain regions of input image. The output dynamic of the first layer activates oscillators from the second layer. Oscillator from the first layer is only able to activate oscillators located in his own column. Active oscillators form separated groups of oscillators that interact with each other and each group of active oscillators is defined by ensemble of synchronous oscillators in the first layer. In other words, oscillators in active areas of the seconds layer interact with each other within own group and do not have any relation with oscillators from other active groups.

Connections between oscillators are formed in line with input image and similarity parameter that is general parameter of the network and that is distributed from 0 to 1, where 0 means no similarity when each fragment of image is unique and only totally similar fragments of image should be taken into account by the network, and 1 means total similarity when image represents one solid segment. In our model we use two similarity parameters: the first for color similarity that is used by the first layer and the second for spatial similarity that is used by the second layer. Connection between oscillators in the first layer is established if similarity between spatial descriptions of these oscillators is less than color similarity. Connections of the first layer are static and they are formed once and never changed during simulation. Connections of the second layer are dynamic because they depend on synchronization in the first layer, but they also depend on spatial similarity in the same way. The primary rule for establishing connection between oscillators in the second layer is synchronization in the first layer i.e. connection can be established only between oscillators that are related to the same active group of oscillators and thus spatial similarity is the secondary rule for forming structure of the second layer. Similarity is often calculated via normalized distance measure such as Euclidian distance, but also Jaccard coefficient or Tanimoto evaluation can be used for that purpose. In our experiments, we have used normalized Euclidian distance as similarity estimation.

The modified Kuramoto model describes dynamic of each oscillator:

$$\dot{\theta}_i = \frac{K}{N_i} \sum_{j \in N_i} \sin(\theta_j - \theta_i) \tag{7}$$

Phase of oscillator is denoted by θ_i and it is key state variable that is distributed from 0 to π. Distribution from 0 to 2π is undesirable because of unstable behavior that follows from the Eq. (7) in case of odd number of oscillators equidistant from each other (not random) in this range. All oscillators have the same value of frequency ω therefore it is omitted in this equation due to permanent phase offset that reduces time of synchronization. Coupling strength K between oscillators is also the same and it affects the convergence rate of synchronization process. High values of coupling strength can adversely affect process output result of the network that has punctured connections and can lead to so-called "races" when oscillators are trying to synchronize with each other. In our experiments, we have used coupling strength from 1 to 3 for achieving adequate results. N_i denotes set of oscillators that have connections with oscillator i.

The first layer receives color description of an input image, for example, it can be RGB representation of pixel or area of pixels. Each oscillator i receives outputs of its neighbors that are defined by set N_i and affect on current phase position in accordance with mean field and thus synchronized with each other. During self-organization of ensembles in the first layer, connections between oscillators are being changed in the second layer. So initially the second layer contains connections in line with the spatial parameter but during simulation connections can be changed by ensembles of oscillators of the first layer until local synchronization is not reached and as a consequence final state of synchronization in the second layer cannot be reach until it is not reached in the first layer, thus segmentation is over when local synchronization is reached in both layers of the network. When synchronization process reaches final state of local synchronization that is described by the degree r_c in the first layer then each synchronous ensemble of oscillators encodes one color cluster and also forms local areas of oscillators in the second layer which do not have connections with oscillators from other areas and interact only with each other. Oscillators of the second layer use their own spatial description in the layer for forming structure in line with similarity parameter for the spatial. Connections in the second layer are formed in line with similarity parameter and activated areas by the first layer. When process of synchronization is over in the second layer each ensemble of oscillators of each activated area corresponds to one separated object.

3.2   Enhancements for Real Image Segmentation

Number of connections in the network can be huge for real images because it depends on number of oscillators and similarity parameters, and as result there are problems with storing these connections in random access memory RAM. Full-interconnected layer in the network contains N_2 connections, where N – number of oscillators in the layer, for instance, in the layer with 4096 oscillators requires 16.7 million connections when one pixel is represented by one oscillator. There are two ways for resolving this issue.

The first is usage of image areas instead of single pixels as it mentioned before when each correspond to area of pixels that can be defined by some equation, for example using usual average or more complexity compression algorithms.

The second way is encoding of connections between oscillators. The easiest way is to use adjacent matrix for connection representation, this way ensures high performance because of fast access to the matrix cell where connection is stored $O(1)$. One cell uses 4 bytes at least for storing state of connection if size of bool type is equal to 32-bit. Thus, the adjacent matrix requires a lot of memory, for example, in case of image size 64 × 64 it requires about 67.1 MB for one layer, and for image 128 × 128 it requires about 1073 MB. Moreover, size of adjacent matrix does not depend on real number of connections. Set of neighbors can be used in this case, but it reduces performance, because in worth case it requires $O(n)$ in case of usual list and $O(\log_2(n))$ in case of balanced binary tree. Set of neighbors is effective method of representation if real number of connections in the network is low, but in worth case it requires the same memory block like usual adjacent matrix, but additionally increases complexity of searching. The last way is bit map that can significantly reduce usage of memory and ensure performance like in adjacency matrix (statement is applicable in case of C/C++ implementation). Bit map represents adjacency matrix where indexes of oscillators are used for access to bit in memory. We have performed experimental study and have found that bit map does not reduce performance of the network during simulation, but it allows to process large images without compression. For example, image with size 64 × 64 requires about 16.7 MB and image 128 × 128 requires about 268 MB.

Segmentation of real images requires a lot computational resources on for calculating of differential equations in both layers. Originally, both layers in the network should be simulated simultaneously, but for reducing numerical computations we propose algorithm that simplifies complexity of simulation of the network.

The differential equation of the phase oscillator is simplified in following way:

$$\theta_i[k+1] = \theta_i[k] + \frac{K}{N_i} \sum_{j \in N_i} \sin\left(\theta_j[k] - \theta_i[k]\right) \tag{8}$$

In this case accuracy is lower in comparison with the family of Runge-Kutta methods, therefore we recommend to use coupling strength $K = 1$. Otherwise it may lead to infinite process of simulation due to lack of convergence.

Both layers are simulated separately: the first layer is simulated until specified stop condition (that is defined by the order of local synchronization, usually $r_c > 0.998$) is not reached. After that connections of the second layer are initialized once in line with synchronous ensembles in the first layer and also in line with the similarity parameter. Thus, there is no need to adjust connections between oscillators in the second layer during simulation. Next, simulation of second layer is executed until stop condition is not reached. Since there are no connections between activated areas of oscillators in the second layer then separate simulation can be performed for each activated area. Another one of the advantages of the separate simulation is reducing memory usage.

The proposed algorithm reduces amount of consumed memory by 8 times and reduces time of processing by 186 times in case of single-core implementation and it allows to use the network for segmentation of real images where memory usage and performance play important role. Results of segmentation of real image by the oscillatory network will be shown later (Sect. 4).

3.3 Parallel Implementation of Oscillatory Network

The ability to be executed in parallel is one of the most important features of biologically plausible systems such as oscillatory networks. In this section we consider implementation of the oscillatory network for multi-core station. Each oscillator is considered as a separate unit that can be executed by processor core and each oscillator has own context in memory where current state of each oscillator is stored (in our case it is current phase, but also it can store frequency and other variables describes state of oscillator). The context is read-only for other oscillators and it is available for writing only for the oscillator that is owner of this context. Thus synchronization between execution units is not required, but it is important to ensure lack of cache coherency using cache write-back and cache invalidate operations if platform does not care about that.

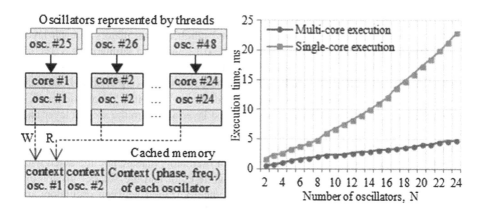

Fig. 1. (Left) The scheme of the multi-core implementation of the oscillatory network. (Right) The comparison of execution times of multi-core implementation and single-core implementation with sizes from 2 to 24 oscillators during 100000 iterations.

We have used HP ProLiant BL460c Generation 7 (G7) workstation that has four processors Intel Xeon X5660 (2.80GHz/6-core/12MB/95W, DDR3-1333, HT, Turbo 2/2/2/2/3/3) with 8 MB shared L3 cache for simulation of the multi-core implementation of the oscillatory network. Each oscillator is simulated by separate thread that can be executed by one of the 24 cores. The scheme of multi-core implementation is shown on Fig. 1 (left). Average execution time of the multi-core implementation of the network with sizes from 2 to 24 oscillators (this range has been used for obtaining results of completely parallel execution) has linear character in comparison with the single-core implementation as it is shown on Fig. 1 (right). In experiments, simulation of the network has been performed with fixed number of iterations using 100000 steps. The multi-core implementation is 3 times faster than the single-core in the small number of oscillators (2–4) and in 5 times faster in case full loading (24 oscillators). In spite of the limit in 24 cores, increase of number of oscillators leads to increase of difference by several times in the execution time between the single and the multi-core implementation – results of simulation of networks with bigger sizes have been presented in Table 1.

Table 1. Comparison of execution times with various number of oscillators.

Oscillators, N	200	300	400	500	600	700	800	900	1000
Single-core., ms	915	2233	3618	5166	7110	10931	14279	16873	20101
Multi-core., ms	123	295	458	628	810	1211	1561	1775	2111
Difference, times	7.43	7.56	7.9	8.22	8.77	9.02	9.14	9.50	9.52

4   Image Segmentation Results

To illustrate general principle of segmentation in comprehended way simple images 32×32 is used for the first example where three black letters are presented on white background. The oscillatory network uses direct projection of the image (one oscillatory column encodes only one pixel, total number of oscillators in each layer $N = 1024$) with similarity between objects $\delta_{obj} = 0.12$ and color similarity $\delta_{color} = 0.5$ due to necessity to allocate black objects from the white background. Input image and result of segmentation are presented on Fig. 2 where four objects are allocated and each of them is denoted by black mask: background, letter 'F', letter 'T' and letter 'K'. Outputs of the network are obtained from the first layer and from the second layer and both are presented on Fig. 3. Dynamic of the second layer is divided into two plots for convenient presentation where dynamic of each plot corresponds to activated group of oscillators. The first layer activates two groups of oscillators that are not interacted with each other, the "white" group corresponds to white color and the "black" group corresponds to black color. At the beginning de-synchronization between oscillators are observed in both groups, but after time process of self-organization is more apparent that is considered as a process of synchronization between oscillators responsible for the similar features. Simulation of the network is stopped when local synchronization reaches $r = 0.999$. Activated group by white color converges to a single point that means there is only one object (white background) and activated group by the black color has three synchronous ensembles of oscillators and it means allocation of three objects (black letters). Since one oscillator corresponds to only one pixel, it is easy to decode allocated objects on the image.

Fig. 2. Initial image 'FTK' at the left and then allocated segments from the image where back masks represent allocated objects: background, letter 'F', letter 'T' and letter 'K' (δcolor = 0.12, δobj = 0.5).

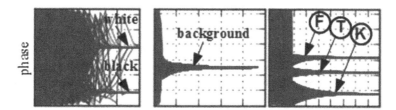

Fig. 3. Output of the oscillatory network (dynamic of the second layer) divided into two groups (middle and right) that are activated by ensembles of synchronous oscillators of the first layer (left). The middle is activated by oscillators that encode white color (background), and the right is activated by oscillators that encode black color (black letters) (Color figure online).

The next example demonstrates results of the network simulation in case of colored image such as satellite image of the White Sea – Fig. 4. Image size is 128 × 128 and the network uses direct projection of oscillator columns, color similarity δ_{color} = 0.15 and object similarity δ_{obj} = 0.1. In this case result of segmentation can be obtained from the first layer due to requirement of sea and land allocation as two segments since there is only color difference. There are only two groups of synchronous oscillators in first layer where one of them corresponds to the sea and another one corresponds to the land. The separate island is not allocated by the second layer due to high level of object similarity and as a consequence there is interconnection between oscillators that responsible for the continental land and for the island. The island allocation can be performed by reducing level of interaction between oscillators in the second layer using object similarity.

Fig. 4. Initial image – map of the White Sea (left) and results of segmentation: the sea and the land with island (δcolor = 0.15, δobj = 0.1) (Color figure online).

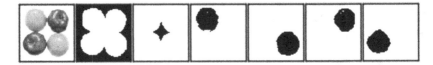

Fig. 5. Initial image (left) and result of segmentation: six allocated objects: two areas of background, two apples and two oranges (δcolor = 0.15, δobj = 0.06) (Color figure online).

The following example demonstrates segmentation of the scene of four fruits where two apples have the same color and two oranges have also the same color – Fig. 5. Size of the image is 128 × 128 and the network uses direct projection with color similarity δ_{color} = 0.06 and object similarity δ_{obj} = 0.15. The color similarity in this example should be much lower due to red color is close to orange color in case of RGB representation. For example, in case of δ_{color} = 0.2 the first layer has only two synchronous groups of oscillators responsible for the white background and for the set of fruits. In case of δ_{color} = 0.06 three synchronous ensembles of oscillators are formed in the first layer that corresponds to allocated objects: white background, set of apples and set of oranges. The second layer performs further segmentation and eventually two synchronous ensembles are formed in each active group: two areas of background, two apples and two oranges, thus six objects are allocated by the network during simulation.

5   Conclusions

In this article we have proposed neural based approach for image segmentation problem (color and object segmentation) using the double-layer oscillatory network based on the modified Kuramoto model that provides result of coloring segmentation from the first layer and result of object segmentation from the second layer. The proposed algorithm uses biologically plausible principles of synchronization where each feature is encoded by synchronous ensemble of oscillators. In addition, we have proposed practical recommendation of the model usage and have described multi-core implementation of the network for simulation on multiprocessor stations. Our experiments demonstrate general capabilities of the network for real image segmentation.

References

1. Arenas, A., DiazGuilera, A., Kurths, Y., Moreno, Y., Changsong, Z.: Synchronization in complex networks. Phys. Rep. **469**, 93–153 (2008)
2. Basar, E.: Brain Function and Oscillations, p. 364. Springer, New York (1998)
3. Bohm, C., Plant, C., Shao, J., Yang, Q.: Clustering by synchronization. In: KDD 2010 Proceeding of the 16th ACM SIGKDD International Conference of Knowledge Discovery and Data Mining, pp. 583–592 (2010)
4. Cumin, D, Unsworth, C.: Generalizing the Kuramoto model for the study of neuronal synchronisation in the brain. Report University of Auckland School of Engineering 638 (2006)

5. Eckhorn, R., Reitbock, H., Arndt, M., Dicke, P.: A neural network for feature linking via synchronous activity: results from cat visual cortex and from simulations. In: Cotterill, R.M.J. (ed.) Models of Brain Function, pp. 255–272. Cambridge University Press, Cambridge (1989)
6. Gu, X.: A new approach to image authentication using local image icon of unit-linking PCNN. In: Proceedings of IJCNN 2006. International Joint Conference on Neural Networks, pp. 1036–1041 (2006)
7. Haken, H.: Brain Dynamics, p. 238. Springer, Heidelberg (2007)
8. Hodgkin, A., Huxley, A.: A quantitative description of membrane current and its application to conduction and excitation in nerve. J. Physiol. **117**, 500–544 (1952)
9. Ishihara, H., Yoshioka, K., Hirose, M.: Proposal on image compression method using synchronization. In: Proceedings of ISOT 2012, International Symposium on Optomechatronic Technologies, Paris, 29–31 October (2012)
10. Johnson, J., Padgett, M.: PCNN models and applications. IEEE Trans. Neural Netw. **10**(3), 480–498 (1999)
11. Kazanovich, Y., Borisyuk, R.: Dynamics of neural networks with a central element. Neural Netw. **12**, 441–454 (1999)
12. Kazanovich, Y., Borisyuk, R., Chik, D., Tikhanoff, V., Cangelosi, A.: A neural model of selective attention and object segmentation in the visual scene: an approach based on partial synchronization and star-like architecture of connections. Neural Netw. **22**, 707–719 (2009). Elsevier
13. Kuramoto, Y.: Chemical Oscillations Waves, and Turbulence, p. 157. Springer, Heidelberg (1984)
14. Li, Z.: A neural model of contour integration in the primary visual cortex. Neural Comput. **10**, 903–940 (1998)
15. Li, Z.: Pre-attentive segmentation in the primary visual cortex. Spat. Vis. **13**, 25–50 (2000)
16. Lindblad, T., Kisner, J.: Image Processing Using Pulse-Coupled Neural Networks, p. 164. Springer, Heidelberg (2005)
17. Miyano, T., Tsutsui, T.: Data synchronization as a method of data mining. In: Proceeding of International Symposium on Nonlinear Theory and its Applications (2007)
18. Novikov, A., Benderskaya, E.: SYNC-SOM Double-layer Oscillatory Network for Cluster Analysis. In: 3rd International Conference on Pattern Recognition Applications and Methods, Proceedings, ESEO, Angers, Loire Valley, France, pp. 305–309, 6–8 March 2014
19. Novikov, A., Benderskaya, E.: Oscillatory neural networks based on the Kuramoto model. Pattern Recogn. Image Anal. **24**(3), 365–371 (2014)
20. Shao, J., He, X., Bohm, C., Yang, Q., Plant, C.: Synchronization-inspired partitioning and hierarchical clustering. IEEE Trans. Knowl. Data Eng. **25**(4), 893–905 (2013)
21. Vassilieva, E., Pinto, G., Acacio, J., Suppes, P.: Learning pattern recognition through quasi-synchronization of phase oscillators. IEEE Trans. Neural Netw. **22**(1), 84–95 (2011)
22. Xiao, Z., Shi, J., Chang, Q.: Image segmentation with simplified PCNN. In: Proceeding of CISP 2009 2nd International Congress on Image and Signal Processing, Tianjin, 17–19 October 2009, pp. 1–4 (2009)
23. Wang, D., Terman, D.: Locally excitatory globally inhibitory oscillator networks. IEEE Trans. Neural Netw. **6**(1), 283–286 (1995)
24. Wang, D., Terman, D.: Image segmentation based on oscillatory correlation. Neural Comput. **9**, 805–836 (1997)
25. Wang, X., Jiao, L., Wu, J.: Extracting hierarchical organization of complex networks by dynamics towards synchronization. Phys. A **388**, 2975–2986 (2009)
26. Wu, J., Jiao, L., Chen, W.: Clustering dynamics of nonlinear oscillator network: application to graph coloring problem. Physica D **240**(2), 1972–1978 (2011)

Using Monte Carlo Method for Searching Partitionings of Hard Variants of Boolean Satisfiability Problem

Alexander Semenov$^{(\boxtimes)}$ and Oleg Zaikin

Institute for System Dynamics and Control Theory SB RAS, Irkutsk, Russia
biclop.rambler@yandex.ru, zaikin.icc@gmail.com

Abstract. In this paper we propose the approach for constructing partitionings of hard variants of the Boolean satisfiability problem (SAT). Such partitionings can be used for solving corresponding SAT instances in parallel. We suggest the approach based on the Monte Carlo method for estimating time of processing of an arbitrary partitioning. We solve the problem of search for a partitioning with good effectiveness via the optimization of the special predictive function over the finite search space. For this purpose we use the tabu search strategy. In our computational experiments we found partitionings for SAT instances encoding problems of inversion of some cryptographic functions. Several of these SAT instances with realistic predicted solving time were successfully solved on a computing cluster and in the volunteer computing project SAT@home. The solving time agrees well with estimations obtained by the proposed method.

Keywords: Monte carlo method · SAT · Partitioning · Tabu search · Cryptanalysis

1 Introduction

The Boolean satisfiability problem (SAT) consists in the following: for an arbitrary Boolean formula (formula of the Propositional Calculus) to decide if it is satisfiable, i.e. if there exists such an assignment of Boolean variables from the formula that makes this formula true. The satisfiability problem for a Boolean formula can be effectively (in polynomial time) reduced to the satisfiability problem for the formula in the conjunctive normal form (CNF). Hereinafter by SAT instance we mean the satisfiability problem for some CNF.

Despite the fact that SAT is NP-complete (NP-hard as a search problem) it is very important because of the wide specter of practical applications. A lot of combinatorial problems from different areas can be effectively reduced to SAT [1]. In the last 10 years there was achieved an impressive progress in the effectiveness of SAT solving algorithms. While these algorithms are exponential in the worst case scenario, they display high effectiveness on various classes of industrial problems.

© Springer International Publishing Switzerland 2015
V. Malyshkin (Ed.): PaCT 2015, LNCS 9251, pp. 222–230, 2015.
DOI: 10.1007/978-3-319-21909-7_21

Because of the high computational complexity of SAT, the development of methods for solving hard SAT instances in parallel is considered to be relevant. Nowadays the most popular approaches to parallel SAT solving are *portfolio* approach and *partitioning* approach [6]. In the portfolio approach several copies of the SAT solver process the same search space in different directions. The partitioning approach implies that the original SAT instance is decomposed into a family of subproblems and this family is then processed in a parallel or in a distributed computing environment. This family is in fact a partitioning of the original SAT instance. The ability to independently process different subproblems makes it possible to employ the systems with thousands of computing nodes for solving the original problem. Such approach allows to solve even some cryptanalysis problems in the SAT form. However, for the same SAT instance one can construct different partitionings. In this context the question arises: if we have two partitionings, how can we know if one is better than the other? Or, if we look at this from the practical point of view, how to find if not best partitioning, then at least the one with more or less realistic time required to process all the subproblems in it? In the present paper we study these two problems.

2   Monte Carlo Approach to Statistical Estimation of Effectiveness of SAT Partitioning

Let us consider the SAT for an arbitrary CNF C. The partitioning of C is a set of formulas

$$C \wedge G_j, j \in \{1, \dots, s\} \tag{1}$$

such that for any $i, j : i \neq j$ formula $C \wedge G_i \wedge G_j$ is unsatisfiable and

$$C \equiv C \wedge G_1 \vee \dots \vee C \wedge G_s.$$

(where "$\equiv$" stands for logical equivalence). It is obvious that when one has a partitioning of the original SAT instance, the satisfiability problems for CNFs (1) can be solved independently in parallel.

There exist various partitioning techniques [6]. The results of the research on estimating the time required to process SAT partitionings can be found in a number of papers on logical cryptanalysis [4,10,11]. In the present paper we propose to construct time estimations for the processing of SAT partitionings using the Monte Carlo method in its classical form [8].

Consider the satisfiability problem for an arbitrary CNF C over a set of Boolean variables $X = \{x_1, \dots, x_n\}$. We call an arbitrary set $\tilde{X} = \{x_{i_1}, \dots, x_{i_d}\}$, $\tilde{X} \subseteq X$ a decomposition set. Consider a partitioning of C that consists of a set of $s = 2^d$ formulas of the kind (1), where G_j, $j \in \{1, \dots, 2^d\}$ are all possible minterms over $\tilde{X}$. Note that an arbitrary formula G_j takes a value of true on a single truth assignment $\left(\alpha_1^j, \dots, \alpha_d^j\right) \in \{0, 1\}^d$. Therefore, an arbitrary formula $C \wedge G_j$ is satisfiable if and only if $C\left[\tilde{X} / \left(\alpha_1^j, \dots, \alpha_d^j\right)\right]$ is satisfiable.

Here $C\left[\tilde{X}/\left(\alpha_1^j,\ldots,\alpha_d^j\right)\right]$ is produced by setting values of variables x_{i_k} to corresponding α_k^j, $k \in \{1,\ldots,d\}$: $x_{i_1} = \alpha_1^j,\ldots,x_{i_d} = \alpha_d^j$. A set of CNFs

$$\Delta_C(\tilde{X}) = \left\{C\left[\tilde{X}/\left(\alpha_1^j,\ldots,\alpha_d^j\right)\right]\right\}_{\left(\alpha_1^j,\ldots,\alpha_d^j\right)\in\{0,1\}^d}$$

is called a decomposition family produced by $\tilde{X}$. It is clear that the decomposition family is the partitioning of the SAT instance C.

Consider some algorithm A solving SAT. In the remainder of the paper we presume that A is complete, i.e. its runtime is finite for an arbitrary input. We also presume that A is a non-randomized deterministic algorithm. We denote the amount of time required for A to solve all the SAT instances from $\Delta_C\left(\tilde{X}\right)$ as $t_{C,A}\left(\tilde{X}\right)$. Below we concentrate mainly on the problem of estimating $t_{C,A}\left(\tilde{X}\right)$.

Define the uniform distribution on the set $\{0,1\}^d$. With each randomly chosen truth assignment $(\alpha_1,\ldots,\alpha_d)$ from $\{0,1\}^d$ we associate a value $\xi_{C,A}(\alpha_1,\ldots,\alpha_d)$ that is equal to the time required for the algorithm A to solve SAT for $C\left[\tilde{X}/(\alpha_1,\ldots,\alpha_d)\right]$. Let $\xi^1,\ldots,\xi^Q$ be all the different values that $\xi_{C,A}(\alpha_1,\ldots,\alpha_d)$ takes on all the possible $(\alpha_1,\ldots,\alpha_d) \in \{0,1\}^d$. Let us denote $\xi_{C,A}\left(\tilde{X}\right) = \{\xi^1,\ldots,\xi^Q\}$, and let $\sharp\xi^j$ be the number of $(\alpha_1,\ldots,\alpha_d)$, such that $\xi_{C,A}(\alpha_1,\ldots,\alpha_d) = \xi^j$. Then $\xi_{C,A}\left(\tilde{X}\right)$ is a random variable with distribution $P\left(\xi_{C,A}\left(\tilde{X}\right)\right) = \{p_1,\ldots,p_Q\}$, where $p_k = \frac{\sharp\xi^k}{2^d}$, $k \in \{1,\ldots,Q\}$. Thus, it is easy to see that

$$t_{C,A}\left(\tilde{X}\right) = \sum_{k=1}^{Q}\left(\xi^k \cdot \sharp\xi^k\right) = 2^d \cdot \mathrm{E}\left[\xi_{C,A}\left(\tilde{X}\right)\right]. \qquad (2)$$

To estimate the expected value $\mathrm{E}\left[\xi_{C,A}\left(\tilde{X}\right)\right]$ we will use the Monte Carlo method [8], according to which, a probabilistic experiment, that consists of N independent observations of values of an arbitrary random variable ξ, is used to approximately calculate $\mathrm{E}[\xi]$. Let $\zeta^1,\ldots,\zeta^N$ be the results of the corresponding observations. From the theoretical basis of the Monte Carlo method it follows that if ξ has finite expected value and finite variance, then the value $\frac{1}{N}\cdot\sum_{j=1}^{N}\zeta^j$ is a good approximation of $\mathrm{E}[\xi]$ when the number of observations is large enough. In our case from the assumption regarding the completeness of the algorithm A it follows that random variable $\xi_{C,A}(\tilde{X})$ has finite expected value and finite variance. We would like to mention that an algorithm A should not use randomization, since if it does then the observed values in the general case will not have the same distribution. The fact that N can be significantly less than 2^d makes it possible to use the preprocessing stage to estimate the effectiveness of the considered partitioning.

So the process of estimating the value (2) for a given $\tilde{X}$ is as follows. We construct a random sample $\alpha^1, \ldots, \alpha^N$, where $\alpha^j = \left(\alpha_1^j, \ldots, \alpha_d^j \right)$, $j \in \{1, \ldots, N\}$ is a truth assignment of variables from $\tilde{X}$. Then consider values $\zeta^j = \xi_{C,A} \left(\alpha^j \right), j = 1, \ldots, N$ and calculate the value

$$F_{C,A} \left(\tilde{X} \right) = 2^d \cdot \left(\frac{1}{N} \cdot \sum_{j=1}^{N} \zeta^j \right). \tag{3}$$

By the above, if N is large enough then the value of $F_{C,A} \left(\tilde{X} \right)$ can be considered as a good approximation of (2). Therefore, instead of searching for a decomposition set with minimal value (2) one can search for a decomposition set with minimal value of $F_{C,A} \left(\cdot \right)$. Below we refer to function $F_{C,A} \left(\cdot \right)$ as *predictive function*.

3   Algorithm for Minimization of Predictive Function

As we already noted above, different partitionings of the same SAT instance can have different values of $t_{C,A} \left(\tilde{X} \right)$. In practice it is important to be able to find partitionings that can be processed in realistic time. Below we will describe the scheme of automatic search for good partitionings that is based on the procedure minimizing the predictive function value in the special search space.

So we consider the satisfiability problem for some CNF C. Let $X = \{x_1, \ldots, x_n\}$ be the set of all Boolean variables in this CNF and $\tilde{X} \subseteq X$ be an arbitrary decomposition set. The set $\tilde{X}$ can be represented by the binary vector $\chi = (\chi_1, \ldots, \chi_n)$. Here

$$\chi_i = \begin{cases} 1, if\ x_i \in \tilde{X} \\ 0, if\ x_i \notin \tilde{X} \end{cases}, i \in \{1, \ldots, n\}$$

With an arbitrary vector $\chi \in \{0,1\}^n$ we associate the value of function $F(\chi)$ computed in the following manner. For vector χ we construct the corresponding set $\tilde{X}$ (it is formed by variables from X that correspond to 1 positions in χ). Then we generate a random sample $\alpha^1, \ldots, \alpha^N$, $\alpha^j \in \{0,1\}^{|\tilde{X}|}$ and solve SAT for CNFs $C \left[\tilde{X}/\alpha^j \right]$. For each of these SAT instances we measure ζ^j — the runtime of algorithm A on the input $C \left[\tilde{X}/\alpha^j \right]$. After this we calculate the value of $F_{C,A} \left(\tilde{X} \right)$ according to (3). As a result we have the value of $F(\chi)$ in the considered point of the search space. Then we solve the problem $F(\chi) \to min$ over the set $\{0,1\}^n$.

The minimization of function $F(\cdot)$ over $\{0,1\}^n$ is considered as an iterative process of transitioning between the points of the search space. By $N_\rho(\chi)$ we denote the neighborhood of point χ of radius ρ in the search space $\{0,1\}^n$. The point from which the search starts we denote as χ_{start}. We will refer to the decomposition set specified by this point as $\tilde{X}_{start}$. The current Best Known Value of $F(\cdot)$ is denoted by F_{best}. The point in which the F_{best} was achieved we denote as χ_{best}. By χ_{center} we denote the point the neighborhood of which is processed at the current moment. We call the point, in which we computed the value $F(\cdot)$, a *checked point*. The neighborhood $N_\rho(\chi)$ in which all the points are checked is called *checked neighborhood*. Otherwise the neighborhood is called *unchecked*.

For the minimization of $F(\cdot)$ we employed the tabu search strategy [5]. According to this approach the points from the search space, in which we already calculated the values of function $F(\cdot)$ are stored in special tabu lists, to which we refer below as to L_1 and L_2. The L_1 list contains only points with checked neighborhoods. The L_2 list contains checked points with unchecked neighborhoods. Below we present the pseudocode of the tabu search algorithm for $F(\cdot)$ minimization.

Algorithm 1. Tabu search altorithm for minimization of the predictive function

---

    Input: CNF C, initial point χ_{start}
    Output: Pair $\langle \chi_{best}, F_{best} \rangle$, where F_{best} is a prediction for C, χ_{best} is a
                 corresponding decomposition set

1   $\langle \chi_{center}, F_{best} \rangle \leftarrow \langle \chi_{start}, F(\chi_{start}) \rangle$
2   $\langle L_1, L_2 \rangle \leftarrow \langle \emptyset, \chi_{start} \rangle$ `// initialize tabu lists`
3   **repeat**
4   bestValueUpdated $\leftarrow$ false
5   **repeat**   `// check neighborhood`
6          $\chi \leftarrow$ any unchecked point from $N_\rho(\chi_{center})$
7   compute $F(\chi)$
8   markPointInTabuLists(χ, L_1, L_2) `// update tabu lists`
9   **if** $F(\chi) < F_{best}$ **then**
10            $\langle \chi_{best}, F_{best} \rangle \leftarrow \langle \chi, F(\chi) \rangle$
11   bestValueUpdated $\leftarrow$ true
12   **until** $N_\rho(\chi_{center})$ *is checked*
13   **if** bestValueUpdated **then** $\chi_{center} \leftarrow \chi_{best}$
14
15   **else** $\chi_{center} \leftarrow$ getNewCenter(L_2)
16
17 **until** timeExceeded$()$ *or* $L_2 = \emptyset$
18 **return** $\langle \chi_{best}, F_{best} \rangle$

In this algorithm the function markPointInTabuLists(χ, L_1, L_2) adds the point χ to L_2 and then marks χ as checked in all neighborhoods of points from L_2 that contain χ. If as a result the neighborhood of some point χ' becomes checked, the point χ' is removed from L_2 and is added to L_1. If we have processed all the points in the neighborhood of χ_{center} but could not improve the F_{best} then as the new point χ_{center} we choose some point from L_2. It is done via the function `getNewCenter`(L_2). To choose the new point in this case one can use various heuristics. At the moment the tabu search algorithm chooses the point for which the total conflict activity [7] of Boolean variables, contained in the corresponding decomposition set, is the largest.

4   Computational Experiments

The algorithms presented in the previous section were implemented as the MPI-program PDSAT[1]. In PDSAT there is one leader process, all the other are computing processes (each process corresponds to 1 CPU core). For every new point $\chi = \chi\left(\tilde{X}\right)$ from the search space the leader process creates a random sample of size N (we use neighborhoods of radius $\rho = 1$). Each assignment from this sample in combination with the original CNF C define the SAT instance from the decomposition family $\Delta_C\left(\tilde{X}\right)$. These SAT instances are solved by computing processes. The value of the predictive function is always computed assuming that the decomposition family will be processed by 1 CPU core. The fact that the processing of $\Delta_C\left(\tilde{X}\right)$ consists in solving independent subproblems makes it possible to extrapolate the estimation obtained to an arbitrary parallel (or distributed) computing system. The computing processes use slightly modified MINISAT solver[2] for solving SAT instances.

Below we present the results of computational experiments in which PDSAT was used on the computing cluster "Academician V.M. Matrosov" to estimate the time required to solve problems of logical cryptanalysis of the A5/1 [2] and Bivium [3] keystream generators. The SAT instances that encode these problems were produced using the TRANSALG system [9]. All the estimations presented below are in seconds.

4.1   Time Estimations for Logical Cryptanalysis of A5/1

For the first time we considered the logical cryptanalysis of the A5/1 keystream generator in [10]. In that paper we described the corresponding algorithm in detail, therefore we will not do it in the present paper. We considered the crypt-analysis problem for the A5/1 keystream generator in the following form: given the 114 bits of keystream we needed to find the secret key of length 64 bits, which produces this keystream (in accordance with the A5/1 algorithm). During predictive function minimization PDSAT used random samples of size $N = 10^4$ SAT instances and worked for 1 day using 5 computing nodes (160 CPU cores in total) within the computing cluster. Using the tabu search algorithm we found the set $S_2 = \{x_2, ..., x_{10}, x_{20}, ..., x_{30}, x_{39}, x_{40}, x_{42}, ..., x_{52}\}$. We compared the time estimations for this set with that of the decomposition set S_1, the structure of which was described in [10]. The S_1 set was constructed manually based on the

---

[1] https://github.com/Nauchnik/pdsat.
[2] http://minisat.se.

analysis of the algorithmic features of the A5/1 keystream generator. The value of predictive function for S_1 is equal to 4.45140e+08, and for S_2 is equal to 4.64428e+08.

Since the obtained estimations turned out to be realistic, we decided to solve non-weakened cryptanalysis instances for A5/1. For this purpose we used the BOINC-based volunteer computing project SAT@home[3]. In total we performed two computational experiments on solving cryptanalysis of A5/1 in SAT@home. In the first experiment we solved 10 cryptanalysis instances using the S_1 set and in the second we solved same 10 instances using the S_2 set. To construct the corresponding tests we used the known rainbow-tables for the A5/1 algorithm. These tables provide about 88 % probability of success when analyzing 8 bursts of keystream (i.e. 914 bits). We randomly generated 1000 instances and applied the rainbow-tables technique to analyze 8 bursts of keystream, generated by A5/1. Among these 1000 instances the rainbow-tables could not find the secret key for 125 problems. From these 125 instances we randomly chose 10 and in the computational experiments applied the SAT approach to the analysis of first bursts of the corresponding keystream fragments (114 bits). In all cases we successfully found the secret keys.

4.2   Time Estimations for Logical Cryptanalysis of Bivium

The Bivium keystream generator [3] uses two shift registers. The first register contains 93 cells and the second contains 84 cells. To initialize the cipher, a secret key of length 80 bit is put to the first register, and a fixed (known) initialization vector of length 80 bit is put to the second register. All remaining cells are filled with zeros. An initialization phase consists of 708 rounds during which keystream output is not released.

In accordance with [11] we considered cryptanalysis problem for Bivium in the following formulation. Based on the known fragment of keystream we search for the values of all registers cells at the end of the initialization phase. Therefore, in our experiments we used the CNF encoding where the initialization phase was omitted. Usually it is believed that to uniquely identify the secret key it is sufficient to consider keystream fragment of length comparable to the total length of shift registers. Here we followed [4,11] and set the keystream fragment length for Bivium cryptanalysis to 200 bits. In the role of $\tilde{X}_{start}$ for the cryptanalysis of Bivium we chose the set formed by the variables encoding the cells of registers of the generator considered at the end of the initialization phase. Further we refer to these variables as *starting* variables. Therefore $\left| \tilde{X}_{start} \right| = 177$. During predictive function minimization PDSAT used random samples of size $N = 10^5$ SAT instances and worked for 1 day using 5 computing nodes (160 CPU cores in total) within the computing cluster. Time estimations obtained for the Bivium cryptanalysis is $F_{best} = 3.769 \times 10^{10}$.

In [4,11] a number of time estimations for logical cryptanalysis of Bivium were proposed. In particular, in [4] several fixed types of decomposition sets were

---

[3] http://sat.isa.ru/pdsat/.

analyzed. Time estimation for the best decomposition set from [4] is equal to 1.637×10^{13}, it was calculated using random samples of size 10^2. Authors of [11] constructed estimations for the sets of variables chosen during the solving process and extrapolated the estimations obtained to time points of the solving process that lay in the distant future. Apparently, as it is described in [11], the random samples of size 10^2 and 10^3 were used. In the Table 1 all three estimations mentioned above are demonstrated. The performance of one core of the processor we used in our experiments is comparable with that of one core of the processor used in [11].

Table 1. Time estimations for the Bivium cryptanalysis problem

Source	Sample size	Time estimation
From [4]	10^2	1.637×10^{13}
From [11]	10^3	9.718×10^{10}
Found by PDSAT	10^5	3.769×10^{10}

To compare obtained time estimations with real solving time we solved several weakened logical cryptanalysis problems for Bivium. Below we use the notation $BiviumK$ to denote a weakened problem for Bivium with known values of K starting variables. We used the volunteer computing project SAT@home to solve 5 instances of $Bivium9$. For all considered instances the time required to solve the corresponding instances agrees well with our estimations. An extended version of this paper can be found online.[4]

Acknowledgements. The authors wish to thank Stepan Kochemazov for numerous valuable comments. This work was partly supported by Russian Foundation for Basic Research (grants 14-07-00403-a and 15-07-07891-a) and by the President of Russian Federation grant for young scientists SP-1184.2015.5.

References

1. Biere, A., Heule, M., van Maaren, H., Walsh, T. (eds.): Handbook of Satisfiability, Frontiers in Artificial Intelligence and Applications, vol. 185. IOS Press, Amsterdam (2009)
2. Biryukov, A., Shamir, A., Wagner, D.: Real time cryptanalysis of A5/1 on a PC. In: Schneier, B. (ed.) FSE 2000. LNCS, vol. 1978, pp. 1–18. Springer, Heidelberg (2001)
3. De Cannière, C.: Trivium: a stream cipher construction inspired by block cipher design principles. In: Katsikas, S.K., López, J., Backes, M., Gritzalis, S., Preneel, B. (eds.) ISC 2006. LNCS, vol. 4176, pp. 171–186. Springer, Heidelberg (2006)
4. Eibach, T., Pilz, E., Völkel, G.: Attacking bivium using SAT solvers. In: Kleine Büning, H., Zhao, X. (eds.) SAT 2008. LNCS, vol. 4996, pp. 63–76. Springer, Heidelberg (2008)

---

[4] http://arxiv.org/abs/1507.00862.

5. Glover, F., Laguna, M.: Tabu Search. Kluwer Academic Publishers, NewYork (1997)
6. Hyvärinen, A.E.J.: Grid Based Propositional Satisfiability Solving. Ph.d. thesis, Aalto University (2011)
7. Marques-Silva, J., Lynce, I., Malik, S.: Conflict-driven clause learning SAT solvers. In: Biere et al. [1], pp. 131–153
8. Metropolis, N., Ulam, S.: The monte carlo method. J. Amer. statistical assoc. **44**(247), 335–341 (1949)
9. Otpuschennikov, I., Semenov, A., Kochemazov, S.: Transalg: a tool for translating procedural descriptions of discrete functions to SAT (tool paper). CoRR abs/1405.1544 (2014)
10. Semenov, A., Zaikin, O., Bespalov, D., Posypkin, M.: Parallel logical cryptanalysis of the generator A5/1 in BNB-grid system. In: Malyshkin, V. (ed.) PaCT 2011. LNCS, vol. 6873, pp. 473–483. Springer, Heidelberg (2011)
11. Soos, M., Nohl, K., Castelluccia, C.: Extending SAT solvers to cryptographic problems. In: Kullmann, O. (ed.) SAT 2009. LNCS, vol. 5584, pp. 244–257. Springer, Heidelberg (2009)

A Class of Non-optimum-time $3n$-Step FSSP Algorithms - A Survey

Hiroshi Umeo$^{(\boxtimes)}$, Masashi Maeda, Akihiro Sousa, and Kiyohisa Taguchi

Univ. of Osaka Electro-Communication,
Hastu-cho, 18-8, Neyagawa-shi, Osaka 572-8530, Japan
umeo@cyt.osakac.ac.jp

Abstract. Synchronization of large-scale networks is an important and fundamental computing primitive in parallel and distributed systems. The synchronization in cellular automata, known as firing squad synchronization problem (FSSP), has been studied extensively for more than fifty years, and a rich variety of synchronization algorithms has been proposed. In the present paper, we give a brief survey on a class of non-optimum-time $3n$-step FSSP algorithms for synchronizing one-dimensional (1D) cellular automata of length n in $3n \pm O(\log n)$ steps and present a comparative study of a relatively large-number of their implementations. We also propose two smallest-state, known at present, implementations of the $3n$-step algorithm. The paper gives the first complete transition rules sets for the class of non-optimum-time $3n$-step FSSP algorithms developed so far.

1 Introduction

The synchronization in ultra-fine-grained parallel computational model of cellular automata has been known as the firing squad synchronization problem (FSSP) since its development, in which it was originally proposed by J. Myhill in the book edited by Moore (1964) to synchronize all/some parts of self-reproducing cellular automata.

In the present paper, we give a brief survey on recent developments in a class of non-optimum-time FSSP algorithms for one-dimensional (1D) cellular automata. Here we focus our attention to the 1D FSSP algorithms having $3n \pm O(\log n)$ synchronization steps and present a comparative study of a relatively large-number of their implementations. We also propose two smallest-state, known at present, implementations included in the same class of the algorithms. A class of $3n$-step algorithms is an interesting class of synchronization algorithms among many variants of FSSP algorithms due to its simplicity and straightforwardness and it is important in its own right in the design of other cellular algorithms. The first optimum-time FSSP algorithm designed by Goto (1962) uses a $3n$-step algorithm in its synchronization phase. This paper gives the first complete transition rule sets for the class of non-optimum-time $3n$-step FSSP algorithms developed so far.

© Springer International Publishing Switzerland 2015
V. Malyshkin (Ed.): PaCT 2015, LNCS 9251, pp. 231–245, 2015.
DOI: 10.1007/978-3-319-21909-7_22

Specifically, we attempt to answer the following questions:

- First, what is the local transition rule set for those FSSP algorithms?
- Are all previously presented transition rule sets correct?
- Do these rule sets contain redundant rules? If so, what is the exact rule set?
- How do the algorithms compare with each other?
- Are there still any new implementations of the non-optimum-time FSSP algorithms?
- Can we generalize those algorithms to a generalized FSSP, where an initial general is located at any position of the array?
- What is the state-change complexity in those algorithms?

In Sect. 2 we give a description of the 1D FSSP and review some basic results on the 1D FSSP algorithms. Section 3 gives a survey on those non-optimum-time FSSP algorithms. We make implementations of those algorithms on a computer, check and compare their transition rule sets.

2   A Class of 3n-Step Synchronization Algorithms

2.1   Firing Squad Synchronization Problem

Fig. 1 shows a finite one-dimensional (1D) cellular array consisting of n cells. Each cell is an identical (except the border cells) finite-state automaton. The array operates in lock-step mode in such a way that the next state of each cell (except border cells) is determined by both its own present state and the present states of its left and right neighbors. All cells (*soldiers*), except the left end cell (*general*), are initially in the quiescent state at time $t = 0$ with the property that the next state of a quiescent cell with quiescent neighbors is the quiescent state again. At time $t = 0$, the left end cell C_1 is in the *fire-when-ready* state, which is the initiation signal for the array. The firing squad synchronization problem is to determine a description (state set and next-state function) for cells that ensures all cells enter the *fire* state at exactly the same time and for the first time. The set of states and the next-state function must be independent of n see Umeo (2009) for details.

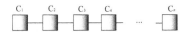

Fig. 1. A one-dimensional (1D) cellular automaton.

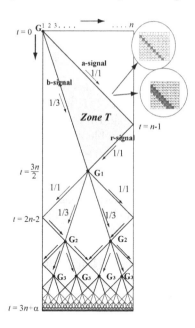

Fig. 2. A space-time diagram for a class of $3n$-step FSSP algorithm and its design parameters: thread-width and *Zone T* in the space-time diagram.

2.2   A Class of $3n$-Step Synchronization Algorithms

A class of $3n$-step algorithms is an interesting class of synchronization algorithms among many variants of FSSP algorithms due to its simplicity and straightforwardness and it is important in its own right in the design of cellular algorithms. Figure 2 shows a space-time diagram for the well-known $3n$-step firing squad synchronization algorithm. The synchronization process can be viewed as a typical *divide-and-conquer strategy* that operates in parallel in the cellular space. An initial *general* G, located at left end of the array of size n, generates two signals, referred to as *a-signal* and *b-signal*, which propagate in the right direction at a speed of $1/1$ and $1/3$, respectively. The a-signal arrives at the right end at time $t = n - 1$, reflects there immediately, and continues to move at the same speed in the left direction. The reflected signal is referred to as *r-signal*. The b- and the r-signals meet at one or two center cells of the arry, depending on the parity of n. In the case that n is odd, the cell $C_{\lceil n/2 \rceil}$ becomes a *general* at time $t = 3\lceil n/2 \rceil - 2$. The *general* is responsible for synchronizing both its left and right halves of the cellular space. Note that the *general* is shared by the two halves. In the case that n is even, two cells $C_{\lceil n/2 \rceil}$ and $C_{\lceil n/2 \rceil+1}$ become the next *general* at time $t = 3\lceil n/2 \rceil$. Each *general* is responsible for synchronizing its left and right halves of the cellular space, respectively.

Thus, at time $t = t_{center}$:

$$t_{center} = \begin{cases} 3\lceil n/2 \rceil - 2 & n: \text{ odd} \\ 3\lceil n/2 \rceil & n: \text{ even}, \end{cases} \tag{1}$$

the array knows its center point and generates one or two new *general(s)* G_1. The new *general(s)* G_1 generates the same 1/1- and 1/3-speed signals in both left and right directions simultaneously and repeat the same procedures as above. Thus, the original synchronization problem of size n is divided into two sub-problems of size $\lceil n/2 \rceil$. In this way, the original array is split into equal two, four, eight, ..., subspaces synchronously. Note that the first general generated at the center G_1 itself is synchronized at time $t = t_{center}$, and the second general G_2 are also synchronized, and the generals generated after that time on are also synchronized. In the last, the original problem of size n can be split into small sub-problems of size 2. In this way, by increasing the synchronized generals step by step, the array can be synchronized. Most of the $3n$-step synchronization algorithms developed so far in Fischer (1965), Herman (1972), Minsky (1967), Umeo, Maeda, and Hongyo (2006), and Yunès (1994, 2007, 2008) are based on the similar scheme. It can be seen that, from the path of the b-signal with or without 1 step delay at the center points at each halving iteration, the time complexity $T(n)$ for synchronizing n cells is $T(n) = 3n \pm O(\log n)$.

A question is **"How can we implement the synchronization diagram above in terms of a small-state finite state automaton?"**.

The three signals a-, b-, and r-signals in the space-time diagram in Fig. 2 play an important role in finding the center cell(s). A triangle area circled by these three signals is also important for its implementation. We call the area *zone T*.

2.3 Complexity Measures and Properties for Synchronization Algorithms

– **Time** Any solution to the original FSSP with a general at one end can be easily shown to require $(2n - 2)$ steps for synchronizing n cells, since signals on the array can propagate no faster than one cell per one step, and the time from the general's instruction until the final synchronization must be at least $2n - 2$.

Theorem 1 Goto (1962) (Lower Bound) The minimum time in which the firing squad synchronization could occur is $2n - 2$ steps, where the general is located on the left end.

Theorem 2 Goto (1962) There exists a cellular automaton that can synchronize any 1D array of length n in optimum $2n - 2$ steps, where the general is located on the left end.

– **Number of States** The following three distinct states: the *quiescent* state, the *general* state, and the *firing* state, are required in order to define any cellular automaton that can solve the FSSP. Note that the boundary state for C_0 and C_{n+1} is not generally counted as an internal state. Balzer (1967) and Sanders (1994) showed that no four-state optimum-time solution exists. Umeo and Yanagihara (2009), Yunès (2008), and Umeo, Kamikawa, and Yunès (2009) gave some 5- and 4-state *partial* solutions that can solve the synchronization problem for infinitely many sizes n, but not all, respectively. The solution is referred to as *partial* solution, which is compared with usual *full* solutions that can solve the problem for all cells.

Theorem 3 Balzer (1967), Sanders (1994) There is no four-state *full* solution that can synchronize n cells.

Yunès (2008) and Umeo, Kamikawa, and Yunès (2009) developed 4-state partial solutions based on Wolfram's rules 60 and 150. They can synchronize any array/ring of length $n = 2^k$ for any positive integer k. Details can be found in Yunès (2008) and Umeo, Kamikawa, and Yunès (2009).

Theorem 4 Yunes (2008), Umeo et al. (2009) There exist 4-state *partial* solutions to the FSSP.

Concerning the optimum-time full solutions, Waksman (1966) presented a 16-state optimum-time synchronization algorithm. Afterward, Balzer (1967) and Gerken (1987) developed an eight-state algorithm and a seven-state synchronization algorithm, respectively, thus decreasing the number of states required for the synchronization. Mazoyer (1987) developed a six-state synchronization algorithm which, at present, is the algorithm having the fewest states for 1D arrays.

Theorem 5 Mazoyer (1987) There exists a 6-state *full* solution to the FSSP.

- **Number of Transition Rules** Any k-state (excluding the boundary state) transition table for the synchronization has at most $(k-1)k^2$ entries in $(k-1)$ matrices of size $k \times k$. The number of transition rules reflects a complexity of synchronization algorithms.

- **Filled-In Ratio** To measure the density of entries in the transition table, we introduce a measure *filled-in ratio* of the state transition table. The filled-in ratio of the state transition table $\mathcal{A}$ is defined as follows: $f_{\mathcal{A}} = e/e_{total}$, where e is the number of exact entries of the next state defined in the table $\mathcal{A}$ and e_{total} is the number of possible entries defined such that $e_{total} = (k-1)k^2$, where k is the number of internal states of the table $\mathcal{A}$.

- **Symmetry vs. Asymmetry** Herman (1971, 1972) investigated the computational power of symmetrical cellular automata, motivated by a biological point of view. Szwerinski (1982) and Kobuchi (1987) considered a computational relation between symmetrical and asymmetrical CAs with von Neumann neighborhood.
  A transition table is said to be *symmetric* if and only if the transition table

$\delta : \mathcal{Q}^3 \to \mathcal{Q}$ such that $\delta(x, y, z) = \delta(z, y, x)$ holds, for any state x, y, z in $\mathcal{Q}$. A symmetrical cellular automaton has a property that the next state of a cell depends on its present state and the states of its two neighbors, but it is same if the states of the left and right neighbors are interchanged. Thus, the symmetrical CA has no ability to distinguish between its left and right neighbors.

- **State-Change Complexity** Vollmar (1982) introduced a *state-change complexity* in order to measure the efficiency of cellular automata, motivated by energy consumption in certain physical memory systems. The state-change complexity is defined as the sum of *proper* state changes of the cellular space during the computations. Vollmar (1982) showed that $\Omega(n \log n)$ state-changes are required by the cellular space for the synchronization of n cells in $(2n - 2)$ steps. Gerken (1987) presented an optimum-time $\Theta(n \log n)$ state-change synchronization algorithm.

Theorem 6 Vollmar (1982) (Lower Bound) $\Omega(n \log n)$ state-change is necessary for synchronizing n cells.

Theorem 7 Gerken (1987) $\Theta(n \log n)$ state-change is sufficient for synchronizing n cells in $2n - 2$ steps.

2.4   A Brief History of the Developments of the $3n$-Step FSSP Algorithms and Their Implementations

The $3n$-step algorithm is a simple and straightforward one that exploits a parallel divide-and-conquer strategy based on an efficient use of 1/1- and 1/3-speed of signals. After Minsky and McCarthy (See Minsky (1967)) gave an idea for designing the $3n$-step synchronization algorithm, Fischer (1965) also presented a $3n$-step synchronization algorithm, yielding a 15-state implementation, respectively. Herman (1972) implemented the $3n$-step algorithm in terms of 10-state finite state automaton. Yunès (1994) developed two seven-state synchronization algorithms. His algorithms were interesting in that he decreased the number of internal states of each cellular automaton by extending the width of signal threads in the space-time diagram. Umeo, Maeda, and Hongyo (2006) presented a 6-state $3n$-step algorithm. Afterward, Yunès (2008) also presented a 6-state $3n$-step algorithm.

3   Implementations of the $3n$-Step FSSP Algorithms

The non-optimum-time $3n$-step FSSP algorithms that we discuss in this paper are as follows:

- **Fischer** (1965) **algorithm,**
- **Minsky-McCarthy** (1967) **algorithm,**
- **Herman** (1972) **10-state algorithm,**

- Yunès (1994) **7-state algorithm**,
- Umeo-Maeda-Hongyo (2006) **6-state algorithm**,
- Yunès (2008) **6-state algorithm**,
- **Two 6-state algorithms proposed in this paper**, and
- **Umeo-Yanagihara (2007) 5-state algorithm**.

In this section, we examine the state transition rule sets for these firing squad synchronization protocols developed so far above. Each state on the first row (column) indicates a state of right (left) neighbor cell, respectively. Each entry of the sub-tables shows a state at the next step. The state "*" that appears in the state transition table is a border state for the left and right end cells. It is noted that, according to the conventions in FSSP, the border state "*" is not counted in the number of states. We have tested the validity of those tables for any array of length n such that $2 \leq n \leq 500$. It reveals that all of the rule tables tested in this paper include no redundant rules. The transition table can also be expressed in a usual 4-tuple $W \ X \ Y \rightarrow Z$ which represents a state transition rule that an automaton in currently in state X, with its left neighbor in state W and the right neighbor in state Y will enter state Z at the next step.

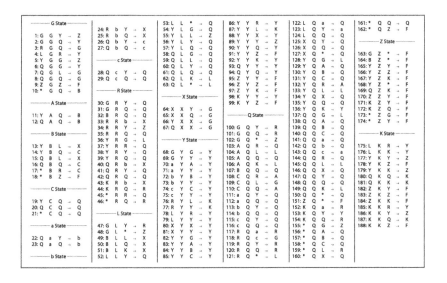

Fig. 3. A 15-state transition table of the Fischer (1965) algorithm.

3.1 Fischer's Algorithm: $\mathcal{A}_1$

Fischer (1965) firstly presented an idea for synchronizing any 1D array in non-optimum-time. We implemented his space-time diagram (See Fig. 1 in Fischer (1965]) for the synchronization in terms of a finite state automaton with 15 states. The set Q of internal states for the Fischer's algorithm is $Q=\{$G, Q, A, B, C, a, b, c, R, L, X, Y, Z, K, F$\}$, where the state G is the initial *general* state, Q is the *quiescent* state, and F is the *firing* state, respectively. The table itself, consisting of 188 4-tuple rules, is constructed newly in this paper. See Fig. 3. The readers find that the table is very sparse in a sense that each table has many empty entries. The filled-in ratio of the implementation is $f_{\text{Fischer [1965]}}$ $= 188/14 \times 15 \times 15 = 5.9$ (%). The time complexity for synchronizing any array of length n is $3n - 4$.

3.2   Minsky-McCarthy Algorithm: $\mathcal{A}_2$

Minsky and McCarthy (See Minsky (1967)) also presented an idea for designing the $3n$-step synchronization algorithm. Yunès (1994) gave an implementation of the algorithm for 14 cells in terms of a 13-state finite state automaton. Figure 4, consisting of 138 rules, is the transition table constructed in this paper based on Fig. 2 in Yunès (1994). The set Q of internal states for the Minsky-McCarthy algorithm is $Q=\{$I, Q, A, B, C, a, b, c, R, L, X, Y, F$\}$, where the state I is the initial *general* state, Q is the *quiescent* state, and F is the *firing* state, respectively. The filled-in ratio of the implementation is $f_{\text{Minsky}-\text{McCarthy [1967]}} = 138/12 \times 13 \times 13 = 6.8$ (%). The time complexity for synchronizing any array of length n is $3n + O(\log n)$.

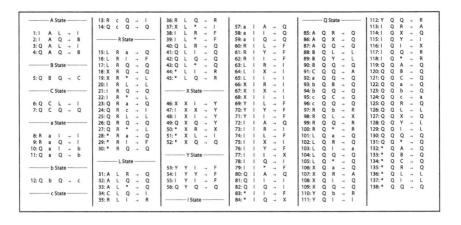

Fig. 4. A 13-state transition table of the Minsky-McCarthy (1967) algorithm.

3.3   Herman's 10-State Algorithm: $\mathcal{A}_3$

Herman (1972) also gave a 10-state implementation for the $3n$-step synchronization algorithm. Figure 5, consisting of 155 rules, is the transition table constructed in this paper based on Figs. 3, 4, and 5 in Herman (1972). The set Q of

──── S State ────				29: V S R → I	56: X I S → G	80: G U R → G	──── X State ────				──── G State ────			
				30: V S S → S	57: X I I → G	81: * U J → V								
				31: G S I → I	58: G I S → J	82: * U R → G	104: S X I → I				129: S G S → G		
1: S S S → S				32: G S J → I	59: G I J → R		105: S X U → X				130: S G G → G		
2: S S I → I				33: G S X → S	60: G I G → G	──── W State ────		106: S X W → S				131: I G I → G		
3: S S J → S				34: * S S → S	61: G I * → G			107: S X V → X				132: I G G → G		
4: S S U → S				35: * S I → I	62: * I S → J	83: S W X → X	108: I X S → I				133: J G J → G		
5: S S W → U				36: * S J → S	63: * I J → R	84: S W G → X	109: I X G → I				134: J G G → G		
6: S S R → I				37: * S X → S	64: * I G → G	85: I W X → X	110: U X S → X				135: U G U → G		
7: S S X → S						86: I W G → X	111: U X G → X				136: U G G → G		
8: S S V → S				──── I State ────		87: X W S → X	112: W X S → S				137: W G W → G		
9: S S G → S					──── J State ────		88: G W S → X	113: W X G → S				138: W G G → G		
10: S S * → S						89: G W I → X	114: V X S → X				139: R G R → G		
11: I S S → I				38: S I I → G	65: S J S → S	90: * W S → X	115: V X G → X				140: R G G → G		
12: I S V → I				39: S I J → J	66: S J I → S	91: * W I → X	116: G X I → I				141: X G X → G		
13: I S G → I				40: S I R → J	67: I J S → S		117: G X U → X				142: X G G → G		
14: I S * → I				41: S I X → G	68: I J U → S		118: G X W → S				143: V G V → G		
15: J S S → S				42: S I G → J	69: I J G → U	──── R State ────		119: G X V → X				144: V G G → G		
16: J S V → S				43: I I S → G	70: U J I → S		120: * X I → I				145: G G S → G		
17: J S G → S				44: I I X → G	71: G J I → U	92: S R G → R	121: * X U → X				146: G G I → G		
18: J S I → S				45: J I S → G	72: * J I → U	93: S R * → R	122: * X W → S				147: G G J → G		
19: U S S → S				46: J I U → I		94: I R G → S	123: * X V → X				148: G G U → G		
20: W S S → U				47: J I W → I		95: I R * → S					149: G G W → G		
21: W S J → I				48: J I G → R	──── U State ────	96: U R G → G					150: G G R → G		
22: R S S → I				49: J I * → R		97: U R * → G	──── V State ────				151: G G X → G		
23: R S V → I				50: U I I → I	73: S U X → V	98: G R S → R					152: G G V → G		
24: X S S → S				51: U I R → I	74: I U X → I	99: G R I → S	124: S V X → W				153: G G G → G		
25: X S G → S				52: W I R → I	75: J U G → G	100: G R U → R	125: S V G → W				154: G G * → F		
26: V S S → S				53: R I S → J	76: R U G → G	101: * R S → R	126: X V S → W				155: * G G → F		
27: V S I → S				54: R I U → I	77: X U S → V	102: * R I → S	127: G V S → W						
28: V S J → S				55: R I W → I	78: X U I → I	103: * R U → G	128: * V S → W						
					79: G U J → V								

Fig. 5. A 10-state transition table of the Herman (1972) algorithm.

──── Q State ────				24: d Q * → Q	──── Z State ────			69: * Z Q → d	88: Q C d → d	110: Q d Z → Q
				25: * Q Q → Q				70: * Z A → A	89: G C G → C	111: Q d A → A
				26: * Q Z → Z	46: Q Z Q → d			71: * Z C → C	90: G C A → C	112: Q d C → Q
1: Q Q Q → Q				27: * Q A → Q	47: Q Z I → d			72: * Z d → Z	91: G C C → C	113: Q d d → Q
2: Q Q G → Z				28: * Q d → Q	48: Q Z Z → d				92: Z C C → C	114: G d Q → C
3: Q Q Z → Z					49: Q Z d → d			──── A State ────	93: Z C d → Z	115: Z d Q → Q
4: Q Q A → Q				──── G State ────		50: Q Z * → d			94: A C G → G	116: Z d Z → Q
5: Q Q C → A						51: Q Z Q → d		73: Q A A → d	95: A C A → C	117: Z d A → Q
6: Q Q d → Q				29: Q G Q → Q	52: G Z A → A			74: Q A C → Q	96: A C d → Z	118: Z d C → A
7: Q Q * → Q				30: Q G G → G	53: G Z C → Z			75: Q A d → d	97: C C G → C	119: Z d d → C
8: G Q Q → Z				31: G G Q → G	54: G Z * → C			76: G A d → C	98: C C Z → C	120: Z d * → Q
9: G Q d → Z				32: G G Z → Q	55: Z Z Q → d			77: Z A d → C	99: C C C → F	121: A d Q → d
10: G Q * → Z				33: G G A → Q	56: Z Z d → d			78: A A Q → C	100: C C d → d	122: A d Z → Q
11: Z Q Q → Z				34: G G C → C	57: A Z G → d			79: C A Z → C	101: C C * → F	123: A d A → G
12: Z Q Z → Z				35: G G d → d	58: A Z d → d			80: C A d → A	102: d C Q → d	124: A d d → G
13: Z Q d → Z				36: Z G G → Q	59: A Z * → A			81: C A * → C	103: d C Z → Z	125: C d Q → Q
14: Z Q * → Z				37: Z G Z → Q	60: C Z G → Z			82: d A Q → A	104: d C C → d	126: C d Z → A
15: A Q Q → Q				38: A G G → G	61: C Z d → C			83: d A G → C	105: d C d → d	127: C d C → Q
16: A Q Z → Z				39: A G A → G	62: C Z * → C			84: d A Z → C	106: * C A → A	128: C d d → Q
17: A Q d → Q				40: C G C → Q	63: d Z Q → d			85: d A d → d	107: * C C → F	129: d d Q → d
18: C Q Q → A				41: C G C → C	64: d Z Z → d			86: * A A → d	108: * C d → C	130: d d Z → d
19: d Q Q → Q				42: d G G → d	65: d Z A → A			87: * A C → C		131: d d A → G
20: d Q G → Z				43: d G d → d	66: d Z C → Z				──── d State ────	132: d d C → Q
21: d Q Z → Z				44: * G Q → G	67: d Z d → G			──── C State ────		133: * d Z → Q
22: d Q A → Q				45: * G Z → C	68: d Z * → Z				109: Q d G → C	134: * d C → Q
23: d Q d → Q											

Fig. 6. A 7-state transition table of the Yunès (1994) algorithm.

internal states for the Herman algorithm is $Q=\{I, S, J, U, W, R, X, V, G, F\}$, where the state I is the initial *general* state, S is the *quiescent* state, and F is the *firing* state, respectively. The filled-in ratio of the implementation is $f_{\text{Herman [1972]}}$ $= 155/9 \times 10 \times 10 = 17.2$ (%). The time complexity for synchronizing any array of length n is $3n + O(\log n)$.

3.4   Yunès Seven-State Algorithm: $\mathcal{A}_4$

Yunès (1994) presented two 7-state implementations for the $3n$-step FSSP algorithms and decreased the number of states required. The set Q of internal states for the first Yunès algorithm is $Q=\{G, Q, A, C, d, Z, F\}$, where the state G is the initial *general* state, Q is the *quiescent* state, and F is the *firing* state, respectively. The following Fig. 6, consisting of 134 rules, is the transition table. The filled-in ratio of the implementation is $f_{\text{Yunès [1994]}} = 134/6 \times 7 \times 7 = 45.6$ (%). The time complexity for synchronizing any array of length n is $3n + O(\log n)$.

Yunès (1994) also gave a different 7-state implementation for the $3n$-step FSSP algorithm. The set Q of internal states for the Yunès algorithm is $Q=\{G, Q, A, C, d, Z, F\}$, where the state Z is the initial *general* state, Q is the *quiescent* state, and F is the *firing* state, respectively. The following Fig. 7, consisting of 134 rules, is the transition table. The filled-in ratio of the implementation is $f_{\text{Yunès [1994]}} = 134/6 \times 7 \times 7 = 45.6$ (%). The time complexity for synchronizing any array of length n is $3n + O(\log n)$. A major difference between these two implementations is a center marking for each splitting.

Q State				G State / Z State				A State / C State				C / d / G State			
1: Q Q Q → Q	25: d Q * → Q	48: * G A → G	71: C Z * → G	—— C State ——	114: G d G → Z
2: Q Q G → Q	—— G State ——	49: * G C → G	72: d Z Q → d	93: Q C G → d	115: G d Z → Q
3: Q Q Z → Z	26: Q G Q → G	50: * G d → G	73: d Z G → Z	94: Q C d → d	116: G d A → A
4: Q Q A → Q	27: Q G G → G	—— Z State ——	74: d Z A → Z	95: G C Q → d	117: G d C → Q
5: Q Q C → A	28: G G Q → G	51: Q Z Q → d	75: d Z C → Z	96: G C Z → Q	118: Z d Q → Q
6: Q Q d → Q	29: G G G → F	52: Q Z G → d	76: d Z * → Z	97: Z C G → Q	119: Z d G → Q
7: Q Q * → Q	30: G G Z → G	53: Q Z A → d	77: * Z Q → d	98: Z C Z → G	120: Z d Z → Q
8: G Q Q → Q	31: G G A → G	54: Q Z d → d	—— A State ——	99: Z C C → G	121: Z d A → Q
9: G Q Z → Z	32: G G C → G	55: Q Z * → d	78: Q A G → C	100: Z C d → A	122: Z d C → A
10: G Q A → Q	33: G G d → G	56: G Z Q → d	79: Q A Z → Z	101: C C Z → G	123: Z d d → C
11: G Q d → Q	34: G G * → F	57: G Z Z → G	80: Q A A → d	102: C C d → G	124: Z d * → G
12: Z Q Q → Z	35: Z G G → G	58: G Z A → Z	81: Q A d → d	103: d C Q → d	125: A d Q → d
13: Z Q G → Z	36: Z G Z → G	59: G Z C → G	82: G A Q → C	104: d C Z → A	126: A d G → d
14: Z Q A → Z	37: Z G d → d	60: G Z d → d	83: G A Z → Z	105: d C C → d	127: A d Z → Q
15: Z Q d → Z	38: A G G → G	61: Z Z Q → d	84: G A A → d	106: d C d → G	128: C d Q → Q
16: Z Q * → Z	39: A G A → G	62: Z Z G → d	85: G A d → A	107: * C Z → G	129: C d G → Q
17: A Q Q → Q	40: C G G → G	63: Z Z * → G	86: Z A Q → Z	108: * C d → G	130: C d Z → A
18: A Q G → Q	41: C G C → G	64: A Z G → d	87: Z A G → Z	—— d State ——	131: d d Q → d
19: A Q Z → Z	42: d G G → G	65: A Z G → Z	88: Z A d → G	109: Q d Z → Q	132: d d G → d
20: A Q d → Q	43: d G A → G	66: A Z Z → d	89: A A Q → C	110: Q d A → A	133: d d Z → C
21: C Q Q → A	44: d G d → G	67: A Z d → Z	90: d A Q → A	111: Q d C → Q	134: * d Z → C
22: d Q Q → Q	45: * G Q → G	68: A Z * → Z	91: d A G → A	112: Q d d → C	
23: d Q G → Q	46: * G G → F	69: C Z G → G	92: d A Z → Z	113: Q d Q → Z	
24: d Q A → Q	47: * G Z → G	70: C Z d → Z			

Fig. 7. A 7-state transition table of the Yunès (1994) algorithm.

3.5   Umeo, Maeda, and Hongyo's 6-State Algorithm: $\mathcal{A}_5$

Umeo, Maeda, and Hongyo (2006) presented a 6-state $3n$-step FSSP algorithm. The implementation was quite different from previous designs. The set Q of internal states for the algorithm is $Q=\{P, Q, R, Z, M, F\}$, where the state P is the initial *general* state, Q is the *quiescent* state, and F is the *firing* state, respectively. The following Fig. 8, consisting of 78 rules, is the transition table. The filled-in ratio of the implementation is $f_{\text{Umeo,Maeda,andHongyo [2006]}} = 78/5 \times 6 \times 6 = 52.0$ (%). The time complexity for synchronizing any array of length

Q State		P State / R State		R State / Z State		Z State / M State	
1: Q Q Q → Q	15: * Q P → P	28: R P * → Z	41: P R M → M	54: Q Z Z → P	70: M Z M → Q
2: Q Q P → P	16: * Q R → Q	29: Z P Q → R	42: R R Q → R	55: Q Z M → Q	71: M Z * → Q
3: Q Q R → Q	17: * Q Z → Q	30: Z P P → Z	43: R R P → M	56: P Z P → Z	72: * Z P → Z
4: Q Q Z → Q	—— P State ——	31: Z P Z → Z	44: R R M → M	57: P Z R → Z	73: * Z R → Q
5: Q Q * → Q	18: Q P Q → Z	32: Z P * → Z	45: Z R Q → P	58: R Z Q → Q	74: * Z Z → F
6: P Q P → Q	19: Q P P → Z	33: * P Q → Z	46: Z R P → R	59: R Z R → Q	75: * Z M → Q
7: P Q P → P	20: Q P R → R	34: * P R → Z	47: Z R M → R	60: R Z Z → Q	—— M State ——
8: P Q * → P	21: Q P Z → R	35: * P Z → Z	48: M R Q → Z	61: R Z * → Q	76: R M R → R
9: R Q Q → Q	22: P P Q → Z	—— R State ——	49: M R P → M	62: Z Z Q → P	77: R M Z → Z
10: R Q R → Q	23: P P R → Z	36: Q R R → R	50: M R R → M	63: Z Z P → Z	78: Z M R → Z
11: Z Q Q → Q	24: P P P → Z	37: Q R Z → P	51: M R Z → R	64: Z Z R → Q	
12: Z Q Z → Q	25: R P Q → R	38: Q R M → Z	52: M R M → M	65: Z Z Z → P	
13: Z Q * → Q	26: R P P → Z	39: P R R → M	—— Z State ——	66: Z Z M → Q	
14: * Q Q → Q	27: R P R → Z	40: P R Z → R	53: Q Z R → Q	67: Z Z * → F	
				68: M Z Q → Q	69: M Z Z → Q	

Fig. 8. A 6-state transition table of the Umeo, Maeda, and Hongyo (2006) algorithm.

n is $3n + \mathrm{O}(\log n)$. The number six was the smallest one known in the class of $3n$−step synchronization algorithms. An important key idea was to increase the number of cells being active during their computation. The algorithm can be extended to a new non-trivial symmetrical six-state $3n$-step generalized firing squad synchronization algorithm. It is seen that the algorithm has $\mathrm{O}(n^2)$ state-change complexity.

Q State					
1:	Q	Q	Q	→	Q
2:	Q	Q	A	→	C
3:	Q	Q	B	→	B
4:	Q	Q	C	→	Q
5:	Q	Q	D	→	Q
6:	Q	Q	*	→	Q
7:	A	Q	Q	→	C
8:	A	Q	A	→	C
9:	A	Q	B	→	B
10:	A	Q	D	→	C
11:	A	Q	*	→	C
12:	B	Q	Q	→	B
13:	B	Q	A	→	B
14:	B	Q	B	→	B
15:	B	Q	C	→	A
16:	B	Q	*	→	B
17:	C	Q	Q	→	Q
18:	C	Q	B	→	A
19:	C	Q	C	→	Q
20:	C	Q	D	→	Q
21:	C	Q	*	→	Q
22:	D	Q	Q	→	Q
23:	D	Q	A	→	C
24:	D	Q	C	→	Q
25:	D	Q	D	→	Q
26:	D	Q	*	→	Q
27:	*	Q	Q	→	Q
28:	*	Q	A	→	C
29:	*	Q	B	→	B
30:	*	Q	C	→	Q
31:	*	Q	D	→	Q
A State					
32:	Q	A	A	→	B
33:	Q	A	C	→	A
34:	A	A	Q	→	B
35:	A	A	A	→	E
36:	A	A	B	→	A
37:	A	A	C	→	C
38:	A	A	D	→	C
39:	A	A	*	→	E
40:	B	A	A	→	A
41:	B	A	B	→	A
42:	B	A	D	→	A
43:	C	A	Q	→	A
44:	C	A	A	→	C
45:	C	A	C	→	C
46:	C	A	D	→	C
47:	D	A	A	→	C
48:	D	A	B	→	A
49:	D	A	C	→	C
50:	D	A	D	→	C
51:	D	A	*	→	C
52:	*	A	Q	→	B
53:	*	A	A	→	E
54:	*	A	B	→	A
55:	*	A	C	→	C
56:	*	A	D	→	C
B State					
57:	Q	B	Q	→	D
58:	Q	B	A	→	C
59:	Q	B	B	→	D
60:	Q	B	C	→	D
61:	Q	B	D	→	D
62:	Q	B	*	→	D
63:	A	B	Q	→	C
64:	A	B	A	→	A
65:	A	B	B	→	A
66:	A	B	D	→	A
67:	A	B	*	→	A
68:	B	B	Q	→	D
69:	B	B	A	→	A
70:	B	B	C	→	A
71:	B	B	D	→	B
72:	C	B	Q	→	D
73:	C	B	B	→	A
74:	C	B	C	→	A
75:	C	B	*	→	A
76:	D	B	Q	→	D
77:	D	B	A	→	A
78:	D	B	B	→	B
79:	D	B	D	→	B
80:	D	B	*	→	B
81:	*	B	Q	→	D
82:	*	B	A	→	A
83:	*	B	C	→	A
84:	*	B	D	→	B
C State					
85:	Q	C	A	→	D
86:	Q	C	B	→	B
87:	Q	C	D	→	C
88:	A	C	Q	→	D
89:	A	C	A	→	D
90:	A	C	B	→	D
91:	A	C	C	→	D
92:	B	C	Q	→	B
93:	B	C	A	→	D
94:	B	C	B	→	B
95:	B	C	C	→	B
96:	B	C	*	→	B
97:	C	C	A	→	D
98:	C	C	B	→	B
99:	C	C	D	→	C
100:	D	C	Q	→	C
101:	D	C	C	→	C
102:	D	C	D	→	C
103:	*	C	A	→	D
104:	*	C	B	→	B
105:	*	C	D	→	C
D State					
106:	Q	D	A	→	Q
107:	Q	D	B	→	Q
108:	Q	D	C	→	A
109:	A	D	Q	→	Q
110:	A	D	A	→	Q
111:	A	D	C	→	B
112:	A	D	D	→	Q
113:	A	D	*	→	Q
114:	B	D	Q	→	Q
115:	B	D	B	→	Q
116:	B	D	D	→	Q
117:	B	D	*	→	Q
118:	C	D	Q	→	A
119:	C	D	A	→	B
120:	C	D	D	→	A
121:	D	D	A	→	Q
122:	D	D	B	→	Q
123:	D	D	C	→	A
124:	*	D	A	→	Q
125:	*	D	B	→	Q

Fig. 9. A 6-state transition table of the Yunès (2008) algorithm.

3.6 Yunès 6-State Algorithm: $\mathcal{A}_6$

Yunès (2008) presented a 6-state implementation for the $3n$-step FSSP algorithm. His implementation was based on wider threads. The set $\mathcal{Q}$ of internal states for the Yunès (2008) algorithm is $\mathcal{Q}$={A, Q, B, C, D, E}, where the state A is the initial *general* state, Q is the *quiescent* state, and E is the *firing* state, respectively. The following Fig. 9, consisting of 125 rules, is the transition table. The filled-in ratio of the implementation is $f_{\text{Yunès [2008]}} = 125/5 \times 6 \times 6 = 69.4$ (%). The time complexity for synchronizing any array of length n is $3n + \mathrm{O}(\log n)$.

3.7 A New 6-State Algorithm: $\mathcal{A}_7$

Here we present a new 6-state implementation for the $3n$-step algorithm. The set $\mathcal{Q}$ of internal states for the algorithm is $\mathcal{Q}$={Z, Q, A, C, d, F}, where the state Z is the initial *general* state, Q is the *quiescent* state, and F is the *firing* state, respectively. The following Fig. 10, consisting of 114 rules, is the 6-state transition table. The filled-in ratio of the implementation is $f_{\text{This Paper}} = 114/5 \times 6 \times 6 = 63.3$ (%). The time complexity for synchronizing any array of length n is $3n + \mathrm{O}(\log n)$.

3.8 A New 6-State Algorithm: $\mathcal{A}_8$

We also present a new 6-state $\mathrm{O}(n^2)$-state-change implementation for the $3n$-step FSSP algorithm. The implementation is quite similar to the algorithm

Q State					
1:	Q	Q	Q	→	Q
2:	Q	Q	Z	→	Z
3:	Q	Q	A	→	Q
4:	Q	Q	C	→	A
5:	Q	Q	d	→	Q
6:	Q	Q	*	→	Q
7:	Z	Q	Q	→	Z
8:	Z	Q	Z	→	Z
9:	Z	Q	d	→	C
10:	Z	Q	*	→	Z
11:	A	Q	Q	→	Q
12:	A	Q	Z	→	C
13:	A	Q	A	→	Q
14:	A	Q	C	→	A
15:	C	Q	Q	→	A
16:	C	Q	A	→	A
17:	d	Q	Q	→	Q
18:	d	Q	A	→	Q
19:	d	Q	d	→	Q
20:	d	Q	*	→	Q
21:	*	Q	Q	→	Q
22:	*	Q	Z	→	Z
23:	*	Q	A	→	Q
24:	*	Q	d	→	Q
Z State					
25:	Q	Z	Q	→	A
26:	Q	Z	Z	→	A
27:	Q	Z	A	→	C
28:	Q	Z	C	→	C
29:	Q	Z	d	→	d
30:	Z	Z	Q	→	A
31:	Z	Z	A	→	d
32:	Z	Z	C	→	Z
33:	A	Z	Q	→	C
34:	A	Z	Z	→	d
35:	A	Z	A	→	d
36:	A	Z	d	→	d
37:	A	Z	*	→	d
38:	C	Z	Q	→	C
39:	C	Z	Z	→	Z
40:	C	Z	C	→	Z
41:	C	Z	*	→	Z
42:	d	Z	Q	→	d
43:	d	Z	A	→	d
44:	*	Z	Q	→	A
45:	*	Z	A	→	d
46:	*	Z	C	→	Z
A State					
47:	Q	A	Z	→	Z
48:	Q	A	A	→	d
49:	Q	A	C	→	Q
50:	Q	A	d	→	d
51:	Z	A	Q	→	Z
52:	Z	A	Z	→	d
53:	Z	A	A	→	d
54:	Z	A	C	→	A
55:	Z	A	d	→	C
56:	Z	A	*	→	d
57:	A	A	Q	→	C
58:	A	A	Z	→	d
59:	A	A	C	→	Q
60:	A	A	d	→	C
61:	C	A	Q	→	Q
62:	C	A	Z	→	A
63:	C	A	A	→	Q
64:	C	A	C	→	Q
65:	d	A	Q	→	A
66:	d	A	Z	→	C
67:	d	A	A	→	C
68:	*	A	Z	→	d
69:	*	A	C	→	Q
C State					
70:	Q	C	A	→	d
71:	Q	C	d	→	d
72:	Z	C	Z	→	A
73:	Z	C	A	→	A
74:	Z	C	C	→	A
75:	Z	C	d	→	A
76:	Z	C	*	→	A
77:	A	C	Q	→	d
78:	A	C	Z	→	A
79:	A	C	d	→	Z
80:	C	C	Z	→	A
81:	C	C	C	→	F
82:	C	C	d	→	Z
83:	C	C	*	→	F
84:	d	C	Q	→	d
85:	d	C	Z	→	A
86:	d	C	A	→	Z
87:	d	C	C	→	Z
88:	d	C	d	→	Z
89:	*	C	Z	→	A
90:	*	C	C	→	F
d State					
91:	Q	d	Z	→	Q
92:	Q	d	A	→	A
93:	Q	d	C	→	Q
94:	Q	d	d	→	C
95:	Z	d	Q	→	Q
96:	Z	d	Z	→	Q
97:	Z	d	d	→	Q
98:	Z	d	*	→	Q
99:	A	d	Q	→	d
100:	A	d	A	→	A
101:	A	d	d	→	A
102:	C	d	Q	→	Q
103:	C	d	C	→	Z
104:	C	d	d	→	Z
105:	d	d	Q	→	d
106:	d	d	Z	→	Q
107:	d	d	A	→	A
108:	d	d	C	→	Z
109:	d	d	d	→	C
110:	d	d	*	→	C
111:	*	d	Z	→	Q
112:	*	d	A	→	A
113:	*	d	C	→	Z
114:	*	d	d	→	C

Fig. 10. A 6-state transition table of the new algorithm.

G State					
1:	G	G	L	→	L
2:	G	G	X	→	L
3:	G	G	Q	→	G
4:	M	G	L	→	G
5:	M	G	X	→	G
6:	M	G	Q	→	L
7:	L	G	G	→	L
8:	L	G	M	→	G
9:	L	G	L	→	L
10:	L	G	X	→	L
11:	L	G	Q	→	M
12:	X	G	G	→	L
13:	X	G	M	→	G
14:	X	G	L	→	L
15:	X	G	Q	→	L
16:	Q	G	G	→	G
17:	Q	G	M	→	L
18:	Q	G	L	→	M
19:	Q	G	X	→	L
20:	*	G	L	→	M
M State					
21:	G	M	G	→	Q
22:	G	M	M	→	Q
23:	G	M	Q	→	Q
24:	G	M	*	→	Q
25:	M	M	G	→	Q
26:	M	M	M	→	F
27:	M	M	L	→	M
28:	M	M	X	→	M
29:	M	M	Q	→	L
30:	M	M	*	→	F
31:	L	M	M	→	M
32:	L	M	L	→	M
33:	L	M	X	→	M
34:	L	M	Q	→	Q
35:	L	M	*	→	Q
36:	X	M	M	→	M
37:	X	M	L	→	M
38:	Q	M	G	→	Q
39:	Q	M	M	→	L
40:	Q	M	L	→	Q
41:	*	M	G	→	Q
42:	*	M	M	→	F
43:	*	M	L	→	Q
L State					
44:	G	L	G	→	G
45:	G	L	M	→	M
46:	G	L	L	→	X
47:	G	L	X	→	M
48:	G	L	Q	→	G
49:	G	L	*	→	M
50:	M	L	G	→	M
51:	M	L	L	→	M
52:	M	L	Q	→	G
53:	M	L	*	→	M
54:	L	L	G	→	X
55:	L	L	M	→	M
56:	L	L	X	→	M
57:	L	L	Q	→	M
58:	X	L	G	→	M
59:	X	L	L	→	M
60:	X	L	Q	→	X
61:	Q	L	G	→	G
62:	Q	L	M	→	G
63:	Q	L	L	→	M
64:	Q	L	X	→	X
65:	Q	L	Q	→	M
66:	*	L	G	→	M
67:	*	L	M	→	M
68:	*	L	Q	→	G
X State					
69:	G	X	G	→	X
70:	G	X	X	→	Q
71:	M	X	M	→	M
72:	M	X	X	→	M
73:	L	X	L	→	Q
74:	L	X	X	→	Q
75:	L	X	Q	→	M
76:	X	X	G	→	Q
77:	X	X	M	→	M
78:	X	X	L	→	Q
79:	X	X	X	→	M
80:	X	X	Q	→	M
81:	Q	X	L	→	M
82:	Q	X	X	→	M
Q State					
83:	G	Q	G	→	Q
84:	G	Q	Q	→	Q
85:	M	Q	M	→	Q
86:	M	Q	Q	→	Q
87:	M	Q	*	→	Q
88:	L	Q	L	→	X
89:	L	Q	Q	→	L
90:	L	Q	*	→	L
91:	X	Q	X	→	M
92:	Q	Q	G	→	Q
93:	Q	Q	M	→	Q
94:	Q	Q	L	→	L
95:	Q	Q	Q	→	Q
96:	Q	Q	*	→	Q
97:	*	Q	G	→	Q
98:	*	Q	M	→	Q
99:	*	Q	L	→	L
100:	*	Q	Q	→	Q

Fig. 11. A transition table of a new 6-state O(n^2)-state-change implementation.

$\mathcal{A}_5$. The set $\mathcal{Q}$ of internal states for the algorithm is $\mathcal{Q}$={L, Q, G, M, X, F}, where the state L is the initial *gene ral* state, Q is the *quiescent* state, and F is the *firing* state, respectively. The following Fig. 11, consisting of 100 rules, is the transition table. The table is nearly symmetric. The filled-in ratio of the implementation is $f_{\text{This Paper}} = 100/5 \times 6 \times 6 = 55.6$ (%). The time complexity for synchronizing any array of length n is $3n + \mathrm{O}(\log n)$. The state-change complexity is $\mathrm{O}(n^2)$.

3.9 Umeo-Yanagihara 5-State Algorithm: $\mathcal{A}_9$

Umeo and Yanagihara (2007) presented a 5-state implementation for the $3n$-step FSSP algorithm. The solution is a *partial* solution that can synchronize any array of length n such that $n = 2^k, k = 1, 2, 3, ...,$. The set $\mathcal{Q}$ of internal states for the implementation is $\mathcal{Q}$={R, Q, S, L, F}, where the state R is the initial *general* state, Q is the *quiescent* state, and F is the *firing* state, respectively. The following Fig. 12, consisting of 125 rules, is the transition table. The filled-in ratio of the implementation is $f_{\text{UmeoandYanagihara [2007]}} = 67/4 \times 5 \times 5 = 67.0$ (%). The time complexity for synchronizing any array of length n is $3n - 3$. Note that the state change complexity is $\mathrm{O}(n^2)$.

Fig. 12. A transition table of the 5-state algorithm.

3.10   State-Change Complexity

Concerning the state-change complexity, the following theorems are established.

Theorem 8. The non-optimum-time algorithms developed by Fischer (1965), Minsky-McCarthy (1967), Herman (1972), Yunès (1994), Yunès (2008), and a new 6-state algorithm in this paper have $O(n \log n)$ optimum state-change complexity for synchronizing n cells in $3n \pm O(\log n)$ steps.

Theorem 9. The non-optimum-time algorithms developed by Umeo-Maeda-Hongyo (2006), a new one in this paper, and Umeo-Yanagihara (2007) have $O(n^2)$ state-change complexity for synchronizing n cells in $3n \pm O(\log n)$ steps.

4   Discussions

We have given a survey on a class of non-optimum-time FSSP algorithms for one-dimensional (1D) cellular automata, focusing our attention to the 1D FSSP

Table 1. Quantitative comparison of transition rule sets for non-optimum-time firing squad synchronization algorithms. The "*" symbol in parenthesis shows the correction and reduction of transition rules made in this paper.

Algorithm	# States	# Rules	Time complexity	State-change complexity	Generals's position	Type	Thread width	Filled-in ratio (%)
Fischer (1965)	15	188*	$3n - 4$	$O(n \log n)$	left	thread	1	5.9
Minsky-McCarthy (1967)	13	138*	$3n + O(\log n)$	$O(n \log n)$	left	thread	1	6.8
Herman (1972)	10	155*	$3n + O(\log n)$	$O(n \log n)$	left	thread	2	17.2
Yunès (1994)	7	134	$3n \pm O(\log n)$	$O(n \log n)$	left	thread	2	45.6
Yunès (1994)	7	134	$3n \pm O(\log n)$	$O(n \log n)$	left	thread	2	45.6
Umeo et al (2006)	6	78	$3n + O(\log n)$	$O(n^2)$	left	plane	–	–
Umeo et al (2006)	6	115	$\max(k, n - k + 1)$ $+2n + O(\log n)$	$O(n^2)$	arbitrary	plane	–	–
Umeo and Yanagihara (2007)	5	67	$3n - 3$ $n = 2^k,$ $k = 1, 2, ..$	$O(n^2)$	left	plane	—	67.0
Yunès (2008)	6	125	$3n + \lceil \log n \rceil - 3$	$O(n \log n)$	left	thread	2, 3	69.4
This Paper	6	114	$3n + O(\log n)$	$O(n \log n)$	left	thread	2, 3	63.3
This Paper	6	100	$3n + O(\log n)$	$O(n^2)$	left	plane	–	55.6

algorithms having $3n\pm O(\log n)$ time complexities. Here, we present a table based on a quantitative comparison of non-optimum-time synchronization algorithms and their transition tables discussed above (Table 1).

References

Balzer, R.: An 8-state minimal time solution to the firing squad synchronization problem. Inf. Control **10**, 22–42 (1967)

Fischer, P.C.: Generation of primes by a one-dimensional real-time iterative array. J. of ACM **12**(3), 388–394 (1965)

Gerken, H.-D.: Über Synchronisations-Probleme bei Zellularautomaten. Diplomarbeit, Institut für Theoretische Informatik, Technische Universität Braunschweig, pp. 50 (1987)

Goto, E.: A minimal time solution of the firing squad problem. Dittoed Course Notes for Applied Mathematics, 298, pp. 52–59. Harvard University (1962)

Herman, G.T.: Models for cellular interactions in development without polarity of individual cells. I. General description and the problem of universal computing ability. Int. J Syst. Sci. **2**(3), 271–289 (1971)

Herman, G.T.: Models for cellular interactions in development without polarity of individual cells. II. Problems of synchronization and regulation. Int. J Syst. Sci. **3**(2), 149–175 (1972)

Kobuchi, Y.: A note on symmetrical cellular spaces. Inf. Process. Lett. **25**, 413–415 (1987)

Mazoyer, J.: A six-state minimal time solution to the firing squad synchronization problem. Theor. Comput. Sci. **50**, 183–238 (1987)

Minsky, M.: Computation: Finite and Infinite Machines, pp. 28–29. Prentice Hall, New Jersey (1967)

Moore, E.F.: The firing squad synchronization problem. In: Moore, F. (ed.) Sequential Machines: Selected Papers, pp. 213–214. Addison-Wesley, Reading MA. (1964)

Sanders, P.: Massively parallel search for transition-tables of polyautomata. In: Proceedings of the VI International Workshop on Parallel Processing by Cellular Automata and Arrays, (Jesshope, C., Jossifov, V., Wilhelmi, W. (eds.) Akademie, pp. 99–108 (1994)

Szwerinski, H.: Symmetrical one-dimensional cellular spaces. Inf. Control **67**, 163–172 (1982)

Umeo, H.: Firing squad synchronization problem in cellular automata. In: Meyers, R.A. (ed.) Encyclopedia of Complexity and System Science, vol. 4, pp. 3537–3574. Springer, Heidelberg (2009)

Umeo, H., Kamikawa, N., Yunès, J.B.: A family of smallest symmetrical four-state firing squad synchronization protocols for ring arrays. Parallel Process. Lett. **19**(2), 299–313 (2009)

Umeo, H., Maeda, M., Hongyo, K.: A design of symmetrical six-state $3n$-step firing squad synchronization algorithms and their implementations. In: El Yacoubi, S., Chopard, B., Bandini, S. (eds.) ACRI 2006. LNCS, vol. 4173, pp. 157–168. Springer, Heidelberg (2006)

Vollmar, R.: Some remarks about the "Efficiency" of polyautomata. Int. J. Theor. Phys. **21**(12), 1007–1015 (1982)

Waksman, A.: An optimum solution to the firing squad synchronization problem. Inf. Control **9**, 66–78 (1966)

Yunès, J.-B.: Seven states solutions to the firing squad synchronization problem. Theor. Comput. Sci. **127**(2), 313–332 (1994)

Yunès, J.-B.: An intrinsically non minimal-time Minsky-like 6-states solution to the firing squad synchronization problem. Theor. Inf. Appl. **42**(1), 55–68 (2008)

Yunès, J.-B.: Simple new algorithms which solve the firing squad synchronization problem: a 7-states $4n$-steps solution. In: Durand-Lose, J., Margenstern, M. (eds.) MCU 2007. LNCS, vol. 4664, pp. 316–324. Springer, Heidelberg (2007)

Yunès, J.B.: A 4-states algebraic solution to linear cellular automata synchronization. Inf. Process. Lett. **19**(2), 71–75 (2008)

CA - Model of Autowaves Formation in the Bacterial MinCDE System

Anton Vitvitsky[✉]

Supercomputer Software Department, ICM&MG, Siberian Branch,
Russian Academy of Sciences, Pr. Lavrentieva, 6, Novosibirsk 630090, Russia
vitvit@ssd.sscc.ru

Abstract. The MinCDE protein system exists in *Escherichia coli* and some other bacteria. It prevents the bacteria from incorrect cell division. Recent studies of MinCDE behavior *in vitro* showed it exhibits self-organization forming protein autowaves and some other patterns. There is a proposition that autowaves arises from an interplay of two opposing mechanisms: cooperative binding of MinD to the membrane, and accelerated MinD detachment due to persistent MinE rebinding. On the basis of this proposition we have developed a cellular automaton model of the process. The behavior of protein concentration, obtained as a result of computer simulations, reveals similarity with the results of experiments *in vitro*. In addition, the protein autowaves resulting from computational experiments are similar to those that emerge *in vitro*.

1   Introduction

Bacterial cell division begins with formation of a ring-like structure on the cell membrane at the midcell (Fig. 1). This ring is called the Z-ring, it consists of FtsZ polymers (tubulin-like protein) and is a framework for downstream cell division proteins [1]. A proper position of Z-ring in the midcell is controlled by the certain self-organization mechanisms. A bright example of these self-organization mechanisms is MinCDE protein system, which is present in the *Escherichia coli*. Currently, the processes leading to a self-organization in this MinCDE system are not quite clear, but are intensively studied [3,4].

In this paper, we propose the Cellular Automata (CA) model of autowaves formation in MinCDE system, simulating results of detailed experimental studies of the protein dynamics MinCDE *in vitro*, [3,4].

2   Oscillations in the Bacterial MinCDE System

The MinCDE system consists of the following interacting proteins: MinC, MinD and MinE. At the beginning of an oscillation cycle (or in the front of the protein wave) MinD starts to bind cooperatively to the membrane. This binding is ATP-dependent, i.e., requires the presence of ATP-nucleotides. MinD also diffuses on

Supported by RFBR under Grant 14-01-31425- mol-a.

V. Malyshkin (Ed.): PaCT 2015, LNCS 9251, pp. 246–250, 2015.
DOI: 10.1007/978-3-319-21909-7_23

Fig. 1. Z-ringassembles at the midcell where the concentration of negative regulators of assembly is lowest [1].

the membrane and periodically detaches back into solution. With rise of MinD density, its residence time on the membrane is increased, and diffusion slows down. MinE from solution also starts to bind to membrane-bound MinD, but at the beginning of an oscillation cycle the ratio of MinE/MinD is still too low and therefore protein detachment, due to ATP-hydrolysis, hardly occurs.

At the middle of an oscillation cycle (or in the middle of the wave), the ratio of MinE/MinD is increased, and correspondingly protein detachment occurs more often. MinE continuously rebinds to a neighboring membrane-bound MinD.

At the end of an oscillation cycle (or in the rear of the wave), the ratio of MinE/MinD reaches its maximum, and MinE, interacting with the membrane, induces a conformational change, resulting in all MinC being displaced back into solution. Finally, MinE stimulates bulk hydrolysis of MinD and all proteins rapidly disappear from the membrane.

3   The Cellular-Automata Model of MinDE Autowaves Formation

A cellular automaton (CA) is a discrete mathematical model consisting of a set of finite state automata called *cells* [4]. Each cell is defined by a pair $(u, \mathbf{x})$, where u is *a cell state* from the finite set of states A, and $\mathbf{x}$ is *a cell name* from the finite set of the name X. The set of cells $\Omega = \{(u, \mathbf{x}) : u \in A, \mathbf{x} \in X\}$ form *a cellular array*.

The CA-model, presented here, simulates self-organization process of MinDE proteins *in vitro*. The CA has a set of names $X = \{\mathbf{x} : \mathbf{x}(i,j), i = 1, .., W, j = 1, ..., H\}$, where $W \times H$ - lattice size, and the set of states $A = \{\emptyset, MinD, MinE, MinDE\}$, where $\emptyset$ - means that the cell is "empty", and the remaining states correspond the presence of one of $MinD, MinE$ proteins or $MinDE$ complex in the cell, respectively. At initial time, all cells $(u, \mathbf{x}) \in \Omega$ are set to $\emptyset$-state.Let us consider the CA-rules in detail:

1. *MinD binding to the membrane:*

$$\emptyset \xrightarrow{\quad P_1 + k \cdot |N^m| \quad} MinD, \tag{1}$$

where P_1 is a basic probability of $MinD$ binding to the membrane, k - coefficient reflecting the strength of cooperative attraction $MinD$ from the solution, $|N^m|$ — the number of cells with the $MinD$-state from the Moore neighborhood.

2. *MinE binding to the membrane-bound MinD*:

$$MinD \xrightarrow{\quad P_2 \quad} MinDE, \qquad (2)$$

where P_2 is a probability of *MinE* binding to *MinD*.
3. *Hydrolysis of ATP and detachment of MinD from the membrane*:

$$MinDE \xrightarrow{\quad P_3 \quad} MinE, \qquad (3)$$

where P_3 is a probability of ATP hydrolysis.
4. *Membrane diffusion of MinD/MinDE*:
   The diffusion is presented by *naive diffusion* [3]. First, choose an empty cell is randomly chosen in the Moore neighborhood. Then, with the probability P_4 the central cell exchangesr states with a chosen one.
5. *Rebinding MinE to an available MinD*: The rule of rebinding is similar to the diffusion but it has some differences. First, we choose a cell with *MinD*-state randomly in the neighborhood (larger than the Moore one). Then, with the probability P_5 the central cell changes its *MinE*-state to the empty-state and the chosen cell changes its *MinD*-state to *MinDE*-state. If the neighborhood does not contain any cell with MinD-state then the central cell becomes empty.

4   Computer Simulation Results

A computational experiment has been carried out. The collected data resulting from the simulation are compared to the data obtained in the experiments *in vitro*. The model parameters of the computational experimenthave been chosen empirically in such a way, that the graph of proteins surface density by time, obtained as a result of computer simulation, matches the graph from the experiments *in vitro* (Fig. 2).

The following parameters have been chosen: rectangular lattice consisting of 600×600 cells, $P_1 = 0.00001$, $P_2 = 0.03$, $P_3 = 1.0$, $P_4 = 1.0$, $P_5 = 0.8$, $k = 0.034375$.

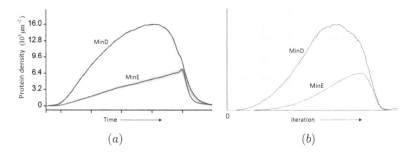

(a)   (b)

Fig. 2. Protein surface densities by time. (a) Estimated surface densities of MinD and MinE *in vitro*. Figure adapted from [3]. (b) Computer simulation result

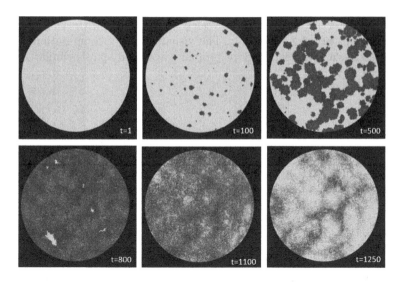

Fig. 3. Visualization of computational experiment

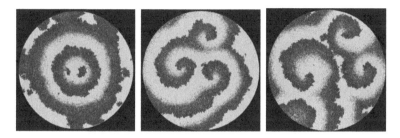

Fig. 4. Emergence of concentric waves (left) and protein spirals (middle and right) in the computational experiment

In Fig. 3 six snapshots of the CA-evolution are shown. On the beginning of simulation (t=1) the membrane is empty. Then, MinD gradually begins to attach to the membrane and attract other MinD dimmers as well as MinE binds to membrane-bound MinD (t=100, t=500). MinE dimmers stimulate ATP hydrolysis by MinD being followed by detachment of the proteins back to the cytosol (t=800, t=1100). As a result, the membrane becomes empty again (t=1250) and the cycle repeats. Self-organization of MinDE proteins is clearly observed after a number of iteration thousands (Fig. 4).

5 Conclusion

The CA-model of the MinDE proteins self-organization, yielding autowaves emergency has been developed. The model is based on the investigation of the process *in vitro* from [3]. The graph of protein concentration, obtained as a result of computer simulations, has revealed the similarity with the graphs from the experiments *in vitro* from [3,4]. The visualization of computational experiments has shown the emergency of protein waves and spirals similar to those that emerge *in vitro*.

References

1. Lutkenhaus, J.: Assembly dynamics of the bacterial MinCDE system and spatial regulation ofthe Z ring. Annu. Rev. Biochem. **76**, 539–562 (2007)
2. Loose, M., Fischer-Friedrich, E., Herold, C., Kruse, K., Schwille, P.: Min protein patterns emerge from rapid rebinding and membrane interaction of MinE. Nat. Struct. Mol. Biol. **18**(5), 577–583 (2011). doi:10.1038/nsmb.2037
3. Ivanov, V., Mizuuchi, K.: Multiple modes of interconverting dynamic pattern formation by bacterial cell division proteins. Proc. Natl. Acad. Sci. **107**(18), 8071–8078 (2010)
4. Bandman, O.: Cellular automata composition techniques for spatial dynamics simulation. In: Hoekstra, A.G., et al. (eds.) Simulating Complex Systems by Cellular Automata: Understanding Complex Systems, pp. 81–115. Springer, Heidelberg (2010)
5. Toffolli, T., Margolus, N.: Cellular Automata Machines: A New Environment for Modeling. MIT Press, Cambridge (1987)

Distributed Computing

Agent-Based Approach to Monitoring and Control of Distributed Computing Environment

Igor Bychkov, Gennady Oparin, Alexei Novopashin,
and Ivan Sidorov[✉]

Matrosov Institute for System Dynamics and Control Theory,
Siberian Branch of Russian Academy of Sciences, Irkutsk, Russia
{bychkov, oparin, apn, ivan. sidorov}@icc. ru

Abstract. This paper discusses a problem of monitoring heterogeneous distributed computing environments which consist of loosely coupled multiplatform computing resources. We propose an approach to the organization of the meta-monitoring system which collects data from existing local monitoring systems and own software sensors, unifies and analyzes data, generates necessary control actions. The approach is based on web-technologies, multi-agent technologies, expert systems, methods of decentralized processing and distributed storage of data.

Keywords: Monitoring · Multi-agent systems · Expert systems · Distributed computing environments · High-performance computer centre

1 Introduction

The paper considers distributed computing environments (DCE) intended for resource-intensive computing experiments. DCE may include high-performance computing (HPC) systems with different architectures and configurations: HPC-servers based on GPU, coprocessors or FPGA; HPC-clusters controlled by various job management systems (PBS/Torque, LSF, SGE, etc.); hybrid computing systems with unequal processor and memory architectures.

To organize DCE it is necessary to solve a number of problems including development of control and monitoring tools. Control tools should provide coordinated work of all DCE components. Examples of control functions are following: load balancing among DCE nodes, command execution in parallel; power on/off compute nodes. Monitoring tools should supply to operator structured and unified information about environment state: values of node hardware sensors (e.g. temperature of CPU and motherboard), load of computing nodes by parallel programs (CPU, RAM, I/O-system, network usage, etc.), load of engineering infrastructure devices and others. With increasing the number of compute nodes in DCE the amount of data which should be analyzed by operator is growing. The risk of human errors is rising simultaneously. Actual graphical and numerical data provided by monitoring system allow operator to respond to any emergency situations in DCE and to identify causes of faults and failures.

© Springer International Publishing Switzerland 2015
V. Malyshkin (Ed.): PaCT 2015, LNCS 9251, pp. 253–257, 2015.
DOI: 10.1007/978-3-319-21909-7_24

2 Related Work

The known monitoring tools have a number of disadvantages which make it very difficult to collect and analyze data in complex computing systems. For example, Ganglia [1] is not able to perform the processing of current values and the notification about the critical and dangerous events. Nagios [2] and Zabbix [3] were initially oriented to network monitoring and are not effective enough for DCE. The main problem of ClustrX Watch [4] is the lack of data visualization means and tools for adding new sensors. LAPTA [5] is designed for in-depth analysis of dynamic characteristics of parallel programs, but it does not allow to control the DCE infrastructure.

Then we propose an approach to the creation of the meta-monitoring system, which differs from the known ones by the set of unique properties: automatic control of software and hardware resources using multi-agent technologies, decentralized scheme of data storage and expert decision-making [6].

3 Model

Let the model of the meta-monitoring system is represented by the following structure:

$$D = <N, L, M, B, V, G, A, R, S>, \tag{1}$$

where N denotes the set of the DCE nodes; L – the set of the DCE links; M – the set of metrics (characteristics); B – the set of metrics values; V – the set of thresholds for metrics values; G – the set of local monitoring systems; A – the set of agents; R – the set of inference rules for an expert subsystem; S – the set of control actions.

The following relations between the sets from (1) are defined: $NM \subseteq N \times M$ – the metrics measured in nodes; $MG \subseteq M \times G$ – the metrics collected from local monitoring systems; $NT \subseteq N \times L$ – the DCE topology; $NA \subseteq N \times A$ – the hierarchical structure of agents; $RM \subseteq R \times M$ – the dependencies between inference rules and measured metrics; $SM \subseteq S \times M$ – the dependencies between control actions and measured metrics.

The meta-monitoring problem is defined as follows. Let $n_i \in N$ is one of the DCE nodes and m_i metrics are measured ($m_i \subset M$) for n_i. Then $B_i(t) = \{b_{ik}(t)\}$ is the set of metrics values for n_i at the time t, where $k = 1, 2, ..., m_i$.

Steady state of a metric at the interval $[t_0, t_1]$ is considered to be the one for which the inequality (2) is satisfied.

$$|b_{ik}(t) - b_{ik}(t_0)| \leq c_{ik}, \forall t \in [t_0, t_1], \forall k \in [1, m_i], \tag{2}$$

where c_{ik} is a positive constant defining the range of values for appropriate metric.

Significant event is a transition from the steady state $B_i(t_0)$ to the steady state $B_i(t_1)$, when there is a change of any metric's value more than relevant threshold $v_{ik} \in V$. The list of the facts for an expert subsystem is formed after each significant event in the interval $[t_0, t_1]$. Further the inference engine generates the control actions S_i on basis of the inference rules R_i for n_i.

4   Architecture

Architecture of the meta-monitoring system for DCE is shown on Fig. 1.

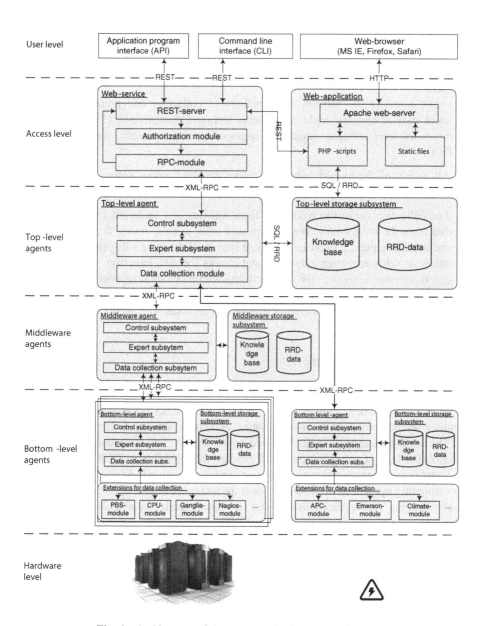

Fig. 1. Architecture of the meta-monitoring system for DCE

The architecture includes the following components:

- *user level* includes the user access tools for interaction with the meta-monitoring system both in batch mode and interactive mode;
- *access-level* includes the data access control subsystem and the server part of a graphical user interface;
- *top-level agent* performs the basic management functions of the meta-monitoring system (in the central node);
- *middleware agents* take the part of the load from top-level agent (in the intermediate nodes);
- *bottom-level agents* perform the functions of collecting and preprocessing data about state of compute node components and other hardware-level devices;
- *hardware-level* includes devices of engineering and computing infrastructure of DCE.

All agents of the meta-monitoring system are autonomous software components [7]. Each agent includes the following subsystems:

- *data collection subsystem* reads data from software and hardware sensors, receives data from the local monitoring systems and other agents, unifies the collected data and translates them into an intermediate format of database subsystem;
- *database subsystem* aggregates data and checks the data integrity;
- *expert subsystem* analyzes the collected data within certain time interval to initialize control actions if appropriate;
- *control subsystem* executes control actions and interacts with agents of upper levels.

5   Experimental Results

The developed model, methods and tools for meta-monitoring were successfully approved in the Supercomputer Centre of ISDCT SB RAS [8] during solving some complex science problems in bioinformatics, chemistry, satisfiability and others.

The meta-monitoring system allowed us to detect ineffective user applications, optimize the load of compute nodes and increase the reliability of DCE. For example, we found dominance of read/write operations to/from a network directory in relation to the computing operations of processor cores for a number of user applications. In these cases changing the applications settings to record results in a local directory has improved computational efficiency up to 30 %.

Besides the meta-monitoring system is used as a power saving software tool for the Matrosov compute cluster [8], which is part of DCE. We have analyzed statistics of cluster usage for the last year and found that the average load of its nodes is less than 75 % (e.g., 90 % – May 2014, 50 % – July 2014, etc.). At the same time 25 % of idle resources consume 3700 kWh per month. In order to save power we have developed an extension for our meta-monitoring system, which includes sensors for PBS, inference rules for an expert subsystem and control actions. As control actions (S from structure (1)) we use operator notifications (e.g. about a huge length of a job queue) and automatic shutdown/start compute nodes by using the IPMI protocol.

6   Conclusions

In this paper the original approach to development of the meta-monitoring system for distributed computing environments has been presented. The novelty of the approach consists in the use of software agents which automatically collect data about state of cluster nodes and infrastructure, perform the data analysis and generate the control actions. The agents are able to make decisions independently or through interactions with other agents. The agent architecture is universal. It makes possible to apply the agents on the different levels of the meta-monitoring system. Using the multi-level hierarchy of agents provides the load balancing between components of the monitoring system and increases the reliability of DCE via decentralization of monitoring and control functions. Furthermore, the agent-based approach guarantees high scalability of the system for DCE with a large number of nodes.

References

1. Massie, M., Li, B., Nicholes, V., Vuksan, V.: Monitoring with Ganglia. O'Reilly Media, Sebastopol (2012)
2. Josephsen, D.: Building a Monitoring Infrastructure with Nagios. Pearson Education, Boston (2007)
3. Zabbix. https://www.zabbix.org
4. ClustrX Watch. http://www.t-platforms.com/products/software/clustrxproductfamily/clustrxwatch.html
5. Adinets, A.V., Bryzgalov, P.A., Voevodin, V.V., Zhumatiy, S.A., Nikitenko, D.A.: About one approach to monitoring, analysis and visualization of jobs on cluster system. In: Numerical Methods and Programming, vol. 12, pp. 90–93. RCC MSU, Moscow (2011)
6. Giarratano, J.C., Riley, G.D.: Expert Systems: Principles and Programming, 4th edn. Course Technology, Boston (2004)
7. Tweedale, J., Ichalkaranje, N.: Innovations in multi-agent systems. J. Netw. Comput. Appl. **30**, 1089–1115 (2007). Elsevier, Melbourne
8. Irkutsk Supercomputer Centre of SB RAS. http://hpc.icc.ru

Virtual Screening in a Desktop Grid: Replication and the Optimal Quorum

Ilya Chernov[✉] and Natalia Nikitina

Institute of Applied Mathematical Research (IAMR),
Pushkinskaya 11, Petrozavodsk 185910, Russia
IAChernov@yandex.ru, nikitina@krc.karelia.ru
http://mathem.krc.karelia.ru

Abstract. We propose a mathematical model of a desktop grid computing system that solves tasks with two possible answers. Replication is used in order to reduce the error risk: wrong answers are returned with some known probabilities and penalty is added to the calculation cost in case of an error. We solve the optimization problems to determine the optimal quorum for tasks of varying duration. Beside the general case, we consider reliable answers of one kind. We apply the model to the problem of virtual screening and show how replication reduces the average cost. Also we demonstrate that when penalties are close to but lower than the critical values, taking different duration of tasks into account significantly reduces the penalty threat at very low additional cost.

Keywords: Grid computing · Virtual screening · Optimal quorum · Replication · Volunteer calculations

1 Introduction

High-performance computing is widely used in science and industry. Recently popularity of desktop grids [2] have increased; such grids connect heterogeneous computers using networks of general purpose. Such grids are useful for problems that demand much computational power due to large amount of independent or weakly dependent tasks. Much effort has been paid recently to optimize the calculation process and improve its reliability [1,3,5–10,12–14]. Batching tasks in order to improve efficiency of a Desktop grid is considered in, e.g., [11], where the author shows that replication is able to reduce the overall time of solving multiple tasks. A distributed computing system with central architecture, but different from Desktop grid, is considered in [3]: there computing nodes are homogenous and computing tasks are strongly connected.

The answer returned by the computing node can be wrong, due to malfunction, errors in transaction, malicious actions; besides, the algorithm can provide a wrong result due to different reasons. For example, gradient methods can converge to local minima and fail to reach the true global minimum. In fact, functions on multi-dimensional space typically have multiple minima; therefore local minimization may fail to find the global solution.

© Springer International Publishing Switzerland 2015
V. Malyshkin (Ed.): PaCT 2015, LNCS 9251, pp. 258–267, 2015.
DOI: 10.1007/978-3-319-21909-7_25

An obvious way to avoid errors is replication: each task is solved until the given number (called the quorum) of identical answers is obtained. However, this means increasing the computational time and/or cost. Another way to reduce the risk is to use higher precision; this also increases the cost of calculation. Provided that estimation of the cost of the wrong answer is known, we are able to evaluate the expected cost and choose the optimal quorum and optimal precision: the optimal strategy of solving the tasks can be obtained.

We consider the problem of optimizing the average cost of a task. The cost consists of time spent on its solution and some penalty paid in case of accepting a wrong answer. Obviously replication does not eliminate the risk, though reduces it significantly. A mathematical model of the computation process and a numerical optimization procedure are applied to the virtual screening problem.

Virtual screening is a computational technique to evaluate the binding energy between a complex protein molecule and a smaller structure called a ligand. Ligands can be substances (medicines, poisons, etc.) that somehow affect functions performed by the protein. The predicted binding energy depends on spatial structure of both molecules. Ligands with lower predicted energy potentially can be used and are to be further tested in a laboratory. However, a protein molecule often has pockets: local minima of the connection energy. If the algorithm finds a false pocket instead of the optimal one, the ligand will be rejected and probably lost. Errors of the other kind are less likely. The algorithm can work with different given precision and uses random numbers; thus a few tries can, possibly, reveal the mistake.

The structure of this article is the following: we give the mathematical description of a problem, then show how to evaluate the optimal quorum; we pay more attention to the special case when one kind of answer is always correct and consider examples based on a piece of statistics from a virtual screening project.

2   The Model

Let us consider the grid computing system (or its part) solving numerous problems with two possible answers: YES or NO. The correct answers have some probabilities q_-, q_+, both greater than 0.5. This means that if the correct answer is YES, it is produced by a node with probability q_+; the wrong answer comes with probability $1 - q_+ = p_-$. A priori probabilities of answers are α_+ and α_-. They are evaluated from the statistics of real calculations. If the YES answer is highly unlikely to be missed, we can assume that $q_+ = 1$. However, $q_- < 1$, i.e., false positive answer can be returned. These probabilities depend on parameters of the algorithm, i.e., random seed or precision.

Of course, different computer nodes can have different probabilities q_+, q_-; however, often the difference is not so significant to be taken into account; even if it is, the nodes can be divided into groups and the results below be applied to a group [5,11]; in case of a large number of nodes, average values of the probabilities can be used; finally, it seems reasonable to use the upper bound of the error probabilites to obtain the robust strategy.

In order to reduce the risk of producing the wrong answer replication or quorum can be used. Under the first approach we understand sending ν copies of each task to randomly chosen computers and expecting equal answers; otherwise the task is re-solved carefully. The second method is sending copies until ν identical answers arrive; possible different answers are discarded. One of the answers can be tested more carefully: so ν answers are enough for e.g. the NO answer while $\mu > \nu$ are needed to take the YES answer. In the $q_+ = 1$ case, obviously, the YES answers are not checked at all: $\nu = 1$, while the NO answer is checked carefully enough: $\mu > 1$.

The obtained answer is given to the user and can possibly still be wrong. The user figures out if the answer is correct or not, in the latter case some kind of penalty F_+ or F_- is added to the time spent on the computation (or the cost of this time) forming the cost function J. Therefore we have an optimization problem: for large redundancy the cost is high, but the risk of the wrong answer and thus of penalty is almost eliminated; while in case of insufficient quorum the calculation cost is low, but instead penalty threat can be too expensive.

Penalties F_+ and F_- are losses due to the unnecessary (possibly expensive) laboratory test, losses in case of a missed good ligand, reputational losses, etc.; they can differ, because missing the correct "YES" answer and missing the correct "NO" answer can be errors of different cost. Obviously false positive result is less costly because laboratory tests would inevitably reveal the mistake; on the other hand, the false negative result means that the good ligand is missed, possibly forever.

We fix the average computation time of a task as a cost unit. Let duration of the considered task be C; some tasks are solved more quickly, so penalty is effectively higher, others are more complex, so penalty seems lower. So possibly short tasks can be tested more carefully. In the docking project time needed to solve a single task varies 10–12 times.

We assume that the penalties are known, and determine the optimal quorums depending on the real cost of a task, and also we show how parameters of the algorithm (e.g., precision) can be chosen such that the expected cost be minimal.

3   Optimal Quorum

Let $p_- = 1 - q_+$, $p_+ = 1 - q_-$. The random cost is given in Table 1. Each column shows the conditional binomial random variable: the number of tries before acception of the correct/wrong YES/NO answer.

Table 1. Random cost, unequal answers

$\nu + i,$	$\nu + i + F_+/C,$	$\mu + j,$	$\mu + j + F_-/C,$
$i = 0 : \mu - 1,$	$i = 0 : \mu - 1$	$j = 0 : \nu - 1$	$j = 0 : \nu - 1$
$\alpha_- \binom{\nu+i-1}{\nu-1} q_-^{\nu} p_+^{i}$	$\alpha_+ \binom{\nu+i-1}{\nu-1} p_-^{\nu} q_+^{i}$	$\alpha_+ \binom{\mu+j-1}{\mu-1} q_+^{\mu} p_-^{j}$	$\alpha_- \binom{\mu+j-1}{\mu-1} p_+^{\mu} q_-^{j}$

Denote its expectation $E(\nu, \gamma)$. We distinguish the basic replication ν that is additional checks for both kinds of answers, and additional checks γ for the answers of greater importance. Firstly we study influence of the basic replication and then that of the additional one.

3.1   Basic Replication

Consider the increment of the expected cost

$$G(\nu, \gamma) = E(\nu, \gamma) - E(\nu + 1, \gamma) = A_+ F_+ + A_- F_- - B.$$

Let us evaluate A_+ (see the details in the Appendix):

$$A_+ = \alpha_+ p_-^\nu q_+^\mu \binom{\nu + \mu - 1}{\nu}\left(1 - p_- \frac{\nu + \mu}{\mu}\right).$$

In the same way A_- is evaluated:

$$A_- = \alpha_- p_+^\mu q_-^\nu \binom{\nu + \mu - 1}{\mu}\left(1 - p_+ \frac{\nu + \mu}{\nu}\right).$$

Note that, as $\mu = \nu + \gamma \geq \nu$, the coefficient $A_+ > 0$ provided that $p_- \leq 0.5$. But $A_- > 0$ only if the node is "reliable":

$$p_+ < \frac{\nu}{\nu + \mu} = \frac{\nu}{2\nu + \gamma} < \frac{1}{2}.$$

This threshold is important: for reliable computing nodes raising both penalties pays, while for unreliable penalty F_- reduces the average cost.

Also, for reliable computers we can choose the penalties in such a way that any basic replication ν is optimal, even more, in a number of ways. While for less reliable ones it can be reasonable (without other conditions) to choose $F_- = 0$, i.e. to forgive the cheap mistakes. Again, any replication ν can be made optimal choosing only F_+.

But in general the condition $G(\nu, \gamma) > 0$ demands (for unreliable nodes)

$$\frac{F_+}{F_-} > -\frac{A_-}{A_+},$$

which in the special case $p_+ = p^+ = p$, $q = 1 - p$ looks like

$$\frac{F_+}{F_-} > \frac{\alpha_-}{\alpha_+} \cdot \frac{p(\nu + \mu) - \nu}{\mu - p(\nu + \mu)} \cdot \left(\frac{p}{q}\right)^\gamma. \tag{1}$$

It is easy to check that

$$0 < \frac{p(\nu + \mu) - \nu}{\mu - p(\nu + \mu)} < 1 \text{ for } \frac{\nu}{\nu + \mu} < p < \frac{1}{2}$$

If also $F_+ = F_-$, computers are always reliable (provided that $p < 0.5$); passing from the quorum ν to $\nu + 1$ pays if

$$F > F_\nu = \frac{E_0(\nu + 1) - E_0(\nu)}{\binom{2\nu-1}{\nu-1}(1 - 2p)p^\nu q^\nu}. \tag{2}$$

The sequence F_ν increases up to $+\infty$. Also note that both low (≈ 0) and high (≈ 0.5) probability of error makes redundancy less useful: in the first case the penalty is unlikely, while in the second it is cheaper to pay the penalty without hopeless costly attempts to avoid it. The quantity F_ν as a function of p (given by (2)) has a minimum in $p^*(\nu) \in (0, 0.5)$; This value is the critical level of reliability: as we have said earlier, for higher values replication is not advantageous because of too high risk of the wrong answer and thus the estimated penalty.

3.2 Additional Replication

Now let us see what the additional replication γ can give. Consider the difference

$$G(\gamma) = E(\nu, \gamma) - E(\nu, \gamma + 1) = a_+ F_+ + a_- F_- - b.$$

As for a_+, it is just an additional term

$$a_+ = -\alpha_+ p_-^\nu q_+^{\nu+\gamma} \binom{2\nu + \gamma - 1}{\nu - 1},$$

which is a chance to miss the correct YES answer when it has been obtained μ times. Let us evaluate a_-. This is done in the similar way as A_- above:

$$a_- = \alpha_- p_+^{\nu+\gamma} q_-^\nu \binom{2\nu + \gamma - 1}{\nu + \gamma} = \alpha_- p_+^{\nu+\gamma} q_-^\nu \binom{2\nu + \gamma - 1}{\nu - 1}.$$

Note that a_- is always positive. Additional checks for the YES answer only improve the chance to get the correct NO answer. So inequality $a_+ F_+ + a_- F_- > 0$ is necessary for $G(\gamma) > 0$, and is sufficient if penalties are large enough. In the special case $p_+ = p_- = p$, $q = 1 - p$ this inequality reduces to

$$\frac{F_+}{F_-} \leq \frac{\alpha_-}{\alpha_+} \left(\frac{p}{q}\right)^\gamma. \tag{3}$$

Provided that F_+, F_- are given and ν, $\mu = \nu + \gamma$ are chosen, inequality

$$A_+ F_+ + A_- F_- \geq B$$

shows if passing to $\nu + 1$, $\mu + 1$ is reasonable, while inequality

$$a_+ F_+ + a_- F_- \geq b$$

shows if passing to ν, $\mu + 1$ pays. The optimal penalties are the solution to the linear optimization problem:

$$Q_+ F_+ + Q_- F_- + Q \to \min,$$

$$a_- F_- + a_+ F_+ \geq b,$$
$$A_- F_- + A_+ F_+ \geq B,$$
$$F_+ \geq 0, \quad F_- \geq 0.$$

Here Q_+, Q_- are probabilities of errors and $Q = E_{0,0}(\nu+1, \gamma+1)$. This problem has a solution: existence of a single point in the domain is sufficient. For reliable nodes such a point is $F_+ = 0$, F_- is high enough; for unreliable ones it is sufficient to find a point such that $a_+ F_+ + a_- F_- > 0$, $A_+ F_+ + A_- F_- > 0$. Here $a_+ < 0$, $A_- < 0$. So we need to check if

$$-\frac{a_+}{a_-} < -\frac{A_+}{A_-}.$$

This inequality reduces to

$$\mu - p_-(\nu + \mu) > p_+(\nu + \mu) - \nu.$$

We have used the unreliability, which makes the right-hand side positive; as both probabilities are less than 0.5, inequality holds for all p_+, p_- if it does for $p_+ = p_- = 0.5$ or at least become an equality. It is easily seen that it indeed becomes an equality, so admissible points exist.

Note that for some cases one of the optimal penalties (F_+) is zero: it is reasonable to forgive one kind of mistakes. For example, the optimal penalties for $\nu = 2$, $\mu = 3$, $\gamma = 1$, $p_+ = p_- = 0.01$ (reliable nodes) are $F_+ = 0$, $F_- = 501.56$, with optimal cost $E = 6.0025$.

Let us consider a complex example with different error probabilities and non-zero penalties for different kinds of answers; assume that $p_+ = 0.004$, $p_- = 0.001$, $\alpha_- = 0.9964$, penalties be $F_+ = 3 \cdot 10^5$, $F_- = 4 \cdot 10^4$. The optimal quorums then are $\nu = 2$, $\mu = 4$ with average cost $E = 2.026$. Table 2 shows how the expected cost E depends on the quorums.

Table 2. Cost E with respect to the quorums.

$\mu - \nu$	0	1	2	3	4	5	6	7
$\nu = 1$:	161.5	3.8	4.25	5.33	6.41	7.49	8.57	9.64
$\nu = 2$:		3.92	2.028	2.026	2.04	2.05	2.06	2.07	2.08
$\nu = 3$:			3.04	3.015	3.019	3.02	3.03	3.03	3.03	3.04

It is interesting that some quorums are not optimal for any penalties. For example, if $(\nu, \mu) = (3, 4)$ is better than $(2, 3)$ and $(3, 3)$, then $\nu = 1$, $\mu = 4$ is optimal. However, $\nu = 3$, $\mu = 5$ is optimal for certain positive penalties.

3.3   Reliable Positive Answers

Let us simplify the results for the case of reliable positive answers, i.e., $q_+ = 1$: the YES answer is for sure obtained. Then $p_- = 0$ and also penalty F_+ is never paid. Also it is obvious that $\nu = 1$: the NO answer is accepted without any check because it can not be false. So from here on $\nu = 1$. The random variable for the cost is in Table 3. Expressions for the coefficients become simpler:

$$A_+ = a_+ = 0, \quad A_- = \alpha_- p_+^\mu q_- \left(1 - p_+(1 + \mu)\right), \quad a_- = \alpha_- p_+^\mu q_-.$$

Table 3. Random cost, unequal answers

$1 + i,\ i = 0 : \mu - 1,\ \mu,\ \ \mu + F_- C,$	
$\alpha_- q_- p_+^i$	$\alpha_+\ \alpha_- p_+^\mu$

4   Examples

Let us consider a few examples of the presented approach; we use a piece of statistics obtained in 2013–2014 in the Luebeck Institute for Experimental Dermatology, University of Luebeck, Germany during virtual screening of ligands for one protein. Open-source software Autodock Vina [15] was used in virtual screening.

A priori probabilities of the answers were $\alpha_+ = 0.036$, $\alpha_- = 0.964$, probability of the error (the false positive answer) was $p_+ = 0.004$, mean duration of a task was 11.37 s. We neglect the risk of false negative answers: $p_- \approx 0$.

Minimal penalties F^* that make quorum μ better than $\mu - 1$ are given in the Table 4.

Table 4. Minimal penalty making quorum μ optimal.

μ	2	3	4	5	6	7
F^*	1.91	$2.28 \cdot 10^2$	$5.67 \cdot 10^4$	$1.42 \cdot 10^7$	$3.54 \cdot 10^9$	$8.86 \cdot 10^{11}$

So, if the penalty is 1.5 units, then the optimal quorum for a task of the mean cost is $\mu = 1$, i.e., no checks are performed. Then the expected cost of a task with penalty taken into account is 1.006: penalty threat is almost eliminated, as well as conflicting answers. However, shorter tasks are relatively cheaper and therefore quorum $\mu = 2$ can become optimal. Taking this difference into account saves 0.14%, so that neglecting variance of duration/cost of tasks seems satisfactory. Total duration of the calculation is 0.2% more, but instead the expected losses for penalties are 44.79% less. In other words, probability of error is reduced from $4.0 \cdot 10^{-3}$ to $2.2 \cdot 10^{-3}$.

If we choose $F^* = 1.92$, then the optimal overall duration would be 0.42% less (because expensive tasks are not checked), but the error probability would grow: from $1.6 \cdot 10^{-5}$ to $1.4 \cdot 10^{-3}$. However, this probability is still less than that of the no-check case.

For high penalties results are the same. For example, for $F = 1.4 \cdot 10^7$ the optimal quorum for a mean task is $\mu = 4$. Taking task cost variance into account saves 0.13%, total duration grows on 0.15%, instead the error probability is twice less: $9.6 \cdot 10^{-11}$ against $2.6 \cdot 10^{-10}$. Note that if no replication is used, overall duration is only 1.5% less, while the average cost per a task is more than $5 \cdot 10^5$ units (compare with optimal cost equal to 1.017).

As the penalty is not known and can not be known precisely, it is reasonable to assume that it is equal to one of these critical values; then it is sufficient to estimate the order of magnitude of the penalty and there is no need to evaluate the precise value.

The highest gain is obtained when $F \approx F^*$ but $F < F^*$ for any F^*: many (about one half) tasks are solved with a higher optimal quorum: they are checked more carefully, thus the error probability is almost eliminated; additional time spent on checks is small in the average, because additional checks are needed rarely. If $F > F^*$, but still is close to it, the situation is vice versa: longer tasks are checked less, so total time is reduced; instead the error probability is higher. The total gain is positive, yet small: about 0.03 %.

Note that replication reduces the error probability significantly, so even great penalty increases the average cost of a task only slightly. Table 5 shows penalties, optimal quorums, and average costs. Note that $E - 1$, i.e., the expected loss for penalty and additional checks, is much less than one unit. Taking variance of task cost into account is not able to improve the optimal solution much.

Table 5. Penalties, optimal quorums μ, expected costs.

$\log_{10} F$	1	2	3	4	5	6	7	8
μ	2	2	3	3	4	4	4	5
$(E - 1) \cdot 10^3$	7.75	9.18	11.27	11.84	14.83	15.06	17.45	18.50

Table 6 shows how the average task cost changes with respect to μ.

Table 6. Expected cost with respect to quorum μ.

μ	1	2	3	4	5	6	7	8	9	10
E	55799	224.2	1.90	1.09	1.02	1.02	1.03	1.03	1.03	1.04

Growth for large μ is linear with gradient equal to α_+ which is small. Therefore the lower bound of the optimal quorum μ is necessary, while choosing too high μ increases the average cost only slightly.

Calculation can be performed with different precision; higher precision demands more time, but reduces the error probability. For example, some precision allows $p_+ = 0.044$ with unit average calculation time, while higher precision provides $p_+ = 0.004$ for eight times longer calculation. However, setting $\mu = 2$ (the optimal value for penalty F between 1.13 and 3.0) or $\mu = 3$ (optimal for $3 < F < 45$) provides error probability at least 10 and 100 times less, respectively. This reduction demands almost no price: the average cost of a task is 1.05 with difference in the third digit; note that this value includes expected penalties, so pure duration per a task is even less. So, most tasks are solved only once, rare additional checks almost do not change the cost, but effectively eliminates the risk.

This shows that it can be reasonable to use lower precision with optimal quorum.

Conclusion

We have described the cost of solving a recognition task in a desktop grid computing system as a random variable that consists of the calculation cost and the penalty in case of an error. The total cost is reduced by redundant checking calculations. We show how to choose the optimal quorum in the general case and how to improve the cost using task cost variance. We show that this gain is maximal when the penalty is close to but lower than a critical value. This theory has been applied to the virtual screening problem and tested on a piece of statistics. It showed that:

- taking cost variance into account pays only when this variance is large;
- using quorums can reduce the average cost drastically;
- overestimation of the quorum is not costly if one kind of answers is reliable;
- the optimal quorum grows exponentially, so a rough estimation is sufficient for evaluation of the quorum.

The results can be used in a task distributing manager of a desktop grid server.

Acknowledgments. The work has been supported by the Russian Foundation for Basic Research (project a-13-07-00008)and the Program of strategic development of Petrozavodsk State University. Perl Data Language [4] was used for calculations.

References

1. Ben-Yehuda, O.A., Schuster, A., Sharov, A., Silberstein, M., Iosup, A.: ExPERT: pareto-efficient task replication on grids and a cloud. In: Parallel and Distributed Processing Symposium (IPDPS), pp. 167–178 (2012)
2. Foster, I., Kesselman, C., Tuecke, S.: The anatomy of the grid: enabling scalable virtual organizations. Int. J. Supercomputer Appl. **15**(3), 200–222 (2001)
3. Ghare, G.D., Leutenegger, S.T.: Improving speedup and response times by replicating parallel programs on a SNOW. In: Feitelson, D.G., Rudolph, L., Schwiegelshohn, U. (eds.) JSSPP 2004. LNCS, vol. 3277, pp. 264–287. Springer, Heidelberg (2005)

4. Glazebrook, K., Economou, F.: PDL: the Perl Data Language. Dr. Dobb's J. 22(9) (1997). http://www.ddj.com/184410442. Accessed 19 September 2013
5. Han, J., Park, D.: Scheduling proxy: enabling adaptive-grained scheduling for global computing system. In: Proceedings of the Fifth IEEE/ACM International Workshop on Grid Computing, pp. 415–420 (2004)
6. Jimènez-Peris, R., Patiño Martìnez, M., Alonso, G., Kemme, B.: Are quorums an alternative for data replication? ACM Transact. Database Syst. 28(3), 257–294 (2003)
7. Kondo, D., Chien, A., Casanova, H.: Scheduling task parallel applications for rapid turnaround on enterprise desktop grids. J. Grid Comput. 5, 379–405 (2007)
8. Kondo, D., Taufer, M., Brooks, C., Casanova, H., Chien, A.: Characterizing and evaluating desktop grids: an empirical study. In: Proceedings of the International Parallel and Distributed Processing Symposium (IPDPS) (2004)
9. Kondo, D., Araujo, F., Malecot, P., Domingues, P., Silva, L.M., Fedak, G., Cappello, F.: Characterizing result errors in internet desktop grids. In: Kermarrec, A.-M., Bougé, L., Priol, T. (eds.) Euro-Par 2007. LNCS, vol. 4641, pp. 361–371. Springer, Heidelberg (2007)
10. Martins, F.S., Andrade, R.M., dos Santos, A.L., Schulze, B., de Souza, J.N.: Detecting misbehaving units on computational grids. Concurr. Comput.: Practice Exp. 22(3), 329–342 (2010). http://dblp.uni-trier.de/db/journals/concurrency/concurrency22.html#MartinsASSS10
11. Rumiantsev, A.S.: Optimizing the execution time of a desktop grid project. Program Syst.: Theory Appl. Online J. 5(1), 175–182 (2014). (in Russian)
12. Sangho, Y. Kondo, D., Bongjae, K.: Using replication and checkpointing for reliable task management in computational grids. In: International Conference on High Performance Computing and Simulation, pp. 125–131 (2010)
13. Silaghi, G.C., Araujo, F., Silva, L.M., Domingues, P., Arenas, A.E.: Defeating colluding nodes in desktop grid computing platforms. J. Grid Comput. 7(4), 555–573 (2009). http://dblp.uni-trier.de/db/journals/grid/grid7.html#SilaghiASDA09
14. Storm, C., Theel, O.: A general approach to analyzing quorum-based heterogeneous dynamic data replication schemes. In: Garg, V., Wattenhofer, R., Kothapalli, K. (eds.) ICDCN 2009. LNCS, vol. 5408, pp. 349–361. Springer, Heidelberg (2008)
15. Trott, O., Olson, A.: AutoDock Vina: improving the speed and accuracy of docking with a new scoring function, efficient optimization, and multithreading. J. Comput. Chem. 31, 455–461 (2010). doi:10.1002/jcc.21334

Partition Algorithm for Association Rules Mining in BOINC–Based Enterprise Desktop Grid

Evgeny Ivashko$^{(\boxtimes)}$ and Alexander Golovin

Institute of Applied Mathematical Research,
Karelian Research Centre of Russian Academy of Sciences,
Petrozavodsk, Russia
{ivashko,golovin}@krc.karelia.ru

Abstract. The paper describes an approach to association rules mining from big data sets using BOINC–based Enterprise Desktop Grid. An algorithm of data analysis and a native BOINC–based application are developed. Several experiments with the aim of validation and performance evaluation of the algorithm implementation are performed. The results of the experiments show that the approach is promising; it could be used by small and medium businesses, scientific groups and organizations.

Keywords: Enterprise Desktop Grid · BOINC · Distributed computing

1 Introduction

Data Mining methods are a popular tool for data analysis. It is also called "knowledge discovery in databases" – the process of discovering interesting and useful patterns and relationships in large volumes of data. The field combines tools from statistics and artificial intelligence to analyze large data sets [11].

There are a number of methods aimed at extracting information from data. One of the popular Data Mining methods is association rules mining [10]. Association rules mining is a well-studied area, especially given its importance in many problems of data analysis. Association rules express the association between observations in a database transaction. The problem of discovering frequent itemsets in a transactional data set (the so called FIM problem) is the first step of association rules mining. A number of algorithms have been suggested to discover frequent itemsets: Apriori [7], FP-Growth [8], Eclat [9], and others.

Development of new algorithms and technologies for mining association rules is vital due to the need to process increasingly large data sets. One of the problems is computational complexity in discovering frequent itemsets, as with the increasing number of elements in the input data exponentially increases the number of potential sets. Therefore such development requires use of parallel data

The work is supported by grants of Russian Fund for Basic Research 13-07-00008 and 15-07-02354.

© Springer International Publishing Switzerland 2015
V. Malyshkin (Ed.): PaCT 2015, LNCS 9251, pp. 268–272, 2015.
DOI: 10.1007/978-3-319-21909-7_26

processing technology. There are algorithms adapted for use in parallel systems, for example: Partition [4], PFP [5], FDM [6].

Analysis of big data sets requires the involvement of high-performance computing systems. One of the high-performance computing technologies is Desktop Grid. It utilizes the power of idle CPU time of desktop computer. There are two concepts of computing resources gathering: Volunteer Desktop Grid aggregating the power of volunteer computers over the Internet, and Enterprise Desktop Grid consisting of computers belonging to the local-area network of an organization. BOINC is one of the most commonly used Desktop Grid software.

The paper describes an approach to association rules mining from big data sets using BOINC–based Enterprise Desktop Grid.

Use of Enterprise Desktop Grid for data processing requires appropriate adaptation of the software. In case of BOINC, it is necessary to develop special software that uses the BOINC API for implementing the interaction between a BOINC client and a running application.

Some previous works have been focused on developing methods and technologies of big data sets analysis based on Data Mining and Desktop Grid environment [14]. There are two main problems appear with the development of the BOINC-based data mining application aimed at big data sets processing [12]. First of all, such applications are not easily decomposable into a great number of small enough independent tasks. Second, these applications are very data-intensive, i.e. every task needs a large portion of data. The first problem is challenging, but it could be solved by development of a special algorithm. Weka4WS [15] extend the Weka toolkit to allow the use of distributed ad-hoc environment to perform data analysis. The project distributedDataMining.org uses BOINC and data mining tool RapidMiner for distributed data analysis [3]; the article gave an overview for integration and interaction of the used tools. However, to solve the second problem it is necessary to develop a specific workaround. For example, the paper [2] describes the approach employing P2P networks and distributed cache servers to workaround the data-intensivity problem. Our research is based on Enterprise Desktop Grid to work with data-intensive applications with speed of local-area networks.

The Berkeley Open Infrastructure for Network Computing (BOINC) is an open source software framework for distributed and grid computing [1]. BOINC is based on the client/server model. It has a central server and project's database storing information about registered users and associated hosts, applications, data of tasks and results of calculations and other information. Also there are special services on the central server, the most important are

- **Work generator** generates workunits and corresponding input files.
- **Scheduler** assigns jobs to a client depending on its characteristics.
- **Validator** decides whether results are correct.
- **Assimilator** periodically checks the completed jobs and processes results according to application-specific rules.

BOINC clients work at the computing nodes. A BOINC application has to communicate to a BOINC-client itself to be runnable in the BOINC environment.

2   Implementation of Partition Algorithm with BOINC

The association rules approach deals with a transactional database. The database consists of a big number of transactions, each of them is a set of several items. For example, such database can be a supermarket's record of purchases for a month: each transaction is a market basket and items are single products. Another example is a server logs database of Web-sites visits. Then each transaction is a single user's work session and items are visited sites.

Association rule is an implication $X \to Y$, where X and Y are itemsets. Such a rule has two main characteristics: support (s) and confidence (c). The rule means that if a transaction contains X then it also contains Y; there are $s\%$ of transactions in the database containing both X and Y; there are $c\%$ of all transactions that contain X also contain Y.

We have chosen the Partition algorithm to solve the problem of finding frequent itemsets in large volumes of data. Partition is a parallel modification of a well-known Apriori algorithm, it has good scalability and performance. The essence of the algorithm is as follows [4].

Below is the description of the BOINC-based Partition algorithm implementation scheme. There are three stages, two of which are executed in parallel on the computing nodes of the grid network. The final stage is the association of intermediate results. Consider this implementation in details.

At the **Prepare stage**, the work generator receives an input source file with the transactional database and the following parameters: the minimum support and confidence, the number which determines into how many parts is the source file divided and some BOINC-related workunit attributes. At **stage I**, the BOINC scheduler distributes jobs to clients (computing nodes of the BOINC-grid). BOINC-clients download input files (which are parts of the original transactional database) from the server. Then the clients run an application that extracts local frequent itemsets from their parts. After that clients upload the output files to the server and report on completing the jobs. At the **Merge stage**, the server side validation service validates the results. At the **Intermediate stage** completed jobs are handled by an assimilator which generates the set of all global candidates based on the received local frequent itemsets. Also this service forms new jobs. At the **stage II** the BOINC scheduler distributes the new jobs to clients. Each BOINC-client calculates support for each global candidate itemset in its part of the transactional database. After receiving the canonical result the assimilator summarizes supports for each candidate and removes the ones whose support is less than the specified minimum. At the same step, the assimilator constructs the association rules.

3   Results of the Experiments

Several experiments with the aim of validation and performance evaluation of the Partition algorithm implementation were performed. We used a BOINC-based Enterprise Desktop Grid with up to 32 computing nodes connected to BOINC-server by local network with a bandwidth of 100 Mb/s.

Table 1. Characteristics of the test datasets

	Filename	Number of transactions	Average length of transaction	Minimum support
I	T10I4D100K.dat	100 000	10	1
II	T25I20D100K.dat	100 000	25	1.5
III	T40I10D100K.dat	100 000	40	5

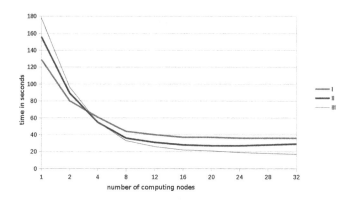

Fig. 1. Results of the experiments on the test datasets.

First of all, we validated the application by test source datasets from the Frequent Itemset Mining Dataset Repository (FIMDR) [13]. Characteristics of the used datasets are presented in Table 1 (filename corresponds dataset of FIMDR). The results of the experiments are presented in the Fig. 1.

The figure shows that use of BOINC can accelerate rules extraction up to 6–9 times. It also shows that the overall time of rules extraction depends on the minimal support and length of transactions. The main performance limitation in the performed experiments is still the network bandwidth. BOINC-server should distribute a database between computing nodes that becomes a very time-consuming operation. But the use of Enterprise Desktop Grid (and local-area network) allows to improve the speed of data analysis.

4   Conclusion and Discussion

BOINC is becoming a more popular tool to perform large-scale computational experiments. BOINC-based Enterprise Desktop Grid allows to small and medium companies or small scientific groups to solve their private problems using their own computing resources. One of such private problems is analysis of big datasets.

This study shows the way to extract association rules from big data sets using BOINC-based Enterprise Desktop Grid. We adapted the Partition algorithm for BOINC and performed the experiments on performance evaluation of association rules extraction. Our results show that Enterprise Desktop Grid allows reducing expended time for data analysis.

Further development of the study will be devoted to adapting other methods of data analysis to BOINC-environment. Also it is important to combine the developed tool with special visualization software.

References

1. Anderson, D.P.: BOINC: a system for public-resource computing and storage. In: Fifth IEEE/ACM International Workshop on Grid Computing, pp. 4–10 (2004)
2. Cesario, E., De Caria, N., Mastroianni, C., Talia, D.: Distributed data mining using a public resource computing framework. In: Desprez, F., Getov, V., Priol, T., Yahyapour, R. (eds.) Grids, P2P and Services Computing, pp. 33–44. Springer, US (2010)
3. Schlitter, N., Laessig, J., Fischer, S., Mierswa, I.: Distributed data analytics using RapidMiner and BOINC. In: Proceedings of the 4th RapidMiner Community Meeting and Conference (RCOMM 2013), pp. 81–95 (2013)
4. Savasere, A., Omiecinski, E., Navathe, S.: An efficient algorithm for mining association rules in large databases. In: Proceedings of 21st International Conference on Very Large Data Bases, pp. 432–444. Morgan Kaufmann, San Francisco (1995)
5. Li, H., Wang, Y., Zhang, D., Zhang, M., Chang, E.Y.: Pfp: parallel fp-growth for query recommendation. In: RecSys 2008 Proceedings of the 2008 ACM conference on Recommender systems, pp. 107–114 (2008)
6. Cheung, D., Han, J., Ng, V.T., Fu, A. W., Fu, Y., Yongjian, A.W.: A fast distributed algorithm for mining association rules. In: Proceedings of International Conference on PDIS 1996, pp. 31–42 (1996)
7. Agrawal, R., Srikant, R.: Fast discovery of association rules. In: Proceedings of the 20th International Conference on VLDB, pp. 307–328. Santiago, Chile (1994)
8. Han, J., Pei, H., Yin, Y.: Mining frequent patterns without candidate generation. In: Proceedings Conference on the Management of Data, pp. 1–12. Dallas, TX (2000)
9. Zaki, M.J.: Scalable algorithms for association mining. IEEE Trans. Knowl. Data Eng. **12**(3), 372–390 (2000)
10. The 5th Annual Rexer Analytics Data Miner Survey. http://www.rexeranalytics.com/Data-Miner-Survey-Results-2011.html
11. Encyclopedia Britannica. http://global.britannica.com/EBchecked/topic/1056150/data-mining
12. Barbalace, D., Lucchese, C., Mastroianni, C., Orlando, S., Talia, D.: Mining@HOME: public resource computing for distributed data mining. Concurrency Comput. Pract. Experience **22**(5), 658–682 (2010)
13. Frequent Itemset Mining Dataset Repository. http://fimi.ua.ac.be
14. Saad, M.K., Abed, R.M.: Distributed data mining on grid environment. Am. Acad. Sch. Res. J. Spec. Iss. **4**(5), 240–243 (2012)
15. Talia, D., Trunfio, P., Verta, O.: Weka4WS: a WSRF-enabled weka toolkit for distributed data mining on grids. In: Jorge, A.M., Torgo, L., Brazdil, P.B., Camacho, R., Gama, J. (eds.) PKDD 2005. LNCS (LNAI), vol. 3721, pp. 309–320. Springer, Heidelberg (2005)

Task Scheduling in a Desktop Grid to Minimize the Server Load

Vladimir V. Mazalov, Natalia N. Nikitina$^{(\boxtimes)}$, and Evgeny E. Ivashko

Institute of Applied Mathematical Research, Pushkinskaya 11,
Petrozavodsk 185910, Russia
{mazalov,nikitina,ivashko}@krc.karelia.ru
http://mathem.krc.karelia.ru

Abstract. Desktop Grids utilize computational resources of desktop computers in their idle time. The BOINC middleware for organizing Desktop Grids has an architecture developed to unite a large number of computing nodes. However, a large flow of server requests may limit the Desktop Grid performance. In the paper we present a game-theoretical model of task scheduling in a Desktop Grid. The model allows to consider the trade-off between the server load and the total time of computations. The solution is illustrated on examples.

Keywords: Desktop grid · BOINC · Virtual screening · Hierarchical game

1 Introduction

In the paper we consider the problem of task scheduling in a Desktop Grid. The term stands for a distributed computing system that collects together desktop computers, servers, cluster nodes and other heterogeneous computational resources connected by the Internet or a local network and working for the Desktop Grid in their idle time. The BOINC middleware [1] can be considered a de-facto standard for organizing Desktop Grids, as it is being actively developed and has been successfully used in several computational projects since 1997.

The BOINC system has server-client architecture: a server distributes independent tasks to the nodes which perform computations and return the results to the server for further processing. Such division of a large resource-demanding computational problem into many independent tasks allows to solve it effectively in a shorter time. Many works ([2–4] etc.) evaluate performance of Desktop Grids by the total computational time or the average throughput rate.

Computational process in a distributed system involves persons with various interests: a system administrator, the users, the owners of computational nodes etc. For this reason, the task of effectively managing distributed resources can be solved using mathematical game theory ([5–7] etc.).

Quite often the independent computational tasks are short-running due to the properties of a computational problem. In such cases Desktop Grid clients

© Springer International Publishing Switzerland 2015
V. Malyshkin (Ed.): PaCT 2015, LNCS 9251, pp. 273–278, 2015.
DOI: 10.1007/978-3-319-21909-7_27

generate an intensive flow of requests to the server, reporting results and asking for more work very frequently. With a fair number of clients, this can cause an effect similar to a DDoS-attack, or at least make increased demands on the server software or hardware. We illustrate the case as follows.

We ran a BOINC server on a virtual machine running CentOS, with eight Intel® Xeon® E5620 CPUs each of 2.4 GHz. A sample short-running BOINC application was used. We launched BOINC clients in parallel on a separate virtual machine, varying their number from 10 to 230. As the number of clients exceeded 70, some of clients' HTTP requests to the server would fail. At the same time, the average number of HTTP requests served by the server per second practically did not change. Consequently, the system software such as web server or the database can limit the scalability of the Desktop Grid or its peak performance. Moreover, when an error occurs, the BOINC client will make a pause before next request to the server. The length of such a pause will increase exponentially if errors repeat. This leads to decreasing both the total number of requests to the server and the average project performance as new clients join.

To reduce the server load we propose to group computational tasks into *parcels*. We treat the overall cost of computing a parcel of tasks as the expenses required to create an input data archive (instead of a single input file), transmit it over the network, compute the tasks on the client, send the results back, unpack and process the results. The expenses can be interpreted, for instance, as the total CPU or wall-clock time consumed at the server, or the total electricity bill. For simplicity, we assume that a computational error in a single task interrupts the computation and causes a whole parcel to be sent to the client again.

Grouping tasks in order to increase work performance of a Desktop Grid is considered, for example, in [3] where the possibility to decrease computational time in the presence of task replication is investigated. In work [8] the authors propose to group the geographically nearby nodes and assign one in each group to be an intermediate scheduler. In this work we present a mathematical model of a Desktop Grid as a two-level hierarchical game. The model describes the problem of choosing an optimal parcel size and allows to reach the balance between the amount of computations and the server load.

Particular performance measurements depend heavily on the server hardware, the system software, the settings of the web server and the database etc. In the examples we used a default installation of system software and BOINC.

2   The Model

Hierarchical games, one of domains of mathematical game theory, allow to model situations in which participants make their choices subsequently. First, the player of the top hierarchy level, or the "center", makes its choice. Then the players of the second level make choice within the restrictions imposed by the choice of the "center", aiming to maximize their own utilities. The "center" knows the rules by which the second level players answer to each of its possible decisions, so its choice maximizes its utility.

Let us describe the computational process in a Desktop Grid as a two-level hierarchical game of $m + 1$ players.

The server M_0 has N computational tasks and wants to distribute them among a set of clients $M_1, \ldots, M_m$ giving each one $N_1, \ldots, N_m$ tasks correspondingly. The server chooses a method of task distribution $u = (N_1, \ldots, N_m)$, where u belongs to U, a set of server strategies,

$$U = \{u = (u_1, \ldots, u_m) : u_i \in \mathbb{Z}, 0 \le u_i \le N, \sum_{i=1}^{m} u_i = N\}$$

As soon as client M_i knows server's decision N_i, it chooses the parcel size n_i which will minimize its expenses. n_i belongs to $V_i(N_i)$, a set of client strategies,

$$V_i(N_i) = \left\{ v_i \in \mathbb{Z} : \begin{cases} 1 \le v_i \le N_i & N_i \ge 1, \\ v_i = 0 & N_i = 0 \end{cases} \right\}$$

The utility function of a client $C_i(N_i, n_i) = -K(n_i) \left\lceil \frac{N_i}{n_i} \right\rceil$ represents its expenses on computing $\left\lceil \frac{N_i}{n_i} \right\rceil$ parcels, where $K(n_i)$ is the cost of computing a parcel of n_i tasks. The optimal parcel size n_i^* will minimize client's expenses — or, equally, maximize its utility function:

$$C_i^* = C_i(N_i, n_i^*) = \max_{n_i \in V_i(N_i)} C_i(N_i, n_i)$$

As the server knows parcel sizes chosen by clients for given N_i, it aims to minimize its own expenses. The utility function of $C_0(N_1, \ldots, N_m, n_1, \ldots, n_m)$ represents server's expenses on creating parcels, waiting for the results and processing them. The optimal tasks distribution $u^* = (N_1^*, \ldots, N_m^*)$ minimizes server's expenses, that is, maximizes its utility function:

$$C_0^* = C_0(N_1^*, \ldots, N_m^*, n_1^*, \ldots, n_m^*) = \max_{u \in U} C_0(N_1, \ldots, N_m, n_1^*, \ldots, n_m^*)$$

We have a game $\Gamma = \langle \{M_0, M_1, \ldots, M_m\}, \{U, V_1, \ldots, V_m\}, \{C_0, C_1, \ldots, C_m\} \rangle$.

In 2014, a joint research project was completed between the IAMR and the Lübeck Institute of Experimental Dermatology (LIED), University of Lübeck (Germany). During the project we created a Desktop Grid infrastructure using available computers and used it to perform virtual drug screening for research interests of the LIED. Virtual drug screening is an important example of the problem being successfully solved on Desktop Grids. It is the computerized evaluation of very large libraries of chemical compounds in order to choose the ones that are most likely to influence the course of disease.

In [9] we proposed analytical expressions of utility functions, optimal parcel sizes and the resulting optimal tasks distribution for a Desktop Grid organized within a group of research institutions. The server expenses expressed the total computational time. The numerical solution was derived using statistical data. The results of computational experiments showed that under all assumptions, in equilibrium situation grouping tasks into parcels would decrease 3.4–fold the total server load in the model. Considering multi-processing, in equilibrium situation we would obtain both 3.5–fold decrease in the server load and 2.1–fold decrease in total computational time in the model.

3   Performance of a Volunteer Computing System

In this section we illustrate how the proposed model may help to improve performance of a BOINC-based volunteer computing system. We use the data published in work [10] and provided by SAT@home project to define coefficients and functions of the model, to find the optimal solution and interpret its effect for a BOINC-based volunteer computing project.

In [10] the authors show that CPU is a bottleneck of the BOINC server. We will interpret cost of one request to the server as the amount of CPU time that the server spends serving this request. The maximal supported number of BOINC clients in a single-server model [10] is 8.8 mln, and the average CPU time that a workunit instance consumes at a client is 12 CPU hours. Hereof we may consider that a single request to a server costs $\frac{8\,800\,000}{12\times3600}$ times more than computing a single task on a client.

We assume that cost functions in the model are linear in parcel size and derive coefficients from [10]. Note that the model [10] does not take into account errors occuring at BOINC clients. We estimate the error probability $1-p = 0.064$ basing on statistical data from a volunteer computing project SAT@home [11]. Under these statements, we obtain the optimal parcel size for a client $n^* = 4$ (Fig. 1).

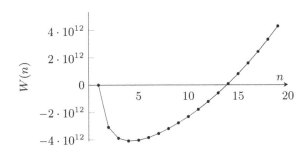

Fig. 1. $W(n)$, the increase in computational cost for a client due to packing $N = 10^8$ tasks into parcels of size n.

Using the model we calculate the total server load to create, distribute and process N tasks to be $1.15214 \times 10^{-2} \times N$ CPU seconds without parcels and $0.581473 \times 10^{-2} \times N$ with parcels of optimal size. Consequently, using parcels the considered BOINC server can support $\frac{1.15214}{0.581473} = 1.98$ times more volunteer clients. The trade-off is that without additional clients, the total time to complete N tasks would increase: in case of no parcels every client spends 524 265 CPU seconds to compute its share of 100 mln tasks, but with parcels of the optimal size this time increases to 638 560 (+22 %). This increase follows from the model assumption that upon failure the whole parcel is sent to the client again.

Conclusion

In our work we present a game-theoretical model for task scheduling in a BOINC-based Desktop Grid. Following the server-client architecture of BOINC, we consider the server load to be minimized at client level and the total computational time to be minimized at server level. As illustrated by computational experiments, the proposed task scheduling discipline allows to significantly decrease the server load with a comparatively small increase of total computational time.

Acknowledgments. The authors would like to thank Dr. Steffen Möller, PhD in the LIED, for supervising the joint research project at Lübeck and all volunteer participants of the computations both in the IAMR and the LIED for providing their computational resources, discussion and help. The work is supported by grants of RFBR 13-07-00008, 15-07-02354 and the program of the Branch of Mathematics of the RAS "Algebraic and combinatorial methods of mathematical cybernetics and information systems of the new generation".

References

1. Anderson, D.P.: BOINC: A system for public-resource computing and storage. In: Proceedings of the 5th IEEE/ACM International Workshop on Grid Computing, pp. 4–10, Washington DC, USA (2004)
2. Kondo, D., Chien, A.A., Casanova, H.: Resource management for rapid application turnaround on enterprise desktop grids. In: Proceedings of the 2004 IEEE/ACM Conference on Supercomputing (SC 2004), pp. 17–30, Washington DC, USA (2004)
3. Rumiantsev, A.S.: Optimizing the execution time of a desktop grid project. Program Syst. Theor. Appl. 5 **1**(19), 175–182 (2014). (in Russian)
4. Chernov, I. A.: Optimal quorum for the model of computational grid with redundancy. applied problems in theory of probabilities and mathematical statistics related to modeling of information systems. In: Proceedings of the Autumn Session of the VIII International Workshop, pp. 648–651 (2014)
5. Zhao, H., Li, X.: Efficient grid task-bundle allocation using bargaining based self-adaptive auction. In: 9th IEEE/ACM International Symposium on Cluster Computing and the Grid, pp. 4–11 (2009)
6. Penmatsa, S., Chronopoulos, A.T.: Job Allocation Schemes in Computational Grids Based on Cost Optimization. In: Proceedings of the IEEE International Parallel and Distributed Processing Symposium, 180a (2005)
7. Donassolo, B., Legrand, A., Geyer, C.: Non-cooperative scheduling considered harmful in collaborative volunteer computing environments. In: 11th IEEE/ACM International Symposium on Cluster, Cloud and Grid Computing (CCGrid), pp. 144–153 (2011)
8. Han, J., Park, D.: Scheduling proxy: enabling adaptive-grained scheduling for global computing system. In: Proceedings of the Fifth IEEE/ACM International Workshop on Grid Computing, pp. 415–420 (2004)

9. Mazalov, V.V., Nikitina, N.N., Ivashko, E.E.: Hierarchical Two-Level Game Model for Tasks Scheduling in a Desktop Grid. Applied Problems in Theory of Probabilities and Mathematical Statistics Related to Modeling of Information Systems, pp. 641–645. IEEE, Leonia, NJ, USA (2014)
10. Anderson, D.P., Korpela, E., Walton, R.: High-performance task distribution for volunteer computing. In: Proceedings of the First International Conference on e-Science and Grid Computing, pp. 196–203. IEEE, Washington DC, USA (2015)
11. Zaikin, O.S., Posypkin, M.A., Semenov, A.A., Khrapov, N.P.: Experience in organizing volunteer computing: a case study of the OPTIMA@home and SAT@home projects. Vestnik of Lobachevsky State University of Nizhni Novgorod, No. 5–2, pp. 340–347 (2012). (in Russian)

An HPC Upgrade/Downgrade
that Provides Workload Stability

Alexander Rumyantsev[✉]

Institute of Applied Mathematical Research of the Karelian Research
Centre of the RAS, Pushkinskaya Str. 11, 185910 Petrozavodsk, Russia
ar0@krc.karelia.ru

Abstract. The workload model of a high-performance cluster is considered in the context of an upgrade (increase the computational power) or
downgrade (save energy) problems. Analytical solutions are found that
provide stochastic stability of the workload. The results of numerical
experiments with log-files of a real workload are presented.

Keywords: HPC workload · Green computing · Stochastic stability

1 Introduction

A high-performance computer cluster (HPC) is a computer system with a high
level of computational capacity, which is mainly built of a large number of
dedicated processors working together as a single homogeneous computational
resource. The HPC system is shared by multiple customers in a competitive
manner, where each customer may (depending on the configuration of the HPC)
partially or fully occupy the processors. When the workload of such a system
exceeds its computational capacity, the *upgrade* problem is basically solved as
follows:

1. either replace each processor with a more powerful one,
2. or increase the number of processors of the same type.

A somewhat similar problem arouses in the context of green computing [1,5,6].
The energy consumption of a typical HPC is roughly the same both in busy and
in idle periods. Hence, it is natural to dynamically decrease the computational
power in idle states, thus lowering the energy consumption of the system. The
downgrade problem has two major solutions:

1. either decrase the frequency and voltage of each processor (e.g. DVFS [5]),
2. or decrease the number of processors in the system (by powering them off).

In this work both the upgrade and downgrade problems are solved in terms of
stochastic stability.

The research is partially supported by Russian Foundation for Basic Research,
projects No. 13-07-00008, 14-07-31007, 15-07-02341, 15-07-02354 and the Program
of Strategic Development of the Petrozavodsk State University.

V. Malyshkin (Ed.): PaCT 2015, LNCS 9251, pp. 279–284, 2015.
DOI: 10.1007/978-3-319-21909-7_28

2   Stability of an HPC Workload Model

Consider a queueing system with s identical processors serving Poisson flow of customers (with intensity λ) arriving at a single queue (with FCFS queueing discipline). A customer i occupies $N_i \leqslant s$ processors simultaneously for the time S_i (i.i.d., exponentially distributed with intensity μ). Let i.i.d. $\{N_i\}, i \geqslant 1$ have a distribution

$$\mathbf{p} = \{p_k := \mathsf{P}\{N = k\}, \quad k = 1, \ldots, s\}. \tag{1}$$

(The indices are omitted for generic elements of a stochastic sequence.) If the number of idle processors at the time of i-th arrival is less than N_i, the customer experiences a delay $D_i > 0$ waiting for resources. We refer the reader to the works [7–9] for a detailed discussion of the HPC workload model. Note that the *driving sequence* $\{T_i, S_i, N_i\}$ for $i \geqslant 0$ may be extracted from the log-file of the queue management software of an HPC (see e.g. the traces in [3]). Another option is distribution sampling [4].

The HPC workload model is *stable* (where the delay D_i converges weakly to a stationary delay $D < \infty$ as $i \to \infty$) iff

$$\rho := \frac{\lambda}{\mu} < C(s), \tag{2}$$

where

$$C(s) := \left[\sum_{m \in \mathcal{M}} \frac{\prod_{i=1}^{s} p_{m_i}}{f(m, s)} \right]^{-1}, \tag{3}$$

$$f(m, s) := \max\{i : \sum_{j=1}^{i} m_j \leqslant s\}, \tag{4}$$

and $\mathcal{M} = \{1, \ldots, s\}^s$ [10].

3   An Upgrade/Downgrade Problem

The aforementioned solutions of the upgrade/downgrade problem need to define the border values

- find μ' s. t. $\lambda/\mu' = C(s)$;
- find s' s. t. $\lambda/\mu = C(s')$.

It is easy to find such m' by

$$\mu' = \frac{\lambda}{C(s)}. \tag{5}$$

To prove the existence of s', we need the following convention:

$$p_i := 0, \quad i = s + 1, \ldots, s'. \tag{6}$$

In case of an upgrade problem, (6) means that after upgrade the tasks still require no more than s processors; for a downgrade problem this is a necessary requirement for the possibility of such a downgrade.

Lemma 1. *For any $s \geq 1$ and $s' > s$, provided* (6),

$$C(s') \geq C(s),\tag{7}$$

and there exists $s_0 > s$, s. t. $C(s_0) > C(s)$.

Proof. Fix $s' > s$ and $m \in \mathcal{M}$. Denote $\mathcal{M}' := \{1,\ldots,s\}^{s'}$ and

$$\mathcal{M}'(m) = \{m' \in \mathcal{M}' : m'_i = m_i, \; i = 1,\ldots,s\}.$$

Then for $m' \in \mathcal{M}'(m)$

$$f(m,s) = \max\{i : \sum_{j=1}^{i} m_j \leq s\} \leq \max\{i : \sum_{j=1}^{i} m'_j \leq s'\} = f(m',s'). \tag{8}$$

Hence,

$$C^{-1}(s') = \sum_{m' \in \mathcal{M}'} \frac{\prod_{i=1}^{s'} p_{m'_i}}{f(m',s')} = \sum_{m \in \mathcal{M}} \sum_{m' \in \mathcal{M}'(m)} \frac{\prod_{i=1}^{s} p_{m_i} \prod_{i=s+1}^{s'} p_{m'_i}}{f(m',s')} \leq$$

$$\leq \sum_{m \in \mathcal{M}} \frac{\prod_{i=1}^{s} p_{m_i}}{f(m,s)} \sum_{m' \in \mathcal{M}'(m)} \prod_{i=s+1}^{s'} p_{m'_i} = C^{-1}(s).$$

Now take $s_0 := s(s+1)$ and note that (8) becomes strict inequality for any $m \in \mathcal{M}$ by the fact that $f(m',s_0) \geq s+1$ for any m'. □

We note that s' can be found numerically by a monotonicity property proven in Lemma 1. However, an approximate solution can be found with the following conjecture, that is true, as preliminary numerical experiments have shown (a strict proof of this result is left for future research).

Conjecture 1. under the assumption (6) for $s' > s$,

$$\lim_{s' \to \infty} [C(s'+1) - C(s')] = \delta := \left[\sum_{k=1}^{2s-1} r_k \right]^{-1}, \tag{9}$$

where

$$r_k := \sum_{1 \leq i, s+i-k \leq s} p_i \left[\sum_{j=k-i+1}^{s} p_j \right], \quad k = 1,\ldots,2s-1. \tag{10}$$

Limit (9) may be treated as $C(s') \approx C(s) + \delta(s'-s)$ for sufficiently large $s' > s$. Thus, the approximate solution is as follows:

$$s' \approx s + \frac{\rho - C(s)}{\delta}. \tag{11}$$

4   Numerical Experiments

To illustrate both solutions, the hpcwld package [2] for the R language [11] was used. The log-file of the Cornell Theory Center (CTC) IBM SP2 cluster was used to illustrate the over-utilized HPC, whereas Ohio Supercomputing Center (OSC) cluster was selected to illustrate the low utilization case. Both log files were taken from [3] and imported into R by using FromSWF function of the aforementioned package. We note that the experiments were held under the assumption of FIFO scheduling discipline, whereas both machines use distinct schedulers (CTC uses EASY Backfill, while OSC uses Maui scheduler). Thus, the results of experiment may serve as some sort of "upper bound" for the system performance. We refer the reader to [9] on details of the package usage for numerical verification of stability.

The plan of both experiments is as follows:

1. parse the log-file and verify the stability by (2);
2. find μ' from (5) and s' from (11);
3. evaluate the delays $D_i(s,\mu), D_i(s,\mu'), D_i(s',\mu)$ (where $D_i(s,\mu)$ is the delay of customer i in the system with s processors with intensity μ) and graphically verify the stability.

4.1   CTC SP2 Cluster

This HPC has $s = 336$ processors, and $\rho = \lambda/\mu = 29.72244 > C(s) = 26.02444$. Evaluating $\delta = 0.0910304$ from (9), one gets $s' = 377$ and $\mu'/\mu = 1.142097$, i. e. either the frequency of each processor should be increased on $\approx 15\,\%$, or the number of processors should be increased to 377. The resulting sample paths of the delays for all the 77221 customers from the log-file are depicted on Picture 1. Interestingly, both upgrade options give nearly the same sample path, that is stable, whereas the delays in the original path (depicted in black) tend to grow unbounded (Fig. 1).

4.2   OSC Cluster

This HPC has 178 processors, however, the maximum number of requested processors per customer is 32, hence, (6) holds for $i = 33, \ldots, 178$. Moreover, $\rho = 16.48379$ is less, than $C(s) = 72.81022$. Evaluating $\delta = 0.417208$ from (9), one has $s' = 43$ and $\mu'/\mu = 0.2263939$. However, this approximate solution needs to be improved by a local search near the value of s', which shows that (2) holds for $s_1 = 47$. The resulting sample paths of the delays for all the 36096 customers from the log-file are depicted on Picture 2. Note that the original delays were near zero (with mean 0.009 and maximum 83 s). Both downgraded systems seems to be stable, although the delays are high enough. However, in case of a demand for power budget reducing, switching off nearly 75 % of the system (or slowing down the processors by a factor of 5) could be an inescapable option (Fig. 2).

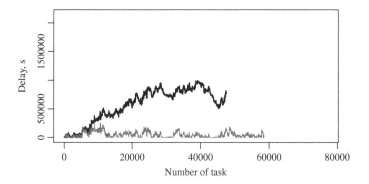

Fig. 1. Sample paths of the delays $D_i(s, \mu)$ evaluated on CTC SP2 data: black $D_i(336, \mu)$, blue $D_i(377, \mu)$, red $D_i(336, 1.142097\mu)$ (Colour figure online)

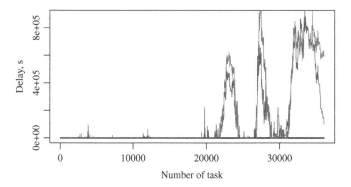

Fig. 2. Sample paths of the delays $D_i(s, \mu)$ evaluated on OSC data: black $D_i(178, \mu)$, blue $D_i(47, \mu)$, red $D_i(178, 0.2377136\mu)$ (Color figure online)

5 Conclusion

As numerical experiments have shown, the presented method of solving the upgrade/downgrade problem may be easily applied to real systems. An extention of this work to the performance of an HPC is left for future research. We note, however, that due to simplicity of model evaluation and stability condition validation, one could consider fine-tuning the parameters of the system numerically to provide reasonable performance measures.

Acknowledgments. Author thanks E. V. Morozov for helpful discussions that allowed to improve the work.

References

1. Alonso, M., et al.: Power saving in regular interconnection networks. Parallel Comput. **36**(12), 696–712 (2010)

2. CRAN - Package hpcwld. http://cran.r-project.org/web/packages/hpcwld/index.html
3. Feitelson D.G.: Parallel Workloads Archive: Logs. http://www.cs.huji.ac.il/labs/parallel/workload/logs.html
4. Feitelson D.G.: Workload modeling for computer systems performance evaluation. http://www.cs.huji.ac.il/feit/wlmod/wlmod.pdf
5. Gandhi, A., et al.: Optimal power allocation in server farms. ACM SIGMETRICS Perform. Eval. Rev. **37**, 157–168 (2009)
6. Gandhi, A., et al.: Power capping via forced idleness. In: Proceedings of Workshop on Energy Efficient Design. pp. 1–6 (2009). http://researcher.watson.ibm.com/researcher/files/us-lefurgy/weed2009_ghandi_paper.pdf
7. Morozov, E.V., Rumyantsev, A.S.: Stability analysis of a multiprocessor model describing a high performance cluster. In: XXIX International Seminar on Stability Problems for Stochastic Models and V International Workshop "Applied Problems in Theory of Probabilities and Mathematical Statistics related to modeling of information systems", pp. 82–83. Institute of Informatics Problems RAS, Moscow (2011)
8. Morozov, E.V., Rumyantsev, A.S.: Stochastic models of multiprocessor systems: stability and moment properties. Inf. Appl. **6**(3), 99–106 (2012)
9. Rumyantsev, A.: Simulating Supercomputer Workload with hpcwld package for R. In: 15th International Conference on Parallel and Distributed Computing, Applications and Technologies, pp. 138–143. IEEE (2014)
10. Rumyantsev, A.: Stabilization of a high performance cluster model. In: 6th International Congress on Ultra Modern Telecommunications and Control Systems and Workshops (ICUMT), pp. 518–521. IEEE (2014)
11. R Foundation for Statistical Computing, Vienna, Austria. ISBN 3-900051-07-0. http://www.R-project.org/

Job Ranking and Scheduling
in Utility Grids VOs

Victor Toporkov[1]([⊠]), Anna Toporkova[2], Alexey Tselishchev[3],
Dmitry Yemelyanov[1], and Petr Potekhin[1]

[1] National Research University "MPEI",
ul. Krasnokazarmennaya, 14, Moscow 111250, Russia
{ToporkovVV, YemelyanovDM, PotekhinPA}@mpei.ru
[2] National Research University Higher School of Economics,
ul. Myasnitskaya, 20, Moscow 101000, Russia
atoporkova@hse.ru
[3] European Organization for Nuclear Research (CERN),
1211 Geneva, 23, Switzerland
Alexey.Tselishchev@cern.ch

Abstract. In this work, we propose approaches to creation of a ranked jobs framework within a model of cycle scheduling in virtual organizations of utility Grids with the decoupling of users from resource providers. Two methods for job selection and scheduling are proposed and compared: the first one is based on the knapsack problem solution, while the second one introduces a heuristic parameter of a job and a computational resource set "compatibility". Along with these methods we present experimental results demonstrating the efficiency of proposed approaches and compare them with random job selection.

Keywords: Grid · Virtual organization · Scheduling · Resource management · Job · Flow · Batch · Knapsack problem

1 Introduction

The complexity of resource management and scheduling in distributed computing environment like Grid is determined by geographical distribution, resource dynamism and inhomogeneity of jobs and execution requirements defined by users of virtual organizations (VO) [1, 2]. A matter of the utmost importance for the VO is to efficiently manage available computational resources while fulfilling requirements of all stakeholders: users, resource owners and VO administrators. The fact that resources of utility Grids are non-dedicated makes the efficient scheduling problem even more complex. In distributed computing with a lot of different participants and contradicting requirements the most efficient approaches are based on economic principles [3–6]. Different approaches to job flow scheduling can be classified based on dispatching methods. When job-dispatching process is decentralized, schedulers usually reside and work on the client side and fulfill end-user requirements (AppLeS [7], PAUA [8]). Centralized job-dispatching implies that a meta-scheduler ensures the efficient usage of all the resources. While managing the scheduling process the meta-scheduler works with

© Springer International Publishing Switzerland 2015
V. Malyshkin (Ed.): PaCT 2015, LNCS 9251, pp. 285–297, 2015.
DOI: 10.1007/978-3-319-21909-7_29

meta-jobs that are accompanied by a resource query, that contains resource charac-teristics required for the job execution (X-Com [9], GrADS [10]). It is also possible to evaluate job resource requirements by other means: statistically or by using expert systems [11]. Generally job-flow scheduling problem is solved using standard methods or algorithms [12–14], which include First-Come-First-Served, backfilling, user rank-ing mechanisms and resource separation. Within these approaches it is important to maintain the queue order and user priorities when executing these jobs. Even more "honest" queue forming is based on economic principles [6], which takes into account single job features and their impact on the queue.

Cycle scheduling scheme (CSS) [15–17] allows fulfilling VO requirements to a greater extent. Such scheduling is based on the set of dynamically updated information about the load of available resources. During the job batch execution the VO policy, as a rule, has higher priority than single batch jobs. This allows optimizing overall job batch execution parameters [18]. However, at the same time the queue order can be affected. There are two main steps in CSS for a single job batch: firstly, several execution options (alternatives) are found for each job for a given scheduling interval [17], and, secondly, the set of alternatives (one alternative for each job) is chosen following the VO policy [15, 16]. In order to fulfill VO user requirements the job batch is populated with the jobs with the highest priority (e.g. those in the beginning of a standard queue). Execution alternatives allocation is also performed sequentially for each job, which, in its turn, guarantees, that the priorities are followed. When additionally, user optimization criteria are used, one can guarantee a "fair" scheduling of the whole job batch [16, 19, 20]. However, it is worth noting, that job selection using simple user priorities can negatively impact the scheduling efficiency of the whole job batch. In other words, in order to increase the whole job batch scheduling efficiency according to the VO requirements one should evaluate different methods of job framework ranking.

In this paper, we review common problems of job batch forming for the cycle scheduling process. Two job batch forming approaches are proposed: the first one is based on the knapsack problem [21], the second one is using a heuristic "compatibility" parameter of a job and a resource domain for the job-flow distribution and job batch selection. The rest of this paper is organized as follows. Section 2 contains brief analysis of related works. In Sect. 3, there are approaches proposed to form a job framework. Section 4 describes the experimental results. Final results and next steps are defined in the summary Sect. 5.

2 Related Works

Many scheduling algorithms and heuristic-based solutions have been proposed for parallel jobs in distributed environments [1–3, 5, 11, 17, 23–30]. In some well-known models [2, 24–26] of distributed computing with non-dedicated resources, only the first fit set of resources is chosen depending on the environment state, while job scheduling optimization mechanisms are usually not supported. In other models [3, 5, 11], the aspects related to the specifics of environments with non-dedicated resources, partic-ularly dynamic resource loading, the competition between independent users, users' global and owners' local job flows, are not presented. In [5], heuristic algorithms for slot selection, based on user-defined utility functions, perform slot window allocation under the maximum total execution cost constraint, but the optimization occurs only on the stage of the best found offer selection. Architecture and an algorithm for performing Grid resources co-allocation without the need for advance reservations based on synchronous queuing of subtasks are introduced in [27]. Advance reservation-based co-allocation algorithms are proposed in [24–26, 28, 29]. First fit resource selection algorithms [2, 24–26] assign any job to the first set of slots matching the resource request conditions without any optimization. Preference-based matchmaking [2] is not focused on the scheduling process. In [23], an approach to resource matchmaking among VOs combining hierarchical and peer-to-peer models of meta-schedulers is proposed. The algorithms described in [28–30] suppose an exhaustive search. Approaches in [29, 30] are based on linear integer programming (IP) or mixed-integer programming (MIP) models. A linear IP-driven model with a genetic algorithm is proposed in [1]. It allows obtaining the best meta-schedule that minimizes the combined cost of all independent users in a coordinated manner. In [30], a MIP model which performs the best scheduling in environments composed of multiple clusters that act collaboratively. The scheduling approaches in [1, 28–30] are efficient under given criteria: the processing cost, the overall makespan, resources utilization, load balancing for related tasks, etc. It is worth remarking here, that complexity of the scheduling process is extremely increased by the resources heterogeneity and the co-allocation process of the tasks of parallel jobs across resource domain boundaries. In this work, algorithms for efficient slot selection based on users', resource owners' and VO administrators' preferences with the linear complexity on the number of all available time-slots are used [15–17]. Our approach takes into account preferences of diverse VO stakeholders. Scheduling optimization is conducted at two levels – when selecting the slots and when executing the job batch.

The CSS model [15–17] has the following basic resource request requirements to computational nodes: the minimal performance p, required for job execution, the maximum total job execution cost (budget) S, a number n of computing nodes needed for the job, and resource reservation time t (estimated for a resource with the performance p). The framework of independent jobs at each scheduling cycle is represented as a job batch in a certain manner formed from the job flow. Such selection makes it possible to increase overall scheduling efficiency in the VO compared to scheduling each job individually due to optimization of the general criterion formalizing the VO policy and fair resource sharing based on preferences of key stakeholders [2–6, 11, 15, 16, 20].

3 Job Framework Forming

3.1 Job Batch Size Restrictions

An important step bearing, at first glance, no relation to job flow cyclic scheduling efficiency is determining the job batch size during each scheduling cycle. By varying the job batch size limit (which can be expressed, for example, in a number of jobs in the batch or their cumulative execution budget) scheduling efficiency can be increased according to one or several different criteria. There are following scheduling efficiency criteria considered in our model: computing nodes utilization level; optimization criterion formalizing the VO policy (for example, job flow execution time minimization, with restriction on the total execution cost); number of execution alternatives found for each job (in CSS a greater number of alternatives means greater scheduling optimization opportunities); number of scheduling cycles required for complete job flow execution (minimizing this factor provides a higher throughput of the distributed computing environment).

Specifying the batch size directly, for instance, by VO administrators, is not reasonable. Under conditions when local schedules of computing nodes change dynamically and parameters of incoming jobs differ significantly and are based on user estimates which are often inaccurate, it is impossible to specify a limit in advance that would allow increasing scheduling efficiency according to the criterion chosen in the VO. A more flexible batch size limiting mechanism can be built based on relation between job requirements and computing environment parameters. In the context of economic principles it is logical to choose resource utilization cost and resource reservation time as base characteristics for such relation. When using time limit, total time of slot occupation is evaluated for each job. This time is normalized to a resource with a base performance ($p = 1$). To execute a job a set of suitable slots has to be allocated. Each of the slots is characterized by start time, length and utilization cost [17]. A slot set forms a "window", for which total time and cost of slot utilization can be calculated. Note that for the purpose of normalization these values need to be calculated to a resource of the base performance. Thus, total time of slot utilization by a single job can be evaluated as $p \cdot t \cdot n$. For a resource domain we evaluate a cumulative slot length. The slot length is also calculated with regard to the base resource performance. The job batch should be composed in a way that total time of the slot utilization by the batch jobs is not greater than the cumulative slot length of the resource domain with a coefficient $L \in (0, 1]$. VO administrators can use the coefficient $L \in (0, 1]$ to control job flow execution process in computing environment. Cost limit is similar to time limit: the maximum job execution budget S is specified by user in the resource request (e.g. using JSDL). At the same time cumulative cost of slots available for use in current scheduling cycle is calculated for resource domain based on the VO pricing policy [4, 14]. As opposed to a batch with a fixed number of jobs, time and cost limits introduction allows adjusting the batch size under conditions of dynamically changing nodes utilization and heterogeneity of the job flow. This is achieved by specifying the value of $L \in (0, 1]$. Experimental study of this approach is conducted and presented in Sect. 4.

3.2   Job and Computing Environment Compatibility Indicator

Job batch grouping schemes proposed in this paper form the batch based on job and computing environment characteristics compatibility. Thus, the batch is composed of jobs which resource requests are most fitted for executing in a current scheduling interval. As compatibility measure of an individual job and a resource domain an empirical coefficient D_Q (Distribution Quality) is proposed. D_Q describes chances for a job to be scheduled and executed successfully during the present resource domain scheduling interval utilization level. It can have positive (high chance to be executed) or negative values (low chance to be executed). To figure out the D_Q coefficient and to find significant parameters of the job and the resource domain experimental studies were conducted.

As a result, the following environment characteristics and resource request parameters that most influence the probability of a successful scheduling outcome were discovered.

- A "price/quality" ratio of the domain computational nodes (Q_0) and user jobs (Q). For an individual computing node Q_0 is calculated as the ratio of the specified utilization cost (per time unit) to its performance factor c/p. Thus the higher the performance and the lower the utilization cost of the node, the lower the value of the coefficient is. For a resource domain an average value of Q_0 by all nodes is taken into account. For an individual job the factor is evaluated in a similar manner: $Q = S/ntp$, where the average value of the maximum acceptable for the user single node price is calculated as $c = S/nt$.
- A number n_0 of the available resources (nodes) in the resource domain and a number n of the computing nodes required for the job execution respectively.
- An average slot length l_s during the scheduling interval with regard to the resource of the base performance and resource reservation time required for job execution, also calculated with regards to the base resource performance: $t \cdot p$.
- Total domain available processor time V_s (cumulative length of available slots) and processor time $t \cdot p \cdot n$ required for the job execution.

Thus, D_Q consists of four summands, corresponding to the mentioned characteristics of the resource domain and the user job. For each of the summands adjusting parameters are introduced: K_q, K_n, K_l, and K_v – weight coefficients of the summands; C_q, C_n, C_l, and C_v – threshold values, approximately determining the value at which at least one alternative for the job is likely to be found.

The D_Q coefficient is defined as the sum of the following terms:

$$D_{Q1} = K_q(Q/Q_0 - C_q)/C_q; \tag{1}$$

$$D_{Q2} = K_n(C_n - n/n_0)/C_n; \tag{2}$$

$$D_{Q3} = K_l(C_l - t \cdot p/l_s)/C_l; \tag{3}$$

$$D_{Q4} = K_v(C_v - t \cdot p \cdot n/V_s)/C_v. \tag{4}$$

The term (1) normalizes the ratio of coefficients Q of the job and Q_0 of the environment. The greater the value of the ratio Q/Q_0, the larger the value of the term is and the higher the probability of successful job execution. For instance, the higher the budget S, allocated by the user to execute a job, the larger the value of the term and entire D_Q is. The term (2) normalizes the ratio of the matching computing nodes number in the domain and the number of nodes required to execute the job n/n_0. The term (3) normalizes the ratio of resource reservation time required to the job and the average slot length. The term (4) characterizes the ratio of slot utilization time required to execute the job and total processor time available during the considered scheduling cycle. Note, that parameters Q_0, n_0, l_s, and V_s are calculated for the set of domain resources that match the job against the minimum performance limit p. If there are no such resources in the domain or the number of such resources is less than the value n from the resource request, D_Q is considered to have the value of negative infinity. Using D_Q coefficient as the sum of (1)–(4) it is possible to form the job batch in different ways. One possible approach consists in selecting jobs with the maximum value of D_Q at each scheduling cycle.

However, in this case a situation similar to job selection by the cost criterion can happen: after successful scheduling of the most "valuable" jobs of the flow at the first cycles, scheduling efficiency may reduce abruptly for jobs left in the queue at subsequent scheduling cycles. Job system generation methods proposed in this paper apply a different policy and are based on selecting jobs with the minimal positive value of D_Q, i.e. the most "problem" jobs out of those that can be executed successfully at current scheduling interval. This policy allows balancing of job flow execution during many cycles and providing the most efficient resource utilization. On the other hand, jobs with very high or negative values of D_Q factor can be moved to other flows and executed in other resource domains.

Two fundamentally different batch generation methods are proposed in the paper. In the first method, a job batch grouping process is reduced to solving the problem of optimal knapsack filling with dynamic programming methods [20, 21]. This approach seems to be natural as it allows formalizing the job selection procedure under characteristics of jobs and resource domains known in advance. The second method is based on D_Q coefficient and allows flexibly adjusting scheduling process to a dynamically changing structure of resources and jobs of the flow. To compare the methods both of them use the job batch size restriction rules while batch job grouping is performed based on D_Q indicator evaluation for each job.

The idea of knapsack problem application for scheduling is not new, however in the known approaches [15, 16, 19, 21] it is usually used for optimal jobs allocation to non-dedicated resources. We propose using it to fill the job batch, as a preparatory step before scheduling. A weight limit of the batch and a weight of an individual job can be either time or cost depending on the chosen limit type. The weight limit is chosen based on summary resource characteristics with some coefficient $L \in (0, 1]$. A value of a job is proposed to be calculated as $1/D_Q$: that is, the lower D_Q coefficient, the higher the value of the job is. This assumption is based on the logic of choosing jobs based on D_Q, and allows achieving a more even resource utilization during several scheduling cycles. Note, that jobs whose value is less than or equal to zero will never be put into the batch, since they make no positive contribution to the total batch value but occupy some "useful weight".

Another approach uses D_Q indicator to select jobs into the batch. However, when forming the batch based on the mentioned job and resource domain compatibility factor, the use of simple D_Q becomes insufficient: when adding jobs to the batch it is necessary to take into account parameters of the jobs already put into this batch.

Thus, it is necessary to slightly modify the D_Q terms. For instance, when time limit is used, total processor time required by the jobs previously put into the batch is added to the term normalizing the ratio of job /domain required /available processor time. In this case, D_{Q4} (4) is modified as follows:

$$D_{Q4} = K_v \left(C_v - \left(t \cdot p \cdot n + \sum_{i=1}^{N} t_i \cdot p_i \cdot n_i \right) / V_s' \right) / C_v. \tag{5}$$

The sum $\sum_{i=1}^{N} t_i \cdot p_i \cdot n_i$ in (5) includes parameters of all N jobs, already put into the batch, and V_s' is total processor time for all the jobs. In case of cost limit, the nominator of the ratio includes the total execution budget of the jobs already put into the batch and the denominator - cumulative cost of all the available slots.

As was mentioned earlier, the jobs with the minimal positive value of D_Q coefficient have the highest priority during the selection process. When the number of jobs in the batch increases, the value of D_{Q4} reduces and may take negative values. Batch generation process continues until there are any jobs with the positive value of D_Q left in the job flow. Note, that in this batch grouping method the limiting coefficient VO administrators operate is represented as adjusting the threshold parameter C_v. Unlike the batch grouping method based on the knapsack problem solution, the batch size limit for this method is not strict. When solving the knapsack problem the limit is strict and cannot be exceeded. When using D_Q, exceeding the limit will result in D_{Q4} taking a negative value while entire D_Q coefficient can still be positive, and then the job will be put into the batch. The approach based on D_Q allows keeping the job selection policy in a more flexible way, as each time the job with the minimal positive value of the coefficient is chosen. The main advantage of the method based on D_Q consists in the ability of taking into account the parameters of jobs already put into the batch. Note, that for both batch grouping methods job ranking is used according to D_Q coefficient in descending order, i.e. the most "problem" jobs are placed in the beginning of the batch. This allows improving scheduling efficiency indicators.

4 Simulation Studies

Many Grid simulators have already been developed [2], e.g. ChicSim, GridSim, SimGrid, OptorSim, etc. We implement our own Grid simulator [22] in order to maximize reuse a code base for CSS with original job flow and application level scheduling algorithms and heuristics [15–17, 22].

Some realistic features are introduced into the experimental setup. The scheduling interval length is assumed to be 600 units of time in simulation steps. The number of nodes in the resource domain is equal to 24. The nodes performance level is given as a uniformly distributed random value in the interval [2, 14]. This configuration provides a sufficient resources diversity level while the difference between the highest and the lowest resource performance levels will not exceed one order within a particular resource domain. Uniform distribution was chosen in the assumption that the node composition is formed by resource selection based on such hard constraints [11] as a computing node type, performance, etc. The node prices are assigned during the pricing stage depending on the node performance level and a random "discount/extra charge" value which is normally distributed. The number of jobs in the queue is assumed to be 150. The jobs budget limit is generated in such a way that the "richest" users can afford to use "expensive" resources with the price formed as a "market value + 60 % extra charge", and the "poorest" users have been forced to rely on 60 % discounts. Jobs expected runtimes are generated on a [50, 150] interval so the whole job queue execution requires several scheduling cycles for every batch grouping method. The following job batch grouping methods were studied:

- Random – each time the batch is filled with a constant number of jobs randomly selected from the job flow;
- KnapsackT – the knapsack problem with a restriction on total reservation time is solved to fill the job batch;
- KnapsackC – the knapsack problem with a restriction on total reservation cost is solved to fill the job batch;
- D_QT – the job batch is filled according to jobs D_Q indicator with a restriction on jobs reservation time;
- D_QC – the job batch is filled according to jobs D_Q indicator with a restriction on jobs reservation cost.

Random grouping method represents a general job queue scheduling policy based on a job submission natural order. The main goal of the experiment is to compare the knapsack and the simple D_Q heuristic job grouping approaches which attempt to reorder a job queue according to the available meta data.

Table 1 contains job batch size limit values used during the simulation series. A job batch size for the Random grouping method was specified accordingly to the average size of the batches formed by other considered approaches.

Table 1. Job batch grouping limit parameters for different experiment series.

Experiment series #	1	2	3	4	5
Job batch size for Random grouping method	6	20	30	40	50
L value for KnapsackT, KnapsackC, D_QT and D_QC grouping algorithms	0.1	0.3	0.5	0.7	0.9

Figure 1 presents a domain nodes utilization level (a) and an average number of execution alternatives (b) found for the batch jobs depending on considered experiment series (see Table 1). As expected, the graphs show that the available resource utilization level increases and the number of possible execution alternatives decreases with increasing the size of the formed job batch.

Figure 2 shows average job execution time (a), a number of alternatives found (b), and an average number of completed scheduling cycles (c), depending on the resulting resource utilization level. Section 3 introduces the coefficient $L \in (0, 1]$ which defines the weight restriction for the knapsack problem and hence the job batch size.

The relation between the coefficient and resource domain nodes utilization level could be graphically seen from Fig. 1 (a). On the other hand, D_QT(C) approaches use D_Q indicator which consists of main terms (1)-(5). Thus, even when one of the terms

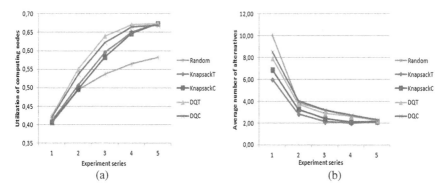

Fig. 1. Average computational nodes utilization level (a) and possible execution alternatives number (b) in different experiment series

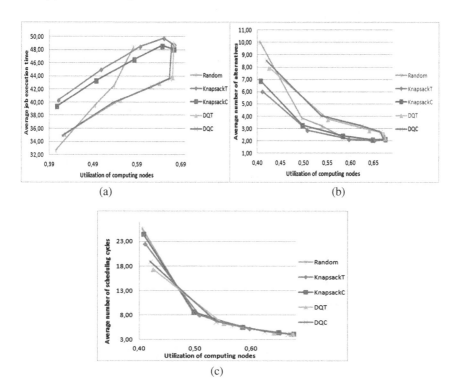

Fig. 2. Average job execution time (a), execution alternatives number (b), and number of required scheduling cycles (c) depending on the nodes utilization level

becomes negative due to the job batch time or cost limit exceeding, the cumulative indicator value may stay positive. Such flexible restriction results in a greater number of potentially executable jobs being selected into the job batch during each scheduling cycle. This explains better scheduling criteria values and a relatively higher level of resources utilization provided by $D_QT(C)$ approaches.

Figure 2 shows a 15 % advantage of $D_QT(C)$ approach over KnapsackT(C) by job execution time (the main VO scheduling criterion), however both job batch grouping algorithms managed roughly the same number of scheduling cycles to complete the whole job flow. As can be seen from Fig. 2 (a), $D_QT(C)$ approach provides the best scheduling results almost for every obtained resource utilization level. Figure 2(b) and (c) are consistent and agree with this conclusion: approaches which provided better VO scheduling criterion values accordingly provided more possible execution alternatives during the scheduling process. The average number of the scheduling cycles required to complete the job flow execution (Fig. 2 (c)) was approximately the same for each grouping method with the same observed resources utilization level.

Figure 3 contains graphical data on the average execution alternatives number, the job batch size and the number of job execution declines and returns to the job flow depending on the scheduling cycle number. The data represents the dynamic job flow scheduling results for every considered job batch grouping algorithm in the experiment series #3 (see Table 1). As was mentioned earlier, $D_QT(C)$ approach aims to select jobs with the minimal positive D_Q compatibility indicator value. The job "weight" (processor time or the total execution budget) is not considered during the selection: the jobs are equally selected from the job flow according to D_Q value until the total weight limit is not exceeded. On the other hand, KnapsackT(C) approach maximizes the sum of the batch jobs D_Q values with a hard restriction on a total weight. As it turns out, this policy tends to select more relatively small jobs since they make a smaller contribution to the total batch weight. Thus, the job flow scheduling is uneven: relatively small jobs are selected for the first scheduling cycles while jobs with higher resource demands remain till the last cycles.

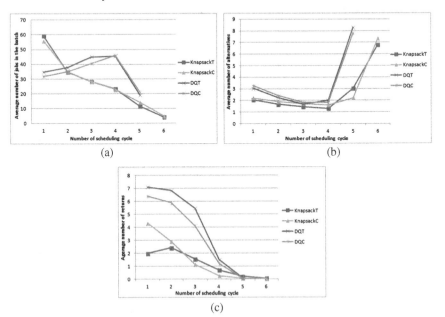

(a)

(b)

(c)

Fig. 3. Average execution alternatives number (a), job batch size (b), and number of job execution declines (c) depending on the scheduling cycle number

The whole picture is presented by Fig. 3. At the first scheduling cycles KnapsackT (C) forms batches of a larger size, while at later cycles with the same weight limit the batch size decreases by several times. Similarly, the number of the alternatives found with KnapsackT(C) approach can be explained. At the first scheduling cycles a large number of relatively small jobs are competing for limited resources and each reserved alternative further "granulates" available processor slots. This granulation makes the execution alternatives search task even more complex, and it becomes difficult to allocate even several execution alternatives for a batch of much more resource demanding jobs. At the same time, the scheduling with $D_QT(C)$ is uniform from cycle to cycle: there is now skew in the batches toward more or less resource demanding jobs. The reduced job execution declines number during the ending scheduling cycles can be explained by the fact that only jobs with a relatively higher D_Q value (the most "suitable" jobs) remain in the job flow. The scheduler tends to form a batch of jobs with resource demands most appropriate for the particular resource domain. The compatibility is defined by the specified values of the D_Q threshold coefficients. The execution of the jobs too small or expensive for the particular resource domain (which should have a higher D_Q value) might granulate available processor time and hint on overall job flow distribution inefficiency.

The obtained simulation results show that KnapsackT(C) job batch forming approach provides better scheduling results with the restriction on the total batch jobs execution budget. D_QT and D_QC approaches showed similar results in all considered experiments and no noticeable advantages observed with cost or time limitation. The Random job batch grouping scheme in some experiments provided the best values of average job execution time and the considered job execution alternatives number. This can be explained by the fact that the job batch was formed without any job-domain "compatibility" rules with the only restriction on the preliminary defined total batch size. As a consequence, the Random approach provided a lower average resource utilization level, a greater number of job execution declines and required a larger number of scheduling cycles for a complete job flow execution compared to other considered approaches.

5   Summary

In this paper, the problem of job selection for resource scheduling in virtual organization in utility Grids is considered. We propose the general compatibility parameter D_Q for a job and a chosen resource domain. Two methods based on different job selection approaches for job batch forming using D_Q are proposed and analyzed. The first method, KnapsackT(C), forms job batches based on the knapsack problem solution for preliminary calculated D_Q for each job and has the strict limit for total execution time or cost of the job batch execution. The second method, $D_QT(C)$, uses D_Q for job selection, which in its turn dynamically changes based on jobs that were already selected and their characteristics, and uses less strict constraints for the job batch size. Experiment results show significant advantage of $D_QT(C)$ over KnapsackT(C).

Further research is aimed at developing methods for job allocation between several resource domains and forming a job framework while fulfilling requirements of all VO participants.

Acknowledgements. This work was partially supported by the Council on Grants of the President of the Russian Federation for State Support of Young Scientists and Leading Scientific Schools (grants YPhD-4148.2015.9 and SS-362.2014.9), RFBR (grants 15-07-02259 and 15-07-03401), the Ministry on Education and Science of the Russian Federation, task no. 2014/123 (project no. 2268), and by the Russian Science Foundation (project no. 15-11-10010).

References

1. Garg, S.K., Konugurthi, P., Buyya, R.: A linear programming-driven genetic algorithm for metascheduling on utility grids. J Par., Emergent and Distr. Systems **26**, 493–517 (2011)
2. Cafaro, M., Mirto, M., Aloisio, G.: Preference-based matchmaking of grid resources with cp-nets. J. Grid Comput. **11**(2), 211–237 (2013)
3. Buyya, R., Abramson, D., Giddy, J.: Economic models for resource management and scheduling in grid computing. J. Concurrency Comput. **14**(5), 1507–1542 (2002)
4. Toporkov, V.V., Yemelyanov, D.M.: Economic model of scheduling and fair resource sharing in distributed computations. J. Program. Comput. Softw. **40**(1), 35–42 (2014)
5. Ernemann, C., Hamscher, V., Yahyapour, R.: Economic scheduling in grid computing. In: Feitelson, D.G., Rudolph, L., Schwiegelshohn, U. (eds.) JSSPP 2002. LNCS, vol. 2537, pp. 128–152. Springer, Heidelberg (2002)
6. Mutz, A., Wolski, R., Brevik, J.: Eliciting honest value information in a batch-queue environment. In: 2007 8th IEEE/ACM International Conference on Grid Computing, pp. 291–297. IEEE Computer Society (2007)
7. Berman, F., Wolski, R., Casanova, H., et al.: Adaptive computing on the grid using appLeS. J. IEEE Trans. On Parallel Distrib. Syst. **14**(4), 369–382 (2003)
8. Cirne, W., Brasileiro, F., Costa, L. et al.: Scheduling in bag-of-task grids: the PAUÁ case. In: 16th Symposium on Computer Architecture and High Performance Computing, pp. 124–131. IEEE (2004)
9. Voevodin, V.: The solution of large problems in distributed computational media. J. Autom. Remote Control **68**(5), 773–786 (2007)
10. Dail, H., Sievert, O., Berman, F., et al.: Scheduling in the grid application development software project. In: Nabrzyski, J., Schopf, J.M., Weglarz, J. (eds.) Grid resource management, pp. 73–98. State of the Art and Future Trends. Kluwer Academic Publishers, Dordrecht (2003)
11. Kurowski, K., Oleksiak, A., Nabrzyski, J., et al.: Multi-criteria grid resource management using performance prediction techniques. In: Gorlatch, S., Danelutto, M. (eds.) JSSPP 2010, pp. 215–225. Springer, Heidelberg (2010)
12. Moab Adaptive Computing Suite. http://www.adaptivecomputing.com/products/moab-adaptive-computing-suite.php. Accessed November 2014
13. Kannan, S., Roberts, M., Mayes, P., et al.: Workload Management with LoadLeveler. IBM, New York (2001)
14. Tsafrir, D., Etsion, Y., Feitelson, D.: Backfilling using system-generated predictions rather than user runtime estimates. J. IEEE Trans. on Parallel Distrib. Sys. **18**(6), 789–803 (2007)
15. Toporkov, V., Toporkova, A., Tselishchev, A., Yemelyanov, D., Potekhin, P.: Preference-based fair resource sharing and scheduling optimization in grid vos. J. Procedia Comput. Sci. **29**, 831–843 (2014)
16. Toporkov, V., Toporkova, A., Tselishchev, A., Yemelyanov, D., Potekhin, P.: Core heuristics for preference-based scheduling in virtual organizations of utility grids. In: Camacho, D., Braubach, L., Venticinque, S., Badica, C. (eds.) IDCVIII. SCI, vol. 570, pp. 309–318. Springer, Heidelberg (2014)

17. Toporkov, V., Toporkova, A., Tselishchev, A., Yemelyanov, D.: Slot selection algorithms in distributed computing. J. of Supercomputing **69**(1), 53–60 (2014)
18. Zhou, Z., Lan, Z., Tang, W., Desai, N.: Reducing energy costs for ibm blue gene/p via power-aware job scheduling. In: 17th Workshop on Job Scheduling Strategies for Parallel Processing, pp. 96–115. Boston (2013)
19. Soner, S., Özturan, C.: Integer programming based heterogeneous cpu-gpu cluster scheduler for slurm resource manager. In: 14th IEEE International Conference on High Performance Computing and Communication and 9th IEEE International Conference on Embedded Software and Systems, pp. 418–424. IEEE, Liverpool (2012)
20. Toporkov, V., Tselishchev, A., Yemelyanov, D., Potekhin, P.: Metascheduling strategies in distributed computing with non-dedicated resources. In: Zamojski, W., Sugier, J. (eds.) DPCIS. AISC, vol. 307, pp. 129–148. Springer, Heidelberg (2014)
21. Vanderster, D.C., Dimopoulos, N.J., Parra-hernandez, R., Sobie, R.J.: Resource allocation on computational grids using a utility model and the knapsack problem. J. Future Gener. Comput. Syst. **25**(1), 35–50 (2009)
22. Toporkov, V., Tselishchev, A., Yemelyanov, D., Bobchenkov, A.: Composite scheduling strategies in distributed computing with non-dedicated resources. J. Procedia Comput. Sci. **9**, 176–185 (2012)
23. Rodero, I., Villegas, D., Bobroff, N., Liu, Y., Fong, L., Sadjadi, S.M.: Enabling interoperability among grid meta-schedulers. J. Grid Comput. **11**(2), 311–336 (2013)
24. Aida, K., Casanova, H.: Scheduling mixed-parallel applications with advance reservations. In: 17th IEEE Int. Symposium on HPDC, pp. 65–74. IEEE CS Press, New York (2008)
25. Ando, S., Aida, K.: Evaluation of scheduling algorithms for advance reservations. In: Information Processing Society of Japan SIG Notes HPC-113, pp. 37–42 (2007)
26. Elmroth, E., Tordsson, J.: A standards-based grid resource brokering service supporting advance reservations, coallocation and cross-grid interoperability. J. of Concurrency Comput. **25**(18), 2298–2335 (2009)
27. Azzedin, F., Maheswaran, M., Arnason, N.: A synchronous co-allocation mechanism for grid computing systems. Cluster Comput. **7**, 39–49 (2004)
28. Castillo, C., Rouskas, G.N., Harfoush, K.: Resource co-allocation for large-scale distributed environments. In: 18th ACM International Symposium on High Performance Distributed Compuing, pp. 137–150. ACM, New York (2009)
29. Takefusa, A., Nakada, H., Kudoh, T., Tanaka, Y.: An advance reservation-based co-allocation algorithm for distributed computers and network bandwidth on QoS-guaranteed grids. In: Frachtenberg, E., Schwiegelshohn, U. (eds.) JSSPP 2010. LNCS, vol. 6253, pp. 16–34. Springer, Heidelberg (2010)
30. Blanco, H., Guirado, F., Lérida, J.L., Albornoz, V.M.: MIP model scheduling for multi-clusters. In: Caragiannis, I., et al. (eds.) Euro-Par Workshops 2012. LNCS, vol. 7640, pp. 196–206. Springer, Heidelberg (2013)

Congestion Elimination on Data Storages Network Interfaces in Datacenters

P.M. Vdovin[(✉)], I.A. Zotov, V.A. Kostenko, and A.V. Plakunov

Moscow State University, Moscow, Russia
{pavel.vdovin,tridcatov,kostmsu}@gmail.com,
artacc@lvk.cs.msu.su

Abstract. In this paper we propose a model of a datacenter that provides an ability to describe a wide class of data center architectures. The resource allocation problem is considered that supports replication of data storage elements. Replication procedure provides a way of congestion elimination on channels connected to data storages with low write rates and high read rates. Experimental investigation that was held with different resource allocation algorithms is showing replication procedure to be able to increase the number of allocated requests.

Keywords: Datacenter · IaaS · SLA · Replication · Data storage · Virtual link

1 Introduction

An opportunity to eliminate congestion on data storages interfaces in datacenters is relevant for databases, which have a low write and high read rates. In this work datacenters with Infrastructure-as-a-Service (IaaS) [1] model are considered with guaranteed service level agreements (SLA) for all resource types: computational resources, data storages and network resources. In order to map resource requests onto physical resources of DCs, it is necessary to solve three interdependent NP-hard problems:

- Assign virtual machines onto computational resources.
- Assign storage elements onto data storages.
- Assign virtual links onto network resources.

All virtual elements have a vector of required resources, and assignment should be performed so that the amount of corresponding physical resources is not exceeded on each physical element.

In [2] the self-organizing cloud platform that is able to deploy administrative virtual networks in datacenters and is compatible with OpenStack [3] platform is described.

The work is performed with financial support of The Ministry of Education and Science of The Russian Federation. Agreement number 14.607.21.0070.

V. Malyshkin (Ed.): PaCT 2015, LNCS 9251, pp. 298–303, 2015.
DOI: 10.1007/978-3-319-21909-7_30

For deploying such networks and achieving high utilization of datacenter physical resources a platform resource scheduler should have the following characteristics:

- Ability to set arbitrary set of SLA requirements.
- Requests assignments on computational resources, data storages and network resources should be performed consistently with respect to requested SLAs.
- Ability to eliminate physical resource segmentation via virtual resources migration in datacenter.
- Ability to balance between the algorithm complexity and the solution quality.
- Ability to add/remove virtual resources in scheduled virtual network (tenant).
- Ability to compactly allocate resources for virtual networks in terms of hop count between network elements.
- Ability to eliminate congestion on storages network interfaces.

The investigation of algorithms, which are compatible with IaaS model [4–17], showed that not all requirements presented above are taken into account. In works [4–9] network resources are not considered, while in [10–14] network resources are considered without SLA requirements (only routing is performed). In [15] data storages are not represented as physical resources, and in [16] only tree topology of DCs is used. None of the algorithms provides an ability to eliminate congestion on storage interfaces.

In [17–20] various algorithms to resource allocation in the datacenters with self-organizing cloud platforms are presented that satisfy all characteristics mentioned above.

In this work we present the replication procedure for data storages network interfaces congestion elimination and the results of its effectiveness.

2 Replication Procedure

There are three algorithms of resource allocation in datacenters, presented in [17–20], that uses the replication procedure for congestion elimination: algorithm with the unified scheduler, algorithm with the individual schedulers and ant colony algorithm. In these algorithms replication procedure is called if virtual link route between virtual machine and storage element is failed to be build with respect to required virtual link bandwidth.

The replication is possible when one of the elements connected by the virtual link is the storage element s for which the replication is available. The replication procedure consists of searching for the data storage m that has sufficient resources for assigning the replica (the replica requires as much resources as the given storage element s) such that total routes length of all virtual links (that include the given storage element s) to this replica of m is minimal. In addition, it is required to create a link l for maintaining the consistency between the replica and the original storage element (the required bandwidth of the link l is determined by the type of the storage element s). If the route of the link l for maintaining the consistency failed to be built, then another data storage s is considered in order of total length increment of virtual links routes.

If the replication procedure returns success, then the virtual links including given storage element s can further be assigned to this replica of m. By rebuilding routes of some virtual links from data storage m to its replica, congestion on interfaces of m (which correspond to data transmission channels connected to m) is eliminated and further virtual links may be assigned to this data storage.

If m with a sufficient number of resources is not found or the route of the link l for maintaining the consistency fails to be built, then the procedure returns failure.

3 Experimental Investigation of Replication Procedure Efficiency

To demonstrate the efficiency of the replication procedure, the FatTree topology [21] was used with total of 85 switches, 60 computational nodes and 60 data storages.

For investigating the effectiveness of replication procedure on congestion elimination, the virtual requests were generated so that the potential possible load of the channels (under the optimal allocation) connected to data storages varied from 0.3 to 1.0. The load of the computation nodes and data storages was fixed to 0.75.

The number of requests in each test set was 100.

All the algorithms were executed once without the replication and once with the replication allowed. The figures show the difference in the number of requests assigned by each of the algorithms with the replication off and on.

Figures 1, 2 and 3 present the results of the algorithm with the individual resource schedulers, unified resource scheduler and ant colony algorithm respectively.

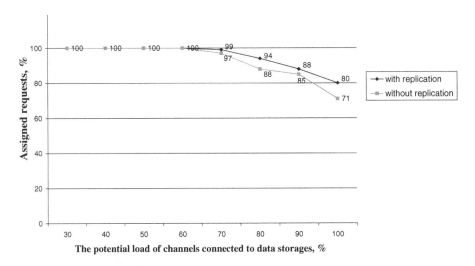

Fig. 1. The algorithm with individual schedulers

The investigation of the algorithms shows that the replication procedure increases the number of the assigned requests in the case when the potentially possible load of channels connected to data storages is 0.7 or higher. Also note that the algorithm with the unified scheduler assigns more requests than the algorithm with individual

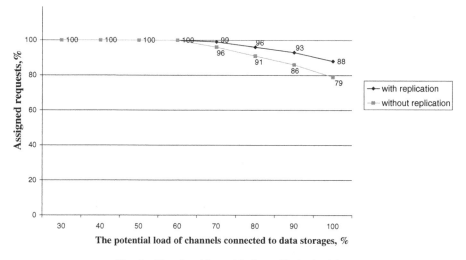

Fig. 2. The algorithm with the unified scheduler

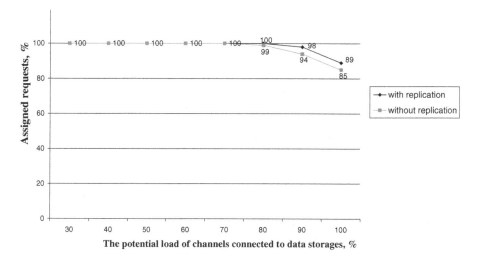

Fig. 3. The ant colony algorithm

schedulers. The ant colony algorithm is more efficient then these algorithms, but it has the largest computational complexity among all three algorithms.

4   Conclusion

The performed experimental investigation of replication procedure efficiency showed that more requests may be allocated on the same physical resources in comparison to algorithms without replication procedure, which is the result of congestion elimination of data storages network interfaces.

References

1. Amies, A., Sluiman, H., Tong, Q.G., et al.: Developing and Hosting Applications on the Cloud. IBM Press, Boston (2012)
2. Vdovin, P.M., Zotov, I.A., Kostenko, V.A., Plakunov, A.V., Smelyansky, R.L.: Comparing various approaches to resource allocating in data centers. J. Comput. Syst Sci. Intern **53**(5), 689–701 (2014)
3. Pepple, K.: Deploying OpenStack. O'Reilly, California (2011)
4. Urgaonkar, B., Rosenberg, A.L., Shenoy, P.: Application placement on a cluster of servers. Int. J. Found. Comput. Sci. **18**, 1023–1041 (2007)
5. Bein D., Bein W., Venigella S.: Cloud storage and online bin packing.In: Proceedings of the 5th International Symposium on Intelligent Distributed Computing Delft: IDC, pp. 63–68 (2011)
6. Nagendram, S., Lakshmi, J.V., Rao, D.V., et al.: Efficient resource scheduling in data centers using MRIS. Indian J. Comput. Sci. Eng. **2**(5), 764–769 (2011)
7. Arzuaga E., Kaeli D.R.: Quantifying load imbalance on virtualized enterprise servers. In: Proceedings of the First Joint WOSP/SIPEW International Conference on Performance Engineering, pp. 235–242. ACM, San Josa (2010)
8. Mishra M., Sahoo A.: On theory of VM placement: anomalies in existing methodologies and their mitigation using a novel vector based approach. In: IEEE International Conference on Cloud Computing (CLOUD), pp. 275–282. IEEE Press, Washington (2011)
9. Zhu, Y., Ammar, M.H.: Algorithms for assigning substrate network resources to virtual network components. In: 25th International Conference on Computer Communications (INFOCOM), Barcelona, pp. 1–12 (2006)
10. Botero, J.F., Hesselbach, X., Fischer, A., et al.: Optimal mapping of virtual networks with hidden hops. Telecommun. Syst. **51**(4), 273–282 (2012)
11. Lischka J., Karl H.: A virtual network mapping algorithm based on subgraph isomorphism detection. In: Proceedings of the 1st ACM Workshop on Virtualized Infrastructure Systems and Architectures, pp. 81–88. ACM, Barcelona (2009)
12. Yu, M., Yi, Y., Rexford, J., et al.: Rethinking virtual network embedding: substrate support for path splitting and migration. ACM SIGCOMM Comput. Commun. Rev. **38**(2), 17–29 (2008)
13. Chowdhury N.M.M.K., Rahman M.R., Boutaba R.: Virtual network embedding with coordinated node and link mapping. In: 28th International Conference on Computer Communications (INFOCOM), Barcelona, pp. 783–791 (2009)
14. Cheng, X., Sen, S., Zhongbao, Z., et al.: Virtual network embedding through topology-aware node ranking. ACM SIGCOMM Comput. Commun. Rev. **41**(2), 38–47 (2011)
15. Jiang J.W., Tian L., Sangtae H., et al.: Joint VM placement and routing for data center traffic engineering. In: 31th International Conference on Computer Communications (INFOCOM), Orlando, pp. 2876–2880 (2012)
16. Korupolu M., Singh A., Bamba B.: Coupled placement in modern data centers. In: IEEE International Symposium on Parallel & Distributed Processing (IPDPS), pp. 1–12, New York (2009)
17. Kostenko, V., Plakunov, A., Nikolaev, A., Tabolin, V., Smeliansky, R., Shakhova, M.: Selforganizing cloud platform. In: Proceedings of the International Science and Technology Conference Modern Networking Technologies (MoNeTec). IEEE, Moscow, Russia, MAKS Pess, pp. 77–82 (2014)

18. Vdovin, P.M., Kostenko, V.A.: Algorithm for resource allocation in data centers with independent schedulers for different types of resources. J. Comput. Syst. Sci. Intern **53**(6), 854–866 (2014)
19. Zotov, I.A., Kostenko, V.A.: Resource allocation algorithm in data centers with a unified scheduler for different types of resources. J. Comput. Syst. Sci. Int. **54**(1), 59–68 (2015)
20. A. Plakunov, V. Kostenko.: Data center resource mapping algorithm based on the ant colony optimization. In: Proceedings of the International Science and Technology Conference Modern Networking Technologies (MoNeTec). IEEE, Moscow, Russia, MAKS Pess, pp. 127–132 (2014)
21. Al-Fares, M., Loukissas, A., Vahdat, A.A.: Commodity data center network architecture. ACM SIGCOMM Comput. Commun. Rev. **38**(4), 63–74 (2008)

Special Processors Programming Techniques

Use of Xeon Phi Coprocessor for Solving Global Optimization Problems

Konstantin Barkalov$^{(\boxtimes)}$, Victor Gergel, and Ilya Lebedev

Lobachevsky State University of Nizhni Novgorod, Nizhni Novgorod, Russia
{barkalov,lebedev}@vmk.unn.ru, gergel@unn.ru

Abstract. This work considers a parallel algorithm for solving multidimensional multiextremal optimization problems. The issue of implementation of the algorithm on state-of-the-art computing systems using Intel Xeon Phi coprocessor is considered. Speed up of the algorithm using Xeon Phi compared to using only CPU is experimentally confirmed. Computational experiments are carried out using a set of a several hundred of multidimensional multiextremal problems.

Keywords: Global optimization · Dimension reduction · Parallel algorithms · Speedup · Intel Xeon Phi

1 Introduction

Optimization problems are of great practical importance. Almost each problem of design of new devices, products or systems includes a stage where optimal variants are selected. Among the most complex optimization problems are problems of global optimization, where the criterion of optimality is multiextremal. While validation of local optimality of a solution requires only analysis of its local neighborhood, global minimum is an integral characteristic of the optimization problem solved and it requires analysis of the whole search domain. As a result, search of a global optimum is reduced to construction of a grid in the parameter domain. It leads to exponential growth of computational effort with more dimensions (the so-called "curse of dimensionality").

A decrease in computational effort can be provided through construction of a non-uniform grid in the search domain: it has to be quite dense in the neighborhood of the global optimum and more sparse farther from the required solution. There is a number of methods allowing to build non-uniform grids of such kind (see, for example, [1–4]). Among those, we note the global search algorithm and its modifications developed within the framework of the information-statistical approach [5–9]. It is experimentally confirmed in [8], that this algorithm is more effective than other known methods of the same purpose.

Use of parallel computing systems significantly expands capabilities for solving global optimization problems. Parallel versions are proposed for almost all existing algorithms (see, for example, [10, 11]). However, the provided versions of algorithms are parallelized in a CPU using MPI and/or OpenMP technologies, whereas currently the main tendency in the field of parallel computing is use of accelerators. Of special

© Springer International Publishing Switzerland 2015
V. Malyshkin (Ed.): PaCT 2015, LNCS 9251, pp. 307–318, 2015.
DOI: 10.1007/978-3-319-21909-7_31

interest in this regard is the Intel Xeon Phi coprocessor. It is based on x86 architecture and standard technologies and libraries can be used in programming for Xeon Phi (unlike specialized and, as a rule, more complex programming technologies for GPU).

The present work contains the results of an analysis of parallel global search algorithm developed within the framework of the information-statistical approach [8], and its implementation using Xeon Phi.

2   Global Search Algorithm with Parallel Trials

Let us consider the problem of global minimum search of an N-dimensional function $\varphi(y)$ in hyperinterval D

$$\varphi(y^*) = \min\{\varphi(y) : y \in D\}, \tag{1}$$

$$D = \{y \in R^N : a_i \leq y_i \leq b_i, 1 \leq i \leq N\}.$$

Let us assume that objective function $\varphi(y)$ satisfies Lipschitz condition

$$|\varphi(y_1) - \varphi(y_2)| \leq L\|y_1 - y_2\|, \quad y_1, y_2 \in D,$$

with constant L, which in the general case is unknown.

The considered approach reduces solving multidimensional problems to solving equivalent one-dimensional problems (*reduction of the dimension*). Thus, use of continuous single-valued mapping like the Peano curve

$$\{y \in R^N : -2^{-1} \leq y_i \leq 2^{-1}, 1 \leq i \leq N\} = \{y(x) : 0 \leq x \leq 1\}$$

allows reduction of the problem of minimization in domain D to a problem of minimization on interval [0,1]

$$\varphi(y^*) = \varphi(y(x^*)) = \min\{\varphi(y(x)) : x \in [0, 1]\}$$

Problems of numerical construction of Peano-type space filling curves and the corresponding theory are considered in detail in [8, 13]. Here we will note that a numerically constructed curve (*evolvent*) is an approximation to a theoretical Peano curve with accuracy 2^{-m}, where m is an evolvent construction parameter. An important property is preservation of boundedness of function relative differences: if function $\varphi(y)$ in domain D satisfies Lipschitz condition, then function $\varphi(y(x))$ on interval [0,1] will satisfy a uniform Hölder condition

$$|\varphi(y(x_1)) - \varphi(y(x_2))| \leq H|x_1 - x_2|^{1/N}, \quad x_1, x_2 \in [0, 1],$$

where Hölder constant H is linked to Lipschitz constant L by the relation

$$H = 2L\sqrt{N+3}.$$

Therefore, it is possible, without limitation of generality, to consider minimization of one-dimensional function

$$f(x) = \varphi(y(x)), \ x \in [0,1],$$

satisfying Hölder condition.

An algorithm for solving problem (1) (let us formulate it here according to [12]) involves constructing a sequence of points x^i, where the values of the minimized function $z^i = f(x^i) = \varphi(y(x^i))$ converging to the solution of the problem are calculated. Let us call the function value calculation process (including construction of image of $y^i = y(x^i)$) the trial, and pair (x^i, z^i) – the result of the trial. At each iteration of the method p of trials is carried out in parallel, and the set of pairs $\{(x^i, z^i)\}$, $1 \leq i \leq k = np$, make up the search information collected by the method after carrying out of n steps. The rules that define the work of a parallel global search algorithm (PGSA) are as follows.

At the first iteration of the method p of arbitrary points $x^1, \ldots, x^p$ in interval $[0,1]$ (for example, these points can be uniformly located), and in these points trials are carried out in parallel. The results of trials $\{(x^i, z^i)\}$, $1 \leq i \leq p$, are saved in the search base of the algorithm.

Suppose, now, that $n \geq 1$ iterations of the method have already been executed. The trial points of the next $(n+1)$-th iteration are then chosen by using the following rules.

Step 1. Renumber points of the set

$$X_k = \{x^1, \ldots, x^k\} \cup \{0\} \cup \{1\}$$

which includes boundary points of interval $[0,1]$, and points $\{x^1, \ldots, x^k\}$ of the previous $k = k(n) = np$ trials, with subscripts in increasing order of coordinate values, i.e.

$$0 = x_0 < x_1 < \ldots < x_{k+1} = 1.$$

Step 2. Supposing that $z_i = f(x_i)$, $1 \leq i \leq k$, calculate values

$$\mu = \max_{1 \leq i \leq k} \frac{|z_i - z_{i-1}|}{\Delta_i}, \ M = \begin{cases} r\mu, \mu > 0, \\ 1, \ \mu = 0, \end{cases} \tag{2}$$

where $r > 1$ is a preset reliability parameter of the method, and $\Delta_i = (x_i - x_{i-1})^{1/N}$.

Step 3. Calculate characteristic for every interval (x_{i-1}, x_i), $1 \leq i \leq k+1$, according to the following formulas

$$R(1) = 2\Delta_1 - 4\frac{z_1}{M},$$

$$R(k+1) = 2\Delta_{k+1} - 4\frac{z_k}{M},$$

$$R(i) = \Delta_i + \frac{(z_i - z_{i-1})^2}{M^2 \Delta_i} - 2\frac{z_i + z_{i-1}}{M}, \quad 1 < i < k+1.$$

Step 4. Arrange characteristics $R(i)$, $1 \leq i \leq k+1$, in decreasing order

$$R(t_1) \geq R(t_2) \geq \ldots \geq R(t_k) \geq R(t_{k+1}) \tag{3}$$

and select p maximum characteristics with interval numbers t_j, $1 \leq j \leq p$.

Step 5. Carry out new trials in points x^{k+j}, $1 \leq j \leq p$, calculated using formulas

$$x^{k+j} = \frac{x_{t_j} + x_{t_j-1}}{2}, \quad t_j = 1, t_j = k+1,$$

$$x^{k+j} = \frac{x_{t_j} + x_{t_j-1}}{2} - \text{sign}(z_{t_j} - z_{t_j-1})\frac{1}{2r}\left[\frac{|z_{t_j} - z_{t_j-1}|}{\mu}\right]^N, \quad 1 < t_j < k+1.$$

The algorithm terminates if the condition $\Delta_{t_j} \leq \varepsilon$ is satisfied at least for one number t_j, $1 \leq j \leq p$; $\varepsilon > 0$ is the preset accuracy.

This method of organizing parallel computing has the following justification [8]. The characteristic $R(i)$ used in the global search algorithm can be considered as probability measure of the global minimum point location in the interval (x_{i-1}, x_i). Inequalities (3) arrange intervals according to their characteristics, and trials are carried out in parallel in p intervals with the largest probabilities.

3   Convergence and Speedup of the Parallel Algorithm

The following theorem form [8] identifies convergence conditions for the algorithm.

Theorem 1. Let $\bar{x}$ be the limit point of the sequence $\{x^k\}$ generated by PGSA during minimization of the Hölder with constant H, $0 < H < \infty$, function $f(x)$, $x \in [0, 1]$, and number p of parallel trials is fixed, $1 \leq p < \infty$, and $\varepsilon = 0$ in the stop condition of the algorithm. Then

- convergence to the internal point $\bar{x} \in (0, 1)$ is bilateral;
- the point $\bar{x}$ is locally optimal if the function $f(x)$ has a finite number of local extremums;
- if, together with $\bar{x}$, another limit point $\hat{x}$ exists then $f(\bar{x}) = f(\hat{x})$;
- for all $k \geq 1$ it follows than $f(x^k) \geq f(\bar{x})$;
- if, at some step, for M from (2) the condition $M > 2^{2-1/N}H$ holds, then the set of limit points of the sequence $\{x^k\}$ will coincide with the set of global minimizers of the function $f(x)$.

More general variants of parallel global search algorithm and corresponding convergence theory are presented in [8].

Let us describe theoretical properties of a parallel algorithm, which characterize its speedup. One of the main indicators of efficiency of parallel algorithms (in any domain, not only in global optimization) is speedup in time

$$S(p) = T(1)/T(p)$$

where $T(1)$ is the time required for solving the problem by a sequential algorithm, and $T(p)$ is the time for solving the same problem by a parallel algorithm in a system with p computing elements. The characteristic of efficiency of parallel algorithms (in relation to algorithms of optimization) is also speedup in number of iterations

$$s(p) = n(1)p/n(p),$$

where $n(1)$ is the number of the trials carried out using the sequential method, and $n(p)$ is the number of the trials carried out using the parallel method with p processors. This characteristic is especially important since in applied problems the time of carrying out of a trial exceeds the time of processing of its results.

It is obvious that number of trials $n(p)$ for sequential and parallel algorithms will differ. Actually, the sequential algorithm when selecting point x^{k+1} of the next $(k+1)$ trial possesses complete information received at the previous k iterations. The parallel algorithm selects not one, but p points x^{k+j}, $1 \leq j \leq p$, at iteration $(k+1)$ based on the same information. It means that selection of point x^{k+j} is carried out in absence of information on the results of the trial in points x^{k+i}, $1 \leq i < j$. Only the first point x^{k+1} will coincide with the point selected by the sequential algorithm. Points of the other trials, generally speaking, can not coincide with the points generated by the sequential algorithm. Carrying out of such trials can reduce efficiency of use of parallel processors. Therefore, let us consider such trials as "redundant", and the value

$$\lambda(p) = \begin{cases} (n(p) - n(1))/n(p), n(p) > n(1) \\ 0, n(p) \leq n(1) \end{cases}$$

as "method redundancy".

Let us set the series of trials $\{x^k\}$ and $\{y^m\}$ generated correspondingly by sequential and parallel algorithms for solving the same problem with $\varepsilon = 0$ in the condition of stopping. The following theorems from [8] determine the number of computing elements p, which can be involved for non-redundant parallelization.

Theorem 2. Suppose x^* is the point of global minimum, x' is the point of the local minimum of function $f(x)$, and the following conditions are fulfilled:

1. Inequality

$$f(x') - f(x^*) \leq \delta, \ \delta > 0, \tag{4}$$

holds.

2. The initial $q(l)$ trials of the sequential and parallel methods coincide, i.e.

$$\{x^1, \ldots, x^{q(l)}\} = \{y^1, \ldots, y^{q(l)}\},$$

Where

$$\left\{x^1, \ldots, x^{q(l)}\right\} \subset \{x^k\}, \ \left\{y^1, \ldots, y^{q(l)}\right\} \subset \{y^m\}.$$

3. There exists a point $y^n \in \{y^m\}$, $n < q(l)$, such that $x' \le y^n \le x^*$ or $x^* \le y^n \le x'$.
4. For value M from (2) the following inequality

$$M > 2^{2-1/N}H$$

holds, where H is Hölder constant of the minimized function.

Then a parallel algorithm of global search using two processors will be non-redundant (i.e. $s(2) = 2$, $\lambda(2) = 0$), while the following condition is satisfied

$$(x_{t_j} - x_{t_j-1})^{1/N} > \frac{4\delta}{M - 2^{2-1/N}H}, \ j = 1, \ 2, \tag{5}$$

where t_j are determined according to (3).

Corollary. Let the objective function $f(x)$ have Q local minimum points $\left\{x_1', \ldots, x_Q'\right\}$, for which condition (4) is fulfilled, and let there exist trial points y^{n_i}, $1 \le i \le Q$, such as

$$y^{n_i} \in \{y^1, \ldots, y^{q(l)}\},$$

$$\alpha_i \le y^{n_i} \le \alpha_{i+1}, \ \alpha_i, \alpha_{i+1} \in \{x^*, x_1', \ldots, x_Q'\}, \ 1 \le i \le Q.$$

Then, if the theorem conditions are satisfied, the parallel algorithm of global search with $Q + 1$ processors will be non-redundant (i.e. $s(Q + 1) = Q + 1$, $\lambda(Q + 1) = 0$), while condition (5) is satisfied.

The theorem conclusion plays a special role for solving multidimensional problems reduced to one-dimensional problems by means of Peano-like evolvent $y(x)$. Evolvent $y(x)$, which is approximation to Peano curve, has the effect of "splitting" of a point of the global minimum $y^* \in D$ to several preimages in interval $[0,1]$. If function $\varphi(y)$ has the only global minimum in D, the "reduced" function $f(x)$ can have up to 2^N local extremum points close (by value) to a global extremum point (see [8]). In the case of applying a parallel global search algorithm for minimization of a similar function it is possible receive non-redundancy when using up to 2^N+1 computing elements.

4   Implementation on Xeon Phi

At the end of 2012 Intel presented Xeon Phi processor with MIC (Many Integral Core) architecture. The basis of MIC is using a large number of x86 computing cores in one processor. As a result, standard technologies, including OpenMP and MPI, can be used for parallel programming. Moreover, a vast number of tools and libraries has been developed for x86 architecture. It is a significant advantage as compared to other accelerators, for which special (usually more complex) technologies of parallel programming (CUDA, OpenCL) are used.

Intel Xeon Phi supports a few modes of coprocessor use, which can be combined to achieve maximum performance depending on characteristics of the solved problem. The process can be started both in the basic operating system or in coprocessor OS. Depending on the mode of use the computing capacity of either basic system processors or coprocessor or basic system processors and coprocessor combined can be used.

In the MPI mode the basic system and each Intel Xeon Phi coprocessor are considered as separate nodes, and MPI processes can be carried out on basic system processors and Xeon Phi coprocessors in any combination.

During operation in the offload mode MPI processes are carried out only on basic system processors, uploading and execution of functions on the coprocessor is used for implementation of Xeon Phi computing capabilities.

Taking into account that peak performance of one Xeon Phi core is comparable to peak CPU core performance (difference can make 5 – 10% depending on exact processor type), it is preferable to use an accelerator to carry out complex operations not requiring transfer of large amounts of data between the CPU and Phi. With regard to the considered parallel global optimization algorithm such a complex operation is parallel calculation of many objective function values. Transfers of data from the CPU to Phi will be minimal: it is only required to transfer to Phi the trial points coordinates, and to receive function values in these points. The functions that determine the trial results processing according to the algorithm and requiring operation with a large volume of collected search information can be efficiently implemented on the CPU. The described organization scheme corresponds well to the accelerator offload mode.

The general scheme of organization of calculations using Xeon Phi is shown in Fig. 1. According to this scheme steps 1–4 of the parallel global search algorithm are performed on the CPU. Coordinates of the p trial points calculated at step 4 of the algorithm are transferred to the accelerator. Calculation of function values in these points is carried out on Xeon Phi, and then the trial results are transferred to the CPU. We use Xeon Phi offload mode for synchronous computing p function values at each iteration. Current implementation of the parallel algorithm supports only one coprocessor.

We note here that parallel calculation of function values in several tens or hundreds of points (up to 240 threads can be launched on Xeon Phi) not always gives speedup of the search process by a factor of tens or hundreds. In this case, the conditions of the theorem of non-redundant parallelization can be violated: the number of local extremums will be less, than the number of computing cores. Then (according to the

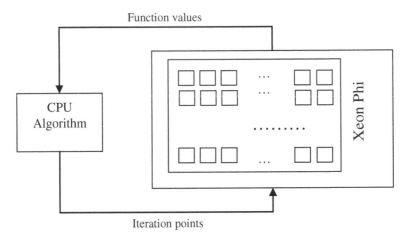

Fig. 1. Scheme of information exchanges

theorem and its corollary) the parallel global search algorithm will generate redundant trial points. Nevertheless, despite some redundancy, use of Xeon Phi will reduce overall algorithm operating time. It is confirmed by computational experiments, results of which are given in Sect. 5.

5   Results of Numerical Experiments

Computing experiments were carried out on one of the nodes of a high-performance cluster of the Nizhny Novgorod State University. The cluster node includes Intel Sandy Bridge E5-2660 2.2 GHz CPUs, 64 Gb RAM, and Intel Xeon Phi 5110P. For implementation of the parallel algorithm Intel C++ Compiler 14.0.2 under CentOS 6.4 was used.

It is significant, that widely known test problems from the field of multidimensional global optimization are characterized by small time of objective function values calculation. Usually, such a calculation is reduced to summation of several (according to problem dimension) values of elementary functions. Therefore, for the purpose of imitation of the computational complexity inherent to applied problems of optimization [14], calculation of the objective function in all performed experiments was complicated by additional calculations without changing the type of function and arrangement of its minimums (series summation from 20 thousand elements).

In work [15] a GKLS generator allowing generation of multiextremal optimization problems with known properties (number of local minimums, size of their domains of attraction, global minimum point, etc.) is described.

The results of numerical comparison of three sequential algorithms – DIRECT [1], DIRECTl [2] and global search algorithm (PGSA from Sect. 2 with $p=1$) – are provided below (results of work of the first two algorithms are given in [3]). Numerical comparison was carried out on function classes *Simple* and *Hard* of dimension 4 and 5 from [3] since solving problems of dimension 2 and 3 requires a small number of iterations and use of accelerator for solving these problems is impractical. Global minimum y^* was considered as found, if the algorithm generated trial point y^k in δ-vicinity of the global minimum, i.e. $\|y^k - y^*\| \leq \delta$. The size of the vicinity was selected (according to [3]) as $\delta = \|b - a\| \sqrt[N]{\Delta}$, N – problem dimension, a and b – borders of search domain D, parameter $\Delta = 10^{-6}$ at $N = 4$ and $\Delta = 10^{-7}$ at $N = 5$. When using the PGSA method for class *Simple* $r = 4.5$ parameter was selected, for class *Hard* $r = 5.6$ was selected; evolvent construction parameter was fixed as $m = 10$. The maximum allowable number of iterations was $K_{max} = 10^6$.

The average number of iterations k_{av} performed by the method for solving a series of problems from these classes is shown in Table 1. Symbol ">" reflects a situation, when not all problems of a class were solved by a method. It means that the algorithm was stopped as the maximum allowable number of iterations K_{max} was achieved. In this case, K_{max} value was used for calculation of the average value of number of iterations k_{av} that corresponds to the lower estimate of this average value. The number of unsolved problems is specified in brackets.

Table 1. Average number of iterations

N	Problem class	DIRECT	DIRECT*l*	PGSA
4	Simple	>47282(4)	18983	11953
	Hard	>95708(7)	68754	25263
5	Simple	>16057 (1)	16758	15920
	Hard	>217215 (16)	>269064 (4)	>148342 (4)

Table 1 shows that PGSA outperforms DIRECT and DIRECT*l* methods on all classes of problems by average number of iterations. And in class *5-Hard* each of the methods solved not all problems: DIRECT did not solve 16 problems, DIRECT*l* and PGSA – 4 problems.

Let us estimate now the speedup when using PGSA implemented on CPU depending on number of used cores p. Tables 2 and 3 show time speedup $S(p)$ and

Table 2. Time speedup $S(p)$ on CPU

p	$N = 4$		$N = 5$	
	Simple	Hard	Simple	Hard
2	2.45	2.20	1.15	1.32
4	4.66	3.90	2.82	2.59
8	7.13	7.35	3.47	5.34

Table 3. Iteration speedup $s(p)$ on CPU

p	$N = 4$		$N = 5$	
	Simple	Hard	Simple	Hard
2	2.51	2.26	1.19	1.36
4	5.04	4.23	3.06	2.86
8	8.58	8.79	4.22	6.56

Table 4. Redundancy $\lambda(p)$ on CPU

p	N = 4		N = 5	
	Simple	Hard	Simple	Hard
2	0.00	0.00	0.23	0.29
4	0.00	0.00	0.47	0.18
8	0.00	0.00	0.00	0.27

iteration speedup $s(p)$ respectively; speedup of parallel algorithm was measured in relation to the sequential one ($p=1$). Table 4 shows the average redundancy $\lambda(p)$ of the method during solving a problem series.

The results of the experiments show considerable acceleration and low redundancy of PGSA when using CPU.

Now let us perform a series of experiments using Xeon Phi. We will measure acceleration and redundancy of an algorithm that uses Phi as compared to a CPU algorithm that fully uses an eight-core CPU. In the experiments we will vary the number of threads p on Xeon Phi. All other parameters of the method will not vary.

Table 5. Time speedup $S(p)$ on Phi

p	N = 4		N = 5	
	Simple	Hard	Simple	Hard
60	0.54	1.02	1.07	1.61
120	0.55	1.17	1.05	2.61
240	0.51	1.06	1.07	4.17

Table 6. Iteration speedup $s(p)$ on Phi

p	N = 4		N = 5	
	Simple	Hard	Simple	Hard
60	8.13	7.32	9.87	6.55
120	16.33	15.82	15.15	17.31
240	33.07	27.79	38.80	59.31

Table 7. Redundancy $\lambda(p)$ on Phi

p	N = 4		N = 5	
	Simple	Hard	Simple	Hard
60	0.00	0.02	0.00	0.13
120	0.00	0.00	0.00	0.00
240	0.00	0.07	0.00	0.00

The results of the experiments (Table 5) show that in most cases an algorithm using Phi is not less fast than a CPU algorithm: acceleration approximately 1.05 is observed. Slowing down with class 4-*Simple* is explained by relative simplicity of the problems solved: the computational load on the coprocessor is not sufficient, the additional costs of transmission of the problem parameters from CPU to Phi produce a significant effect. With class 5-*Hard*, which is characterized by a high computational effort, a quadruple acceleration is observed; additional costs produce no decisive impact here.

An important additional feature is also acceleration on number of iterations, which goes up to a several dozens, if Phi is fully used (see Table 6). For example, solving a problem from class 5-*Hard* required on average only 633 parallel iterations on Phi, whereas when using all computing cores of the CPU the number of iterations was more than 37 thousand. At the same time, an algorithm using Phi is almost non-redundant in comparison to a CPU algorithm (see Table 7).

6 Conclusions

The work considers a parallel algorithm of global search developed within the framework of the information statistical approach to multiextremal optimization. This algorithm can be used for solving time-consuming optimization problems on state-of-the-art multiprocessor systems as it allows efficient implementation through use of Intel Xeon Phi coprocessor. The results of computational experiments confirm a high efficiency and low redundancy of the parallel algorithm. This very well correlates with the theoretical statements provided above.

Acknowledgements. The research is supported by the grant of the Ministry of education and science of the Russian Federation (the agreement of August 27, 2013, № 02.B.49.21.0003).

References

1. Jones, D.R., Perttunen, C.D., Stuckman, B.E.: Lipschitzian optimization without the Lipschitz constant. J. Optim. Theory Appl. **79**(1), 157–181 (1993)
2. Gablonsky, J.M., Kelley, C.T.: A locally-biased form of the DIRECT algorithm. J. of Glob. Optim. **21**(1), 27–37 (2001)
3. Sergeyev, Y.D., Kvasov, D.E.: Global search based on efficient diagonal partitions and a set of Lipschitz constants. SIAM J. Optim. **16**(3), 910–937 (2006)
4. Žilinskas, J.: Branch and bound with simplicial partitions for global optimization. Math. Model. Anal. **13**(1), 145–159 (2008)
5. Gergel, V.P.: A method of using derivatives in the minimization of multiextremum functions. Comput. Math. Math. Phys. **36**(6), 729–742 (1996)
6. Gergel, V.P.: A global optimization algorithm for multivariate functions with lipschitzian first derivatives. J. Glob. Optim. **10**(3), 257–281 (1997)
7. Gergel, V.P., Sergeyev, Y.D.: Sequential and parallel algorithms for global minimizing functions with lipschitzian derivatives. Comput. Math Appl. **37**(4–5), 163–179 (1999)
8. Strongin, R.G., Sergeyev, Y.D.: Global optimization with non-convex constraints. Sequential and Parallel Algorithms. Kluwer Academic Publishers, Dordrecht (2000)
9. Barkalov, K.A., Strongin, R.G.: A global optimization technique with an adaptive order of checking for constraints. Comput. Math. Math. Phys. **42**(9), 1289–1300 (2002)

10. Evtushenko, Y., Malkova, V.U., Stanevichyus, A.A.: Parallel global optimization of functions of several variables. Comput. Math. Math. Phys. **49**(2), 246–260 (2009)
11. Paulavicius, R., Zilinskas, J., Grothey, A.: Parallel branch and bound for global optimization with combination of Lipschitz bounds. Optim. Meth. Softw. **26**(3), 487–498 (2011)
12. Grishagin, V.A., Sergeyev, Y.D., Strongin, R.G.: Parallel characteristical algorithms for solving problems of global optimization. J. Glob. Optim. **10**(2), 185–206 (1997)
13. Sergeyev, Y.D., Strongin, R.G., Lera, D.: Introduction to global optimization exploiting space-filling curves. Springer, Heidelberg (2013)
14. Barkalov, K., Polovinkin, A., Meyerov, I., Sidorov, S., Zolotykh, N.: SVM regression parameters optimization using parallel global search algorithm. In: Malyshkin, V. (ed.) PaCT 2013. LNCS, vol. 7979, pp. 154–166. Springer, Heidelberg (2013)
15. Gaviano, M., Lera, D., Kvasov, D.E., Sergeyev, Y.D.: Software for generation of classes of test functions with known local and global minima for global optimization. ACM Trans. Math. Softw. **29**, 469–480 (2003)

Increasing Efficiency of Data Transfer Between Main Memory and Intel Xeon Phi Coprocessor or NVIDIA GPUS with Data Compression

Konstantin Y. Besedin$^{(\boxtimes)}$, Pavel S. Kostenetskiy, and Stepan O. Prikazchikov

South Ural State University, Chelyabinsk, Russia
besedin.k@gmail.com, {kostenetskiy,prikazchikovso}@susu.ru

Abstract. Efficient data transfer between main memory and Intel Xeon Phi coprocessor or GPU plays crucial role in using this devices for database processing. This paper addresses this problem by using data compression methods such as RLE, Null Suppression, LZSS and combination of RLE and Null Suppression. The chosen compression methods were implemented for Intel Xeon Phi coprocessors and NVIDIA GPUs. It is shown experimentally that these compression methods can be used to increase the efficiency of database processing using Intel Xeon Phi coprocessor and NVIDIA GPUs under certain conditions imposed on the data under treatment. It is also shown that, when a compression method allows one to process data without decompression, such a processing procedure can additionally increase the efficiency of this method.

1 Introduction

Processing of big amounts of data is an important part of modern scientific world. This processing introduces a lot of challenges for scientists and software engineers. Some of this challenges can be addressed by using parallel database management systems (DBMS) [1–4]. Database community shows a growing interest to increasing the performance of database processing by using devices like Intel Xeon Phi or GPU [5,6]. Both GPUs and manycore coprocessors have specific characteristics that need to be considered for developing high performance algorithms for these devices. The need to transfer data from main memory to device via PCI-E bus is considered as one of main bottlenecks for GPU and manycore coprocessors programming.

In this paper, we focus on three compression methods, used in DBMSs: RLE (Run Length Encoding), Null Suppression and LZSS. We implemented decompression algorithms for these compression methods for two hardware platforms: for Intel Xeon Phi coprocessor and for NVIDIA GPUs. A number of experiments have been done using these implementations. We also evaluate the effectiveness of combinations of different compression methods. At this moment, only combination of RLE and Null Suppression is evaluated.

© Springer International Publishing Switzerland 2015
V. Malyshkin (Ed.): PaCT 2015, LNCS 9251, pp. 319–323, 2015.
DOI: 10.1007/978-3-319-21909-7_32

2   Compression Methods

Data compression is one of possible ways to increase efficiency of data transfer between main memory and coprocessor or GPU. Data compression is already used by database systems to decrease the amount of data involved in I/O operations [7,8].

There are a number of papers devoted to implementation of different compressions methods for GPU [9,10]. Paper [6] shows that combining several simple schemas like RLE and Null Suppression can additionally increase the efficiency of compression.

To the best of our knowledge, there are no prior researches in using data compression for database processing on Intel Xeon Phi coprocessors.

At this stage of research, we've implemented all chosen compression methods for Intel Xeon Phi coprocessor and for GPU NVIDIA Tesla K40m we implemented RLE and Null Suppression methods. For Intel Xeon Phi we used OpenMP technology and used coprocessor in offload mode. Implementation for NVIDIA Tesla GPU was done with CUDA.

3   Experiments

Compression ratio of each compression method depends on characteristics of its input method [6]. For each evaluated compression method we vary such characteristic and show its impact on compression effectiveness. For RLE, LZSS and combination of RLE and Null suppression we use number of runs to number of elements ratio. For Null Suppression we use "value size" — the number of bytes needed to represent compressed element without data loss.

For some compression methods, it is possible to process compressed data without prior decompression [11]. In this paper, we evaluate three such methods: RLE, Null Suppression and combination of Null Suppression and RLE.

We consider all experiments on 1500 Mb of input data. This data consist of 64-bit unsigned integers. To generate it, we have implemented random generators that are able to generate test data with required value of one of data characteristics described above. Experiments for LZSS and RLE methods were using the same input data.

For each compression method we evaluate processing time for uncompressed data, processing time for compressed data with prior decompression and processing time for compressed data without prior decompression (if allowed by compression method). By "processing time" we mean the sum of time of data transfer to device memory, the time of decompression (if needed) and the time of applying an aggregate function to data. In our experiments we use sum as aggregate function. We had to implement compression-aware version of aggregate function to allow it to operate on data compressed with RLE compression method or combination of RLE and Null Suppression. Null suppression method does not require specific version of this aggregate function. During our experiments we imply that data is already in main memory in compressed form.

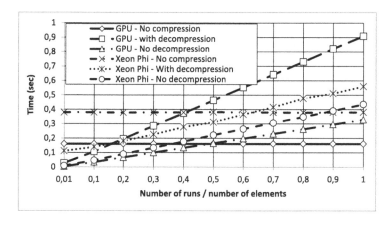

Fig. 1. Processing time for data, compressed with RLE

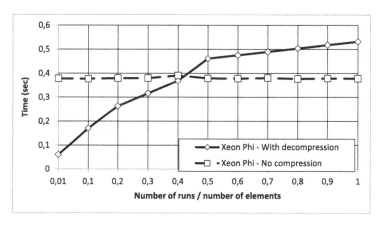

Fig. 2. Processing time for data, compressed with LZSS

Figure 1 show results of experiments with RLE compression method. Results for experiments with LZSS compression method are presented on Fig. 2. Results for Null Suppression experiments are summarized on Fig. 3. Results for combination of Null Suppression and RLE are shown on Fig. 4.

It can be seen that evaluated compression methods can be used to increase the efficiency of database processing using Intel Xeon Phi coprocessor and NVIDIA GPUs under certain conditions imposed on the data under treatment. Experiments also show that processing data in compressed form can additionally increase the efficiency of compression method.

4   Conclusion

This paper focuses on using data compression methods, such as RLE, Null Suppression, LZSS and combination of RLE and Null Suppression to increase effi-

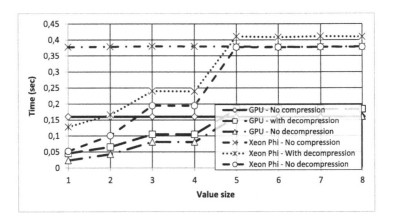

Fig. 3. Processing time for data, compressed with Null Suppression

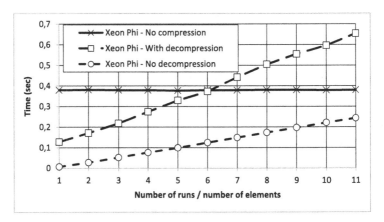

Fig. 4. Processing time for data, compressed with combination of RLE and Null Suppression

ciency of data transfer between main memory and coprocessor. We implemented decompression algorithms for these compression methods for two hardware platforms: for Intel Xeon Phi coprocessor and for NVIDIA GPUs. We used these implementations to show experimentally that these compression methods can be used to increase the efficiency of data transfer between main memory and Intel Xeon Phi coprocessor or NVIDIA GPUs under certain conditions imposed on the data under treatment. We expect that these results can be applied not only for data transfer through PCI-E bus, but also for other situations where efficient data transfer is required, like network communications with Ethernet/Infiniband. It is also shown that, when a compression method allows one to process data without decompression, such a processing procedure can additionally increase the efficiency of this method.

Acknowledgment. This work was supported in part by the Ministry of Education and Science of Russia under the Federal Targeted Programme for Research and Development in Priority Areas of Development of the Russian Scientific and Technological Complex in 2014–2020 (Agreement No. 14.574.21.0035).

References

1. Kostenetskii, P.S., Sokolinsky, L.B.: Simulation of hierarchical multiprocessor database systems. Program. Comput. Softw. **39**(1), 10–24 (2013)
2. Kostenetskiy, P.S., Sokolinsky, L.B.: Analysis of hierarchical multiprocessor database systems. In: International Conference on High Performance Computing, Networking and Communication Systems, HPCNCS-07, Orlando, Florida, USA, pp. 245–251, 9–12 July 2007
3. Pan, C.S., Zymbler, M.L.: Taming elephants, or how to embed parallelism into postgreSQL. In: Decker, H., Lhotská, L., Link, S., Basl, J., Tjoa, A.M. (eds.) DEXA 2013, Part I. LNCS, vol. 8055, pp. 153–164. Springer, Heidelberg (2013)
4. Sokolinskiy, L.B., Ivanova, E.V.: Using distributed column indexes for query execution over very large databases. In: The International conference "Advanced Mathematics, Computations and Applications– 2014" Institute of Computational Mathematics and Mathematical Geophysics of Siberian Branch of Russian Academy of Sciences, pp. 51–52. Academizdat, Novosibirsk, June 2014
5. Besedin, K.Y., Kostenetskiy, P.S.: Simulating of query processing on multiprocessor database systems with modern coprocessors. In: 37th International Convention on Information and Communication Technology, Electronics and Microelectronics, MIPRO 2014, Opatija, Croatia, pp. 1614–1616, 26–30 May 2014
6. Fang, W., He, B., Luo, Q.: Database compression on graphics processors. Proc. VLDB Endowment **3**(1–2), 670–680 (2010)
7. Ng, W.K., Ravishankar, C.V.: Block-oriented compression techniques for large statistical databases. IEEE Trans. Knowl. Data Eng **9**(2), 314–328 (1997)
8. Roth, M.A., Horn, S.J.V.: Database compression. SIGMOD Rec. **22**(3), 31–39 (1993)
9. Wu, L., Storus, M., Cross, D.: Final project cuda wuda shuda: cuda compression project. Technical report Cs315a, Stanford University (2009)
10. Ozsoy, A., Swany, M.: Culzss: Lzss lossless data compression on cuda. In: Proceedings of the 2011 IEEE International Conference on Cluster Computing. CLUSTER 2011, pp. 403–411. IEEE Computer Society, Washington (2011)
11. Abadi, D.J., Madden, S., Ferreira, M.: Integrating compression and execution in column-oriented database systems. In: Proceedings of the ACM SIGMOD International Conference on Management of Data, Chicago, Illinois, USA, pp. 671–682, 27–29 June 2006

Parallelizing Branch-and-Bound on GPUs for Optimization of Multiproduct Batch Plants

Andrey Borisenko[1]([✉]), Michael Haidl[2], and Sergei Gorlatch[2,3]

[1] Tambov State Technical University, Tambov, Russia
borisenko@mail.gaps.tstu.ru
[2] University of Muenster, Muenster, Germany
{michael.haidl,gorlatch}@uni-muenster.de
[3] Cells-in-Motion Cluster of Excellence (EXC 1003 – CiM), Muenster, Germany

Abstract. Parallel implementation of the Branch-and-Bound (B&B) technique for optimization problems is a promising approach to accelerating their solution, but it remains challenging on Graphics Processing Units (GPUs) due to B&B's irregular data structures and poor computation/communication ratio. The contributions of this paper are as follows: (1) we develop two basic implementations (iterative and recursive) of B&B on systems with GPUs for a practical application scenario - optimal design of multi-product batch plants; (2) we propose and implement several optimizations of our CUDA code using both algorithmic techniques of reducing branch divergence and GPU-specific properties of the memory hierarchy; and (3) we evaluate our implementations and optimizations on a modern GPU-based system and we report our experimental results.

Keywords: GPU computing · CUDA · Branch-and-bound · Combinatorial optimization · Multi-product batch plant design

1 Motivation and Related Work

Combinatorial optimization [8] is often very time-consuming due to "combinatorial explosion" – the number of combinations to be examined grows exponentially, such that even the fastest supercomputers would require an intolerable amount of time. A common approach in applications is to formulate a mixed-integer nonlinear programing (MINLP) model [6,15] and to exploit the Branch-and-bound (B&B) technique for solving it. In B&B, the search space is represented as a tree whose root is the original problem, the internal nodes are partially solved subproblems, and the leaves are the potential solutions. B&B proceeds in several iterations where the best solution found so far (upper bound) is progressively improved: a bounding mechanism is used to eliminate the subproblems that are not likely to lead to optimal solutions and to cut their corresponding sub-trees. This reduces the size of the explored search space, but can be still time-consuming in practice and requires acceleration, for example using parallel computing.

© Springer International Publishing Switzerland 2015
V. Malyshkin (Ed.): PaCT 2015, LNCS 9251, pp. 324–337, 2015.
DOI: 10.1007/978-3-319-21909-7_33

Parallelization of B&B has been extensively studied, recently with a focus on systems comprising Graphics Processing Units (GPUs). Recent approaches usually address the most time consuming bounding mechanism of B&B [10]. The main difficulty in B&B are irregular data structures not well suited for GPU computing and the low computation/communication ratio. In [2], a hybrid implementation of B&B for the knapsack problem demonstrates that for small problem sizes it is not efficient to launch the B&B computation kernels on GPU. A parallel implementation in [3] with CUDA makes use of data y compression. In [11], a hybrid CPU-GPU implementation is presented, and [16] studies the design of parallel B&B in large-scale heterogeneous compute environments with multiple shared memory cores, multiple distributed CPUs and GPU devices.

In this paper, we parallelize B&B and illustrate it with a practical application – the optimal selection of equipment for multi-product batch plants [12]. We develop and evaluate an implementation of the B&B method on a CPU-GPU system using the CUDA programming environment [13] in two versions – an iterative and a recursive one – and we describe their optimizations, as well as compare them to each other. We report experimental results about the speedup of our GPU-based implementations as compared to the sequential CPU version.

2 Problem Formulation

Our application use case is a *Chemical-Engineering System* (CES) – a set of equipment (reactors, tanks, filters, dryers etc.) which implement the processing stages for manufacturing certain products. CES comprises I processing stages; i-th stage is equipped with units from a finite set X_i, with J_i being the number of equipment variants in X_i. All equipment variants of a CES are described as $X_i = \{x_{i,j}\}, i = \overline{1, I}, j = \overline{1, J_i}$, where $x_{i,j}$ is the main size j (working volume) of the unit suitable for stage i. A variant $\Omega_e, e = \overline{1, E}$ of a CES, where $E = \prod_{i=1}^{I} J_i$ is the number of all possible system variants, is an ordered set of equipment unit variants, selected from the respective sets. Each variant Ω_e of a system must be in an operable condition (*compatibility constraint*) i.e., it must satisfy the conditions of a joint action for all its processing stages: $S(\Omega_e) = 0$ if compatibility constraint is satisfied. An operable variant of a CES must run at a given production rate in a given period of time (*processing time constraint*), such that it satisfies the restrictions for the duration of its operating period $T(\Omega_e) \leq T_{max}$.

Thus, designing an optimal CES can be formulated as the following optimization problem [9]: to find a variant $\Omega^* \in \Omega_e, e = \overline{1, E}$ of a CES, where the optimality criterion – equipment costs $Cost(\Omega_e)$ – reaches a minimum, and both compatibility constraint and processing time constraint are satisfied:

$$\Omega^* = argmin \ Cost(\Omega_e), e = \overline{1, E} \tag{1}$$

$$\Omega_e = \{x_{1,j_1}, x_{2,j_2}, \ldots, x_{I,j_I} | j_i = \overline{1, J_i}, i = \overline{1, I}\}, e = \overline{1, E} \tag{2}$$

$$x_{i,j} \in X_i, i = \overline{1, I}, j = \overline{1, J_i} \tag{3}$$

$$S(\Omega_e) = 0, e = \overline{1, E} \tag{4}$$

$$T(\Omega_e) \leq T_{max}, e = \overline{1, E} \tag{5}$$

Figure 1 shows the search space as a tree: all possible variants of a CES with I stages are represented by a tree of height I (see Fig. 1). Each tree level corresponds to one processing stage of the CES, each edge corresponds to a selected equipment variant taken from X_i, where X_i is the set of possible variants at stage i of the CES. For example, the edges from level 0 correspond to elements of X_1. Each node $n_{i,k}$ at the tree layer $N_i = \{n_{i,1}, n_{i,2}, \ldots, n_{i,k}\}$, $i = \overline{1, I}$, $k = \overline{1, K_i}$, $K_i = \prod_{l=1}^{i}(J_l)$ corresponds to a variant of a beginning part of the CES, composed of equipment units for stages 1 to i. Each path from the root to one of the leaves thus represents a complete variant of the CES. To enumerate all possible variants of a CES, a depth-first traversal of the tree is performed as in Fig. 1: starting at level 0, all device variants of the CES at a given level are enumerated and appended to the valid beginning parts of the CES obtained at previous levels, starting with an empty beginning part at level 0. This process continues recursively for all valid beginning parts resulting from appending device variants of the current level to the valid beginning parts from previous levels. When a leaf is reached, the recursive process stops and the new solution is compared to the current optimal solution, possibly replacing it.

A complete tree traversal (selecting a device on each edge traversal) and checking constraints (Eqs. 4 and 5) would result in a considerable computational effort. E.g., for a CES consisting of 16 stages where each process stage can be equipped with devices of 5 to 12 standard sizes [9], the number of choices is 5^{16}–12^{16} (approximately 10^{11}–10^{17}). Hence, performing an exhaustive search (pure brute-force solution) for finding a global optimum is usually impractical.

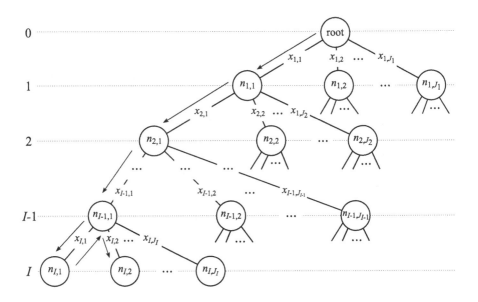

Fig. 1. Tree traversal in depth-first search.

3   Parallelization for GPU

Figure 2 illustrates our strategy for dividing the initial search tree into subtrees for parallels processing: the sequential *host* process on the CPU dispatches computations to multiple *device* threads on the GPU and then gathers the results from these threads. The tree-like organization of B&B provides a potential for parallelization, as all branches of the tree can be processed simultaneously.

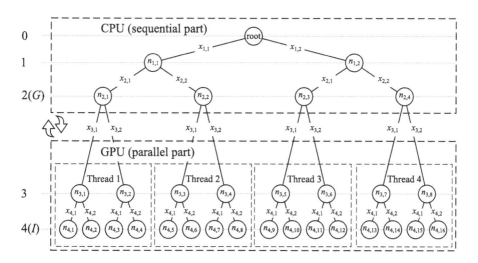

Fig. 2. Dividing the search tree into subtrees for parallel processing.

All nodes $N_i = \{n_{i,1}, n_{i,2}, \ldots, n_{i,k}\}$, $i = \overline{G+1, I}$, $k = \overline{1, K_i}$, $K_i = \prod_{l=1}^{i} J_l$ at each layer i below G are traversed in an independent *thread*. The total number of threads is $N_{threads} = \sum_{i=1}^{G} K_i$, $1 \leq G \leq I$. The *granularity* parameter G limits the number of threads: each subtree below the granularity level will be processed as one thread on the GPU. E.g., Fig. 2 shows a CES consisting of 4 stages ($I = 4$) where each stage can be equipped with 2 devices ($J_1 = J_2 = J_3 = J_4 = 2$); the number of all possible system variants is $2^4 = 16$. We use granularity $G = 2$, so the threads number is $2^2 = 4$. All three nodes at layers from 0 to 2 are processed on CPU, then partial solutions are transferred to GPU, and all other nodes at layers from 3 to 4 are processed in parallel on GPU.

Host Code. We use the CUDA Runtime API [5,14]. The host (see Listing 1) begins its work by loading *input data* from file by calling `ReadInputData()` (line 2). The `inData` is a pointer to the structure whose fields store all necessary input data for the calculation. The values of these data are not changed during calculations and are the same for all threads, i.e. constant. Based on the input data `inData` and *granularity* parameter `G`, the required number of threads `numThreads` is computed by `ThreadsNumber()` (line 4) and used in

PrepOperationalData() for preparing *operational data* (line 6). The oprData is a pointer to the structure whose values are changed in the calculations independently by each thread with *thread index* threadID.

```
 1   main() {      /* prepare all necessary data */
 2     ReadInputData(inData);
 3     /* number of threads */
 4     numThreads = ThreadsNumber(inData, G);
 5     /* prepare all operational data */
 6     PrepOperationalData(oprData, inData, numThreads);
 7     /* start tree traversal for dividing tree into */
 8     /* subtrees and creating beginning parts of the CES */
 9     EnumerateHost(0, 0);
10     /* send all necessary data to device */
11     cudaMemcpyHtoD(inData,oprData,W, G);
12     /* define parameters for kernel launch */
13     blocksPerGrid = numThreads / MAX_THREADS_PER_BLOCK + 1;
14     threadsPerBlock = MAX_THREADS_PER_BLOCK;
15     /* starting kernel function on device */
16     FindSolution<<<blocksPerGrid, threadsPerBlock>>>(G);
17     /* synchronize device */
18     cudaDeviceSynchronize();
19     /* copy the results from device to host */
20     cudaMemcpyDtoH(W, Cost, minCost);
21     /* find global optimal solution */
22     for (n=1; n<=numThreads; n++) {
23         if(Cost[n] < minCost) { Wopt = W[n];
24             minCost = Cost[n]; }}}
25   EnumerateHost(threadID, level) {
26     for (j=1; j <= J[level]; j++) {
27         Wloc[level] = X[level, j];
28         if (level < G) { /* check granularity */
29             EnumerateHost(threadID, level + 1); }
30         else{ W[threadID] = Wloc;
31             threadID++; }}}
```

Listing 1. The host pseudo-code.

Both ReadInputData() and PrepOperationalData() make all necessary memory allocations and variables initialization. Then host performs a depth-first traversal of the tree using recursive function EnumerateHost() to the level defined by G (line 9). Here W is a two-dimensional array, represented as an array of length numThreads each element of this is a vector of length I specifying the device variant at each stage of the solution. In lines 25–31, the host creates beginning parts of CES at levels from 0 to G and saves them in W. The host neither checks constraints (Eqs. 4 and 5) nor evaluates upper and lower bounds of the objective function (Eq. 1); this will be performed on the GPU. After the host sends all necessary data to the device (line 11), it defines the number of blocks (line 13) and threads (line 14) in one block and then starts the kernel function FindSolution() (line 16).

A CUDA kernel launch is asynchronous and returns control to the CPU immediately after starting the GPU process. By calling cudaDeviceSynchronize() (line 18), the CPU is forced to idle until the GPU work has completed and the host receives the results from the device (line 20). Here Cost is an array of size numThreads; its elements are the local optimal costs of CES which were found by each thread. The minCost is the global minimal cost of CES, whose value is used to seek for the best optimal solution. The host compares local minimal costs with global minimum and searches for the best CES-variant (lines 22–24). Here, vector Wopt specifies the unit variant at each stage of the optimal solution.

Kernel Code. We have developed a recursive (Listing 2) and an iterative (Listing 3) tree traversal implementations. The recursive approach can be used on NVIDIA GPU devices of Compute Capability 2.0 and higher. On the older NVIDIA devices, the iterative approach should be used.

```
1    __global__ void FindSolution (G)
2    {    /* obtaining thread identifier */
3         threadID = blockDim.x * blockIdx.x + threadIdx.x;
4         /* if threadID not greater maximal thread numbers */
5         if (threadID <= numThreads) {
6             /* start subtree traversal */
7             EnumerateDevice(threadID, G + 1); }}
8    __device__ EnumerateDevice(threadID, level)
9    {    for (j=1; j <= J[level]; j++) {
10            /* append device variant to beginning part */
11            Wloc[level] = X[level, j];
12            if (level < I) {
13                /* check compatibility constraint and upper bound */
14                if (S(Wloc) == 0 && DefineCost(Wloc, level) < minCost){
15                    /* search recursively */
16                    EnumerateDevice(threadID, level + 1); }}
17            else {/* leaf node */
18                /* check optimality criterion */
19                if (Cost (Wloc) < minCost)) {
20                    /* check processing time constraint */
21                    if (T(Wloc) <= Tmax) {
22                        /* make solution new (local) optimal solution */
23                        W[threadID] = Wloc;
24                        Cost[threadID] = DefineCost(Wloc, I);
25                        atomicMinDbl(minCost, Cost[threadID]); }}}}}
```

Listing 2. The kernel pseudo-code using recursive approach.

The kernel function FindSolution() (Listing 2) calculates a global thread identifier threadID (line 3) which is compared with the required number of threads numThreads (line 5). This is necessary because of rounding of kernel launch parameters (see Listing 1, line 13): the actual number of running threads can be greater than numThreads, but all memory allocations for oprData are done for numThreads. For all threads with valid threadID, the kernel function calls EnumerateDevice() at level G+1 (line 7). Within this function, the device traverses the remaining sub-trees at levels from G+1 to I of the received CES' beginning parts to find solutions. All unit variants of the CES at a given level are enumerated and appended to the beginning parts of the CES. Valid beginning parts are obtained at previous levels, starting at level G+1. This process continues recursively for all valid beginning parts that result from appending unit variants of the current level to the valid beginning parts from previous levels.

When a leaf node is reached, the recursive process stops and the current solution is compared to the current optimal solution, possibly replacing it. When traversing the tree, the compatibility constraint (Eq. 4, function S()) is checked for the corresponding part of the CES. We also compare the cost for the current beginning part of the CES, consisting of the first level stages (function DefineCost()) with a global upper bound (variable minCost) (line 14). The initialization of the upper bound is done with the sum of all maximum units costs for each production stage. If the current beginning part of the CES fulfills the compatibility constraint and its costs do not exceed the global upper bound, we recursively continue tree traversal to the next level (line 16). If a leaf node of the tree is reached (17), a new full solution has been found and its costs (Eq. 1, function Cost()) are compared to the cost of the previous optimal solution minCost (line 19). If a better solution is obtained, the processing time constraint (Eq. 5, function T()) is checked for the corresponding CES (line 21). If this constraint is fulfilled, a new valid optimal solution replaces the previous optimal solution (line 23), saves it value (line 24) and its costs are taken as the new upper bound and saved with atomic function atomicMinDbl (line 25).

The iterative stack-based traversal algorithm is presented in Listing 3. The kernel function is the same like in Listing 2 (lines 1 – 7), so we omit it. In our implementation, stack, which must be not smaller than the number of tree levels, is a one-dimensional array stack of size MAX_STACK_SIZE (line 10). The stack is accessed for adding and removing data elements through its *top*, variable stackTop (line 11). The standard elementary stack operations are push (line 13), pop (line 14), and empty (line 15). At first stack is empty and therefore the first node (units variant for level 1) of the tree is added (line 17). While the stack is not empty, we pop the stack, find all possible choices after the previous one and push these choices onto the stack (line 19). At each loop iteration, the last item on the stack is popped (line 20). If at the current level there are no other nodes, we return to the previous level (line 22). Otherwise we push the next node at the current level into the stack (line 24). If the current beginning part of the CES fulfills the compatibility constraint and its costs do not exceed the global upper bound (line 26), we append unit variants at the current level to the valid beginning parts of CES from previous levels (line 28). Otherwise we discard deeper levels of the tree and go to the next loop iteration. If the last level is not reached (line 29), we continue tree traversal to the next level (line 30) and push the first node at the next level into the stack (line 32). If a leaf of the tree is reached (line 33), the algorithm works like recursive version (see Listing 2). The solution cost is compared to the global cost of the previous optimal solution (line 35). If a better solution is obtained, then processing time constraint is checked for the corresponding CES (line 37). If this constraint is fulfilled, this mean that a new optimal solution has been found which replaces the previous optimal solution and its cost is taken as the new upper bound (lines 39 – 41).

```
7    ...
8    __device__ EnumerateDevice(threadID, level)
9    {    /* declaration of standard last-in, first-out stack */
10       stack[MAX_STACK_SIZE];
11       stackTop = 0;
12       /* defining standard stack operations */
13   #define push(x)  stack[stackTop++] = (x)
14   #define pop()    stack[--stackTop]
15   #define empty    (stackTop == 0)
16       /* add to stack first item on level 1 */
17       push(1);
18       /* loop while stack is not empty */
19       while (!empty) {  /* retrieve the top node from stack */
20           j = pop();
21           /* if on this level are not other nodes */
22           if(j == J[level] + 1) {level--;}
23           else {  /* add to stack next trees node on current level */
24               push(j + 1);
25               /* check compatibility constraint and upper bound */
26               if (S(Wloc) == 0 && DefineCost(Wloc, level) < minCost){
27                   /* append device variant to beginning part */
28                   Wloc[level] = X[level, j];
29                   if (level < I) {  /* go to next trees level */
30                       level++;
31                       /* add to stack first node on current level */
32                       push(1); }
33                   else {/* leaf node */
34                       /* check optimality criterion */
35                       if (Cost (Wloc) < minCost)) {
36                           /* check processing time constraint */
37                           if (T(Wloc) <= Tmax) {
38                               /* make solution new (local) optimal solution */
39                               W[threadID] = Wloc;
40                               Cost[threadID] = DefineCost(Wloc, I);
41                               atomicMinDbl(minCost, Cost[threadID]); }}}}}}}
```

Listing 3. The kernel pseudo-code using iterative approach.

4 Optimizations

In our previous work [1] we have used Standard Template Library (STL), in particular container class *std::vector*, for implementing the mathematical model of CES. Since there is no implementation of STL containers in CUDA, we revised the program code, using many (about 50) multidimensional (from 1- to 5-dimensional) arrays. To keep the arrays' contents contiguous, we simulate a multidimensional array with a one-dimensional array which guarantees that all array elements are in a flat chunk of memory. This is convenient for data transfer with standard **memcpy**-like functions and faster for the memory access as compared to fragmented memory (e.g., there are fewer cache misses and better performance). On the test platform presented in Sect. 5, this new sequential CPU version without STL vectors is approximately 8 – 9 times faster than the STL-based sequential version. Our experiments in the sequel are carried out for this new CPU version of the program.

We perform and evaluate two optimizations: (a) shared memory utilization, and (b)reducing branching in the kernel function.

Our target architecture (GPU) comprises *Streaming Multiprocessors* (SMs) contains some *Streaming Processors* (SP) (since Fermi microarchitecture NVIDIA changes the name SP to *CUDA cores*). The DRAM or *Global memory* is the biggest memory region (several gigabytes) on the graphic device. It is off-SM, therefore relatively slow and can be accessed both by host and device. In addition, every SM has a small on-chip memory that can be configured as *Shared memory* and as *L1 cache*. In addition to the *L1 cache*, Kepler introduces a *Read-only data cache* for data for the duration of the function. The *L2 cache* is the primary point of data unification between the SM units, servicing all load and store requests and providing data sharing across the GPU.

```
2   ...
3       threadID = blockDim.x * blockIdx.x + threadIdx.x;
4       __shared__ sharedMemory[];
5       /* send data by shared memory pointer */
6       __device__ SendDataByMemoryPtr(Wloc, sharedMemory);
7   #ifdef ITERATIVE
8       __device__ SendDataByMemoryPtr(stack, sharedMemory);
9   #endif
10      /* if threadID  not greater maximal thread numbers */
11      if(threadID <= numThreads){
12          /* start subtree traversal */
13          EnumerateDevice(threadID, G + 1);}
14  ...
```

Listing 4. The kernel pseudo-code with optimization *O1*.

Listing 4 shows our first memory optimization: we move the local array Wloc (line 6) from global memory to shared memory using device function SendDataByMemoryPtr() and for the iterative approach we move the stack too (line 8) (optimization *O1*). Moreover we move inData from the constant to the shared memory, as shown in Listing 5 line 6 (optimization *O2*), using function SendDataByMemoryPtr() which is a device function for sending data to the specified memory address.

The SMs of an NVIDIA GPU only get one instruction at a time and all CUDA cores execute the same instruction. Threads within a *warp* (a group of 32 thread, which are used in the hardware implementation to coalesce memory access and instruction dispatch) execute the same instruction in each cycle, disabling threads that are not on the same path of control-flow; this is also known as *thread or branch divergence* [4,7]. The most common code construct that can cause thread divergence is branching for conditionals in an *if-then-else* statement. Branch divergence can hurt performance due to lower utilization of the processing elements, which cannot be compensated for via increased amount of parallelism [7].

```
2   ...
3       threadID = blockDim.x * blockIdx.x + threadIdx.x;
4       __shared__ sharedMemory[];
5       /* send data by shared memory pointer */
6       SendDataByMemoryPtr(inData, sharedMemory);
7       /* if threadID not greater maximal thread numbers */
8       if(threadID <= numThreads){
9           /* start subtree traversal */
10          EnumerateDevice(threadID, G + 1);}
11  ...
```

Listing 5. The kernel pseudo-code with optimization *O2*.

In addition to shared memory utilization, we reduce the branch divergence by removing the checking of the compatibility constraint and the upper bound (Listing 2 line 14, Listing 3 line 26). In this case all restrictions are checked only at the last tree level, such that code paths have fewer branches (this effectively transforms B&B into an exhaustive search). The implementations optimized this way get suffix 'm' (e.g., Recursion *O1m*, Iteration *O2m*) in our evaluation.

As an additional advantage, this optimization also reduces the stall time of GPU threads which finish their work earlier and remain idle while waiting for the last thread of the same warp to finish. The stall times of threads could alternatively be minimized by a more complex work distribution across threads. However this would arguably introduce more thread divergence, and also require additional information exchanges across threads, as well as higher load on the registers and shared memory, with more atomic operations, which together would negatively affect the kernel's runtime.

5   Experimental Results

Our experiments are conducted on a hybrid system comprising: (1) CPU: Intel Xeon E5-1620 v2, 4 cores with Hyper-Threading, 3.7 GHz with 16 GB RAM, and (2) GPU: NVIDIA Tesla K20c with 13 SMs, each with 192 CUDA Cores (total 2496 CUDA Cores), 5 GB of global memory and up to 48 KB of shared memory per SM. We use Ubuntu 14.04.1, NVIDIA Driver version 340.29, CUDA version 6.5 and GNU C++ Compiler version 4.8.2. We study the design of a CES consisting of 16 processing stages with 3 variants of devices at every stage as test case (total $3^{16} = 43\,046\,721$ CES variants).

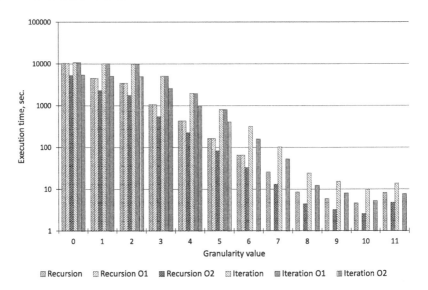

Fig. 3. Results of memory optimization. Run-time depending on granularity.

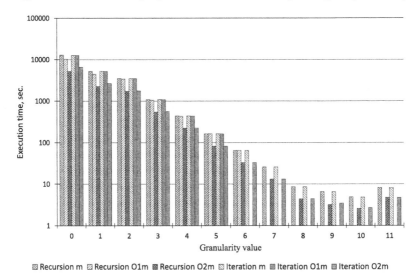

Fig. 4. Results of branch optimization. Run-time depending on granularity.

Our first experiment (see Fig. 3) concerns selecting a suitable granularity value which sets the level of parallelism and, therefore, is important for parallel program performance. On the one hand, increasing the number of threads increases the performance of the parallel program on the GPU, but on the other hand, it needs more memory: each thread needs own memory in `oprData`. We run our CUDA-based implementation setting granularity values from 1 to 11.

We observe in Fig. 3 that the runtime is reduced with increasing granularity, but for larger granularity values, too many threads are created and too much memory is required for operational data `oprData`, such that the implementation runs out of global memory for certain versions, and for granularity greater than 11 it runs out of memory for all versions.

The results of our memory optimization are also shown in Fig. 3 (the vertical axis has a logarithmic scale). For optimization *O1*, we use shared memory for array `Wloc` (recursive approach) and for `Wloc` and `stack` (iterative approach). For granularity greater than 6 for recursive and greater than 5 for iterative approach, we run out of shared memory. The total runtime of our program consists of the runtimes of the CPU-part and the GPU-part. Our measurements show that the CPU runtime affects insignificantly the total runtime (CPU takes about 0.1 % of the total runtime). While optimization *O1* does not have a significant impact on the runtime, optimization *O2* significantly reduces the execution time, on average by 48 %. We observe that the recursive implementation (Listing 2) is faster than the iterative one (Listing 3) on average by about 2.2 times.

The results of branch optimization are presented in Fig. 4: in addition to the shared memory utilization, we remove the checking of the compatibility constraint and the upper bound (Listing 2 line 14, Listing 3 line 26). We observe in Fig. 4 that the branch optimization has significantly different effects for recursion and iteration: for recursion the runtime was slightly increased within 5 %; for iteration the runtime was decreased on average by 60 %. Based on the experimental data, we conclude that for the recursive version the fastest is the *O2*-optimization, and for the iterative version the *O2m*-optimization is the fastest.

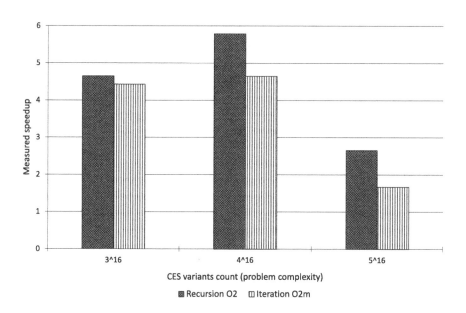

Fig. 5. Experimental results. Measured speedup.

The fastest recursive version for our test example is faster than the fastest iterative version by about 5 %.

In Fig. 5, the speedups of our parallel GPU versions for Recursion $O2$ and Iteration $O2m$ tree traversal implementations are presented as compared to the sequential CPU-program.

We have run several test cases for problems of larger dimension and we measured the speedup of the parallel GPU-based implementation vs. sequential CPU-program. Our experiments were carried out for the above example (16 processing stages) but with 4 (total $16^4 = 4\,294\,967\,296$ CES variants) and 5 (total $16^5 = 152\,587\,890\,625$ CES variants) units, correspondingly.

We observe in Fig. 5 that the speedup value for the recursive version is between 2.66 and 5.79, and for the iterative version it is between 1.67 and 4.43.

6   Conclusion

We proposed two parallelization approaches to implement a parallel B&B algorithm for solving the optimization problem for multi-product batch plants on a CPU-GPU platform.

Two basic implementations – based on recursive and iterative approaches to the tree traversal – have been presented. We analyzed the impact of the degree of parallelism controlled by the granularity parameter, and conducted GPU-specific optimizations of our programs: using shared memory and reducing branch divergence. Our results show that the recursion-based implementation in general is faster than the iteration-based on our test platform, and that our optimizations significantly reduce the total runtime. Our future work will extend the optimization space using the most modern features of the GPU programming approaches.

Acknowledgement. This work was partially supported by the Deutsche Forschungsgemeinschaft (DFG), Cells-in-Motion Cluster of Excellence (EXC 1003 – CiM), University of Muenster, Germany. Andrey Borisenko was supported by the DAAD (German Academic Exchange Service) and by the Ministry of Education and Science of the Russian Federation under the "Mikhail Lomonosov II"-Programme.

References

1. Borisenko, A., Kegel, P., Gorlatch, S.: Optimal design of multi-product batch plants using a parallel branch-and-bound method. In: Malyshkin, V. (ed.) PaCT 2011. LNCS, vol. 6873, pp. 417–430. Springer, Heidelberg (2011)
2. Boukedjar, A., Lalami, M.E., El Baz, D.: Parallel branch and bound on a CPU-GPU system. In: PDP, pp. 392–398. Citeseer (2012)
3. Boyer, V., El Baz, D., Elkihel, M.: Solving knapsack problems on GPU. Comput. Oper. Res. **39**(1), 42–47 (2012)
4. Chakroun, I., Mezmaz, M., Melab, N., Bendjoudi, A.: Reducing thread divergence in a GPU-accelerated branch-and-bound algorithm. Concurr. Comput. Pract. Exp. **25**(8), 1121–1136 (2013)

5. Farber, R.: CUDA Application Design and Development. Elsevier, Amsterdam (2011)

6. Fumero, Y., Corsano, G., Montagna, J.M.: A mixed integer linear programming model for simultaneous design and scheduling of flowshop plants. Appl. Math. Model. **37**(4), 1652–1664 (2013)

7. Han, T.D., Abdelrahman, T.S.: Reducing branch divergence in GPU programs. In: Proceedings of the Fourth Workshop on General Purpose Processing on Graphics Processing Units, p. 3. ACM (2011)

8. Hoffman, K., Padberg, M.: Combinatorial and integer optimization. In: Hoffman, K.L., Padberg, M. (eds.) Encyclopedia of Operations Research and Management Science, pp. 94–102. Springer, Heidelberg (2001)

9. Malygin, E., Karpushkin, S., Borisenko, A.: A mathematical model of the functioning of multiproduct chemical engineering systems. Theo. Found. Chem. Eng. **39**(4), 429–439 (2005)

10. Melab, N., Chakroun, I., Mezmaz, M., Tuyttens, D.: A GPU-accelerated branch-and-bound algorithm for the flow-shop scheduling problem. In: IEEE International Conference on Cluster Computing (CLUSTER), pp. 10–17. IEEE (2012)

11. Meyer, X., Chopard, B., Albuquerque, P.: A branch-and-boundalgorithm using multiple GPU-based LP solvers. In: 20th International Conference on HighPerformance Computing (HiPC), pp. 129–138. IEEE (2013)

12. Mokeddem, D., Khellaf, A.: Optimal solutions of multiproduct batch chemical process using multiobjective genetic algorithm with expert decision system. J. Anal. Meth. Chem. **2009**, 1–9 (2009)

13. NVIDIA Corporation: CUDA C programming guide 6.5, August 2014. http://docs.nvidia.com/cuda/pdf/CUDA_C_Programming_Guide.pdf

14. Sanders, J., Kandrot, E.: CUDA by Example: An Introduction to General-Purpose Gpu Programming. Addison-Wesley Professional, Boston (2010)

15. Terrazas-Moreno, S., Grossmann, I.E., Wassick, J.M.: A mixed-integer linear programming model for optimizing the scheduling and assignment of tank farm operations. Ind. Eng. Chem. Res. **51**(18), 6441–6454 (2012)

16. Vu, T., Derbel, B.: Parallel branch-and-bound in multi-core multi-CPU multi-GPU heterogeneous environments (2014). https://hal.inria.fr/hal-01067662

Optimal Dynamic Data Layouts for 2D FFT on 3D Memory Integrated FPGA

Ren Chen$^{(\boxtimes)}$, Shreyas G. Singapura, and Viktor K. Prasanna

University of Southern California, Los Angeles, CA 90089, USA
{renchen,singapur,prasanna}@usc.edu

Abstract. FPGAs have been widely used for accelerating various applications. For many data intensive applications, the memory bandwidth can limit the performance. 3D memories with through-silicon-via connections provide potential solutions to the latency and bandwidth issues. In this paper, we revisit the classic 2D FFT problem to evaluate the performance of 3D memory integrated FPGA. To fully utilize the fine grained parallelism in 3D memory, optimal data layouts so as to effectively utilize the peak bandwidth of the device are needed. Thus, we propose dynamic data layouts specifically for optimizing the performance of the 3D architecture. In 2D FFT, data is accessed in row major order in the first phase whereas, the data is accessed in column major order in the second phase. This column major order results in high memory latency and low bandwidth due to high row activation overhead of memory. Therefore, we develop dynamic data layouts to improve memory access performance in the second phase. With parallelism employed in the third dimension of the memory, data parallelism can be increased to further improve the performance. We adopt a model based approach for 3D memory and we perform experiments on the FPGA to validate our analysis and evaluate the performance. Our experimental results demonstrate up to **40**x peak memory bandwidth utilization for column-wise FFT, thus resulting in approximately **97 %** improvement in throughput for the complete 2D FFT application, compared to the baseline architecture.

1 Introduction

FPGAs have been used as accelerators for many applications such as Signal Processing, Image Processing, Packet classification etc. The general purpose processors cannot keep up with the demands of these applications in terms of performance. Even with the high performance of FPGAs, meeting the throughput requirement of these applications is a challenging task. Most of the applications are data intensive and this translates to frequent accesses to the memory. The bottleneck in these cases is the low bandwidth and high latency of the memory.

3D memory has been widely studied in the research community with the high bandwidth and short latency access being the important parameters. 3D memories consist of stack of layers connected using Through Silicon Vias (TSVs) [9].

This material is based in part upon work supported by the National Science Foundation under Grant Number ACI-1339756.

© Springer International Publishing Switzerland 2015
V. Malyshkin (Ed.): PaCT 2015, LNCS 9251, pp. 338–348, 2015.
DOI: 10.1007/978-3-319-21909-7_34

The high speed vertical TSVs along with the third dimension of memory result in short latencies and packs in large memory sizes compared to the conventional 2D memories. Although 3D memories are expected to provide 10× bandwidth compared to 2D memory, this is subject to the ideal conditions. These include data layouts which reduce row activation overhead, high page hit rate for stride access, etc. These problems are similar to the issues in the conventional planar memories. But, employing the solutions in the context of 3D memory is not trivial due to the structure and organization of 3D memory.

In this paper, we target 2D FFT application on 3D memory integrated FPGA and evaluate its performance with throughput and latency as the target metrics. 2D FFT is a data intensive application with stride memory access patterns. 2D FFT consists of two phases and the access patterns in the two phases require mutually conflicting data layouts. The ideal data layout in the first phase is row major data layout whereas, the second phase requires a column major data layout. Therefore, a static data layout trying to improve the performance in one phase will lower the performance in the other phase. The main reasons for this low performance are high number of row activations and low page hit rate. Therefore, with a static data layout the true capability of 3D memory cannot be realized. We address this problem by extending our solution of dynamic data layouts [6] to 3D memory. The main contributions in this paper are:

1. Model the 2D FFT application on 3D memory integrated FPGA.
2. Develop optimal dynamic data layouts to optimize performance of 2D FFT on 3D memory.
3. Evaluation of optimized and baseline implementation with throughout and latency as the performance metrics.

2   Related Work

As the well-known simplest multidimensional FFT algorithm, the row-column algorithm has been commonly used to implement 2D FFT by performing a sequence of 1D FFTs [10,15]. In this algorithm, input elements hold by an $N \times N$ array are stored in row-major order in the external memory such as DRAM. One major issue in the implementation of the 2D FFT architecture is the considerable delay caused by DRAM row activation which are mainly introduced by the strided memory access in the column wise 1D FFTs. To solve this problem, the authors in [2] propose a tiled data mapping method to improve the external memory bandwidth utilization. They logically divide the input $N \times N$ input array into $\frac{N}{k} \times \frac{N}{k}$ tiles and map the elements in each tile to consecutive memory locations. They conclude that the DRAM bandwidth utilization is maximized when the size of each tile is set to be the size of the DRAM row buffer. However, this solution introduces non-trivial on-chip hardware resource cost for local transposition. Various traditional 2D memory based 2D FFT architectures achieving high throughput performance have been developed in [10,16]. In [10], the authors propose a 2D decomposition algorithm which enables local 2D FFT on sub-blocks. In this way, the times of DRAM row activation is minimized. Vector radix 2D FFT in [16] presents a general structure theorem to construct a multi-dimensional FFT architecture by decomposing a large size problem into small size 2D FFTs. The external memory row activation overhead is not considered.

3D memory is expected to provide bandwidth higher than the 2D memory by an order of magnitude. There have been many works which have focused on this aspect of 3D memory. Reference[17] implements matrix multiplication and 2D FFT on a Logic-in-Memory architecture. The architecture consists of a logic layer is interleaved between two segments of memory layers to form a 3D architecture. The performance metrics are energy efficiency and bandwidth. In [8], the authors develop power efficient FFT on an architecture consisting of memory layers stacked on multiple FPGA layers. The authors focus on energy efficiency improvement while moving to a 3D architecture from a 2D architecture.

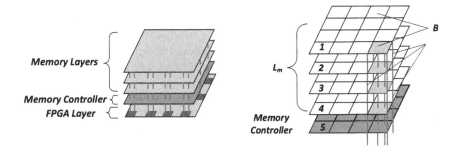

Fig. 1. (a) 3D MI-FPGA architecture (b) 3D memory

3   3D Memory Integrated FPGA (3D MI-FPGA)

Our model of 3D architecture consists of 3D memory integrated with FPGA interacting through TSVs. We extend our previous work on 3D architectures [13,14]. Here, we provide a brief overview of 3D Memory Integrated FPGA (3D MI-FPGA). The architecture consists of three components: 3D memory, FPGA and TSVs. Figure 1 illustrates the architecture of 3D memory integrated FPGA. The memory is composed of several layers (L) vertically stacked one above the other. Each of these layers is partitioned into several banks. Vaults are defined as the group of banks (1, 2, 3, 4 in Fig. 1) across layers which share a set of interconnects (TSVs). This set of banks residing on one layer which belong to the same vault (B) is analogous to the number of banks in a chip in the 2D memory. The reason being these set of banks share the bus in 2D memory and they share the TSVs in the 3D memory architecture. This set of TSVs shared by the banks in a vault is denoted by N_{tsv}. Each vault has a dedicated memory controller which handles the memory accesses to that particular vault. These memory controllers form a separate layer in the memory. Vaults can be activated at the same time as they do not share the TSVs. On the other hand, the banks in a given vault share the TSVs and the activation of these banks has to be pipelined or interleaved as in the case of 2D memory. Denoting by BW_{vault} the bandwidth of a vault, the total bandwidth of 3D memory is $V \times BW_{vault}$. The FPGA architecture is similar to that of the conventional FPGA consisting reconfigurable logic, DSP blocks, on-chip memory (Block RAM and Distributed RAM) and memory controllers. The difference is that we model the FPGA to interact with the memory through the set of TSVs connecting the FPGA and the memory. These TSVs are between memory controllers on FPGA and those in the memory. FPGA accesses the data in the memory through the TSVs which are high speed, low latency vertical interconnects. The TSVs are characterized by the number of TSVs and latency of data transfer across them. These two parameters affect the amount of data that can be transferred between memory and FPGA in a given unit of time. Each TSV can transfer 1 bit of data at a time. Therefore, higher the number of TSVs, higher the bandwidth.

3.1 Timing Parameters

Bandwidth and latency of accesses to the 3D memory depend on a certain set of timing parameters and we discuss these in this section. Data in the 3D memory is stored in rows which combine to form a bank and which group together to form a vault. Therefore, each row belongs to a specific bank and vault. When memory is accessed, depending on the address a specific row, bank and vault are activated. Therefore, although some of the parameters overlap with that of the 2D memory, certain additional parameters have to be defined taking into account the architecture and different accesses possible in the context of a 3D memory. We model the 3D memory using the following parameters:

1. $t_{diff\text{-}row}$: minimum time required between issuing two successive activate commands to different rows in the same bank
2. $t_{diff\text{-}bank}$: minimum time required between successive activate commands to different rows in different banks in same or different vaults
3. $t_{in\text{-}row}$: minimum time required between successive accesses to elements in the same row in the same bank
4. $t_{in\text{-}vault}$: minimum time required between accesses to different rows in different banks in the same vault

The values of the above parameters have a significant impact on latency and bandwidth of the 3D memory. In general, accessing data from different vaults causes zero latency. Hence, a parameter such as $t_{diff\text{-}vault}$ is not defined. This is because, since vaults are completely independent and can be active at the same time, this parameter is equal to zero. Since the banks located in different layers but belonging to the same vault can be activated in a pipeline, this latency ($t_{in\text{-}vault}$) is lower than that of accessing data from banks belonging to the same layer and same vault. Other parameters are similar to the parameters of 2D memory. Therefore, accessing data from the same row in a bank $(t_{in\text{-}row})$ is faster than accessing data from two rows in different banks $(t_{diff\text{-}bank})$. The highest latency is seen when we access data from two different rows in the same bank in the same vault denoted by $t_{diff\text{-}row}$.

4   2D FFT Architecture

4.1   1D FFT Kernel

An N-point (floating-point) 1D FFT kernel is implemented by concatenating several basic components including radix block, data path permutation (DPP) unit, and twiddle factor computation (TFC) unit. The design of each architecture component relies on the FFT algorithm in use. Implementation details of those components will be introduced next. We applied several energy optimizations discussed in [3–5] onto the design components to reduce their energy consumption. The 1D FFT kernel supports processing continuous data streams so as to maximize design throughput and the memory bandwidth utilization.

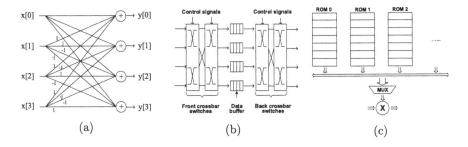

Fig. 2. (a) Radix-4 block (b) Data permutation unit (c) Twidlle factor coeffcient unit

Radix Block. The radix block is used to perform a butterfly computation on some input samples. For example, the radix block for radix-4 FFT takes four input samples, performs the butterfly computation and then generates four results in parallel. Each radix block is composed of complex adders and subtractors. The structure of a radix block is determined by the FFT algorithm in use. Figure 2a shows the structure of radix block for radix-4 FFT.

DPP Unit. DPP unit is used for data permutation between butterfly computation stages in FFT. A DPP unit is composed of multiplexers and data buffers. In subsequent clock cycles, data from previous butterfly computation stage are first multiplexed and written into several data buffers. Each stored data element will be buffered with a certain number of clock cycles and then read out. Outputs from data buffers will also be multiplexed and fed into the next butterfly computation stage. Figure 2b shows the DPP unit used for a radix-4 based FFT design. Each DPP unit consists of eight 4-to-1 multiplexers and four data buffers. In each cycle, a data buffer may be read and written simultaneously on different addresses. The size of each data buffer depends on the ordinal number of its present butterfly computation stage and the FFT problem size. Note that each data element is a complex number including both its real part and imaginary part, hence the data width is 64 bit.

TFC Unit. A TFC unit consists of two parts: the TFC generation logic and the complex number multiplier. As shown in Fig. 2c, the TFC generation logic includes several lookup tables (functional ROMs) for storing twiddle factor coefficients, where the data read addresses will be updated with the control signals. The size of each lookup table is determined by the ordinal number of its present butterfly computation stage and the FFT problem size. Each lookup table can be implemented using a BRAM or distributed RAM (dist. RAM) on FPGA [1]. Each complex number multiplier consists of four real number multipliers and two real number adders/subtractors.

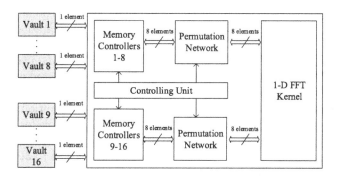

Fig. 3. 2D FFT processor architecture

The proposed 2D FFT architecture is shown in Fig. 3, in which a controlling unit (CU) and a permutation network are introduced. The permutation network is developed based on our work in [7]. The CU is responsible for reconfiguring the permutation network to achieve the dynamic data layout.

4.2   Baseline Architecture

In baseline architecture, when performing column-wise 1D FFTs, memory address is increased with a stride equals to FFT problem size N after each memory access. However, a minimum activate-to-activate delay exists when successively accessing two rows in the same bank, same vault or accessing in two banks in the same vault. This delay results in a decline in 3D memory bandwidth utilization, thus the entire system throughput is impaired.

4.3   Optimized Architecture

In the optimized architecture, the controlling unit is responsible for reconfiguring the permutation network dynamically to ensure data results of row-wise 1D FFTs are mapped onto the different vaults using the optimal dynamic data layout. Through this data remapping, vault row activation will be only needed after several successive accesses on the same row rather than every memory access. Thus, the impact of vault row activations on the entire system throughput will almost be minimized. Furthermore, to reduce the times of vault row activation, data inputs of several consecutive column-wise 1D FFTs will be moved from vaults to local memory together, without waiting for the completion of the current executed 1D FFT.

4.4   Optimal Dynamic Data Layouts

Our work in this paper is based on the dynamic data layouts (DDL) developed for the traditional 2D external memory in [12]. In this approach, the data layout in memory is dynamically reorganized during computation. After reorganizations, non-unit stride accesses are converted to unit stride accesses, thereby reducing cache misses. The data layout is *optimal* from the performance point of view as it maximizes the memory bandwidth utilization. However, the data reorganization overhead with regarding to latency and on-chip SRAM buffer consumption has not been considered. In [6] , we proposed the *optimal dynamic data layouts* for 2D memory such that peak memory bandwidth utilization is achieved with minimal data reorganization overhead. The data reorganization overhead is evaluated using the reorganizing latency and the on-chip buffer consumption. We further optimized our approach in [6] so that this technique is applicable for 3D memory based architecture. In the baseline, row major order data layouts are employed. In our approach, instead of mapping results of row-wise FFTs to 3D memory in row major order, we employ block-based dynamic data layout, and the results are read block-by-block by the column-wise 1D FFT. The dynamic data layout is organized into blocks, each of size $w \times h$. w and h represent the width and height, respectively. w is dynamically determined by the stride permutation to be performed in 1D FFTs. We assume the row buffer size in each 3D memory vault is s, the number of memory banks in each vault is b, the number of valuts to be accessed in parallel is n_v. To achieve the *optimal dynamic data layout*, h is calculated based on the equation below:

$$h = \begin{cases} n_v \cdot sb/m & \text{if } 0 < m < sb\frac{t_{diff\_row}}{t_{in\_row}}; \\ n_v \cdot t_{diff\_bank}/t_{in\_row} & \text{if } sb\frac{t_{diff\_row}}{t_{in\_row}} \leq m < sb ; \\ n_v \cdot t_{diff\_row}/t_{in\_row} & \text{if } m \geq sb. \end{cases} \quad (1)$$

Note that $w = s/h$. The permutation network will be employed for permuting the data in these blocks locally. Due to the limitation of space, we cannot give all the relevant details. For more information, please refer our previous work in [6].

4.5   Metrics of Evaluation

We evaluate the performance of 2D FFT on 3D memory integrated FPGA with respect to the metrics throughput and latency for the entire application.

Throughput: defined as the maximum bandwidth of the memory supported by the application. It is measured in Giga Bytes per second (GB/s). Since our architecture is streaming data every cycle, the bandwidth at which memory operates determines the total execution time of the application. Therefore, higher the throughput, lower the execution time.

Latency: defined as the time elapsed between accessing first input from the memory and the time at which the first output is generated by the FFT kernel. We measure latency in the unit of ns. This penalty is paid just once and at the beginning of the processing. As we employ a streaming architecture, after the first output is generated, the subsequent outputs are generated every cycle of operation.

5   Experimental Results

Before evaluating the performance of the entire design, we separately estimate the throughput for both the baseline and the optimized architecture for the 3D architecture described in Sect. 3. Table 1 shows the throughput performance of the 3D memory before and after the proposed optimization. There is no much performance difference between the baseline architecture and the optimized architecture regarding memory access by row-wise 1D FFTs. The reason for that is the system throughput is almost not affected by row-wise 1D FFTs in both architectures. From the Table 1, it shows that performance loss for column-wise FFT increases with a larger problem size. Through the proposed optimization, the peak bandwidth utilization is improved to 40.0 %, 32.0 %, and 28.8 % for **1024 × 1024**, **4096 × 4096** and **8192 × 8192** size 2D FFTs respectively.

In order to give a thorough view of the performance of the complete 2D FFT implementation, we evaluate the entire system architecture based on our memory model and our actual implementation of 2D FFT design on FPGA. Table 2 presents the throughput and latency performance comparison between

Table 1. Throughput comparison: column-wise FFT

	1024 × 1024	4096 × 4096	8192 × 8192
	2D FFT	2D FFT	2D FFT
Throughput of column-wise FFT (Baseline)	6.4 Gb/s	3.2 Gb/s	3.2 Gb/s
Peak bandwidth utilization	1.00 %	0.5 %	0.5 %
Throughput of column-wise FFT (Optimized)	32 GB/s	25.6 GB/s	23.04 GB/s
Peak bandwidth utilization	40.0 %	32.0 %	28.8 %

the baseline 2D FFT architecture and the optimized 2D FFT architecture. It shows that the optimized 2D FFT architecture achieves 32.0, 25.6 and 23.0 GB/s in throughput for 1024×1024, 4096×4096 and 8192×8192 problem sizes, respectively. The throughput performance is improved by 95.1 %, 97.0 %, 96.6 % for 1024×1024, 4096×4096 and 8192×8192 point 2D FFT, respectively. The latency is reduced by up to 3x by using our proposed optimizations. Comparing the results of the throughput in the 1D FFT kernel and the entire 2D FFT architecture, we observe that the optimization for 3D memory access makes a major contribution in the performance improvement. Moreover, the sustained throughput of the optimized 2D FFT architecture achieves up to 40 % of the *peak memory bandwidth*, which is an upper bound on the performance of the chosen FFT algorithm and 3D system architecture. Note that when calculating the *peak memory bandwidth*, we ignored the run-time behavior of the target applications.

Table 2. Performance Comparison: Entire 2D FFT application

FFT size	Baseline architecture			Optimized architecture			Performance improvement (throughput)
	Throughput (GB/s)	Latency (ns)	Data Parallelism # elements	Throughput (GB/s)	Latency (ns)	Data Parallelism # elements	
1024×1024	16.4	1.60 ms	32	32.0	524 μs	32	95.1%
4096×4096	13.0	7.48 ms	32	25.6	2.4 ms	32	97.0%
8192×8192	11.7	145.4 ms	32	23.0	46.6 ms	32	96.6%

6 Conclusion

In this paper, we proposed dynamic data layout optimizations to obtain a high throughput 2D FFT architecture on 3D memory integrated architecture. The proposed architecture achieves high throughput by maximizing and balancing the bandwidth between the external memory and FFT kernel on FPGA. By proposing the dynamic data layouts realized with the on-chip permutation network, the delay caused due to row activation overhead is highly reduced, thus leading to significant performance improvement. The experimental results comparing with the baseline architecture show that our implementation outperforms in throughput and latency. In the future, we plan to build a design framework targeted at throughput-oriented signal processing kernels, which enables automatic data layout optimizations addressing new 3D memory technologies.

References

1. Virtex-7 FPGA Family. http://www.xilinx.com/products/virtex7
2. Akin, B., Milder, P., Franchetti, F., Hoe, J.: Memory bandwidth efficient two-dimensional fast fourier transform algorithm and implementation for large problem sizes. In: 20th International Symposium on Field-Programmable Custom Computing Machines, pp. 188–191. IEEE, April 2012
3. Chen, R., Park, N., Prasanna, V.K.: High throughput energy efficient parallel FFT architecture on FPGAs. In: IEEE High Performance Extreme Computing Conference (HPEC), pp. 1–6. IEEE (2013)
4. Chen, R., Prasanna, V.K.: Energy-efficient architecture for stride permutation on streaming data. In: International Conference on Reconfigurable Computing and FPGAs, pp. 1–7 (2013)
5. Chen, R., Prasanna, V.K.: Energy efficient parameterized FFT architecture. In: International Conference on Field-programmable Logic and Application. pp. 1–7. IEEE (2013)
6. Chen, R., Prasanna, V.K.: DRAM row activation energy optimization for stride memory access on FPGA-based systems. In: Sano, K., Soudris, D., Hübner, M., Diniz, P.C. (eds.) Applied Reconfigurable Computing. LNCS, vol. 9040, pp. 349–356. Springer, Switzerland (2015)
7. Chen, R., Prasanna, V.K.: Energy and memory efficient bitonic sorting on FPGA. In: International Symposium on Field-Programmable Gate Arrays, pp. 45–54. ACM/SIGDA (2015)
8. Gadfort, P., Dasu, A., Akoglu, A., Leow, Y.K., Fritze, M.: A power efficient reconfigurable system-in-stack: 3D integration of accelerators, FPGAs, and DRAM. In: International Conference on System-on-Chip Conference (SOCC), pp. 11–16. IEEE (2014)
9. Hybrid Memory Cube Consortium: Hybrid Memory Cube Specification. http://hybridmemorycube.org/files/SiteDownloads/HMC_Specification%201_0.pdf
10. Kim, J.S., Yu, C.L., Deng, L., Kestur, S., Narayanan, V., Chakrabarti, C.: FPGA architecture for 2D discrete fourier transform based on 2D decomposition for large-sized data. In: IEEE Workshop on Signal Processing Systems, pp. 121–126. IEEE, October 2009
11. Langemeyer, S., Pirsch, P., Blume, H.: Using SDRAMs for two-dimensional accesses of long $2^n \times 2^m$-point FFTs and transposing. In: International Conference on Embedded Computer Systems (SAMOS), pp. 242–248. IEEE, July 2011
12. Park, N., Prasanna, V.: Dynamic data layouts for cache-conscious implementation of a class of signal transforms. IEEE Trans. Signal Process. **52**(7), 2120–2134 (2004)
13. Singapura, S.G., Panangadan, A., Prasanna, V.K.: Performance modeling of matrix multiplication on 3D memory integrated FPGA. In: 22nd Reconfigurable Architectures Workshop, IPDPDS. IEEE (2015) (to appear)
14. Singapura, S.G., Panangadan, A., Prasanna, V.K.: Towards performance modeling of 3D memory integrated FPGA architectures. In: Sano, K., Soudris, D., Hübner, M., Diniz, P.C. (eds.) Applied Reconfigurable Computing. LNCS, vol. 9040, pp. 443–450. Springer, Switzerland (2015)
15. Wang, W., Duan, B., Zhang, C., Zhang, P., Sun, N.: Accelerating 2D FFT with non-power-of-two problem size on FPGA. In: International Conference on Reconfigurable Computing and FPGAs, pp. 208–213. IEEE, December 2010

16. Wu, H., Paoloni, F.: The structure of vector radix fast fourier transforms. IEEE Trans. Acoust. Speech Signal Process. **37**(9), 1415–1424 (1989)

17. Zhu, Q., Akin, B., Sumbul, H.E., Sadi, F., Hoe, J.C., Pileggi, L., Franchetti, F.: A 3D-stacked logic-in-memory accelerator for application-specific data intensive computing. In: IEEE International Conference on 3D Systems Integration Conference (3DIC). pp. 1–7. IEEE (2013)

High-Performance Reconfigurable Computer Systems Based on Virtex FPGAs

Alexey I. Dordopulo[1]([⊠]), Ilya I. Levin[2], Yuri I. Doronchenko[2], and Maxim K. Raskladkin[2]

[1] Southern Scientific Centre of the Russian Academy of Sciences,
Rostov-on-Don, Russia
scorpio@mvs.tsure.ru
[2] Scientific Research Centre of Supercomputers and Neurocomputers Co. Ltd.,
Taganrog, Russia
levin@mvs.sfedu.ru, doronchenko@mvs.tsure.ru,
raskladkin@mail.ru

Abstract. The paper covers architectures and comparison characteristics of reconfigurable computer systems (RCS) based on field programmable gate arrays (FPGAs) of the Xilinx Virtex-7 family and technologies of task solving by means of software development tools. In the paper we also consider architecture and assembly of next-generation RCS with a liquid cooling system and give results of calculations and prototyping of principal technical solutions which provide the performance of 1 PFlops for a standard computational 47U rack with the power of 150 kWatt. This is promising approach because of RCS with a liquid cooling system have a considerable advantage for lot of engineering and economical parameters such as real and specific performance, power efficiency, mass and dimension characteristics, etc., in comparison with similar systems.

Keywords: FPGA · High-level programming language for reconfigurable computer systems · Programmable soft-architecture · Architecture description language

1 Introduction

One of the promising approaches which provide achievement of high real performance of a computer system is adaptation of its architecture to the structure of the solving task and creation of a computing device which performs structural and procedural fragments of calculations with the same efficiency. That is why domestic [1] and foreign vendors of computers use field programmable gate arrays (FPGAs) more and more frequently. FPGAs speedup calculations of computationally laborious fragments. It is possible to create stand-alone accelerators which contain one or two FPGAs, or computing

This work was financially supported in part by the Russian Ministry of Education under Grant RFMEFI57814X0006

© Springer International Publishing Switzerland 2015
V. Malyshkin (Ed.): PaCT 2015, LNCS 9251, pp. 349–362, 2015.
DOI: 10.1007/978-3-319-21909-7_35

complexes. Such corporations as Nallatech [2] and Pico Computing, Inc. [3] produce accelerators and base boards with small number (up to 4) of FPGAs which are used as components of servers and heterogeneous clusters of such vendors as Hewlett-Packard and IBM.

Convey Computer [4] and Maxeler Technologies [5] create hybrid supercomputers based on their own heterogeneous cluster nodes, Each node can contain 1-4 FPGAs and some general-purpose processors. The company SRC Computers [6] uses a similar solution. It produces nodes such as MAP processor for 1U, 2U and 4U MAPstation rack. MAPstation 1U contains one MAP processor. MAPstation 2U contains up to three MAP processors. MAPstation 4U can contain up to 10 different nodes such as MAP processor, a node with a general-purpose microprocessor or a memory node.

Scientific team of Scientific Research Centre of Supercomputers and Neurocomputers (SRC SC & NC Co. Ltd., Taganrog, Russia) together with scientists of Scientific Research Institute of Multiprocessor Computer Systems at Southern Federal University (SRI MCS SFU, Taganrog, Russia) and Southern Scientific Centre of the Russian Academy of Sciences (SSC RAS, Rostov-on-Don, Russia) design and produce reconfigurable computer systems, similar to supercomputers which contain a set of FPGAs, united into computational fields by high-speed data transfer channels, and considered as the principal computational resource.

The range of designed and produced items is rather various: from completely stand-alone small-size reconfigurable accelerators (computational blocks), desk-top computational modules or computational modules which are to be placed into a rack, to computer systems, which can consist of several computational racks and must be placed in a specially equipped computer room.

Reconfigurable computer systems which contain large computational FPGA-fields, are usually used for solving computationally laborious problems in various fields of science and technology, because of several considerable advantages in comparison with cluster-type multiprocessor computer systems such as high real and specific performance, high power efficiency, etc.

2   RCS Based on Xilinx Virtex-7 FPGAs

Reconfigurable computational block Celaeno. On basis of Xilinx Virtex-7 FPGAs we have designed a reconfigurable computational 1U block (RCB) Celaeno intended for data processing. Data is received via Gigabit Ethernet channel with no support of IP-protocols.

Figure 1a shows the RCB Celaeno. Figure 1b shows the open block (no top cover) and its printed circuit board. The RCB Celaeno is produced in two modifications: Celaeno-K based on Kintex-7 XC7K160T FPGAs and Celaeno-V based on Virtex-7 XC7VX485TFPGAs. Specifications of the RCB Celaeno of these two modifications are given in Table 1.

The RCB Celaeno contains 6 FPGAs of the computational field, an embedded host-computer, a power supply system, a control system, a cooling system and other subsystems. All FPGAs of the computational field are connected according a

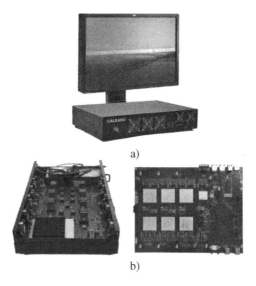

Fig. 1. RCB Celaeno and its components

Table 1. Specifications of the RCB Celaeno-K and Celaeno-V

Parameter	Celaeno-K	Celaeno-V
Number and type of FPGAs	6	6
Total number of equivalent gates in the FPGAs of the computational field, million	96	288
RAM size, Gb	3	3
Performance of the computational module Pi_{32}/Pi_{64}, GFlops	150/75	440/220
Working frequency, MHz	330	400
Rate of data exchange via Ethernet channel, Gb/sec	1	1
Rate of data transfer via LVDS between FPGAs of the computational field, MHz	900	1200
Power, Watt	200	320
Dimensions, mm	480 × 270 × 70	480 × 270 × 70
Cost, million rubles	1.3	2.0

lattice-like structure by LVDS-channels, and each FPGA is connected to its own units of dynamic memory of 256 Mbyte each.

To control and configure the computational field of the RCB an embedded computer (computer-on-module of the Kontron COM-Express family) is used. It is placed on the printed circuit board of the computational module. It provides connection with peripheral devices, development and debugging of parallel applications of computationally laborious tasks, generation of initial data files, which, together with the executable file of the application, are loaded into the computational field via PCI-Express bus and LVDS-channel. When the task is done, its results are transferred into the COM-Express processor unit.

Possible areas of application of the RCB Celaeno are symbolic processing, mathematical physics, simulation and computational experiment, digital signal processing, linear algebra, etc.

Reconfigurable. According to the state contract №14.527.12.0004 from 03.10.2011 the scientific team of SRI MCS SFU designed a reconfigurable computer system RCS-7 based on Virtex-7 FPGAs, which contains a computational field of 576 Virtex-7 XC7V585T-FFG1761 FPGAs (58 million of equivalent gates each), assembled into one 47U computational rack with the peak performance of 10^{15} fixed-point operations per second. The principal structural component of the RCS-7, intended for placement into a standard 19" computational rack, is a computational module (CM) 24V7-750 (CM Pleiad), which contains 4 boards of the computational module (BCM) 6V7-180 (see Fig. 2); a control unit CU-7; a power supply subsystem; a cooling subsystem, and other subsystems. Figure 2 shows the CM 24V7-750.

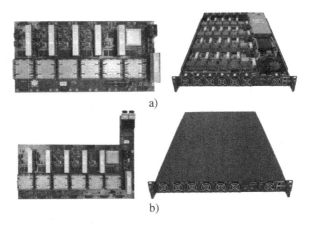

a)

b)

Fig. 2. Computational module (CM) 24V7-750 (a – boards of the CM Pleiad, b – CM Pleiad with no top cover/with a top cover)

Each board of the CM 24V7-750 contains 6 Virtex-7 XC7V585T-1FFG1761 FPGAs of the computational field, connected sequentially, and 12 MT47H128M16HR-25E chips of distributed dynamic memory, organized as 128 M*16 with read/write frequency up to 400 MHz. The total size of distributed dynamic memory is 12 GByte. Data can be transferred between the FPGAs via 144 LVDS differential lines at frequency of 800 MHz.

The performance of the one board is 645.9 GFlops for processing of 32-digit floating point data, and the performance of the CM 24V7-750 is 2.58 TFlops for processing of 32-digit floating point data.

Reconfigurable computational module Taygeta. The scientific team of SRC SC & NC has designed a 19" 2U computational module Taygeta, based on Virtex-7 FPGAs and intended for high-performance multirack RCSs. Figure 3a shows the CM Taygeta, which contains 4 boards 8V7-200, an embedded host-computer, a power supply system, a control system, a cooling system, and other subsystems. The boards of the CM Taygeta are connected by LVDS-channels, running at frequencies up to 1000 MHz. Figure 3b shows the board 8V7-200.

a)   b)

Fig. 3. The CM Taygeta (a – the CM Taygeta without top cover, b – BCM 8V7-200)

The board of the computational module (BCM) 8V7-200 is a 20-layer printed circuit board with double-side mounting of elements. It contains 8 XC7VX485T-1FFG1761 FPGAs (48.5 million equivalent gates each), 16 chips of distributed memory DDR2 SDRAM with total capacity of 2 GByte, LVDS and Ethernet interfaces, and other components.

The performance of one BCM 8V7-200 is 667 GFlops for processing of 32-digit floating point data, and the performance of the CM Taygeta is 2.66 TFlops, respectively.

RCS based on CM Pleiad and CM Taygeta. On the base of already considered CM Pleiad, in 2013 we had designed a reconfigurable computer system RCS-7 (Fig. 4a), which contained 24 computational modules, and which can be extended up to 36 computational modules. The performance of RCS-7, when it contains from 24 to 36 24V7-750 CMs is from 62 to 93 TFlops for processing of 32-digit floating point data, and 19.4 ÷ 29.4 TFlops for processing of 64-digit floating point data, respectively. Fields of application of RCS-7 and RCS-7-based computer complexes are digital signal processing and multichannel digital filtering (Ali M. Reza, 2013; Mazher et al., 2013).

Figure 4b shows an RCS, designed on the base of the CM Taygeta. The performance of its one rack, which contains 18 CMs Taygeta is 48 TFlops for processing of floating point data with single precision, and 23 TFlops for processing of 64-digit floating point data.

High-performance RCSs based on the CM Taygeta are intended for solving computationally laborious problems of science and industry, drug design and symbolic processing, and for such problems they provide a significant advantage of the majority of technical and economical parameters such as specific performance, power efficiency, etc., in comparison with cluster-type multiprocessor computer systems.

a)   b)

Fig. 4. RCS based on Xilinx Virtex-7 FPGAs (a – RCS-7 on the base of the CM Pleiad, b – RCS-7 on the base of the CM Taygeta)

3   Next-Generation Reconfigurable Systems Based on Xilinx UltraScale FPGAs

Further development of open scalable architecture (Levin, 2010), used for design of RCSs based on Xilinx Virtex-7 FPGAs, is a variety of next-generation components for new designed products – Xilinx FPGAs of a new generation family UltraScale, based on 20 nm technol. In comparison with FPGAs of Virtex-7 family they have lower power consumption and higher performance.

Reconfigurable Computational Block Based on UltraScale FPGAs. The designed RCB Celaeno-U will also be produced as a 1U block, but in contrast to its precursors Celaeno-K and Celaeno-V, it will contain 4 Xilinx UltraScale XCVU095 FPGAs (95 million equivalent gates each), which will create a computational field of 380 million equivalent gates in total. Figure 5 shows the structure chart of the RCB Celaeno-U and assembly outline of the board.

The keys of Fig. 5:

- DD1-DD4 – the computational Xilinx UltraScale XCVU095 FPGAs;
- DD5 – Xilinx UltraScale XCKU040 FPGA of the BCM controller;
- A1-A9 – distributed memory modules;
- X2-X4, X7-X12 – connectors of different types of interfaces.
- In comparison with the previous version of the RCB Celaeno-V the performance of the RCB Celaeno-U will increase in 1.7–1.8 times while its power will grow not more than in 1.3 times.

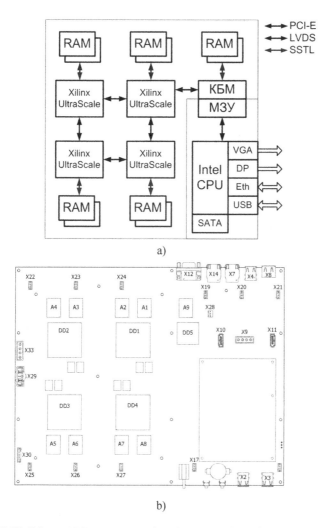

Fig. 5. RCB Celaeno-U (a – structure chart, b – assembly outline of the RCB board)

RCS with Liquid Cooling Based on UltraScale FPGAs. The time of air cooling systems, used in modern high-performance computer systems and supercomputers, designed on their basis, including reconfigurable supercomputers, is practically over. The majority of computer designers are oriented to liquid cooling systems which will help to solve problems of cooling of the designed computer complexes. It is reasonable to use liquid cooling, particularly submersion of boards of computational modules into a liquid cooling agent (mineral oil), for computational modules of RCSs designed on the base of next-generation FPGA families.

The direction of design of next-generation RCSs based on liquid cooling is actively developed in SRC SC & NC. New designs of printed boards and computational modules with high board density are designed. Specifically, next-generation computational modules Scate-8 for multirack RCSs of super-high performance are designed at present.

The board of the next-generation computational module contains 8 Virtex Ultra-Scale FPGAs (not less than 100 million equivalent gates each). The computational module consists of two sections: the first section contains 16 boards of the computational module with the power of up to 800 Watt each, completely submerged into electrically neutral liquid cooling agent. The second section contains a pump system and a heat-transfer device, which provide flow and cooling of the cooling agent. Figure 6a shows the 3U CM outline.

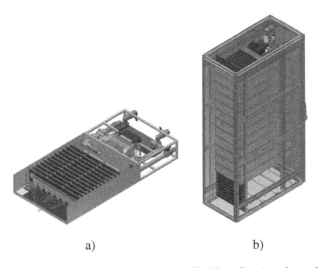

a)   b)

Fig. 6. The outline of the computer system based on liquid cooling (a – the outline of the CM Scate-8, b – the outline of the Scate-8 based computational rack)

According to performed analysis, use of liquid cooling and creation of computer systems on the base of the CM Scate-8 provide more than petaflops-like performance of a single computational rack of the RCS. The computational 19" rack of the super-computer can contain up to 12 CM Scate-8 with liquid cooling. Figure 6b shows the outline of the rack. Table 2 contains the performance and the power of the next-generation RCS.

Table 2. The Performance and Power of the Next-Generation RCS on the Base Xilinx UltraScale FPGAs

Parameter	Value
Performance of the CM Scate-8	105 TFlops
Performance of the computational rack based on the CM Scate-8	1 PFlops
Power of the CM Scate-8	13 kWatt
Power of the computational rack based on the CM Scate-8	150 kWatt

Table 3. Performance of Reconfigurable Supercomputers

Product, year of production, FPGA family	Board performance Pi_{32}/Pi_{64}, GFlops	CM performance Pi_{32}/Pi_{64}, GFlops	Performance of 47U rack Pi_{64}, TFlops
Taygeta, years 2012/2013, Virtex-7	900/300	3600/1200	68–100
Scate, years 2015/2016, UltraScale	7250/2500	82500/30000	1000–1250

On basis of reconfigurable systems, produced in SRC SC & NC, it is possible to watch growth rates of RCS performance when the FPGA family is changed.

In 2015–2016 on base of the described design we will create super-high-performance computer complexes with effective cooling of computational FPGAs both of the UltraScale family and of the next-generation FPGA family.

4 RCS Software

At present there are plenty of various development suits for development of structural solutions of applied tasks for FPGAs. The most popular suits which can be used as separate development tools and as parts of some complexes are synthesizers, developed by FPGA vendors: ISE and Vivado (Xilinx, Inc.) [8], Quartus II (Altera Corporation) [9] and Actel Libero IDE (Actel Corporation) [10]. These software tools, besides the development environment of digital devices, contain a number of utilities: analyzers of timing characteristics, placing editors, FPGA programming units, systems of digital device simulation, etc. Owing to a wide range of tools these development suits provide a complete cycle of digital device development within single FPGA: development of the initial description of the project, synthesis, simulation, placement, tracing, chip programming.

Continuous growth of FPGA capacity makes design of applied task solutions for FPGAs by means of hardware description languages (VHDL, AHDL, Verilog, etc.) [11] and design of digital devices by means of graphic editors more and more laborious. That is why at present the leading vendors of FPGAs and reconfigurable computers are oriented to high-level languages. As a result, the new development environment Vivado by Xilinx, Inc. contains a new design tool Vivado HLS, based on a high-level language. The development kit Altera SDK [12], used for Altera FPGA design, contains tools for a new standard OpenCL of parallel programming of heterogeneous systems. These solutions use translators of C-like languages, which generate code in the hardware description languages on the register transfer level (RTL, C-to-RTL) from the program in some C-like high-level programming language.

In spite of similarity of syntaxes of C-like languages with the C language, such approach does not mean that initial C-code, developed for a PC or a cluster computer system will be correctly interpreted by C-to-RTL translators. The language C was chosen as a basic one because of its wide popularity, which makes mastering of new FPGA application development and design tools much easier.

In addition, when we use C-to-RTL translators the whole application or its explicitly selected procedures are translated into RTL-descriptions of single FPGAs. Such development suits have no tools of automatic decomposing of the parallel program into fragments for a set of interconnected FPGAs.

When we use Vivado HLS, the project is designed within one FPGA, and if the application developer needs hardware resource more than the resource of one FPGA, then he himself must distribute calculations between several projects for each FPGA and synchronize control and data streams between them.

The OpenCL standard is used by the company Nallatech (vendor of reconfigurable computers) and allows use of several FPGAs in one project. In this case solutions in FPGAs are programmed by means of functions, called from the library of tools of Altera SDK. Each FPGA involved in computational process performs calculations described by a certain fragment of code. So, the program written according to the OpenCL standard is a basic code, written for traditional processors, and some separate fragments of code, written for FPGAs, involved into computational process as co-processors. In this case the problem of data synchronization is responsibility of the programmer.

Another well-known FPGA programming tool is a complex created by the company Mitrionics Inc., which contains a Mitrion Virtual Processor (MVP), programmed by means of the high-level programming language Mitrion-C, and a library of functions MithalAPI included in the development kit Mitrion SDK [13] for development of host-programs. The developed Mitrion-C program must be completely realized on a single virtual processor MVP. It is impossible to program multichip RCSs, and as a result, it considerably reduces effectiveness of the software complex of the company Mitrionics Inc. To program multichip RCSs which consist of interconnected FPGAs the programmer himself must realize an interface (protocol) of data exchange between FPGAs and solve problems concerning data flow synchronization. In this case the RCS program degenerates into a program for a cluster (a set of MVP), implemented in FPGAs, and it considerably reduces effectiveness of tasks realized on multichip RCSs.

5   Language COLAMO and Software Complex for Multichip RCS

An alternative approach to RCS programming is suggested in SRI MCS SFU which deals with design of multichip reconfigurable computer systems of various architectures and configurations for more than 15 years.

The experience of SRI MCS SFU in solving problems of various types has proved that effective solving of modern laborious problems requires programming tools which can provide:

– programming in a high-level programming language;
– support of multichip programming;
– high operating frequency of FPGAs;
– high density of placement in FPGAs;

- support of pipeline and macropipeline organization of calculations. Specialists of SRI MCS SFU have developed and widely used a software complex, which consists of:
- a translator of the programming language COLAMO, which translates of the initial code written in COLAMO into an information graph of a parallel application;
- a synthesizer Fire!Constructor of scalable circuit solutions on the level of FPGA logic gates, which maps the information graph, generated by the translator of the COLAMO-language, on an RCS architecture, places the mapped solution into FPGAs and provides automatic synchronization of the fragments of the information graph in different FPGAs;
- a library of IP-cores, which correspond to operators of the COLAMO-language (self-contained structurally implemented hardware devices) for various problem domains, and interfaces which match the rate of data processing and connect all components into a single computing structure;
- debugging tools, access tools, and tools of monitoring of RCS condition.

The high-level language COLAMO is intended for description of the parallel algorithm and creation of a special-purpose computing structure, generated according to the principles of structural procedural organization of calculations [1, 14, 15], within the RCS architecture. Such computing structure implies sequential change of structurally (hardwarily) implemented fragments of the information graph of the task. Each fragment is a computational data flow pipeline. So, the application (applied task) for the RCS consists of the structural component, represented as a set of hardwarily implemented fragments of calculations, and of the procedural component, represented as a control program of sequential change of computing structures and organization of dataflows. The control component is, one and the same for all structural fragments. To provide such organization of calculations the programming language contains such structure as "cadr". A cadr is a program-indivisible component, a set of operators implemented as arithmetic-logic instructions and read/write instructions, performed on various functional devices, interconnected according to the information structure of the algorithm.

The language COLAMO has no explicit forms of parallelism description. Parallelization is provided by declaration of types of access to variables and by indexing of array items, which is typical for data flow languages. To address to data it is possible to use two principal access methods: parallel access (declared by Vector type) and sequential access (declared by Stream type). The degree of parallelism is defined according to the minimal value of the parameter of parallelization. For Stream type the degree of parallelism is 1, and for Vector type it is defined according to the minimal value of Vector type of each array, involved in computing process. For parallel type of access it is possible to process concurrently all dimensions of arrays, declared as Vector. In this case the hardware resource, needed for calculations, will grow, but the processing time will drop down.

Multidimensional data arrays can have plenty of dimensions. Each dimension can have sequential or parallel access type, declared by keywords Stream or Vector, respectively. Change of access type allows very simple control of the degree of parallelism of calculations on the level of data structure description, the processing rate, and the occupied resource. Owing to this, the programmer can describe various types of parallelism in a rather short form.

Besides the access type, the variable in the language COLAMO also has type of storage: memory (Mem), register (Reg) and commutation (Com).

The memory variable is stored in a cell of distributed memory, and hence it keeps its value till the next reassignment. For the memory variable it is possible to perform only one process at the same time. That is why, according to semantics of the COLAMO language, in any cadr any memory variable complies with two rules: the single-assignment rule and the rule of single substitution. The single-assignment rule means that the memory variable changes its value only once in the cadr. The rule of single substitution means that the variable in the cadr can be used for only one process of reading or writing.

To describe connections between the elements of the information graph of the task the COLAMO language has switching variables. Since the switching variable describes information connections, it requires no computational hardware resource for itself. It is impossible to get access to the value of the switching variable when the cadr is done. The translator needs switching variables to define information dependencies during generation of the computing structure of the task. As memory variables, switching variables comply the single-assignment rule, but not the rule of single substitution. Owing to use of switching variables data flows can be easily forked and duplicated, but it is impossible to create recursion.

To realise recursion the COLAMO language has a register variable, which is a hardware register used to store intermediate data, received during computational process. The single-assignment rule is the only restriction for register variables in the cadr.

To translate the program written in the high-level language COLAMO means to generate a circuit configuration of the computer system (a structural component) and a parallel program which controls data flows (a stream component and a procedural component) [1,14,15]. To generate the structural component means to create a computational graph which corresponds to information dependencies between results of calculations. In this case for each operation, used in the program, a specialized computing unit is substituted according to data access, data types, their capacity, etc. The synthesized information graph of the task is transferred to the synthesizer Fire!Constructor for mapping on the multichip RCS hardware resource [16].

The problem of automatic mapping of the parallel program on the multichip RCS hardware resource consists of three steps: partition of the information graph into disjoint subgraphs, placement of the subgraphs into RCS FPGAs, and tracing of external connections of the placed subgraphs within the RCS communication system.

The result of the synthesizer Fire!Constructor is a set of files of VHDL-descriptions, time constraints, and user constraints. VHDL-files describe structural implementation of the fragments of the parallel program. These files and the library of circuit components are the basis for projects, created by the synthesizer ISE for each single FPGA. Then the synthesizer ISE generates FPGA bitstream files which are loaded into the RCS.

The COLAMO-application is developed within a single project and can be translated for any RCS, which has a description and all required libraries, included into the RCS software suit. In contrast to other existing RCS application development suits, the programmer has no need to define in the text of the program, which fragments and in which FPGAs will be performed. The synthesizer Fire!Constructor provides automatic splitting of the computing structure of the COLAMO-program into several projects by means of the synthesizer Xilinx ISE, and, in addition, it provides synchronization of data flows both inside each FPGA and between them.

6 Conclusion

According to Table 3, FPGAs as principal components of reconfigurable supercomputers provide a permanent, practically linear growth of the RCS performance and give new prospects of creation of supercomputers of petaflops performance. It is possible to claim that design solutions used for the next-generation computational modules, based on Xilinx Virtex UltraScale FPGAs, will help to concentrate a powerful computational resource in a single 47U computational rack and to provide the specific performance of the RCS, based on Xilinx Virtex UltraScale FPGAs, on the level of the best world characteristics for cluster supercomputers. Owing to this, UltraScale-based RCSs can be considered as a basis for the next-generation high-performance computer complexes, which provide high efficiency of calculations and practically linear growth of performance for extending computational resource.

References

1. Kalyaev, I.A., Levin, I.I., Semernikov, E.A., Shmoilov, V.I.: Reconfigurable multipipeline computing structures. Nova Science Publishers, New York (2012)
2. Nallatech, a subsidiary of Interconnect Systems Inc.,http://www.nallatech.com/
3. Picocomputing. http://picocomputing.com/
4. Convey computer. http://www.conveycomputer.com
5. Maxeler Technologies. http://www.maxeler.com/
6. SRC computers. http://www.srccomp.com/
7. Levin, I.I.: Reconfigurable computer systems with open scalable architecture. In: Proceedings of the 5th International Conference Parallel calculations and Control Problems, PACO 2010, pp. 83–95. V.A. Trapeznikov Institute of control problems of the Russian Academy of Sciences, Moscow (2010)
8. Zotov, V.I.: Design of digital devices based on XILINX FPGAs using WebPACK ISE. Goryachaya liniya-Telekom, Moscow (2003)
9. Quartus II Handbook Version 10.1 Volume 1: Design and Synthesis. Altera Corporation (2010)
10. Libero IDE v9.1 User's Guide. Actel Corporation (2010)
11. Design for Xilinx FPGAs using high-level languages and Vivado HLS. J. Components and technologies. 12, (2013)
12. Altera measurable advantage. http://www.altera.com/literature/lit-opencl-sdk.jsp
13. Mitrionics. http://www.mitrionics.com/
14. Kalyaev, I.A., Levin, I.I., Dordopulo, A.I., Slasten, L.M.: Reconfigurable Computer Systems Based on Virtex-6 and Virtex-7 FPGAs. In: IFAC Proceedings Volumes, Programmable Devices and Embedded Systems, vol. 12, part 1, pp. 210–214 (**ISSN** 14746670) (2013)
15. Kalyaev, I.A., Levin, I.I., Dordopulo, A.I., Slasten, L.M.: FPGA-based reconfigurable computer systems. In: Science and Information Conference (SAI), pp. 148–155. London, 7–9 October 2013

16. Gudkov, V.A., Gulenok, A.A., Kovalenko, V.B., Slasten, L.M.: Multi-level Programming of FPGA-based Computer Systems with Reconfigurable Macroobject Architecture. In: IFAC Proceedings Volumes Programmable Devices and Embedded Systems, vol. 12, part 1, pp. 204–209 (**ISSN** 14746670) (2013)
17. Gudkov, V.A., Gulenok, A.A., Dordopulo, A.I., Slasten, L.M.: programming tools of reconfigurable multiprocessor computer systems. In: TSURE Proceedings "Intelligent and multiprocessor systems", 16(71), pp. 16–20. TSURE Publishing, Taganrog (2006)
18. Semernikov, E.A., Kovalenko, V.B.: Organization of multilevel programming of reconfigurable computer systems. Herald of computer and information technologies. 9, pp. 3–10. Mashinostroyeniye, Moscow (2011)
19. Gudkov, V.A., Gulenok, A.A., Kovalenko, V.B., Slasten, L.M.: Preprints of the 12th IFAC Conference on Programmable Devices and Embedded Systems PDES, pp. 65–70. Technical University of Ostrava, Czech Republic (2013)

Parallelizing Biochemical Stochastic Simulations: A Comparison of GPUs and Intel Xeon Phi Processors

P. Cazzaniga[1], F. Ferrara[2], M.S. Nobile[2], D. Besozzi[3(✉)], and G. Mauri[2]

[1] Dipartimento di Scienze Umane e Sociali, Università degli Studi di Bergamo,
Piazzale S. Agostino 2, 24129 Bergamo, Italy
paolo.cazzaniga@unibg.it
[2] Dipartimento di Informatica, Sistemistica e Comunicazione,
Università degli Studi di Milano-Bicocca, Viale Sarca 336, 20126 Milano, Italy
{nobile,mauri}@disco.unimib.it
[3] Dipartimento di Informatica, Università degli Studi di Milano,
Via Comelico 39, 20135 Milano, Italy
besozzi@di.unimi.it

Abstract. Stochastic simulations of biochemical reaction networks can be computationally expensive on Central Processing Units (CPUs), especially when a large number of simulations is required to compute the system states distribution or to carry out advanced model analysis. Anyway, since all simulations are independent, parallel architectures can be exploited to reduce the overall running time. The purpose of this work is to compare the computational performance of CPUs, general-purpose Graphics Processing Units (GPUs) and Intel Xeon Phi coprocessors based on the Many Integrated Core (MIC) architecture, for the execution of Gillespie's Stochastic Simulation Algorithm (SSA). To this aim, we consider an *ad hoc* implementation of SSA on GPUs, while exploiting the peculiar capability of MICs of reusing existing CPUs source code. We measure the running time needed to execute several batches of simulations, for various biochemical models of increasing size. Our results show that in all tested cases GPUs outperform the other architectures, and that reusing available code with the MICs does not represent a clever strategy to fully leverage Xeon Phi horsepower.

1 Introduction

Mechanistic models of biochemical reaction networks are more and more becoming a valuable mean to elucidate the functioning of complex biological systems [1,2], thanks to their capability of achieving a global-level understanding of emergent dynamics. For this purpose, efficient and reliable algorithms are needed to simulate the temporal evolution of biochemical reaction networks. In the case of *stochastic* mechanistic models, one of the most applied approach relies on Gillespie's Stochastic Simulation Algorithm (SSA), based on the stochastic formulation of chemical kinetics [3], that was proven to be equivalent to the Chemical Master Equation

© Springer International Publishing Switzerland 2015
V. Malyshkin (Ed.): PaCT 2015, LNCS 9251, pp. 363–374, 2015.
DOI: 10.1007/978-3-319-21909-7_36

[4]. However, SSA is computationally very expensive, and the same holds for other stochastic simulation algorithms thereafter introduced to improve the computational efficiency (see, e.g., [5]). Moreover, a typical analysis of stochastic models usually requires the execution of a large number of independent simulations, in order to explore the multi-dimensional space of kinetic parameters (see, e.g., the work presented in [6] concerning the problem of parameter estimation). Therefore, an efficient strategy for the parallelization of SSA is essential to achieve an effective reduction of the computational costs.

The traditional methods to parallelize algorithms consist in multi-threading [7], distributed computing on clusters [8], custom circuitry produced with Field Programmable Gate Array (FPGA) [9], or general-purpose Graphics Processing Unit (GPGPU) computing [10–12]. Anyway, these technologies require either a custom implementation of the code, that cannot be directly ported on a parallel architecture, or a scheduler to manage the parallel execution of processes over a distributed architecture. An alternative solution is nowadays represented by the family of Intel Xeon Phi coprocessors, based on the Many Integrated Core (MIC) architecture. One of the most important features of the MIC architecture is that the code implemented for Intel Central Processing Units (CPUs) can be directly compiled and executed on the Xeon Phi coprocessors.

The aim of this work is to compare the computational performance of CPUs, GPUs and Intel Xeon Phi coprocessors for the execution of increasing numbers of SSA simulations, for the analysis of stochastic models of different biochemical systems. To the best of our knowledge, this is the first attempt in the evaluation of the Xeon Phi coprocessors performance for such kind of tasks.

Several works previously investigated the performance of the Intel Xeon Phi coprocessors and compared its speed-up with respect to other parallel architectures, showing different outcomes according to the investigated problem. For instance, [13] presented a comparison between Xeon Phi 5100P and Nvidia Tesla K20s video card for the the simulation of spin systems, highlighting that a careful implementation of the C code allows the Xeon Phi to compete with the GPU. The parallelization of non-bonded electrostatic computation for Virtual Screening was introduced in [14], showing that Nvidia Tesla K20x outperforms Xeon Phi 5100. In particular, this work emphasized that OpenMP source code must be optimized to improve performance. In [15] the authors presented the performance comparison for a tracking algorithm based on the Hough transform, executed on a multi-core Intel CPU, an Nvidia Tesla K20c GPU, and an Intel Xeon Phi 7120 coprocessor. Their results highlighted that, for this particular problem, the CPU performs better than both GPU and Xeon Phi, and the authors conclude that an implementation with offloaded calculations to the coprocessors might be desirable to achieve better performances. A multi-threaded version of an algorithm to tackle the tensor transpose problem was given in [16], implemented for a multicore CPU, the Intel Xeon Phi and the Nvidia Tesla K20x. The results showed a significant speed-up achieved with the multicore CPU and the Xeon Phi coprocessor with respect to the Nvidia GPU, since the optimization of L1 cache is easier than the implementation of a coalesced global memory access on the GPU. Finally, [17] presented a comparison of the acceleration on Xeon Phi 5110P and Nvidia Tesla K20x for protein docking calculation based on the fast Fourier transform. In this case, the GPU proved to be 5 times faster than the Xeon Phi coprocessor, considering that they required comparable implementation costs.

In this work, we compare the performance of MIC, GPUs and CPUs for the execution of different batches of parallel simulations of a mechanistic model of prokaryotic gene regulation [18], as well as different synthetic models of increasing size. So doing, we can also evaluate the impact of the model size on the computational performance, irrespective of any actual dynamical properties (i.e., oscillations, bistability, etc.) that the biochemical system could present. In addition to the computational time, we evaluate the costs and power consumption of the hardware employed, and discuss the effort to port the existing code on parallel architectures.

The paper is structured as follows. In Sect. 2 we briefly introduce SSA, and describe the GPU and MIC parallel architectures. In Sect. 3 we show the speedup obtained by Xeon Phi and Nvidia Tesla K20s with respect to CPU and discuss the comparison of their computational performance. Finally, in Sect. 4 we conclude with some final remarks and future directions of this work.

2   Methodology

2.1   Stochastic Simulation of Biochemical Reaction Networks

A classic mathematical approach for the investigation of biochemical reaction networks consists in defining a system of coupled Ordinary Differential Equations (ODEs), which describe the rate of change of each molecular species according to all the chemical reactions where species appear either as reactant or product. The dynamics of this system can then be simulated by means of numeric solvers [19]. However, when the biological system is characterized by molecular species occurring in low amounts, the randomness affecting the temporal evolution cannot be neglected. In this case, stochastic algorithms represent a valuable alternative to ODEs, which are not appropriate to capture the effects of stochastic processes [2]. One of the most used procedures is the Stochastic Simulation Algorithm (SSA) [3], which allows to achieve an exact reproduction of the temporal evolution of biochemical networks, under the following assumptions: the reactions occur within a *single volume*, whose physical conditions (e.g., pressure, temperature) remain constant during the whole simulation time; the volume is *well-stirred*, that is, molecules are uniformly distributed inside the reaction volume; the amount of each molecular species is represented by the (integer) number of molecules.

According to the stochastic formulation of chemical kinetics [3], a mechanistic model of a biochemical network can be defined by specifying the set of N molecular species, $\mathcal{S} = \{S_1, \ldots, S_N\}$, interacting through a set of M chemical reactions, $\mathcal{R} = \{R_1, \ldots, R_M\}$. A reaction is formally defined as

$R_j \colon \sum_{i=1}^{N} \alpha_{ji} S_i \xrightarrow{c_j} \sum_{i=1}^{N} \beta_{ji} S_i$, where c_j is a stochastic constant encompassing the physical and chemical properties of R_j, and $\alpha_{ji}, \beta_{ji} \in \mathbb{N}$ are the stoichiometric coefficients associated, respectively, to the i-th reactant and to the i-th product of the j-th reaction, for $i = 1, \ldots, N$ and $j = 1, \ldots, M$. The state of the system at time t is represented by the vector $\mathbf{x} = \mathbf{x}(t) \equiv (x_1(t), \ldots, x_N(t))$, where $x_i(t) \in \mathbb{N}$ denotes the number of molecules of species S_i occurring in the system at time t. Given the system state $\mathbf{x}$, SSA identifies the reaction to execute in the next time interval $[t, t + \tau]$, by calculating the probability of each reaction R_j to occur in the next infinitesimal time step $[t, t + dt]$. This probability is proportional to $a_j(\mathbf{x})dt$, being $a_j(\mathbf{x}) = c_j \cdot d_j(\mathbf{x})$ the *propensity function* of reaction R_j, where $d_j(\mathbf{x})$ is the number of distinct combinations of the reactant molecules in R_j occurring in state $\mathbf{x}$. The time τ before a reaction takes place is computed as $\tau = \frac{1}{a_0(\mathbf{x})} \ln\left(\frac{1}{\rho_1}\right)$, where $a_0(\mathbf{x}) = \sum_{j=1}^{M} a_j(\mathbf{x})$. The index j of the reaction to be executed is the smallest integer in $[1, M]$ such that $\sum_{j'=1}^{j} a_{j'}(\mathbf{x}) > \rho_2 \cdot a_0(\mathbf{x})$. Finally, ρ_1, ρ_2 are random numbers sampled in $[0,1]$ with a uniform probability. We refer to [3] for further details.

The simulation of a stochastic model by means of SSA can be very time consuming. Moreover, a single SSA run is in general not sufficient to properly investigate the dynamics of a biochemical system, which usually requires the collection of multiple temporal evolutions to assess the probability distribution of chemical species at a given time. Both these issues motivate the development of accelerated stochastic simulators.

2.2   Graphics Processing Units

The emerging field of GPGPU computing allows developers to exploit the great computational power of modern multi-core GPUs, by giving access to the underlying parallel architecture that was conceived for speeding up real-time three-dimensional computer graphics [20]. GPUs represent a valuable alternative to traditional high-performance computing infrastructures, since they are characterized by low costs and a reduced energy consumption, allowing the access to tera-scale computing on common workstations of mid-range price. Nevertheless, a direct porting of sequential code on the GPU is most of the times unfeasible, due to the innovative architecture and the intrinsic limitations of this technology. As a consequence, the full exploitation of GPUs computational power and massive parallelism is still challenging [21].

Among the existing libraries for GPGPU computing, Nvidia's Compute Unified Device Architecture (CUDA) is probably the most mature. CUDA is a parallel computing platform and programming model introduced by Nvidia in 2006, which combines the Single Instruction Multiple Data (SIMD) architecture with multi-threading and automatically handles the conditional divergence between threads. However, this flexibility has a drawback, since any divergence of the execution flow among threads results in a serialization of instructions execution, affecting the overall performances.

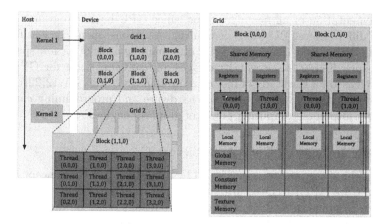

Fig. 1. Architecture of CUDA's threads and memory hierarchy. *Left side.* Threads organization: a kernel is invoked from the CPU (the host) and is executed in multiple threads on the GPU (the device). Threads are organized in three-dimensional structures named blocks, which are organized in three-dimensional grids. The programmer must decide the dimensions of blocks and grids before the kernel launch. *Right side.* Memory hierarchy: threads can access data from multiple kind of memories, all with different scopes and features. Registers and local memories are private for each thread; shared memory lets threads belonging to the same block communicate, and has low access latency; all threads can access the global memory, which suffers of high latencies, but it is cached since the introduction of the Fermi architecture; texture and constant memory can be read from any thread and are equipped with a cache as well.

Following the naming conventions used in CUDA, a C/C++ function, called *kernel*, is loaded from the host (the CPU) to the devices (one or more GPUs) and replicated in many copies named *threads*. Threads can be organized in three-dimensional structures named *blocks* which, in turn, are contained in three-dimensional *grids* (Fig. 1, left side). Whenever the host runs a kernel function, the GPU creates the corresponding grid and automatically schedules each block of threads on an available streaming multi-processor of the GPU, thus allowing a transparent scaling of performances on different devices. GPUs are equipped with different types of memory (Fig. 1, right side). The GPU memory hierarchy consists in the *global memory* (accessible from all threads), the *shared memory* (accessible from threads of the same block), the *local memory* (registers and arrays, accessible from owner thread), and the *constant memory* (cached and not modifiable). The best performances in the execution of CUDA code are achieved by exploiting the constant and shared memories as much as possible. Unfortunately, they are very limited resources; on the contrary, the global memory is very large (thousands of MBs), but suffers of high latencies. In order to mitigate this issue, starting from the Fermi architecture the global memory has been equipped with a L2 cache.

Despite the remarkable advantages concerning the computational speed-up, computing with GPUs usually requires either the re-design or the development and implementation of *ad hoc* algorithms, since GPU-based programming substantially differs from CPU-based computing. As a consequence, scientific applications of GPUs might undergo the risk of remaining a niche for few specialists. To avoid such limitations, several packages and software tools for GPUs were recently released for the study of biological systems (see, e.g., [11,12,22]), so that also users with no knowledge of GPUs hardware and programming can access the high-performance computing power of graphics engines.

2.3 Many Integrated Core Architecture

The Xeon Phi is a coprocessor developed by Intel, based on the concept of Many Integrated Core (MIC) architecture. A single MIC integrates multiple cores based on the x86 instruction set, interconnected by means of an on-die bidirectional ring, which also connects the GDDR memory controllers (GDDR MC) and the PCI express (PCIe) interface logic (Fig. 2). Each core of the MIC can run up to 4 simultaneous threads in hardware and is equipped with a 512-bit wide vector unit (VPU) with 32 vector registers per thread. VPUs allow a further level of parallelism, by executing up to 8 double-precision operations per cycle. Each core is also equipped with a L2 cache, which is kept coherent by a global-distributed tag directory (TD). Similarly to GPUs, Xeon Phi coprocessors are connected to the main host computer through a PCIe system interface; differently from GPUs, Xeon Phi machines do not rely on a SIMD paradigm and run an embedded Linux μOS. Thanks to the OS and the x86 technology, the main advantage of Xeon Phi is the capability of leveraging the existing software for regular Intel (and compatible) processors. Specifically, the Xeon Phi can be programmed in two ways: the *native mode*, in which the application runs directly on the Xeon Phi and communicates with the host by means of the system bus, exploiting multi-threading libraries like openMP and MPI; the *offload mode*, in which the main application on the host offloads a portion of highly parallel code, defined by means of compiler directives.

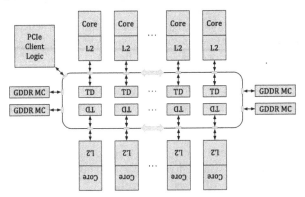

Fig. 2. Scheme of the Intel Xeon Phi coprocessor. It is composed of a bidirectional ring interconnecting multiple processing cores, caches, memory controllers and a PCI express interface.

In this work we rely on the native mode to investigate the computational speed-up that can be realized without the need of reimplementing or modifying the source code.

3   Results

To compare CPUs, GPUs and Xeon Phi coprocessors, we exploited EURORA (EURopean many integrated cOre Architecture [23]), a supercomputer created by the Italian consortium CINECA. This machine—designed to satisfy the most relevant HPC constraints (sustainable performance, space occupancy and costs)—combines multiple state-of-the-art accelerators, i.e., Intel Xeon Phi coprocessors and Nvidia GPUs based on the Kepler architecture. Specifically, EURORA consists of 64 compute nodes, and is equipped with a total of 64 Intel Xeon Phi 5120D coprocessors (60 cores, clock 1.05 GHz) and 64 Nvidia Tesla K20 GPUs (2496 cores, clock 706 MHz). Half of the compute nodes are equipped with Intel Xeon Sandy-Bridge E5-2658 (8 cores, clock 2.10 GHz); the other half of the compute nodes are equipped with Intel Xeon SandyBridge E5-2687W (8 cores, clock 3.10 GHz). In our tests, we considered the nodes equipped with the E5-2687W processors as reference CPUs. Thanks to its peculiar hybrid architecture, EURORA represents an ideal machine for a direct comparison between CPUs, GPUs and Xeon Phi coprocessors, exploited for the same intensive computation. In particular, we compare the computational performance of these three architectures for the specific task of executing stochastic simulations of biochemical systems, as described hereafter.

3.1   Experimental Setting

We implemented two different versions of SSA: one compatible with the Intel Xeon SandyBridge E5-2687W and the Intel Xeon Phi 5120D, and one specifically conceived for the Nvidia Tesla K20 architecture. Both versions were implemented in C language, with the exception of CUDA kernels. CPU and MIC exploit the Intel Math Kernel Library for the generation of random deviates. The CUDA version exploits peculiar GPU memories for the state of the system (which is stored in the shared memory) and the matrices of stoichiometric coefficients (which are stored in the constant memory).

To compare the performances of the three architectures for SSA execution, in the first batch of tests we performed an increasing number of simulations on the CPU, GPU and MIC of a biological model describing a prokaryotic gene regulatory network (PGN) [18]. In all simulations we stored 1000 time points of the dynamics of all species.

In the PGN, a gene (DNA), transcribed into the messenger RNA $(mRNA)$ and translated into a protein (P), is inhibited by the binding with a dimer (P_2) of the protein itself $(DNA \cdot P_2)$. The model consists in 8 chemical reactions: $R_1: DNA+P_2 \xrightarrow{c_1} DNA \cdot P_2$; $R_2: DNA \cdot P_2 \xrightarrow{c_2} DNA+P_2$; $R_3: DNA \xrightarrow{c_3} DNA+ mRNA$; $R_4: mRNA \xrightarrow{c_4} \lambda$; $R_5: 2P \xrightarrow{c_5} P_2$; $R_6: P_2 \xrightarrow{c_6} 2P$; $R_7: mRNA \xrightarrow{c_7} mRNA + P$; $R_8: P \xrightarrow{c_8} \lambda$, where λ denotes the degradation of the chemical

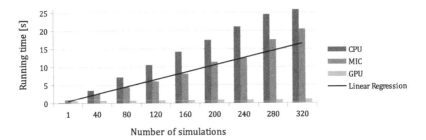

Fig. 3. Comparison of the running time to execute an increasing number of stochastic simulations of the PGN model on the three different architectures: CPU (left bars), MIC (middle bars) and GPU (right bars). The solid line represents the estimation of the running time on the MIC assuming a number of cores larger than 60: a linear regression highlights the degradation of performances when the number of parallel threads is larger than 240. In all tested cases, the GPU largely outperforms the other architectures. Moreover, the running time of the GPU remains basically constant up to 320 parallel simulations, thanks to the large number of available cores.

species. The stochastic constants used in the following tests are $(c_1, \ldots, c_8) = (0.1, 0.7, 0.35, 0.3, 0.1, 0.9, 0.2, 0.1)$. The initial state of the system assumes 250 molecules of DNA and 0 molecules of the other species.

In the second batch of tests we evaluated the impact of the size of the biochemical reaction network on the computational performance of the Intel Xeon Phi coprocessor. Since for these tests we were not interested in any dynamical properties (i.e., oscillations, bistability, etc.) that the system could present, we exploited the methodology used in [10,24] to randomly generate different synthetic models, having a number of species N and of reactions M equal to $(20 \times 20), (40 \times 40), (80 \times 80), (160 \times 160)$, and stochastic constants randomly sampled in the uniform interval $(0, 1)$.

3.2   Computational Results

In Fig. 3 we show the results concerning the first batch of tests, related to the comparison of the running times for the simulation of the PGN model on the CPU (left bars), MIC (middle bars) and GPU (right bars). These results clearly highlight that GPUs outperform the other architectures. It is worth noting that the running time of GPU remains constant throughout all tests, thanks to the high number of cores it contains, which allow to completely distribute all the simulations on this architecture. As a matter of fact, during the simulation of the PGN model, the K20 was far from a full usage of resources. In the case of 320 parallel simulations of the PGN model, the GPU achieves a speed-up of 26× with respect to the CPU and of 18.5× with respect to Xeon Phi. Concerning the MIC, we obtained a speed-up of 1.4× with respect to the CPU.

The acceleration provided by MIC increases up to 240 simulations, thanks to the fact that Xeon Phi executes up to 4 concurrent threads on each one of the 60 available cores. In order to highlight this trend, we estimated by linear

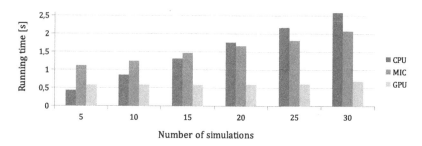

Fig. 4. Break-even of the running time to execute stochastic simulations of the PGN model on the three different architectures: CPU (left bars), MIC (middle bars) and GPU (right bars). When a few simulations are executed, the CPU outperforms the other architectures thanks to its higher clock frequency. The GPU become advantageous when more than 10 simulations are executed. The break-even for the MIC is around 20 simulations.

regression the running time of MIC in the case of 280 and 320 simulations, using the running times obtained in the case of $1, \ldots, 240$ simulations (Fig. 3, solid line). We observe that the measured running times are higher than the expected running time of MIC in the case of 280 and 320 simulations, proving that launching more than 240 simulations (4 threads $\times$ 60 cores) reduces the speed-up provided by the MIC (from 1.7$\times$ to 1.4$\times$, in these tests), since 240 simulations fully occupy the computing resources.

It is also worth noting that when only a few simulations need to be executed, the CPU outperforms both GPU and MIC thanks to its higher clock frequency. Figure 4 shows that the break-even between CPU and MIC is around 20 simulations, while the break-even between CPU and GPU is around 10 simulations. However, a further increase of the speed-up achieved by MIC could be obtained by exploiting MIC's vectorial instruction set, which is currently not considered in our SSA implementation.

In the second batch of tests, we investigated the influence of the size of the simulated model (i.e., number of species and reactions) on the performance of the Xeon Phi coprocessor. Figure 5 shows the results obtained by increasing both the number of parallel simulations and the size of the model. As expected, the running time increases with the number of parallel simulations; however, a 4-fold increase of the size of the synthetic model leads to a 2-fold increase of the running time.

Finally, we highlight that the performance of GPUs (since the introduction of the Fermi architecture) and MICs are affected by the Error Correcting Code (ECC), used to avoid any error caused by natural radiations [25]. This functionality—which is enabled on both accelerators on the EURORA supercomputer—introduces a relevant overhead due to bits verification. Tests performed by Fang *et al.* revealed that ECC causes a bandwidth reduction greater than 20 % on the MIC [25]. GPUs support ECC over the whole memory hierarchy, including global memory, L1 and L2 caches, and registers [20].

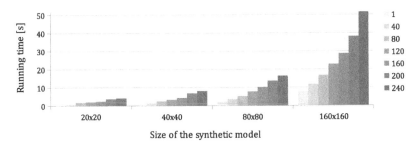

Fig. 5. Comparison of MIC running time to execute an increasing number of stochastic simulations (from 1 to 240) of synthetic models having a number of species N and of reactions M equal to $(20 \times 20), (40 \times 40), (80 \times 80), (160 \times 160)$, and stochastic constants randomly sampled in the uniform interval $(0, 1)$.

According to Kraus *et al.* [26], also for GPUs the bandwidth reduction is around 20%. Thus, the speed-up that we obtained using GPUs and MIC for stochastic simulation ($26\times$ and $1.4\times$, respectively) could still increase by disabling the ECC functionality.

4   Conclusion

We investigated the performances of different architectures—focusing on the Intel Xeon Phi coprocessors—to perform stochastic simulations of the dynamics of mechanistic models of biological and synthetic systems. Our results highlight how the GPU outperforms the CPU and the MIC when more than 10 simulations are required. We believe that this analysis will facilitate the choice of a proper parallelization methodology when a large number of independent simulations are needed, as is the case of many computationally expensive tasks that are typical in the study of biological systems (e.g., parameter sweep, parameter estimation, sensitivity analysis [6, 27, 28]).

An additional issue that should be considered during the selection of a proper parallel architecture concerns the costs required to port the code onto GPU and MIC. In particular, in the case of GPGPU computing on Nvidia video cards, the effort necessary to re-implement the source code using the CUDA programming technique is relevant and must be taken into account. On the contrary, Xeon Phi coprocessors should be fully compatible with CPUs; however, according to our experience, the code has to be adapted in order to be correctly executed on MIC (considering the *native mode*).

We also tested the *offload* capability of the Xeon Phi coprocessor, that is, the possibility of automatically distributing independent calculations over multiple threads (e.g., the instructions contained in a `for` cycle). However, our preliminary offload tests highlighted that data structures such as matrices, defined by means of multiple pointers to the memory, should be avoided in order to automatically exploit this parallelization mode; otherwise, the source code must be re-implemented. As a future development of this work, we plan to modify our implementation of the SSA algorithm, by using linear data structures which can leverage the offload compiler directives, necessary to parallelize the appropriate regions of the source code (e.g., the evaluation of propensity functions and the update of the system state) with Xeon Phi coprocessors.

An additional comparison between different architectures might be based on the evaluation of the cost, power consumption and theoretical peak performance. In the case of the devices tested in this work, the CPU Intel Xeon SandyBridge E5-2687W has a cost of around \$1800, with a power consumption of 150 W and theoretical peak performance of 198.4 GFlops. The characteristics of the other devices are: Intel Xeon Phi 5120D \$ 2700, 245W, 2022 GFlops, and Nvidia Tesla K20 \$ 2700, 225 W, 3520 GFlops. Considering this information, to achieve the theoretical peak of the Tesla K20 GPU, either 18 CPUs or 2 Xeon Phi 5120D would be required (with a consequent increase in terms of cost and power consumption). However, to fully leverage the computational power of these devices, CPU's multi-threading and MIC's vectorial instruction set should be exploited, both requiring further relevant modifications of the existing source code.

References

1. Aldridge, B.B., Burke, J.M., Lauffenburger, D.A., Sorger, P.K.: Physicochemical modelling of cell signalling pathways. Nat. Cell Biol. **8**, 1195–1203 (2006)
2. Wilkinson, D.: Stochastic modelling for quantitative description of heterogeneous biological systems. Nat. Rev. Genet. **10**, 122–133 (2009)
3. Gillespie, D.T.: Exact stochastic simulation of coupled chemical reactions. J. Phys. Chem. **81**, 2340–2361 (1977)
4. Gillespie, D.T.: A rigorous derivation of the chemical master equation. Physica A **188**, 404–425 (1992)
5. Cao, Y., Gillespie, D.T., Petzold, L.R.: Efficient step size selection for the tau-leaping simulation method. J. Chem. Phys. **124**, 044109 (2006)
6. Nobile, M.S., Cazzaniga, P., Besozzi, D., Pescini, D., Mauri, G.: Reverse engineering of kinetic reaction networks by means of Cartesian Genetic Programming and Particle Swarm Optimization. In: IEEE Congress of Evolutionary Computation, pp. 1594–1601 (2013)
7. Tian, T., Burrage, K.: Parallel implementation of stochastic simulation of large-scale cellular processes. In: 8th International Conference on High-Performance Computing in Asia-Pacific Region, pp. 621–626 (2005)
8. Kent, E., Hoops, S., Mendes, P.: Condor-COPASI: high-throughput computing for biochemical networks. BMC Syst. Biol. **6**, 91 (2012)
9. Macchiarulo, L.: A massively parallel implementation of Gillespie algorithm on FPGAs. In: International Conference of the IEEE on Engineering in Medicine and Biology Society, pp. 1343–1346 (2008)
10. Nobile, M.S., Cazzaniga, P., Besozzi, D., Pescini, D., Mauri, G.: cuTauLeaping: A GPU-powered tau-leaping stochastic simulator for massive parallel analyses of biological systems. PLoS ONE **9**, e91963 (2014)
11. Nobile, M.S., Besozzi, D., Cazzaniga, P., Mauri, G., Pescini, D.: cupSODA: A CUDA-powered simulator of mass-action kinetics. In: Malyshkin, V. (ed.) PaCT 2013. LNCS, vol. 7979, pp. 344–357. Springer, Heidelberg (2013)

12. Nobile, M.S., Cazzaniga, P., Besozzi, D., Mauri, G.: GPU-accelerated simulations of mass-action kinetics models with cupSODA. J. Supercomput. **69**, 17–24 (2014)
13. Bernaschi, M., Bisson, M., Salvadore, F.: Multi-Kepler GPU vs. multi-Intel MIC for spin systems simulations. Comput. Phys. Commun. **185**, 2495–2503 (2014)
14. Fang, J., Varbanescu, A.L., Imbernon, B., Cecilia, J.M., Perez-Sanchez, H.: Parallel computation of non-bonded interactions in drug discovery: NVidia GPUs vs. Intel Xeon Phi. In: Proceedings of the 2nd International Work-Conference on Bioinformatics and Biomedical Engineering. pp. 579–588 (2014)
15. Halyo, V., LeGresley, P., Lujan, P., Karpusenko, V., Vladimirov, A.: First evaluation of the CPU, GPGPU and MIC architectures for real time particle tracking based on Hough transform at the LHC. J. Instrum. **9**, P04005 (2014)
16. Lyakh, D.I.: An efficient tensor transpose algorithm for multicore CPU, Intel Xeon Phi, and NVidia Tesla GPU. Comput. Phys. Commun. **189**, 84–91 (2015)
17. Shimoda, T., Suzuki, S., Ohue, M., Ishida, T., Akiyama, Y.: Protein-protein docking on hardware accelerators: comparison of GPU and MIC architectures. BMC Syst. Biol. **9**, S6 (2015)
18. Nobile, M.S., Besozzi, D., Cazzaniga, P., Mauri, G., Pescini, D.: A GPU-based multi-swarm PSO method for parameter estimation in stochastic biological systems exploiting discrete-time target series. In: Giacobini, M., Vanneschi, L., Bush, W.S. (eds.) EvoBIO 2012. LNCS, vol. 7246, pp. 74–85. Springer, Heidelberg (2012)
19. Butcher, J.C.: Numerical Methods for Ordinary Differential Equations. John Wiley & Sons, New York (2003)
20. Nickolls, J., Dally, W.J.: The GPU computing era. Micro IEEE **30**, 56–69 (2010)
21. Farber, R.M.: Topical perspective on massive threading and parallelism. J. Mol. Graph. Model. **30**, 82–89 (2011)
22. Harvey, M.J., Fabritiis, G.D.: A survey of computational molecular science using graphics processing units. WIREs Comput. Mol. Sci. **2**, 734–742 (2012)
23. Cavazzoni, C.: EURORA: a European architecture toward exascale. In: Proceedings of the Future HPC Systems: The Challenges of Power-Constrained Performance, 1, ACM (2012)
24. Komarov, I., D'Souza, R.M., Tapia, J.J.: Accelerating the Gillespie τ-leaping method using graphics processing units. PLoS ONE **7**, e37370 (2012)
25. Fang, J., Varbanescu, A.L., Sips, H., Zhang, L., Che, Y., Xu, C.: Benchmarking Intel Xeon Phi to guide kernel design. Technical report, Delft University of Technology, Netherlands (2013)
26. Kraus, J., Pivanti, M., Schifano, S.F., Tripiccione, R., Zanella, M.: Benchmarking GPUswith a parallel Lattice-Boltzmann code. In: IEEE 25th International Symposium on ComputerArchitecture and High Performance Computing, pp. 160–167 (2013)
27. Besozzi, D., Cazzaniga, P., Pescini, D., Mauri, G., Colombo, S., Martegani, E.: The role of feedback control mechanisms on the establishment of oscillatory regimes in the Ras/cAMP/PKA pathway in *S. cerevisiae*. EURASIP J. Bioinform. Syst. Biol. 2012 (2012)
28. Gunawan, R., Cao, Y., Petzold, L.R., Doyle, F.J.: Sensitivity analysis of discrete stochastic systems. Biophys. J. **88**, 2530–2540 (2005)

Cost of Bandwidth-Optimized Sparse Mesh Layouts

Martti Forsell[1,2,3], Ville Leppänen[1,2,3]([✉]), and Martti Penttonen[1,2,3]

[1] VTT, Computing Platforms, Oulu, Finland
[2] Department of Information Technology, University of Turku, Turku, Finland
[3] Department of Computer Science, University of Eastern Finland, Joensuu, Finland
Ville.Leppanen@utu.fj

Abstract. The requirements of interconnection networks for shared memory chip multiprocessors (CMP) differ from those used in traditional application-specific networks on chip (NOC). This is because modern CMP cores tend to inject memory references to the network frequently (up to once per clock cycle) and the latency of references should be as low as possible. The throughput computing paradigm is a mechanism to trade the low latency requirement to high throughput in CMPs by overlapping memory references from processors with a help of multithreading. To meet the bandwidth requirements of throughput computing CMPs we have studied using d-dimensional sparse meshes and tori. Unfortunately it has turned out that either there is too much bandwidth leading to high silicon area and energy consumption of the links get longer decreasing the clock rate. In this paper we study the cost of bandwidth-optimized 2-dimensional meshes and tori for CMPs using the throughput computing paradigm. We present the layout as well as determine link length, degree of node and compare them to those of d-dimensional meshes and tori. For area and power efficiency considerations, we also give estimates on silicon area and power consumption.

Keywords: NOC · Under-populated networks · Layout · Sparse networks · Throughput computing

1 Introduction

Network-on-chip (NOC) is an extensively studied design paradigm for communication subsystems of highly integrated multi-resource systems like e.g. chip multiprocessors (CMP) and Systems-on-Chip (SOC) [4,12]. Majority of NOC investigations focus on connecting a set of heterogeneous resources (application-specific) [20]. Such interconnection designs are often done without supporting scalability with some fixed communication throughput requirement. We consider that there exists growing importance for supporting high-performance systems aimed at general purpose use. Moreover, we consider it likely that such systems are based on high-bandwidth networks and rather homogeneous resource structures due to the usefulness of re-programmability and design re-use.

© Springer International Publishing Switzerland 2015
V. Malyshkin (Ed.): PaCT 2015, LNCS 9251, pp. 375–389, 2015.
DOI: 10.1007/978-3-319-21909-7_37

The purpose of this paper is to consider so-called underpopulated – or sparse – NOCs supporting high-throughput computing. Specifically, we aim to estimate the efficiency of proposed layouts by using an analytical methodology. For throughput computing, it is characteristic the processing nodes are enabled to useful operations, although the executed computations as such might contain suboperations with long delays like loading/storing data to/from the memory [17]. The essential "trick" of efficient throughput computing is to hide the latency of such long operations simply by interleaving elementary instructions (or sequences) from several threads at each computing node. The latency hiding succeeds we there is enough work to distribute (enough executable threads per node) and enough data moving capacity (enough network bandwidth). In this paper we only consider sparse networks with enough bandwidth.

Most NOC related studies consider networks, where all nodes can participate into communication by being original sources and sinks. In this paper, we call these "ordinary" networks as *dense networks* (*fully populated*). Dense networks have problems with scalability under constant throughput assumption. Consider that nodes are assumed to inject (and received) messages/packets at some constant rate $1/\alpha$, meaning one message/packet per α (logical program) steps. It is easy to see that dense constant degree networks cannot be scaled up while still preserving the injection rate $1/\alpha$, since scaling means longer expected route for the messages and thus higher expected needed message moving capacity per node (but that remains constant for constant degree networks). Consequently, dense networks have insufficient communication bandwidth considering scalability, and the only practical possibilities are to either increase communication locality or limit the communication frequency of nodes (that is decrease the α).

An an opposite to dense networks are so-called *underpopulated networks* [2] or *sparse networks* [1,5–8,11,15,22–25]. In sparse networks, all nodes are not sources and sinks, only some fraction. A large fraction of the nodes are intermediate nodes enabling sufficiently increased communication capacity to meet the needs of throughput computing. The focus of this paper is in presenting layouts for mesh-based scalable sparse network topologies that are capable to support some constant (sustained) injection rate $1/\alpha$, independent of the network size.

Despite the rather large interest towards various kind of the sparse networks, only a few studies consider the NOCs context. Moreover, comparisons regarding the efficiency of layouts for sparse network has previously studied only by us in [6]. In this paper, we extend the analysis with new kinds of sparse mesh-based topologies and new layouts.

In Sect. 2 we present on the basics of cost metrics for layouts of sparse networks. Actual used metrics are then defined based on the basic framework in Sect. 4. In all cases, layout is made for plane (2D). In Sect. 3, we present definitions for various sparse networks (both sparse meshes and tori) and naturally also consider layouts for them. Layouts are given for 2D, 3D and 4D versions of mesh-based networks. Moreover, we also describe a recursive layout construction method for higher dimensions. In Sect. 4 we make a comparison of actual layouts. Our results vary network sizes and provide information on the power consumption, maximum frequency, chip area, frequency/power and frequency/area. Finally, conclusions are given in Sect. 5.

2 Basics of Layout Structures

2.1 Setting for Layouts

In the following, we denote by p the NOC size meaning that there exists p source/sink nodes communicating with each other – exchanging messages by using the network connections. Moreover, we assume that each such node of the NOC can communicate with any other node. We make no particular assumption about the nature of communication. Considering the efficiency of routing, the distribution of routing distances of course matters. Naturally low latency is a desired property, but we assume no means to increase locality of communication and thus decrease expected latency – rather we consider there to exist rather high average latency L, and require that the following bandwidth and slackness conditions are met. By the *slackness* condition we assume that in a time period of L steps, a node produces at most L packets to be sent. The *bandwidth* condition on the other hand requires that the network capacity makes it possible to move $p\phi$ messages per time step, where ϕ is the network diameter. Thus, we set the latency L to be proportional to 2ϕ.

The meaning of bandwidth condition above is that the network enables throughput computing. In general, without any assumptions on communication patterns, it can be proved that having constant degree network and the requirement to move $\Omega(p\phi)$ messages per logical step implies that at most a fraction of $O(1/\phi)$'th of nodes can be sinks/sources. Ordinary dense meshes are unable to support sustained throughput computing and thus those are not considered in this paper.

For simplicity, we make the assumption that the NOC layout is a grid-like cell structure consisting of $\sqrt{p} \times \sqrt{p}$ similar *cells*. As shown in Fig. 1, a cell consists of a *slot* for processing node, and an *infrastructure* for intermediate network, consisting of wiring and some intermediate router nodes which are considered to be very much simpler components (in size) than the processing nodes. Typically, the memory of processing node requires a lot of chip area. See also Fig. 2 for a more concrete illustration.

In this paper, we assume uniform size $w \times w$ for all the cells so that $w = W/\sqrt{p}$ and the side length of chip is W. For the total area of each cell, we also include (besides processing node located into the cell) the area related to the wiring that connects the nodes to each other, and thus the area is related to W and w. For the distance from a node within (i, j) to a node in another cell (k, l) we use the Manhattan metrics, i.e., it is $(|k - i| + |l - j|) \times w$. However, the connections between intermediate nodes inside a cell are considered to have a small non-zero length independent of W. We consider only regular networks, we all nodes have a uniform (topology dependent) in-degree and out-degree.

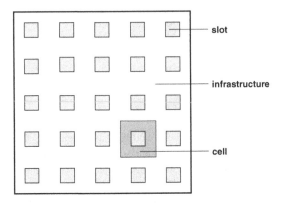

Fig. 1. An illustration on grid-noc structure.

2.2   On Cost Metrics

As the purpose of this paper is to compare various sparse networks and their layouts, we need a cost metrics that is normalized with respect to p, the number of processing nodes. As individual metrics we will use e.g. normalized frequency/chip area and frequency/power. It is tempting to consider the bandwidth or routing capacity as a parameter for sparse network cost metric. However, the bandwidth depends on p and thus it is implicitly taken into account for area and power).

In Sect. 3, we will describe layouts for sparse networks defined in the following section. The purpose of describing layouts exactly is to calculate values for a set of properties as the basis of cost evaluation of a given sparse network. We assume that in all cases a sparse network N has p sink/source nodes with resources and $I_N(p)$ intermediate routing nodes. For layout descriptions, we assume that one *slot* is reserved for each sink/source node + its resources and such slots are distributed evenly over the layout area. As a practical simplification, we assume the overall chip area to consists of $\sqrt{p} \times \sqrt{p}$ equally sized rectangular *cells* so that all cell consists of one slot, a set of allocated intermediate routing nodes, and the needed crossing wiring. Despite the regularity of sparse network, due to the topology of sparse network, allocation of nodes to cells, and consequently due to the layout of wiring, the actual cells can be quite different. For the above reasons, we make no common assumption on the slot position within each cell. In the following, we call the above layout as *grid-noc layout*, see Fig. 1.

Finally, for being able to compare grid-noc layouts, we denote by $w_L(p)$ the width of cells for given layout L. Estimated wire length distribution is denoted by a function of $dist_L : 0 \ldots 2\sqrt{p} \mapsto R$ whereas $dist_L(i) = x$ simply means that L has x wires of length $\approx w_L(p) \times i$. In $dist_L$, the wires placed fully inside a cell are considered to have wire length 0 (yet, we use value 0.5 in the evaluations of Sect. 4). For non-local wires between two nodes in different cells, we use $y \times w_L(p)$ as wire length distance estimation, where y is the Manhattan distance.

3 Definitions and Layouts for Mesh-Based Networks

Many sparse network topologies have been defined in the literature. In fact, it is possible to define a sparse network using almost any dense network as the basis and setting only a fraction of the nodes as "processors". As there are plenty of such networks, we are interested only of some of them: On sparse networks providing some fixed sustained throughput between the "processors". More precisely, we expect that there exists a constant c_N so that each "processor" of N can send/receive a message every c_N'th steps on the average.

We call *scalable* a sparse network topology N, if c_N is independent of the size of the network (of p). Such *scalable sparse networks* have special characteristics. For a network with p sources and an average routing distance (often of the same order as the *diameter*) ϕ, the network must be able to move $\Omega(p\phi)$ packets in each step. For a network with degree δ, it means that the network must have $\Omega(p\phi/\delta)$ intermediate nodes. Scalable sustained throughput is not possible otherwise. Notice that this is sensible when such intermediate nodes are much simpler than the sources. Also, although a packet has an average latency $\Omega(\phi)$ to arrive to its target, the sources and sinks can send and receive packets at constant rate. Assuming that computations on the processors tolerate this latency (as is with throughput computing), then the processors do no busy waiting and are fully employed by computations.

Next, we present sparse mesh-based networks – most of which are previously defined in the literature. For the networks, we give a corresponding grid-noc layout that we believe to be the best and state properties of such layouts (e.g. wire length distributions). We should mention that non-mesh-based sparse solutions have also been proposed: for the butterfly network (used e.g. in the SB-PRAM [1] and Fluent machine [18,19] constructions); for the cube-connected-cycles (CCC) in [15] as it can be seen as a sparse version of the hypercube (constant-degree version). Sparse version of mesh of trees network is quite natural – it has been studied in [13] and also used in the Paraleap [3]. However, we focus on sparse meshes/tori (e.g. since we have used those e.g. in Eclipse [5,8] and investigated in [11]).

3.1 Definition of Sparse Meshes

In the literature, there are several definitions for sparse meshes/tori. In [7,14], the processing nodes (source/sink) are placed on the outer surface of mesh – this is problematic considering tori, since there is no natural outer surface. Thus, we adopt the most natural definition used in several publications [5,10,11,21]: The p processor nodes are placed as a $(d-1)$-dimensional plane. This definition is also natural considering tori.

Definition 1. *A regular d-dimensional n-sided mesh is a graph* $G^{n,d}_{mesh} = (V, E)$, *where*

$$V = \left\{ V_{a_1, a_2, \ldots, a_d} \,\middle|\, 0 \leq a_i \leq n-1, 1 \leq i \leq d \right\}$$

is a set of $p = n^d$ nodes, and

$$E = \left\{ (V_{a_1,a_2,...,a_d}, V_{b_1,b_2,...,b_d}) \right|$$

$$\sum_{i=1}^{d} |a_i - b_i| = 1, 0 \le a_i, b_i \le n - 1, 1 \le i \le d \right\}$$

defines the connections between the nodes (no wrap-around connections allowed). The degree of a mesh is 2d, and the diameter is $d\sqrt[d]{p} - d$.

Definition 2. *A regular d-dimensional n-sided toroidal mesh (torus) is a graph $G_{torus}^{n,d} = (V, E)$, where*

$$V = \left\{ V_{a_1,a_2,...,a_d} \middle| 0 \le a_i \le n - 1, 1 \le i \le d \right\}$$

is a set of n^d nodes, and

$$E = \left\{ (V_{a_1,a_2,...,a_d}, V_{b_1,b_2,...,b_d}) \right.$$

$$\left| \sum_{i=1}^{d} (a_i - b_i) \bmod n = 1, 0 \le a_i, b_i \le n - 1, 1 \le i \le d \right\}$$

defines the connections between the nodes. The degree of a torus is 2d, and the diameter is $\frac{d}{2}\sqrt[d]{p}$ (if n mod 2 = 0). The torus is completely symmetric.

3.2   Layouts

The layouts user here for sparse meshes/tori are defined already in [6]. We review the definition next. A layout for a 2-d $n \times n$ sparse mesh is constructed by mapping all nodes $V_{j,0}, \ldots V_{j,n-1}$ to cell C_j. The $n - 1$ connections along the Y-axis are all within cells, and thus from the view point of this study, the actual layout of the nodes within the cells is not seen an interesting issue. Only, the connections along X-axis are implemented as purely between the cells. Notice that there are only n connections between each pair of cells C_j and C_{j+1}, for $0 \le j < n - 1$. In order to minimize the length of such connections, the cells should be organized in form of a "snake", where the cell C_{j+1} is always a neighbor of the cell C_j, for $0 \le j < n - 1$, in the underlying grid. It is easy to see that there exists a plenty of such "snakes" for all values of n.

Let $dist_{sm}^{d,n}(x)$ denote the number of connections of length x cells in a d-dimensional n-sided sparse mesh. Connections that are internal to a cell are considered to have length 0. Thus, $dist_{sm}^{2,n}(0) = n \times (n-1)$, and also $dist_{sm}^{2,n}(1) = n \times (n-1)$. Clearly, $dist_{sm}^{2,n}(x) = 0$, for $x > 1$.

The case of 2-d $n \times n$ tori is almost identical. The toroidal connections within each cell are easy to arrange. The cells C_{n-1} and C_0 can also be placed next to each other, if n is even. The existence of such a Hamiltonian cycle is easy to see.

However for an odd n, no such cycle seems to exist – yet, only one pair of cells needs to placed at distance 2 cells from each other.

Let $dist_{st}^{d,n}(x)$ denote the number of connections of length x cells in d-dimensional n-sided sparse tori. Thus, $dist_{st}^{2,n}(0) = n^2$. Also $dist_{st}^{2,n}(1) = n^2$, if $2|n$. If n is odd, then $dist_{st}^{2,n}(1) = n(n-1)$ and $dist_{st}^{2,n}(2) = n$.

When constructing layouts for higher dimensional sparse meshes and tori, the above constructions are useful. E.g., the layout for a 3D sparse mesh is a generalization of the 2D case: Simply pack each pile of nodes along z-axis as a cell. We illustrate the situation in Fig. 2 and omit the the more detailed descriptions of layouts.

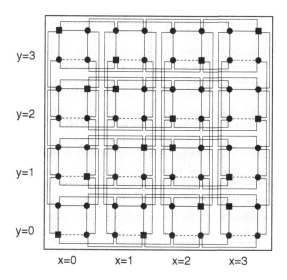

Fig. 2. 3D sparse mesh: A grid-noc layout.

3.3   Bandwidth-Optimized 2-Dimensional Meshes and Tori

The bandwidth or data moving capacity of previously discussed sparse meshes and tori is in fact unnecessarily large. For example, if in the d-dimensional sparse torus packets would be routed by using the axes only in one direction, the average routing distance between processor nodes would be $d(n-1)/2$. Since a torus is symmetric, the average routing distance from any node can be evaluated by considering the distances from the corner node $(0, 0, \ldots, 0)$. It is easy to observe that the sum of such routing distances to all processor nodes from the corner node is $d \times n^{d-2} \times n(n-1)/2$. Since there are n^{d-1} processors, the claim follows. If routing is done to both directions along each axes, the average routing distance is halved to $d(n-1)/4$. As the d-dimensional sparse torus can move d packets (or $2d$) per intermediate node at each step, the data moving capacity exceeds the minimal data moving requirement by a factor

$$\frac{dn^d}{n^{d-1} \times d(n-1)/4} \approx 4.$$

Similar calculations can be presented for the sparse meshes, see e.g. [5].

Since, the bandwidth can be regarded sub-optimal in the presented sparse torus and mesh network, we consider bandwidth-optimized variants of the 2-dimensional meshes and tori by organizing the processor and routing nodes in the way done e.g. in [8].

Definition 3. *A bandwidth-optimized 2-dimensional sparse mesh (torus) is as a regular 2-dimensional $p/4 \times p/4$-sided mesh (torus) where all the nodes are intermediate nodes. The processors are extra nodes that are evenly distributed across the network and attached to groups of $\sqrt{p}/4 \times \sqrt{p}/4$ adjacent intermediate nodes so that the processor at processor-wise position (a_i, a_j) is attached to the intermediate nodes at intermediate node-wise positions (a_k, a_l), where $i\sqrt{p}/4 \leq k \leq (i+1)\sqrt{p}/4 - 1$ and $j\sqrt{p}/4 \leq l \leq (j+1)\sqrt{p}/4 - 1$.*

As an illustration, in a bandwidth-optimized 2-dimensional 8×8 processor sparse mesh, each processor node is attached to 2×2 group of intermediate nodes. Both 2-dimensional sparse mesh and torus have $p^2/16$ intermediate nodes.

Definition 4. *A bandwidth-optimized 2-dimensional multimesh (torus) is as a collection of $\sqrt{p}/4$ interleaved regular 2-dimensional $\sqrt{p}$-sided meshes (tori) where all the nodes are intermediate nodes. The processors are extra nodes that are evenly distributed across the network and attached to groups of corresponding intermediate nodes from the interleaved meshes (tori) so that the processor at processor-wise position (a_i, a_j) is attached to the intermediate nodes at intermediate node-wise positions (a_i, a_j) at each mesh (torus).*

Figure 3 shows a bandwidth optimized 2-dimensional 8×8 processor multimesh, where the small white squares are intermediate nodes grouped as groups of 2 nodes and large grey squares are processors. The number of intermediate nodes for 2D multimesh (and torus) is $\sqrt{p} \times \sqrt{p} \times \sqrt{p}/4 = p^{3/2}/4$.

The layouts of the 2-dimensional bandwidth optimized sparse meshes and tori are straightforward since the processors and attached intermediate nodes are already organized as cells.

3.4   Summary of Properties for Comparison

To compare the layouts presented above, we have calculated a set of characteristic values: I: amount of intermediate routers, Δ: the max degree of nodes, len_{avg}: the average wire length of links, wt: the max amount of parallel wires (X-axis or Y-axis) crossing a cell, len_{max}: max length of links, and logical diameter of sparse network. These characteristic values are shown in Table 1.

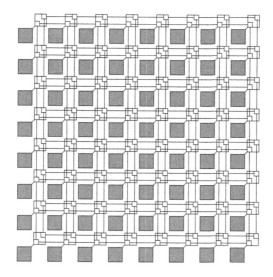

Fig. 3. A layout of a 2-dimensional bandwidth-optimized multi mesh. All connections between the nodes are shown.

4 Comparison

4.1 Preliminaries

We need to link Table 1 to known properties of silicon technology. Assume a CMP with P processors based on the layouts of Table 1. Besides a processor, each cell consists of SRAM with S_d/P bytes data capacity, an SRAM to hold S_i/P bytes instruction, and $I + 1$ intermediate routers, where S_d and S_i are amounts of data / instruction SRAM per CMP, respectively.

We assume synchronous NOC operation, and therefore for interconnects the minimum clock cycle is $D_c = D_s + 2D_l$, where delays D_s and D_l are for the intercommunication switch and the signal of longest interconnect (that is $len_{max} \times w_L$). Due to handshaking, we count the signal delay twice. The max clock frequency is $1/D_c$.

Typically in NOCs, an interconnect is implemented as multiple layers of parallel wires connecting adjacent switches. The signal delay in such a setting can be approximated using so-called parasitic wire model and employing an optimal number of scaled repeaters as shown in [16] and in our 100+ parameter performance-area-power model [9]. The details are not discussed here.

As discussed in [6], based on the models, the silicon area of a cell is $A_c = (\sqrt{A_p + A_m + A_s(I + P)/P} + W_l \times wt)^2$, where A_p is the area of the processor, A_m is the area taken by the local instruction and data SRAM blocks, A_s is the area of the interconnection switch, and W_l is the width of a link and wt is the maximum number of parallel links per row and column. Similarly in [6], it is argued that the power consumption of CMP is $P_{cmp} = P(P_p + P_m + P_s(I + P)/P) + P_w$, where P_p is the power consumption of a processor, P_m is the

Table 1. Properties of layouts for sparse networks.

Network layout	p	I	Δ	wt	len_{avg}	len_{max}	logical diameter
Sparse mesh, 2d	n	$p^2 - p$	4	$2p$	0.75	1	$2p$
Sparse torus, 2d	n	$p^2 - p$	4	$3p$	0.75	1	p
Sparse mesh, 3d	n^2	$p^{3/2} - p$	6	$2\sqrt{p}$	5/6	1	$3\sqrt{p}$
Sparse torus, 3d	n^2	$p^{3/2} - p$	6	$3\sqrt{p}$	1.5	2	$3/2 \times \sqrt{p}$
Sparse mesh, 4d	n^3	$p^{4/3} - p$	8	$\approx p^{1/2}$	$\approx p^{1/6}$	$p^{1/6}$	$4p^{1/3}$
Sparse torus, 4d	n^3	$p^{4/3} - p$	8	$\approx 2p^{1/2}$	$\approx 2p^{1/6}$	$2p^{1/6}$	$2p^{1/3}$
Optimized sparse mesh, 2d	$4n$	$(p/4)^2$	4	$\sqrt{p}/4$	$4/\sqrt{p}$	1	$p/2$
Optimized sparse torus, 2d	$4n$	$(p/4)^2$	4	$\sqrt{p}/4$	$8/\sqrt{p}$	2	$p/4$
Multimesh, 2d	$(4n^2)^{2/3}$	$p^{3/2}/4$	4	$\sqrt{p}/4$	1	1	$2\sqrt{p}$
Multitorus, 2d	$(4n^2)^{2/3}$	$p^{3/2}/4$	4	$\sqrt{p}/4$	2	2	$\sqrt{p}$

power consumption of local SRAM blocks, P_s is the power consumption of an interconnect switch, and P_w is the power consumption of the interconnect links. We can approximate P_w by summing the power consumption of individual links together and taking the link length distribution into account. For individual links of length L_w the power consumption can be obtained from the equation $P_l = W_{link} A_w K (H C_{drv} + C_s/K) V_p^2 / D_{cycle}$, where W_{link} is the number of parallel wires per link, A_w is the average activity factor of the link, K is the number of repeaters, H is the size of the repeaters, C_{drv} is the input capacitance of a minimum sized inverter, C_s is the self capacitance of the wire, V_p is the voltage, and D_{cycle} is the clock cycle.

4.2   Results

Our results are shown in Figs. 4, 5, 6, 7, 8, 9, 10 and 11. To yield actual values for the Figures, we applied standard 65 nm silicon parameters for the models, assumed the CMP to have 8 . . . 128 processor cores, 1 MB and 0.7 MB data/instruction SRAM, and 105 nm as the minimal global wire width. We consider that the practical goodness is best illustrated with the frequency/power and frequency/area figures.

Analysing the figure above and Table 1, we make the following observations:

- Based on Table 1, the 2-dimensional sparse meshes and tori are most costly in terms of area for intermediate routers.
- Layouts with constant maximum length of links is possible, but only for 2D and 3D sparse meshes and tori.
- The shortest, only $2p^{1/3}$ logical diameter is for 4D sparse torus whereas it is longest for 2D sparse mesh (proportional to p).

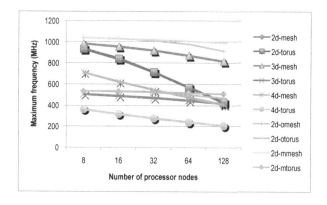

Fig. 4. Results on max clock frequencies.

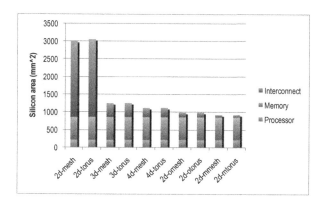

Fig. 5. Results on silicon area for $P = 64$.

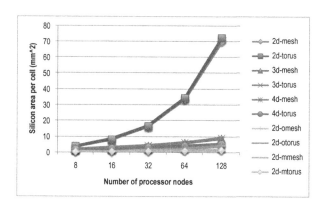

Fig. 6. Results on silicon area per cell.

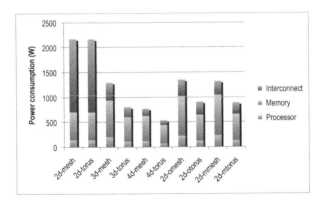

Fig. 7. Results on power consumption for $P = 64$.

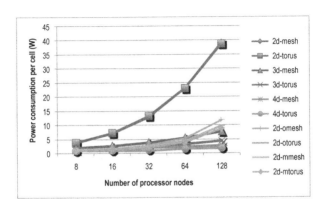

Fig. 8. Results on power consumption per cell.

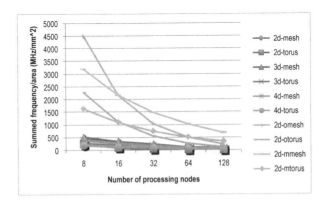

Fig. 9. Results on summed frequency/silicon area.

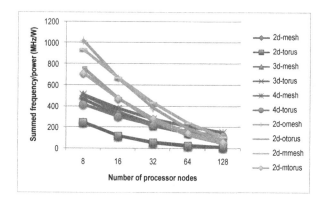

Fig. 10. Results on summed frequency/power consumption.

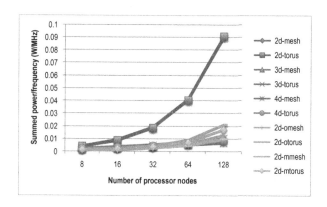

Fig. 11. Results on summed power consumption/frequency.

- Tori provide worse maximum clock rates than meshes. Our 2D bandwidth-optimized multimesh has the highest clock rate whereas 4D torus has the lowest.
- The higher dimensional sparse meshes and tori have clearly smaller silicon area requirement for the interconnection network – 2D is the worst. However, even for $P = 64$ the layout area for interconnections requires a large fraction of total area. The bandwidth-optimized 2-dimensional networks have the smallest silicon area. The same is trend is true for power consumption but bandwidth-optimized 2-dimensional networks do not perform equally well.
- Scaling the dimension of the network up will have a positive effect on both the area and power costs. Scaling up the number of processor nodes will increase the relative area and power costs per cell due to the additional intermediate router nodes.
- The most illustrating summed frequency/power and frequency/area figures will decrease as the amount of processors is increased. Overall based on these metrics, 4D and 3D meshes perform the best whereas 2D torus is the weakest.

5   Conclusions

In this paper, we have given a set of layouts for sparse NOCs for the purpose of supporting high-throughput CMPs. We calculated characteristic values on layouts and applied power and area cost models to compare the goodness of layouts in case of certain kind of 2D silicon implementation technology. According to our comparison, the bandwidth-optimized multimesh and sparse mesh perform the best in most tests. If we do not count the performance-optimized networks, the 3D sparse mesh is seen as the best considering clock frequency and frequency per area figures. However, perhaps surprisingly 4D sparse mesh outperforms the others in the frequency/power consumption comparison. In the comparisons, all meshes are found to perform better than tori. It appears that in the model the effect of higher achievable maximum frequency (toroidal connection increase link length) is more dominating than having shorter logical diameter.

In future work, we plan to consider also 3D layouts for the sparse structures as co-called through-silicon-via's effectively enable 3D stacking of 2D layouts. We also consider ways to make the comparisons normalized with respect to the routing capacity/bandwidth. The bandwidth-optimized constructions we found better essentially because the ordinary sparse meshes/tori simply have too much routing capacity. We also consider studying higher than 4-dimensional sparse structures, and model the effect of routing algorithms.

References

1. Abolhassan, F., Drefenstedt, R., Keller, J., Paul, W., Scheerer, D.: On the physical design of PRAMs. Comput. J. **36**(8), 756–762 (1993)
2. Azizoglu, M., Egecioglu, Ö.: Lower bounds on communication loads and optimal placements in torus networks. IEEE Trans. Comput. **49**(3), 259–266 (2000)
3. Balkan, A.O., Qu, G., Vishkin, U.: An area-efficient high-throughput hybrid interconnection network for single-chip parallel processing. In DAC 2008: Proceedings of the 45th annual Design Automation Conference, pp. 435–440 (2008)
4. Benini, L., Micheli, G.D.: Networks on chips: a new SoC paradigm. Computer **35**(1), 70–78 (2002)
5. Forsell, M.: A scalable high-performance computing solution for network on chips. IEEE Micro **22**(5), 46–55 (2002)
6. Forsell, M., Leppänen, V., Penttonen, M.: Cost of sparse mesh layouts supporting throughput computing. In: Proceedings of 14th Euromicro Conference on Digital System Design, DSD 2011, pp. 316–323. IEEE Computer Society, August 2011
7. Forsell, M., Leppänen, V., Penttonen, M.: Efficient two-level mesh based simulation of PRAMs. In: Proceedings of International Symposium on Parallel Architectures, Algorithms and Networks, ISPAN 1996, pp. 29–35. IEEE (1996)
8. Forsell, M., Leppänen, V.: High-bandwidth on-chip communication architecture for general purpose computing. In: Proceedings of the 9th World Multiconference on Systemics, Cybernetics and Informatics, pp. 1–6, vol. IV (2005)
9. Forsell, M.: On the performance and cost of some PRAM models on CMP hardware. Int. J. Found. Comput. Sci. **21**(3), 387–404 (2010)

10. Honkanen, R., Leppänen, V., Penttonen, M.: Hot-potato routing algorithms for sparse optical torus. In: International Conference on Parallel Processing, ICPP 2001, pp. 302–307 (2001)

11. Honkanen, R.T., Leppänen, V., Penttonen, M.: Address-free all-to-all routing in sparse torus. In: Malyshkin, V.E. (ed.) PaCT 2007. LNCS, vol. 4671, pp. 200–205. Springer, Heidelberg (2007)

12. Jantsch, A., Hannu, T. (eds.): Networks on Chip. Kluwer Academic Publishers, San Francisco (2003)

13. Leppänen, V.: On implementing EREW work-optimally on mesh of trees. J. Univ. Comput. Sci. $1(1)$, 23–34 (1995)

14. Leppänen, V., Penttonen, M.: Work-optimal simulation of PRAM models on meshes. Nordic J. Comput. $2(1)$, 51–69 (1995)

15. Leppänen, V., Penttonen, M., Forsell, F.: A layout for sparse cube-connected cycles network. In: Procceedings of 12th International Conference on Computer Systems and Technologies, ICPS, vol. 578, pp. 32–37. ACM Press (2011)

16. Pamunuwa, D., Zheng, L.-R., Tenhunen, H.: Maximizing throughput over parallel wire structures in the deep submicrometer regime. IEEE Trans. VLSI Syst. $11(2)$, 224–243 (2003)

17. Sun Microsystems.: Throughput Computing: Changing the Economics and Ecology of the Data Center with Innovative SPARC Technology (2005)

18. Ranade, A.: How to emulate shared memory. J. Comput. Syst. Sci. $42(3)$, 307–326 (1991)

19. Ranade, A., Bhatt, S., Johnsson, S.: The fluent abstract machine. In: Proceedings of 5th MIT Conference on Advanced Research in VLSI, pp. 71–93 (1988)

20. Salminen, E., Kulmala, A., Hämäläinen, D.: Survey of Network-on-chip Proposals, March 2008. White paper OCP-IP

21. Sibeyn, J.F.: Solving fundamental problems on sparse-meshes. In: Arnborg, S. (ed.) SWAT 1998. LNCS, vol. 1432, pp. 288–300. Springer, Heidelberg (1998)

22. Valiant, L.: General purpose parallel architectures. In: Leeuwen, J. (ed.) Handbook of Theoretical Computer Science: Algorithms and Complexity, vol. A, pp. 943–971. Elseiver, Amsterdam (1990)

23. Xu, T., Leppänen, V., Forsell, M.: DSNOC: a hybrid dense-sparse network-on-chip architecture for efficient scalable computing. In: Proceedings of ScalCom 2013 - The 13th IEEE International Conference on Scalable Computing and Communication, pp. 528–535. IEEE (2013)

24. Xu, T., Leppänen, V., Forsell, M.: Exploration of a heterogeneous concentrated-sparse on-chip interconnect for energy efficient multicore architecture. In: Proceedings of 14th IEEE International Conference on Computer and Information Technology (CIT 2014), pp. 204–211. IEEE (2014)

25. Xu, T.C., Leppänen, V.: Cache- and communication-aware application mapping for shared-cache multicore processors. In: Pinho, L.M.P., Karl, W., Cohen, A., Brinkschulte, U. (eds.) ARCS 2015. LNCS, vol. 9017, pp. 55–67. Springer, Heidelberg (2015)

Toward a Core Design to Distribute an Execution on a Manycore Processor

Bernard Goossens[1,2](✉), David Parello[1,2], Katarzyna Porada[1,2], and Djallal Rahmoune[1,2]

[1] DALI, UPVD, 66860 Perpignan Cedex 9, France
[2] LIRMM, CNRS: UMR 5506 - UM2, 34095 Montpellier Cedex 5, France
{bernard.goossens,david.parello,katarzyna.porada,
djallal.rahmoune}@univ-perp.fr

Abstract. This paper presents a parallel execution model and a core design to run C programs in parallel. The model automatically builds parallel flows of machine instructions from the run trace. It parallelizes instruction fetch, renaming, execution and retirement. Predictor based fetch is replaced by a fetch-decode-and-partly-execute stage able to compute in-order most of the control instructions. Tomasulo's register renaming is extended to memory with a technique to match consumer/producer pairs. The Reorder Buffer is adapted to parallel retirement. A *sum* reduction code is used to illustrate the model and to give a short analytical evaluation of its performance potential.

Keywords: Microarchitecture · Parallelism · Manycore · Automatic parallelization

1 Introduction

Every parallel machine programmer dreams he can run his unchanged C programs on a parallel computer.

Figure 1 shows a C version and a *pthread* version of a sum reduction function. The difference does not lie in the code text (based on the same algorithm) but in its execution. The C code is run sequentially using a stack and the *pthread* code is run in parallel with the help of the *pthread* system primitives.

This paper aims to show that if we change the execution model, the C code run can have the same behaviour as the *pthread* run, i.e. parallel execution. Section 2 explains how to run a C program in parallel. Section 3 evaluates the Instruction Level Parallelism (ILP) in benchmarks based on parallel algorithms and lists the main published works on ILP. Section 4 describes the parallel execution model and its core microarchitecture. Section 5 gives an analytical evaluation of the performance potential of the proposed model and core design. It also mentions the on-going developments of simulators and concludes.

© Springer International Publishing Switzerland 2015
V. Malyshkin (Ed.): PaCT 2015, LNCS 9251, pp. 390–404, 2015.
DOI: 10.1007/978-3-319-21909-7_38

```
typedef struct{unsigned long *p; unsigned long i;} ST;
void *sum(void *st){
  ST str1,str2; unsigned long s,s1,s2;
  pthread_t tid1, tid2;
  if (((ST *)st)->i>2){
    str1.p=((ST *)st)->p; str1.i=((ST *)st)->i/2;
    pthread_create(&tid1,NULL,sum,(void *)&str1);
    str2.p=((ST *)st)->p + ((ST *)st)->i/2;
    str2.i=((ST *)st)->i - ((ST *)st)->i/2;
    pthread_create(&tid2,NULL,sum,(void *)&str2);
    pthread_join(tid1,(void *)&s1);
    pthread_join(tid2,(void *)&s2);
  }
  else if(((ST *)st)->i==1){s1=((ST *)st)->p[0];s2=0;}
  else {s1=((ST *)st)->p[0];s2=((ST *)st)->p[1];}
  s=s1+s2; pthread_exit((void *)s);
}
```

```
unsigned long
sum(unsigned long t[], unsigned long n){
  if (n==1) return t[0];
  else if (n==2) return t[0]+t[1];
  else return sum(t,n/2) + sum(&t[n/2],n-n/2);
}
```

(a) C implementation

(b) pthread implementation

Fig. 1. A vector sum reduction: C and pthread implementations

```
1  sum:                      //sum(t,n)
2        cmpq    $2,%rsi     //n>2
3        ja      .L2         //if n>2 goto .L2
4        movq    (%rdi),%rax //rax=t[0]
5        jne     .L1         //if n!=2 goto .L1
6        addq    8(%rdi),%rax//rax+=t[1]
7  .L1:  ret                 //return(rax)
8  .L2:  pushq   %rbx        //save rbx
9        pushq   %rdi        //save t
10       pushq   %rsi        //save n
11       shrq    %rsi        //rsi=n/2
12       call    sum         //sum(t,n/2)
13       popq    %rbx        //rbx=n
```

```
14       pushq   %rbx                //save n
15       subq    $8,%rsp             //allocate temp
16       movq    %rax, 0(%rsp)       //temp=sum(t,n/2)
17       leaq    (%rdi,%rsi,8),%rdi  //rdi=&t[n/2]
18       subq    %rsi,%rbx           //rbx=n-n/2
19       movq    %rbx,%rsi           //rsi=n-n/2
20       call    sum                 //sum(&t[n/2],n-n/2)
21       addq    0(%rsp),%rax        //rax+=temp
22       addq    $8,%rsp             //free temp
23       popq    %rsi                //restore rsi (n)
24       popq    %rdi                //restore rdi (t)
25       popq    %rbx                //restore rbx
26       ret                         //return rax
```

Fig. 2. The *sum* function in X86

2 Running a C Program in Parallel

Figure 2 shows the *sum* function translation into x86 (*AT&T* syntax; rightmost operand is the destination). The code is run sequentially because the hardware is unable to *fork* at lines 12 and 20. The control flow travels along the binary tree of calls depth first, leading to a 59 instructions run trace shown on Fig. 3 (Fig. 4 left part shows the call tree for *sum(t,5)*).

Figure 5 shows a modified code for the *sum* function. The hardware is assumed to be able to fork, i.e. start a second instruction flow or *section* which occurs on lines 10 and 16 (*fork* instructions replace *call* instructions). Unlike a *call* instruction, a *fork* instruction does not save a return address.

Non volatile registers (i.e. *rbx*, *rdi* and *rsi* in this example) are copied to the forked section, replacing the stack save/restore pair. Hence *push* and *pop* are removed. The stack pointer (SP) itself (*rsp*) is copied to the forked section[1].

The *endfork* instruction ends a section. Unlike a return, the *endfork* does not give control back to a return address.

Figure 6 shows the parallel run. It starts on core 1 which fetches and executes instructions 1-1 to 1-5. The *fork* instruction starts a new section on core 2. The new section matches the resume path after fork, i.e. instruction *subq* on line 11 (Fig. 5). Core 1 continues its own section (callee path back to line 1,

---

[1] The stack in each section keeps its local variables, e.g. *temp* on Fig. 5.

```
1 sum:   cmpq $2, %rsi  //sum(t,5)          30           ja    .L2
2         ja    .L2                          31           movq  (%rdi), %rax
3 .L2:    pushq %rbx                         32           jne   .L1
4         pushq %rdi                         33    .L1:   ret
5         pushq %rsi                         34    popq   %rbx
6         shrq  %rsi                         35    pushq  %rbx
7         call  sum                          36    subq   $8, %rsp
8         sum:  cmpq $2, %rsi  //sum(t,2)    37    movq   %rax, 0(%rsp)
9               ja    .L2                     38    leaq   (%rdi,%rsi,8), %rdi
10              movq  (%rdi), %rax            39    subq   %rsi, %rbx
11              jne   .L1                     40    movq   %rbx, %rsi
12              addq  8(%rdi), %rax           41    call   sum
13        .L1:  ret                           42    sum:  cmpq $2, %rsi  //sum(&t[3],2)
14        popq  %rbx                          43          ja    .L2
15        pushq %rbx                          44          movq  (%rdi), %rax
16        subq  $8,%rsp                       45          jne   .L1
17        movq  %rax, 0(%rsp)                 46          addq  8(%rdi), %rax
18        leaq  (%rdi,%rsi,8), %rdi           47    .L1:  ret
19        subq  %rsi, %rbx                     48    addq  0(%rsp), %rax
20        movq  %rbx, %rsi                     49    addq  $8,%rsp
21        call  sum                            50    popq  %rsi
22        sum:  cmpq $2, %rsi  //sum(&t[2],3) 51    popq  %rdi
23              ja    .L2                      52    popq  %rbx
24        .L2:  pushq %rbx                     53    ret
25              pushq %rdi                     54    addq  0(%rsp), %rax
26              pushq %rsi                     55    addq  $8,%rsp
27              shrq  %rsi                     56    popq  %rsi
28              call  sum                      57    popq  %rdi
29              sum:  cmpq $2, %rsi //sum(&t[2],1)  58  popq  %rbx
                                               59    ret
```

Fig. 3. The instruction trace for the run of *sum(t,5)*.

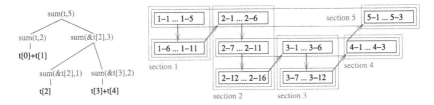

Fig. 4. The call tree (left) for the run of *sum(t,5)* and its sections (right).

instruction *cmpq*). Both sections are run in parallel, leading to the fetch and execution of instructions 1-6 to 1-11 (core 1 section) and 2-1 to 2-16 (core 2 section).

As core 2 receives valid copies of registers *rdi*, *rsi*, *rbx* and *rsp*, instructions 2-1 and 2-3 to 2-6 can be executed. Only instruction 2-2 must wait until register *rax* is set by core 1 section. The synchronisation need is easy to detect, thanks to register renaming. Instruction 2-2 consumes a source *rax* produced by the closest instruction writing to *rax* on the sequential path. As soon as instruction 1-10 writes to *rax*, the written value is forwarded to instruction 2-2.

```
1  sum:              //sum(t,n)              11  subq   $8, %rsp            //allocate temp
2  cmpq   $2, %rsi   //n>2                   12  movq   %rax, 0(%rsp)       //temp=sum(t,n/2)
3  ja     .L2        //if n>2 goto .L2       13  leaq   (%rdi,%rsi,8), %rdi //rdi=&t[n/2]
4  movq   (%rdi), %rax //rax=t[0]            14  subq   %rsi, %rbx          //rbx=n-n/2
5  jne    .L1        //if n!=2 goto .L1      15  movq   %rbx, %rsi          //rsi=n-n/2
6  addq   8(%rdi), %rax //rax+=t[1]          16  fork   sum                 //sum(&t[n/2],n-n/2)
7  .L1:   endfork    //return(rax)           17  addq   0(%rsp), %rax       //rax+=temp
8  .L2:   movq   %rsi, %rbx //rbx=n          18  addq   $8, %rsp            //free temp
9         shrq   %rsi        //rsi=n/2       19  endfork                    //return rax
10        fork   sum         //sum(t,n/2)
```

Fig. 5. The *sum* function in X86 modified by *fork* instructions.

The full run is divided by forks into 5 sections (Fig. 4, right part). Each section is framed by a red rectangle. The longest section is composed of 16 instructions (sections 2, from 2-1 to 2-16). Sections are numbered in execution trace order as indicated by the green arrows. Instructions framed by a blue rectangle belong to the same call level (e.g. instructions 1-1 to 1-5, 2-1 to 2-6 and 5-1 to 5-3 form the same call level). A section is a full recursive descent (e.g. section 1 combines 1-1 to 1-5 for $n = 5$ and 1-6 to 1-11 for $n = 2$).

Out-of-order execution is crucial to parallelize fetch. As instruction 2-2 does not block 2-6, the second call can be run in parallel with the first one.

This example shows that if the hardware is changed, the *sum* function can run in parallel as in the *pthread* implementation. However in the *pthread* or *MPI* models, any link between threads must be explicitly added to the code through OS communication primitives (e.g. MPI_Send and MPI_Recv or socket based communications in *pthread*).

In the parallel model, the sections are totally ordered. New sections are inserted in place in the list of existing sections, possibly in parallel, building the sequential trace of the run. This structure and the renaming process (which assigns a new location to each write of each instruction in the sequential trace) ensure that each read can match the most recent preceding write. In the *pthread* or *MPI* models, this sequential structuration of threads is not available.

For example in *MPI*, if x is local to task t_x and y is local to task t_y, to copy y into x task t_y sends y to task t_x, which receives it (rendezvous). In *pthread*, if x is global, threads t_x and t_y can communicate through x but they must synchronize writes and reads with *pthread* mutex. The OS must be invoked to link the sender and the receiver or to synchronize multiple writers and readers.

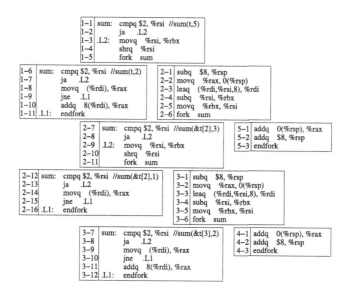

Fig. 6. The instruction trace for the parallel run of *sum(t,5)*.

In our parallel model, the equivalent of thread t_y is section s_y writing to y and the equivalent of thread t_x is section s_x reading y to copy it into x. Section s_y has instruction i_y writing to y (say i_y is $addq$ %rbx, %rax, with y in rax, i.e. $y = y + z$ for some z in rbx) and section s_x has instruction i_x reading y (say i_x is $movq$ %rax, %rcx, with rcx being x, i.e. $x = y$). Instruction i_y allocates rax_0 to rename rax destination. Instruction i_x renames its source rax. As sections s_y and s_x are ordered and no instruction updates rax between i_y and i_x, the renaming of rax in i_x matches rax_0. Moreover, the hardware synchonizes the reader i_x with the writer i_y until rax_0 is full. Hence, rendezvous or mutex are not necessary and the OS need not be sollicited.

Renaming is the key to synchronization and communication between dependent sections. Renaming should be extended to all hardware locations. For example, instruction 5-1 reads the top of stack word $0(rsp)$. This memory location is written by instruction 2-2. If instruction 2-2 destination $a = 0(rsp)$ is renamed r, instruction 5-1 renames the same address a with the same name r, exhibiting the dependency with instruction 2-2. Instruction 5-1, which computes the final sum, executes after it has received rax from section 4 (second half of the sum) and a from section 2 (first half of the sum).

3 ILP in Programs

Figure 7 displays the ILP of ten benchmarks of the PBBS suite [1]. The PBBS benchmarks implement various classical parallel algorithms (see Table 1).

On Fig. 7, for each of the 10 benchmarks, the 11 leftmost bars (those with numbered keys) match eleven parallel runs of the benchmark with increasing datasets. The rightmost bar (blue colour, $seq11$ key) matches sequential runs with the same dataset as key 11 parallel runs.

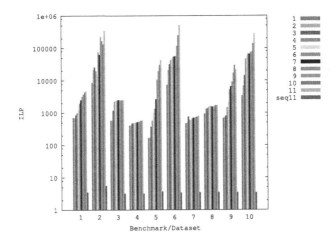

Fig. 7. ILP of ten benchmarks parallel and sequential runs (Colour figure online)

Table 1. Ten benchmarks of the PBBS suite

Benchmark	
01 : breadthFirstSearch/ndBFS	02 : comparisonSort/quickSort
03 : convexHull/quickHull	04 : dictionary/deterministicHash
05 : integerSort/blockRadixSort	06 : maximalIndependentSet/ndMIS
07 : maximalMatching/ndMatching	08 : minSpanningTree/parallelKruskal
09 : nearestNeighbors/octTree2Neighbors	10 : removeDuplicates/deterministicHash

The sequential runs consider all the dependencies excluding the register false ones (Write After Read and Write After Write), assuming an unlimited register renaming capacity, and excluding the control flow ones, assuming perfect branch prediction. The sequential runs ILP measures the ultimate performance of actual out-of-order speculative processors.

The parallel runs assume the trace is available when the run starts (no fetch delay) and in the same time all the destinations (including memory) are renamed. The SP dependencies are not considered. The parallel runs ILP measures the ultimate performance an ideal parallel machine achieves when the run order only depends on the producer to consumer dependencies, excluding the SP. Each instruction on the trace is run at the cycle next to the last source reception. The processor is assumed to run all the ready instructions in the same cycle with a single cycle latency (as in the sequential runs).

All runs are continued until completion. For each benchmark, the 11 parallel runs vary from 1M to 1G instructions (increasing factor 2) and the sequential runs are 1G instructions long. The figure shows that sequential runs have a very low ILP (ranging from 3.2 to 5.6) and parallel runs have a very high ILP (ranging from 600 to 508 K for dataset 11). The difference comes from the dominating distant ILP. Moreover, when a benchmark is data parallel its parallel run ILP increases proportionally to the dataset (e.g. benchmarks 1, 2, 5, 6, 9 and 10).

The sequential run ILP we have measured confirms ILP reported values (such as [2]). Since 50 years many successive research works on ILP were published.

In 1967, Tomasulo [3] presented an algorithm to run floating point instructions out-of-order. He introduced register renaming which is still used in today's speculative cores to parallelize on-the-fly instructions. In 1970, Tjaden and Flynn [4] measured the available parallelism in a 10 instructions window. They ran their test programs at 1.86 instructions per cycle.

In 1984, Nicolau and Fisher [5] measured the available parallelism to feed a VLIW processor. In their experiments, they included a measure of ILP from runs on an ideal machine with infinite resources. They discovered that scientific codes present a high ILP, over 1000.

In 1991, David Wall presented the first study centered on ILP [2]. He measured that the available parallelism a "real" processor finds in 13 benchmarks is

5 on average, ranging from 3 to 45^2. In an "ideal" processor[3], ILP ranges from 6 to 60 with an average at 25. From this study, we know that there is ILP but it seems impossible to catch more than 5 independent instructions per cycle.

In 1992, Austin and Sohi [6] measured the SPEC89 suite ILP and analyzed its distribution. They showed that ILP is arbitrarily distant from the instruction pointer. They also pointed out the serializing effect of the stack manipulations. The same year, Lam and Wilson [7] studied the impact of control on ILP. Their measures showed that a processor with a perfect branch predictor could dramatically improve its performance. As in Austin and Sohi work, distant ILP was detected. To capture this distant ILP, a processor must be able to speculate on the control flow and use multiple instruction pointers. In 1997, Moshovos and Sohi have proposed memory renaming in [8], using a predictor to find the store renaming a load. In 1999, Postiff et al. [9] measured SPEC95 suite ILP. They pointed out that the stack introduces many parasitic dependencies. To capture distant ILP, the application should be multi-threaded.

In 2004, Cristal et al. [10] described a kilo-instructions microarchitecture. The authors suggested that to capture more ILP, the processor must have access to instructions far from the fetch point. They gave solutions to allocate later and free sooner the needed resources to optimize their usage and so, take care of more "on-the-fly" instructions with the same resources. In 2012, Sharafeddine et al. [11] proposed an architecture to partition a run into parallel threads, forking the leading thread at call. In the sum example this leads to fork on both of the highest levels calls but not on the lower levels, capturing only a small part of the distant ILP. In 2013, Goossens and Parello [12] analyzed distant ILP and showed that ILP could be highly increased when removing stack pointer updates and false memory dependencies.

From these works, we deduce that (i) high ILP is available, (ii) most of it comes from very distant instructions and (iii) sequential fetch and stack are the main obstacles on the ILP capture. Two ideas are suggested to help capture distant ILP: following multiple instruction flows [7] and renaming memory [8].

4   An Execution Model to Run Programs in Parallel and Its Core Implementation

In Sect. 3 we assumed the full trace is available at run start and all the destinations are pre-renamed. This is not realistic. However, code fetch and destinations renamings should occur as soon as possible to allow distant ILP capture.

4.1   Parallelizing Fetch

A section is composed of dynamically contiguous instructions. A section starts when a *fork* instruction creates it. It ends when an *endfork* instruction is reached.

---

[2] "Good" model with a 2 K instructions window size, 64 instructions issued per cycle, 256 renaming registers, a branch predictor based on an infinite number of 2-bits counters and a perfect memory aliasing disambiguation.

[3] "Perfect" model enhances "good" model: infinite renaming, perfect branch predictor.

A control flow instruction (jump, call or branch) does not end a section. The same section continues after the control flow instruction.

When a new section is forked, a message is sent to a hosting core[4]. The message contains the forked Instruction Pointer (IP), its SP and the set of non volatile registers. The registers copies remove stack push/pop, i.e. stack and SP dependencies. The message also contains the identification of the neighbour sections (e.g. the current creating section). The choosen core queues the message while it fetches another section. When the section creation message is dequeued, it fills the register file local to the fetch pipeline stage. The IP, the SP and the non volatile registers are initialized and other registers are emptied.

For example, when instruction 1-5 forks, a section creation message is sent to core 2 (say), including register rdi value t, register rsi value 2 and register rbx value 5. When instruction 2-6 forks, the SP is transmitted to section 5, pointing on the same stack word as section 2. Hence, sections 2 and 5 share the same stack portion. Thanks to memory renaming[5], when instruction 5-1 reads stack word 0, it matches with instruction 2-2 write to stack word 0. Both instructions compute the same address $a = rsp + 0$.

The fetch pipeline stage fetches along the section pointed to by IP. The fetch stage has no branch predictor. There are two reasons for such a choice. First, moving fast along the flow is better obtained by a parallel fetch along multiple control-computed sections than by a sequential fetch along a single predicted path, even if the prediction is perfect. Second, a predictor is less cost-effective in a core if the flow is divided into sections and distributed on multiple cores. For these reasons, the fetch stage computes its control rather than predicting it. To keep the stage hardware simple, each cycle fetches and computes a single instruction. As a result, each core fetches more slowly than an actual speculative core but the cores fetch much faster altogether.

Figure 8 shows the fetch-and-decode pipeline stage. The IP addresses the Instruction Memory Hierarchy (IMH, i.e. L1 instruction cache). The fetched instruction addresses the Register File (RF) to read full registers sources. If all the needed sources are full, the instruction is computed in the ALU. Floating point instructions, memory accesses, complex integer instructions and instructions having empty sources are not computed in the fetch stage but later[6]. Computed instructions results are written back to RF, setting the destination register to full. Uncomputed instructions set their destination register(s) to empty.

The fetch stage includes instruction decoding (not shown). When a *fork* instruction is decoded, it generates a section creation message. The created section starts at the next instruction. The current section continues at the *fork* instruction target. It ends when an *endfork* instruction is decoded. Then, IP

---

[4] Hosting core choice to optimize load balancing is out of the scope of this paper.

[5] Memory renaming duplicates same address based stack frames. This allows multiple sections to update their local variables in their frames in parallel.

[6] In the *sum* example, the conditional branches are all computed in the fetch stage, allowing the parallelization of the fetch by fetching fastly the fork instructions.

register is set to empty and at the next cycle the fifo head message is dequeued and IP and RF are initialized, which starts the fetch of a new section.

As mentioned on Fig. 8, the stage critical path is longer than in a speculative out-of-order pipeline, including a L1 cache traversal, an instruction decoding, a register file read (2 read ports), an ALU (Arithmetic and Logic Unit) computation and a register file write (1 write port). This leads to a slow frequency processor. Core slowness is to be compensated by parallelism.

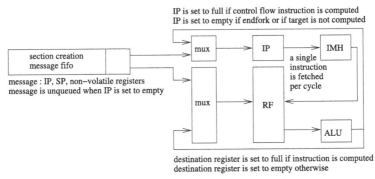

Fig. 8. Fetch-decode pipeline stage

4.2   Core Pipeline Microarchitecture

Figure 9 shows the six-stages pipeline building the core microarchitecture. On the bottom part of the design we find a full size rectangle dedicated to communications with other cores in the processor chip (assumed to be connected by a Network-on-Chip). The forking request unit (FRU) handles the income/outcome of section creation messages. The register renaming request unit (RRRU) handles the income/outcome of source registers renamings. The register exporting request unit (RERU) handles the import/export of renamed registers values. The address renaming request unit (ARRU) handles the income/outcome of source memory addresses renamings. The memory exporting request unit (MERU) handles the import/export of renamed memory values. The instruction exporting request unit (IERU) handles the outcome of retired instructions.

The fetch-decode and register-rename stages follow a single section up to its end. Renamed instructions enter in-order in a Reorder Buffer (ROB in the retire stage) and in the Instruction Queue (IQ in the execute-write-back stage). Load/store instructions enter in-order in the address renaming queue (ARQ in the address-rename stage). Register-register instructions from multiple sections are mixed in the execute-write-back stage. They read sources in a memory keeping the core renamed registers (register renaming memory or RRM). Load/store instructions compute the access address in the execute-write-back stage and save

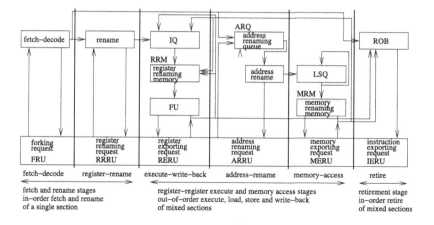

Fig. 9. Six-stages core pipeline

it in the ARQ. Memory addresses in ARQ are renamed in-order and renamed memory access instructions enter the Load/Store Queue (LSQ). As these instructions are renamed they can be run out-of-order.

Register Renaming. Each instruction in the core can be uniquely identified by its section identifier and its ordinal number in the section. If we assume the number of sections hosted by a core is bounded by *max_section* and the number of instructions in a section is bounded by *max_instruction*, a core can host at most *max_section* * *max_instruction* instructions, i.e. as many renamed destinations. Each renamed destination can be uniquely identified by a pair (#section, #instruction) (or (s,i) in short).

The fetch-decode stage delivers a partially evaluated instruction to the rename stage which renames the empty sources, i.e. either find their local (s,i) renaming or, if not hosted by the local core, look for the producing core.

If a source *s* may not be locally renamed by an instruction *inst* (no instruction previously fetched in the same section has written to *s*), its value is requested to the preceding section, i.e. to another hosting core through the RRRU. In the same time, a destination *d* is allocated in RRM for the missing register, as if *s* would be locally written. This destination *d* serves as a caching of the missing source *s*. Later references to *s* in the same section are renamed *d*.

The renaming request travels from section to section until a producer is found (i.e. an instruction writing to *s*). In the *sum* function example, the only register to be renamed is register *rax* in instructions 12 and 17. In both cases, for any size of the data, the producer is the section just preceding the renaming one.

Each core on the travel receives the renaming request in its RRRU. It renames source *s*. If the renaming misses, the request is propagated through the RRRU. If it hits, an export instruction is added to the IQ where it waits for the requested value. When it is written, the export instruction is notified in IQ and run. It reads the value in RRM and send it to the requester through the RERU.

The value reaches the requesting core through its RERU. It is written in RRM, entry d. The IQ is notified that destination d is ready, which allows the waiting instruction $inst$ to start execution and read d in RRM as source s.

Renaming seems very sequential. To find the producer of source s, the trace of executed instructions must be travelled backward from the consumer down to the first instruction writing to s. However, (i) only the portion of code ranging from the producer to the consumer is to be visited and (ii) SP based variables with a positive offset (e.g. $0(rsp)$) benefit from a shortcut eliminating instructions belonging to a call level deeper than the consumer. Statement (i) implies that if a producer is close from a consumer, the portion of code to consider is short. This is the case for function results used by the resume code (register rax in the sum function example). Statement (ii) implies that if a consumer and a producer address the same stack frame, the portion of code to consider is also short, excluding in between function calls. This is the case for local variables set at function start and later used (stack location $0(rsp)$ in the sum example).

Only for global variables and heap pointers the travel from producer to consumer can represent a long path, as all the in between sections must be visited to make sure they do not contain any more recent producer of the consumed address. However, the caching feature ensures that the high price is rarely paid. Once renamed in an intermediary consuming section, a global or heap variable is cached and it can be consumed by neighbour sections for cheap.

Instruction 2-2 on Fig. 6 illustrates fast renaming applying statement (i). After the renaming of register rax misses in section 2, a request is sent to the core hosting section 1. The renaming hits in section 1 (instruction 1-10) and $t[0] + t[1]$ in rax is sent to the core hosting section 2.

Instruction 5-1 illustrates fast renaming applying statement (ii). After the renaming of stack location $0(rsp)$ at address a misses in section 5, a request is sent to section 2, bypassing sections 3 and 4 which are at a lower call level than instruction 5-1. The renaming hits in section 2 (instruction 2-2) and $t[0] + t[1]$ in $0(rsp)$ is sent to the core hosting section 5.

Instruction 1-8 illustrates high price renaming of global variable $t[0]$. The request travels back to the loader which installs code and global initialized data. The hardware can (i) access to full cache lines instead of single words and (ii) cache the accessed lines along the return path. From statement (i), instruction 1-8 gets its own word $t[0]$ but also instruction 1-10 word $t[1]$. Moreover, from statement (ii), core 1 caches the memory line containing $t[0]$ up to $t[4]$ which can be consumed cheaply by sections 2 ($t[2]$) and 3 ($t[3]$ and $t[4]$)[7].

Memory Renaming. Memory renaming is done like register renaming. Instead of a Register Alias Table (RAT), the address-rename stage uses a Memory Address Alias Table (MAAT). There is one MAAT per section, each MAAT

---

[7] Stores update full lines. The loader sets a cleared line and loops to update it successively with $t[0]$ up to $t[4]$. The full line right padded with zeros is exported to its first consumer, i.e. section 1. Sections 2 and 3 get section 1 cached copy.

having one entry per instruction in the section. Each MAAT is a fully associative cache. Renaming address a in section s means looking for a in section s MAAT. If the search misses, it indicates that section s does not write to a and the renaming should be looked for in the section preceding s.

A memory renaming request works like a register one. When renaming address a misses, a memory line is allocated in the MRM to host line l_a containing a (it caches l_a). The renaming request travels along contiguous sections until a producer of l_a is found. Each visited core receives the request in its ARRU. The renaming request is enqueued in the ARQ to avoid bypassing renamings of addresses of the same section not yet done. When the request is dequeued, if the renaming misses, it is propagated to the preceding section. When it hits, an instruction to export l_a is added to the LSQ. The exported memory line travels back to the requesting core where it is received in the MERU. From there, it is written to the MRM and the LSQ is notified that a is ready.

Memory renaming transforms the code at run time into a single assignment form. Synchronisation of consumers with their producers and single assignment ensure sequential consistency without any coherency protocol requirement. Hence, the cores distributed memory is coherent.

Parallelizing Retirement. Sections are created in parallel by *fork* instructions. To keep cores loads acceptable (at most *max_section* hosted in a core), terminated sections should retire at the same speed, i.e. retirement should be parallelized. Retirement frees the sections in the cores to allow new sections in.

Instructions retire in-order (within their section) by exporting their result to the successor section[8]. To be retired, an instruction must be terminated. An instruction is not exported if it holds a result useless for successors, i.e. if (i) its destination is updated later in the section or (ii) it writes to a non volatile register or (iii) it is a control flow instruction or (iv) it writes to stack or heap in a location freed later in the section or (v) it has exported its computation to a consumer. The successor discards exported instructions if their production may not be consumed anymore. An exported instruction i is discarded by the successor section s' if i writes to a destination renamed or freed in s'.

In the *sum* function example, no instruction is exported. For example, the instruction i consuming the final sum s to be displayed renames s and receives the value v exported from instruction 5-1. Statement (v) says that instruction 5-1 retirement does not export v to the next section. Instruction i renaming caches v which can be consumed by later instructions.

An exported instruction is sent to the successor section through the IERU. It is received in the RRRU (register write) or in the ARRU (memory write). The destination is tentatively renamed and in case of a hit the exported instruction is discarded (already renamed destination). It is also discarded if it writes to

---

[8] The oldest section, i.e. the only one with no predecessor, dumps its renamings to the data memory hierarchy (DMH). When it receives a renaming request which misses, it loads from DMH and exports the loaded line.

stack or heap in a location later freed by the section. Otherwise, the instruction gets a new local renaming and is saved in its new section ROB.

What is New in the Proposed Core Design? The core shown on Fig. 9 is much simpler than actual speculative cores. The core is as small as possible to maximize the number of cores on the die. As the core is not speculative, there is no predictor nor renaming repair unit (checkpoints). The LSQ is simpler than the usual Load-Store Queue as loads are not speculative and store-to-load forwarding is not necessary. The data memory hierarchy is kept coherent as only the oldest section can write to and read from it. There is no memory coherency hardware (e.g. MESI protocol handler). There is no vector computing unit (e.g. XMM-like) as vectorization is better obtained through parallelism. Each core implements a single-issue pipeline (no superscalar or VLIW).

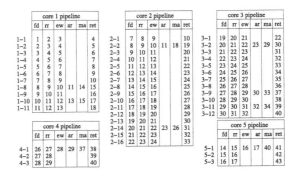

Fig. 10. Execution timing of the *sum(t,5)* run.

5   Analytical Performance Evaluation of the Parallel Execution Model on the *sum* Example and Conclusion

Figure 10 shows 5 tables (one per core) giving the execution timing of the *sum(t,5)* run. The instructions numbers are given on the left of each table. The columns of a table match the 6 pipeline stages. A value in a column represents the cycle at which the instruction is treated by the corresponding pipeline stage. For example, instruction 1-8 (load) is fetched at cycle 8, register renamed at cycle 9, load address is computed at cycle 10 and renamed at cycle 11, renamed memory is accessed at cycle 14 (counting 3 cycles to reach the producer and return the $t[0]$ value) and retired at cycle 15. We assume the sections can be hosted in different cores. We also assume a single-issue pipeline. Instruction cache L1 is assumed to always hit. From this example we see that the code is fetched in 30 cycles (last fetched instruction is 3-12), i.e. 1.5 instructions per cycle. If the data size is doubled, the fetch time is 42 cycles (104 instructions fetched, i.e. 2.5

instructions per cycle). Fetch latency (i.e. IL1 miss rate) can impact the fetch time which impacts the total run time. The more the code is parallel, the more the total run time is independent of renaming and execution latencies.

The number of instructions is $45 * 2^n + 14 * (2^n - 1)$ for the sum of a $5 * 2^n$ elements array (i.e. 45 for $sum(t,5)$, 104 for $sum(t,10)$). The fetch time is $30 + 12 * n$ (i.e. 30 for $sum(t,5)$, 42 for $sum(t,10)$). For 1280 elements, 15090 instructions are fetched in 126 cycles, i.e. 120 instructions per cycle. This shows that even though one instruction is fetched per cycle per core and the control is computed rather than predicted, fetching in parallel is efficient. Even for modest data sizes, it outperforms any speculative fetching hardware.

Renaming is not penalized by the code distribution. Most of the sources are provided by register value copy at fork. Function results and stack local variables are obtained from the predecessor section. Global variables are reached quickly from the first section (which starts at the *main* function entrance) and then fastly propagated to other sections, thanks to full memory line caching. Eventually, instruction retirement frees the resources at the same rate they are allocated, avoiding cores saturations. The retirement time is[9] $43 + 15 * n$. For 1280 elements, the 15090 instructions are retired in 163 cycles, i.e. 92 instructions per cycle.

Devil is in the details. The estimation of the *sum* run timing is not precise enough to prove that the model is successful. However, it shows that distant ILP is captured: the run is parallelized in a divide-and-conquer way. To fix the Instruction Per Cycle (IPC) of a manycore processor based on our core design, an elaborate simulation is necessary. Two such simulators are on-going projects: a VHDL implementation of the core pipeline to prove hardware feasability and a *qemu* and *simplescalar* based simulator to quantify IPC.

The model presented focuses on functions. In the same way, loops can be parallelized. *For* loops can be vectorized, each iteration forming a separate section with no control. It heritates its iteration counter that can be saved in a register and used in the iteration body. *While* loops can be parallelized, launching each iteration in sequence (no speculation) but parallelizing their bodies.

With the introduction of manycore chips as general purpose processors, the time has come to produce parallel programs automatically. This paper suggests that we are not so far from the goal and the hardware can greatly help.

References

1. Shun, J., Blelloch, G.E., Fineman, J.T., Gibbons, P.B., Kyrola, A., Simhadri, H.V., Tangwongsan, K.: Brief announcement: the problem based benchmark suite. In: Proceedings of the 24th ACM Symposium on Parallelism in Algorithms and Architectures, SPAA 2012, pp. 68–70 (2012)

---

[9] 15 cycles is the fetch time of instructions (Fig. 5) 2, 3, 8-10 (5 cycles), the creation time of the forked section (2 cycles), the fetch time of instructions 11-16 (5 cycles) and the retirement of instructions 17-19 (3 cycles).

2. Wall, D.W.: Limits of instruction-level parallelism. In: WRL Technical Note TN-15 (1990)
3. Tomasulo, R.M.: An efficient algorithm for exploiting multiple arithmetic units. IBM J. Res. Dev. **11**, 25–33 (1967)
4. Tjaden, G.S., Flynn, M.J.: Detection and parallel execution of independent instructions. IEEE Trans. Comput. **19**, 889–895 (1970)
5. Nicolau, A., Fisher, J.: Measuring the parallelism available for very long instruction word architectures. IEEE Trans. Comput. **C–33**, 968–976 (1984)
6. Austin, T.M., Sohi, G.S.: Dynamic dependency analysis of ordinary programs. In: Proceedings of the 19th Annual International Symposium on Computer Architecture, ISCA 1992, pp. 342–351 (1992)
7. Lam, M.S., Wilson, R.P.: Limits of control flow on parallelism. In: Proceedings of the 19th Annual International Symposium on Computer Architecture, ISCA 1992, pp. 46–57 (1992)
8. Moshovos, A., Breach, S.E., Vijaykumar, T.N., Sohi, G.S.: Dynamic speculation and synchronization of data dependences. In: Proceedings of the 24th Annual International Symposium on Computer Architecture, ISCA 1997, pp. 181–193 (1997)
9. Postiff, M.A., Greene, D.A., Tyson, G.S., Mudge, T.N.: The limits of instruction level parallelism in SPEC95 applications. In: CAN, vol. 27, pp. 31–34 (1999)
10. Cristal, A., Santana, O.J., Valero, M., Martínez, J.F.: Toward kilo-instruction processors. ACM Trans. Archit. Code Optim. **1**, 389–417 (2004)
11. Sharafeddine, M., Jothi, K., Akkary, H.: Disjoint out-of-order execution processor. ACM Trans. Archit. Code Optim. (TACO) **9**, 19:1–19:32 (2012)
12. Goossens, B., Parello, D.: Limits of instruction-level parallelism capture. Procedia Comput. Sci. **18**, 1664–1673 (2013). 2013 International Conference on Computational Science

Heuristic Algorithms for Optimizing Array Operations in Parallel PGAS-programs

Ivan Kulagin[1][✉], Alexey Paznikov[1,2], and Mikhail Kurnosov[1,3]

[1] Siberian State University of Telecommunications and Information Sciences,
86 Kirova street, Novosibirsk, Russia 630102
ikulagin@sibsutis.ru

[2] Rzhanov Institute of Semiconductor Physics of the Siberian Branch of the RAS,
13 Lavrentev avenue, Novosibirsk, Russia 630090
apaznikov@isp.nsc.ru

[3] Saint Petersburg Electrotechnical University "LETI", 5 Professor Popov street,
Saint-Petersburg, Russia 197376
mkurnosov@gmail.com

Abstract. The algorithms for optimizing array operations in PGAS-Programs are represented. They minimize execution time by taking into account hierarchical structure of computer systems in reduction and by preloading of remote elements to nodes while accessing distributed arrays. Algorithms are implemented for Cray Chapel and IBM X10.

Keywords: PGAS · Compiler optimization · Reduction · Scalar replacement

1 Introduction

The main approach to the parallel programs development in modern distributed computer systems (CS) is message-passing interface (MPICH2, Open MPI, Intel MPI). The major challenge for modern CS is their lack of programmability. To exploit all the resources of modern systems, we need to use diverse technologies (OpenMP/Intel TBB/Intel Cilk Plus, NVIDIA CUDA/OpenCL, SSE/AVX) in conjunction with the MPI. While this model provides a great deal of flexibility and performance potential, it burdens programmers with the complexity of utilizing multiple programming systems in the same applications.

Need of simplification parallel programming has lead the development of high-level tools, e.g. the languages that implement the model of a partitioned global address space (PGAS), including Cray Chapel, IBM X10, UPC. PGAS-programs does not explicitly call communication functions unlike MPI; instead they operate with distributed structures and instructions for parallel tasks management (threads, activities) and synchronization. All the communications are

The reported study was partially supported by RFBR, research projects 15-37-20113, 15-07-02693 and by Ministry of Education and Science of the Russian Federation (02.G25.31.0058 from 12.02.2013).

scheduled by the compiler and performed by the runtime-system which provides the transparent access to remote nodes' memory. High abstraction level of PGAS allows to reduce the complexity of parallel programs development, but requires the development of effective methods for optimizing compilation.

One can emphasize the two most common patterns in parallel PGAS-programs: (i) iteration by distributed arrays, (ii) specified reduction operation for distributed arrays' elements (reduce, reduction). The existing algorithms of operations on distributed arrays [3,4] does not take into account the features of PGAS, such as high intensity of one-side communications, memory consistency, multithreading, etc. The compiler optimization algorithms implemented in IBM X10 [1], UPC [2] do not minimize overheads in PGAS-programs that perform cyclic access to the array's elements, located on the remote nodes.

In this paper, we propose the algorithms for optimizing the communications in operations on distributed arrays. The algorithms are implemented for Cray Chapel and IBM X10.

2   Communications Optimization

2.1   PGAS Model

Let $P = \{1, 2, \ldots, N\}$ is the set of SMP/NUMA-nodes of a distributed CS. Each node $i \in P$ consists of n CPU cores and the local memory.

PGAS model realizes the abstraction of a multicore node – locale (region, place). Each locale manages its own local memory segment. Dynamically spawned tasks (activities, threads) run within the locale. A task can access the global address space comprised nodes' local memory segments. The local segment access performs much faster, because the access to the remote ones demands the communications. The design units required for developing of PGAS-programs: *begin S* – performs the instructions S asynchronously on the separate thread, *on i S* – performs the instructions S on the node i, *on x S* – performs the instructions S on the node which owns the object x, *coforall S* – performs each iteration of the loop body S in the independent thread, *sync T* – the synchronization variable.

2.2   Parallel Reduction Algorithm

Reduction is the collective operation, which performs some associative operation $\bigotimes$ with the distributed array $V[1 : D]$. The result r of the operation is placed in the memory of the thread initialized reduction: $r = V[1] \bigotimes V[2] \bigotimes \ldots \bigotimes V[D]$.

This paper offers the algorithm *BlockReduce* of reduction in PGAS-programs (Fig. 1). In Fig. 2 you can see the algorithm for Cray Chapel.

Each node $i \in P$ is aware of the set V_i of array's V elements storing in its local memory. In the first stage (Fig. 2, lines 3–17) of the sub-arrays V_i are splitted into n parts (by number of cores) (Fig. 2, line 6). Then these parts are processed in parallel. The threads $t = 1, 2, \ldots, n$ of each node i perform reduction with their sub-array V_{it} (Fig. 2, lines 7–11).

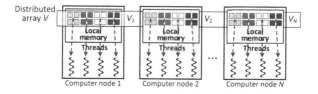

Fig. 1. Distribution of array elements in the algorithm *BlockReduce*

Input: $V[1 : D]$ – distributed array, $\bigotimes$ – operation for reduction.
Output: r – reduction result for the array V.
1: **procedure** BLOCKREDUCE($V[1 : D]$, $\bigotimes$)
2:   ▷ Parallel computation of operation $\bigotimes$ over the local elements in the locales
3:   **coforall** i **in** $[1, 2, \ldots, N]$ **do**
4:     **on** i
5:       ▷ Split V_i to n (cores number) blocks V_{it}
6:       SPLITARRAY(V_i, n)
7:       **coforall** t **in** $[1, 2, \ldots, n]$ **do**
8:         **forall** x **in** V_{it} **do**
9:           $r[i][t] \leftarrow r[i][t] \bigotimes x$
10:          **end for**
11:        **end coforall**
12:        **forall** t **in** $[1, 2, \ldots, n]$ **do**
13:          ▷ $r[i]$ is the reduction of elements V_i located on the locale i
14:          $r[i] \leftarrow r[i] \bigotimes r[i][t]$
15:        **end for**
16:      **end on**
17:    **end coforall**
18:    **return** $r \leftarrow$ BINTREE($r[1 : N]$)
19: **end procedure**

Fig. 2. Algorithm *BlockReduce*

On the second stage the nodes organize the binary tree with the root is first locale. Each operation $\bigotimes$ for the pair of values $r[first], r[second]$ is performed in the separate thread on that node, wherein the value $r[first]$ is located. After the reduction of all the values the barrier is performed.

Barrier may be implemented e.g. by *Dissemination barrier* algorithm ($O(\log N)$). Then *BlockReduce* complexity equals $O(T = O(|V|/N + \log N)$. In the current implementation we used the *Centralized barrier* with the time $O(N)$.

2.3 Arrays Access Optimization

Another common pattern in PGAS-programs is the looping through the elements, wherein the threads access the elements in the memory of other nodes (Fig. 3a). In this case, the runtime-system provides required elements.

In current days PGAS-compilers use relatively straightforward heuristics. Accessing to a remote array's element causes the copying entire array to the

local memory. Though copying the whole array is redundant and incurs communication overheads. *Scalar replacement* algorithm [1,2] reduces these overheads.
While the looping through remote array's elements, runtime-system copies to
local memory the entire array at each iteration (Fig. 3a). That scheme is highly
inefficient. *Scalar replacement* also may cause the redundant copying in loops
because the total number of sent elements exceeds the entire array.

We propose the *ArrayPreload* algorithm optimizing the looping access to
remote arrays for minimizing communication time. *ArrayPreload* prevents multiple copying of remote arrays by preemptive copying the array once before loop
iterations (Fig. 3b). Figure 3 shows the example of optimization of array A access
for IBM X10 language. Unoptimized version (Fig. 3a) incurs passing the array A
to the id node on each iteration. The optimization (Fig. 3b) involves the copying
array A in advance to every node to the local array $localA$. The statement at
used in IBM X10 corresponds the statement on.

```
for (i in 0..R) {
    val id: Long = i % Places.MAX_PLACES;
    at (Place.place(id)) {
        // The using one element of A copies the
        // entire array to the node placeId
        var a: Long = A(i % Size);
    }
}
```

```
// Loop prologue copies A on every node one time
// saving the A to distributed array LocalA
val localA: DistArray[Array[Long]] = ...
for (i in 0..R) {
    val id: Long = i % Places.MAX_PLACES;
    at (Place.place(id)) {
        // Using the local copy LocalA
        // of array A
        var a: Long = localA(id)(i % Size);
    }
}
```

(a) Version without optimization   (b) Optimized version (*ArrayPreload*)

Fig. 3. Example of optimization by passing the array A in a IBM X10 program

The *ArrayPreload* algorithm is based on static analyze by Abstract Syntax
Tree (AST) traversal. The first stage realizes the search of the loops with accessing remote array elements. The second checks if array is not changed during the
loop iterations so as to avoid violation the original program during optimization.
The way this examination depends on compiler implementation, e.g. this check
may be implemented on base of previously built loop context.

The third stage makes AST transformation which includes loop prologues for
each found arrays. The prologue performs coping a remote array to local memory
once before iterations. The remote array access is replaced by access to the
local one copied by prologue loop. Computational complexity of the algorithm
ArrayPreload is determined of the AST height.

3   Experiments and Results

Experiments are carried out on the cluster A (16 nodes: 2 x Quad-Core Intel
Xeon E5420, Gigabit Ethernet) and cluster B (6 nodes: 2 x Quad-Core Intel Xeon
E5420, Infiniband QDR). The algorithms are implemented for the languages
Cray Chapel (*BlockReduce*) and IBM X10 (*ArrayPreload*).

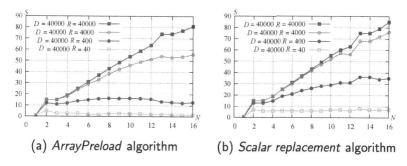

(a) *ArrayPreload* algorithm   (b) *Scalar replacement* algorithm

Fig. 4. Speedup of test program (cluster A)

The evaluation of reduction algorithms was done on the basis of microbenchmarks (reduction of distributed array of length $D = 4000, \ldots, 20000$) and Chapel programs PTRANS (transposition of distributed matrices) and miniMD (molecular dynamics). Node number N was varying from 1 to 16.

BlockReduce efficiency depends on the N and D. Algorithm outperforms by 10–30 % the default algorithm *DefaultReduce*. Slight benefit on the real programs is due to the reduce computation time is much less than the total execution time.

For the efficiency evaluation of *ArrayPreload* and *Scalar replacement* the microbenchmark was used. The benchmark performs the looping through the array's elements placed in the memory of remote nodes.

Both *ArrayPreload* and *Scalar replacement* perform the speed-up from 5 to 82 times (Fig. 4). Generally the efficiency depends on the interconnect performance, the number of nodes, the size of array, the number of iterations.

4   Conclusion

The proposed algorithms reduces the execution time of PGAS programs by means of minimizing the communication overheads. It has been achieved by preemptive copying of remote arrays and taking into account the computer system structure. The algorithms may be used for the wide range of PGAS languages.

References

1. Barik, R., Zhao, J., Grove, D., Peshansky, I., Budimlic, Z., Sarkar, V.: Communication optimizations for distirbuted-memory X10 programs. In: IEEE International Parallel and Distributed Processing Symposium, pp. 1–13 (2011)
2. Chen, W., Iancu, C., Yelick, K.: Communication optimizations for fine-grained UPC applications. In: 14th International Conference on Parallel Architectures and Compilation Techniques (PACT), pp. 267–278 (2005)
3. Kurnosov, M.: All-to-all broadcast algorithms in hierarchical distributed computer systems. Vestnik of computer and information technologies **5**, 27–34 (2011). [in Russian]
4. Rabenseifner, R.: Optimization of collective reduction operations. In: Bubak, M., van Albada, G.D., Sloot, P.M.A., Dongarra, J. (eds.) ICCS 2004. LNCS, vol. 3036, pp. 1–9. Springer, Heidelberg (2004)

Progressive Transactional Memory in Time and Space

Petr Kuznetsov[1] and Srivatsan Ravi[2]([✉])

[1] Télécom ParisTech, Paris, France
petr.kuznetsov@telecom-paristech.fr
[2] TU Berlin, Berlin, Germany
srivatsan.ravi@inet.tu-berlin.de

Abstract. Transactional memory (TM) allows concurrent processes to organize sequences of operations on shared *data items* into atomic transactions. A transaction may commit, in which case it appears to have executed sequentially or it may *abort*, in which case no data item is updated.

The TM programming paradigm emerged as an alternative to conventional fine-grained locking techniques, offering ease of programming and compositionality. Though typically themselves implemented using locks, TMs hide the inherent issues of lock-based synchronization behind a nice transactional programming interface.

In this paper, we explore inherent time and space complexity of lock-based TMs, with a focus of the most popular class of *progressive* lock-based TMs. We derive that a progressive TM might enforce a read-only transaction to perform a quadratic (in the number of the data items it reads) number of steps and access a linear number of distinct memory locations, closing the question of inherent cost of *read validation* in TMs. We then show that the total number of *remote memory references* (RMRs) that take place in an execution of a progressive TM in which n concurrent processes perform transactions on a single data item might reach $\Omega(n \log n)$, which appears to be the first RMR complexity lower bound for transactional memory.

Keywords: Transactional memory · Mutual exclusion · Step complexity

1 Introduction

Transactional memory (TM) allows concurrent processes to organize sequences of operations on shared *data items* into atomic transactions. A transaction may *commit*, in which case it appears to have executed sequentially or it may *abort*, in which case no data item is updated. The user can therefore design software having only sequential semantics in mind and let the TM take care of handling

---

Petr Kuznetsov—The author is supported by the Agence Nationale de la Recherche, ANR-14-CE35-0010-01, project DISCMAT.

© Springer International Publishing Switzerland 2015
V. Malyshkin (Ed.): PaCT 2015, LNCS 9251, pp. 410–425, 2015.
DOI: 10.1007/978-3-319-21909-7_40

conflicts (concurrent reading and writing to the same data item) resulting from concurrent executions. Another benefit of transactional memory over conventional lock-based concurrent programming is *compositionality*: it allows the programmer to easily compose multiple operations on multiple objects into atomic units, which is very hard to achieve using locks directly. Therefore, while still typically *implemented* using locks, TMs hide the inherent issues of lock-based programming behind an easy-to-use and compositional transactional interface.

At a high level, a TM implementation must ensure that transactions are *consistent* with some sequential execution. A natural consistency criterion is *strict serializability* [19]: all committed transactions appear to execute sequentially in some total order respecting the timing of non-overlapping transactions. The stronger criterion of *opacity* [13], guarantees that *every* transaction (including aborted and incomplete ones) observes a view that is consistent with the *same* sequential execution, which implies that no transaction would expose a pathological behavior, not predicted by the sequential program, such as division-by-zero or infinite loop.

Notice that a TM implementation in which every transaction is aborted is trivially opaque, but not very useful. Hence, the TM must satisfy some *progress* guarantee specifying the conditions under which a transaction is allowed to abort. It is typically expected that a transaction aborts only because of *data conflicts* with a concurrent one, e.g., when they are both trying to access the same data item and at least one of the transactions is trying to update it. This progress guarantee, captured formally by the criterion of *progressiveness* [12], is satisfied by most TM implementations today [6,7,14].

There are two design principles which state-of-the-art TM [6–8,11,14,21] implementations adhere to: *read invisibility* [4,9] and *disjoint-access parallelism* [5,16]. Both are assumed to decrease the chances of a transaction to encounter a data conflict and, thus, improve performance of progressive TMs. Intuitively, reads performed by a TM are invisible if they do not modify the shared memory used by the TM implementation and, thus, do not affect other transactions. A disjoint-access parallel (DAP) TM ensures that transaction accessing disjoint data sets do not contend on the shared memory and, thus, may proceed independently. As was earlier observed [13], the combination of these principles incurs some inherent costs, and the main motivation of this paper is to explore these costs.

Intuitively, the overhead invisible read may incur comes from the need of *validation*, *i.e.*, ensuring that read data items have not been updated when the transaction completes. Our first result (Sect. 4) is that a read-only transaction in an opaque TM featured with *weak* DAP and *weak* invisible reads must *incrementally* validate every next read operation. This results in a quadratic (in the size of the transaction's read set) step-complexity lower bound. Informally, weak DAP means that two transactions encounter a memory race only if their data sets are connected in the *conflict graph*, capturing data-set overlaps among all concurrent transactions. Weak read invisibility allows read operations of a transaction T to be "visible" only if T is concurrent with another transaction. The lower bound is derived for *minimal* progressiveness, where transactions are guaranteed to commit only if they run sequentially. Our result improves the lower bound [12,13] derived for *strict-data partitioning* (a very strong version of DAP) and (strong) invisible reads.

Our second result is that, under weak DAP and weak read invisibility, a strictly serializable TM must have a read-only transaction that accesses a linear (in the size of the transaction's read set) number of distinct memory locations in the course of performing its last read operation. Naturally, this space lower bound also applies to opaque TMs.

We then turn our focus to *strongly progressive* TMs [13] that, in addition to progressiveness, ensures that *not all* concurrent transactions conflicting over a single data item abort. In Sect. 5, we prove that in any strongly progressive strictly serializable TM implementation that accesses the shared memory with *read, write* and *conditional* primitives, such as *compare-and-swap* and *load-linked/store-conditional*, the total number of *remote memory references* (RMRs) that take place in an execution of a progressive TM in which n concurrent processes perform transactions on a single data item might reach $\Omega(n \log n)$. The result is obtained via a reduction to an analogous lower bound for mutual exclusion [3]. In the reduction, we show that any TM with the above properties can be used to implement a *deadlock-free* mutual exclusion, employing transactional operations on only one data item and incurring a constant RMR overhead. The lower bound applies to RMRs in both the *cache-coherent (CC)* and *distributed shared memory (DSM)* models, and it appears to be the first RMR complexity lower bound for transactional memory.

2   Model

TM Interface. A *transactional memory* (in short, *TM*) supports *transactions* for reading and writing on a finite set of data items, referred to as *t-objects*. Every transaction T_k has a unique identifier k. We assume no bound on the size of a t-object, *i.e.*, the cardinality on the set V of possible different values a t-object can have. A transaction T_k may contain the following *t-operations*, each being a matching pair of an *invocation* and a *response*: $read_k(X)$ returns a value in some domain V (denoted $read_k(X) \rightarrow v$) or a special value $A_k \notin V$ (*abort*); $write_k(X, v)$, for a value $v \in V$, returns *ok* or A_k; $tryC_k$ returns $C_k \notin V$ (*commit*) or A_k.

Implementations. We assume an asynchronous shared-memory system in which a set of $n > 1$ processes $p_1, \ldots, p_n$ communicate by applying *operations* on shared *objects*. An object is an instance of an *abstract data type* which specifies a set of operations that provide the only means to manipulate the object. An *implementation* of an object type τ provides a specific data-representation of τ by applying *primitives* on shared *base objects*, each of which is assigned an initial value and a set of algorithms $I_1(\tau), \ldots, I_n(\tau)$, one for each process. We assume that these primitives are *deterministic*. Specifically, a TM *implementation* provides processes with algorithms for implementing $read_k$, $write_k$ and $tryC_k()$ of a transaction T_k by *applying primitives* from a set of shared *base objects*.

We assume that processes issue transactions sequentially, *i.e.*, a process starts a new transaction only after the previous transaction is committed or aborted. A primitive is a generic *read-modify-write* (*RMW*) procedure applied to a base object [10]. It is characterized by a pair of functions $\langle g, h \rangle$: given the current state of the base object, g is an *update function* that computes its state after the primitive is applied, while h is a *response function* that specifies the outcome of the primitive returned to the process. A RMW primitive is *trivial* if it never changes the value of the base object to which it is applied. Otherwise, it is *nontrivial*. An RMW primitive $\langle g, h \rangle$ is *conditional* if there exists v, w such that $g(v, w) = v$ and there exists v, w such that $g(v, w) \neq v$. For *e.g*, *compare-and-swap (CAS)* and *load-linked/store-conditional (LL/SC)* are nontrivial conditional RMW primitives while *fetch-and-add* is an example of a nontrivial RMW primitive that is not conditional.

Executions and Configurations. An *event* of a process p_i (sometimes we say *step* of p_i) is an invocation or response of an operation performed by p_i or a rmw primitive $\langle g, h \rangle$ applied by p_i to a base object b along with its response r (we call it a *rmw event* and write $(b, \langle g, h \rangle, r, i)$). A *configuration* specifies the value of each base object and the state of each process. The *initial configuration* is the configuration in which all base objects have their initial values and all processes are in their initial states.

An *execution fragment* is a (finite or infinite) sequence of events. An *execution* of an implementation I is an execution fragment where, starting from the initial configuration, each event is issued according to I and each response of a rmw event $(b, \langle g, h \rangle, r, i)$ matches the state of b resulting from all preceding events. An execution $E \cdot E'$, denoting the concatenation of E and E', is an *extension* of E and we say that E' *extends* E.

Let E be an execution fragment. For every transaction identifier k, $E|k$ denotes the subsequence of E restricted to events of transaction T_k. If $E|k$ is non-empty, we say that T_k *participates* in E, else we say E is T_k-*free*. Two executions E and E' are *indistinguishable* to a set $\mathcal{T}$ of transactions, if for each transaction $T_k \in \mathcal{T}$, $E|k = E'|k$. A TM *history* is the subsequence of an execution consisting of the invocation and response events of t-operations.

The *read set* (resp., the *write set*) of a transaction T_k in an execution E, denoted $Rset(T_k)$ (and resp. $Wset(T_k)$), is the set of t-objects on which T_k invokes reads (and resp. writes) in E. The *data set* of T_k is $Dset(T_k) = Rset(T_k) \cup Wset(T_k)$. A transaction is called *read-only* if $Wset(T_k) = \emptyset$; *write-only* if $Rset(T_k) = \emptyset$ and *updating* if $Wset(T_k) \neq \emptyset$. Note that, in our TM model, the data set of a transaction is not known apriori and it is identifiable only by the set of data items the transaction has invoked a read or write on in the given execution.

Transaction Orders. Let $txns(E)$ denote the set of transactions that participate in E. An execution E is *sequential* if every invocation of a t-operation is either the last event in the history H exported by E or is immediately followed by a matching response. We assume that executions are *well-formed*: no process invokes a new operation before the previous operation returns. Specifically, we assume that for all T_k, $E|k$ begins with the invocation of a t-operation, is sequential and has no events after A_k or C_k. A transaction $T_k \in txns(E)$ is *complete in* E if $E|k$ ends with a response event. The execution E is *complete* if all transactions in $txns(E)$ are complete in E. A transaction $T_k \in txns(E)$ is *t-complete* if $E|k$ ends with A_k or C_k; otherwise, T_k is *t-incomplete*. T_k is *committed* (resp., *aborted*) in E if the last event of T_k is C_k (resp., A_k). The execution E is *t-complete* if all transactions in $txns(E)$ are t-complete.

For transactions $\{T_k, T_m\} \in txns(E)$, we say that T_k *precedes* T_m in the *real-time order* of E, denoted $T_k \prec_E^{RT} T_m$, if T_k is t-complete in E and the last event of T_k precedes the first event of T_m in E. If neither $T_k \prec_E^{RT} T_m$ nor $T_m \prec_E^{RT} T_k$, then T_k and T_m are *concurrent* in E. An execution E is *t-sequential* if there are no concurrent transactions in E.

Contention. We say that a configuration C after an execution E is *quiescent* (and resp. *t-quiescent*) if every transaction $T_k \in txns(E)$ is complete (and resp. t-complete) in C. If a transaction T is incomplete in an execution E, it has exactly one *enabled* event, which is the next event the transaction will perform according to the TM implementation. Events e and e' of an execution E *contend* on a base object b if they are both events on b in E and at least one of them is nontrivial (the event is trivial (and resp. nontrivial) if it is the application of a trivial (and resp. nontrivial) primitive). We say that a transaction T is *poised to apply an event e after E* if e is the next enabled event for T in E. We say that transactions T and T' *concurrently contend on b in E* if they are each poised to apply contending events on b after E.

We say that an execution fragment E is *step contention-free for t-operation* op_k if the events of $E|op_k$ are contiguous in E. We say that an execution fragment E is *step contention-free for T_k* if the events of $E|k$ are contiguous in E. We say that E is *step contention-free* if E is step contention-free for all transactions that participate in E.

3   TM Classes

TM-correctness. We say that $read_k(X)$ is *legal* in a t-sequential execution E if it returns the *latest written value* of X, and E is *legal* if every $read_k(X)$ in H that does not return A_k is legal in E.

A finite history H is *opaque* if there is a legal t-complete t-sequential history S, such that (1) for any two transactions $T_k, T_m \in txns(H)$, if $T_k \prec_H^{RT} T_m$, then T_k precedes T_m in S, and (2) S is equivalent to a *completion* of H.

A finite history H is *strictly serializable* if there is a legal t-complete t-sequential history S, such that (1) for any two transactions $T_k, T_m \in txns(H)$, if $T_k \prec_H^{RT} T_m$, then T_k precedes T_m in S, and (2) S is equivalent to $cseq(\bar{H})$, where $\bar{H}$ is some completion of H and $cseq(\bar{H})$ is the subsequence of $\bar{H}$ reduced to committed transactions in $\bar{H}$.

We refer to S as an opaque (and resp. strictly serializable) *serialization* of H.

TM-liveness. We say that a TM implementation M provides *interval-contention free (ICF)* TM-liveness if for every finite execution E of M such that the configuration after E is quiescent, and every transaction T_k that applies the invocation of a t-operation op_k immediately after E, the finite step contention-free extension for op_k contains a matching response.

A TM implementation M provides *wait-free TM-liveness* if in every execution of M, every t-operation returns a matching response in a finite number of its steps.

TM-progress. We say that a TM implementation provides *sequential TM-progress* (also called *minimal progressiveness* [13]) if every transaction running step contention-free from a t-quiescent configuration commits within a finite number of steps.

We say that transactions T_i, T_j *conflict* in an execution E on a t-object X if $X \in Dset(T_i) \cap Dset(T_j)$, and $X \in Wset(T_i) \cup Wset(T_j)$.

A TM implementation M provides *progressive* TM-progress (or *progressiveness*) if for every execution E of M and every transaction $T_i \in txns(E)$ that returns A_i in E, there exists a transaction $T_k \in txns(E)$ such that T_k and T_i are concurrent and conflict in E [13].

Let $CObj_H(T_i)$ denote the set of t-objects over which transaction $T_i \in txns(H)$ conflicts with any other transaction in history H, i.e., $X \in CObj_H(T_i)$, iff there exist transactions T_i and T_k that conflict on X in H. Let $Q \subseteq txns(H)$ and $CObj_H(Q) = \bigcup_{T_i \in Q} CObj_H(T_i)$.

Let $CTrans(H)$ denote the set of non-empty subsets of $txns(H)$ such that a set Q is in $CTrans(H)$ if no transaction in Q conflicts with a transaction not in Q.

Definition 1. *A TM implementation M is* strongly progressive *if M is weakly progressive and for every history H of M and for every set $Q \in CTrans(H)$ such that $|CObj_H(Q)| \leq 1$, some transaction in Q is not aborted in H.*

Invisible Reads. A TM implementation M uses *invisible reads* if for every execution E of M and for every read-only transaction $T_k \in txns(E)$, $E|k$ does not contain any nontrivial events.

In this paper, we introduce a definition of *weak* invisible reads. For any execution E and any t-operation π_k invoked by some transaction $T_k \in txns(E)$, let $E|\pi_k$ denote the subsequence of E restricted to events of π_k in E.

We say that a TM implementation M satisfies *weak invisible reads* if for any execution E of M and every transaction $T_k \in txns(E)$; $Rset(T_k) \neq \emptyset$ that is not concurrent with any transaction $T_m \in txns(E)$, $E|\pi_k$ does not contain any nontrivial events, where π_k is any t-read operation invoked by T_k in E.

Disjoint-Access Parallelism (DAP). Let $\tau_E(T_i, T_j)$ be the set of transactions (T_i and T_j included) that are concurrent to at least one of T_i and T_j in E. Let $G(T_i, T_j, E)$ be an undirected graph whose vertex set is $\bigcup_{T \in \tau_E(T_i, T_j)} Dset(T)$ and there is an edge between t-objects X and Y *iff* there exists $T \in \tau_E(T_i, T_j)$ such that $\{X, Y\} \in Dset(T)$. We say that T_i and T_j are *disjoint-access* in E if there is no path between a t-object in $Dset(T_i)$ and a t-object in $Dset(T_j)$ in $G(T_i, T_j, E)$.

A TM implementation M is *weak disjoint-access parallel (weak DAP)* if, for all executions E of M, transactions T_i and T_j concurrently contend on the same base object in E only if T_i and T_j are not disjoint-access in E or there exists a t-object $X \in Dset(T_i) \cap Dset(T_j)$ [5,20].

Lemma 1. *([5,18]) Let M be any weak DAP TM implementation. Let $\alpha \cdot \rho_1 \cdot \rho_2$ be any execution of M where ρ_1 (and resp. ρ_2) is the step contention-free execution fragment of transaction $T_1 \notin txns(\alpha)$ (and resp. $T_2 \notin txns(\alpha)$) and transactions T_1, T_2 are disjoint-access in $\alpha \cdot \rho_1 \cdot \rho_2$. Then, T_1 and T_2 do not contend on any base object in $\alpha \cdot \rho_1 \cdot \rho_2$.*

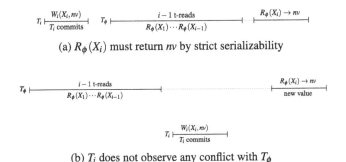

(a) $R_\phi(X_i)$ must return nv by strict serializability

(b) T_i does not observe any conflict with T_ϕ

Fig. 1. Executions in the proof of Lemma 2; By weak DAP, T_ϕ cannot distinguish this from the execution in Fig. 1(a)

4   Time and Space Complexity of Sequential TMs

In this section, we prove that (1) that a read-only transaction in an opaque TM featured with *weak* DAP and *weak* invisible reads must *incrementally* validate every next read operation, and (2) a strictly serializable TM (under weak DAP and weak read invisibility), must have a read-only transaction that accesses a linear (in the size of the transaction's read set) number of distinct base objects in the course of performing its last t-read and tryCommit operations.

We first prove the following lemma concerning strictly serializable weak DAP TM implementations.

Lemma 2. *Let M be any strictly serializable, weak DAP TM implementation that provides sequential TM-progress. Then, for all $i \in \mathbb{N}$, M has an execution of the form $\pi^{i-1} \cdot \rho^i \cdot \alpha^i$ where,*

- *π^{i-1} is the complete step contention-free execution of read-only transaction T_ϕ that performs $(i-1)$ t-reads: $read_\phi(X_1) \cdots read_\phi(X_{i-1})$,*
- *ρ^i is the t-complete step contention-free execution of a transaction T_i that writes $nv_i \neq v_i$ to X_i and commits,*
- *α_i is the complete step contention-free execution fragment of T_ϕ that performs its ith t-read: $read_\phi(X_i) \to nv_i$.*

Proof. By sequential TM-progress, M has an execution of the form $\rho^i \cdot \pi^{i-1}$. Since $Dset(T_k) \cap Dset(T_i) = \emptyset$ in $\rho^i \cdot \pi^{i-1}$, by Lemma 1, transactions T_ϕ and T_i do not contend on any base object in execution $\rho^i \cdot \pi^{i-1}$. Thus, $\rho^i \cdot \pi^{i-1}$ is also an execution of M.

By assumption of strict serializability, $\rho^i \cdot \pi^{i-1} \cdot \alpha_i$ is an execution of M in which the t-read of X_i performed by T_ϕ must return nv_i. But $\rho^i \cdot \pi^{i-1} \cdot \alpha_i$ is indistinguishable to T_ϕ from $\pi^{i-1} \cdot \rho^i \cdot \alpha_i$. Thus, M has an execution of the form $\pi^{i-1} \cdot \rho^i \cdot \alpha_i$.

Theorem 1. *For every weak DAP TM implementation M that provides ICF TM-liveness, sequential TM-progress and uses weak invisible reads,*

(1) If M is opaque, for every $m \in \mathbb{N}$, there exists an execution E of M such that some transaction $T \in txns(E)$ performs $\Omega(m^2)$ steps, where $m = |Rset(T_k)|$.
(2) if M is strictly serializable, for every $m \in \mathbb{N}$, there exists an execution E of M such that some transaction $T_k \in txns(E)$ accesses at least $m-1$ distinct base objects during the executions of the mth t-read operation and $tryC_k()$, where $m = |Rset(T_k)|$.

Proof. For all $i \in \{1, \ldots, m\}$, let v be the initial value of t-object X_i.

(1) Suppose that M is opaque. Let π^m denote the complete step contention-free execution of a transaction T_ϕ that performs m t-reads: $read_\phi(X_1) \cdots read_\phi(X_m)$ such that for all $i \in \{1, \ldots, m\}$, $read_\phi(X_i) \to v$.

By Lemma 2, for all $i \in \{2, \ldots, m\}$, M has an execution of the form $E^i = \pi^{i-1} \cdot \rho^i \cdot \alpha_i$.

For each $i \in \{2, \ldots, m\}$, $j \in \{1, 2\}$ and $\ell \le (i-1)$, we now define an execution of the form $\mathbb{E}^i_{j\ell} = \pi^{i-1} \cdot \beta^\ell \cdot \rho^i \cdot \alpha^i_j$ as follows:

- β^ℓ is the t-complete step contention-free execution fragment of a transaction T_ℓ that writes $nv_\ell \ne v$ to X_ℓ and commits
- α^i_1 (and resp. α^i_2) is the complete step contention-free execution fragment of $read_\phi(X_i) \to v$ (and resp. $read_\phi(X_i) \to A_\phi$).

Claim 1. *For all $i \in \{2, \ldots, m\}$ and $\ell \le (i-1)$, M has an execution of the form $\mathbb{E}^i_{1\ell}$ or $\mathbb{E}^i_{2\ell}$.*

Proof. For all $i \in \{2, \ldots, m\}$, π^{i-1} is an execution of M. By assumption of weak invisible reads and sequential TM-progress, T_ℓ must be committed in $\pi^{i-1} \cdot \rho^\ell$ and M has an execution of the form $\pi^{i-1} \cdot \beta^\ell$. By the same reasoning, since T_i and T_ℓ have disjoint data sets, M has an execution of the form $\pi^{i-1} \cdot \beta^\ell \cdot \rho^i$.

Since the configuration after $\pi^{i-1} \cdot \beta^\ell \cdot \rho^i$ is quiescent, by ICF TM-liveness, $\pi^{i-1} \cdot \beta^\ell \cdot \rho^i$ extended with $read_\phi(X_i)$ must return a matching response. If $read_\phi(X_i) \to v_i$, then clearly $\mathbb{E}^i_1$ is an execution of M with T_ϕ, T_{i-1}, T_i being a valid serialization of transactions. If $read_\phi(X_i) \to A_\phi$, the same serialization justifies an opaque execution.

Suppose by contradiction that there exists an execution of M such that $\pi^{i-1} \cdot \beta^\ell \cdot \rho^i$ is extended with the complete execution of $read_\phi(X_i) \to r$; $r \notin \{A_\phi, v\}$.

The only plausible case to analyse is when $r = nv$. Since $read_\phi(X_i)$ returns the value of X_i updated by T_i, the only possible serialization for transactions is T_ℓ, T_i, T_ϕ; but $read_\phi(X_\ell)$ performed by T_k that returns the initial value v is not legal in this serialization—contradiction.

We now prove that, for all $i \in \{2, \ldots, m\}$, $j \in \{1, 2\}$ and $\ell \le (i - 1)$, transaction T_ϕ must access $(i - 1)$ different base objects during the execution of $read_\phi(X_i)$ in the execution $\pi^{i-1} \cdot \beta^\ell \cdot \rho^i \cdot \alpha_j^i$.

By the assumption of weak invisible reads, the execution $\pi^{i-1} \cdot \beta^\ell \cdot \rho^i \cdot \alpha_j^i$ is indistinguishable to transactions T_ℓ and T_i from the execution $\tilde{\pi}^{i-1} \cdot \beta^\ell \cdot \rho^i \cdot \alpha_j^i$, where $Rset(T_\phi) = \emptyset$ in $\tilde{\pi}^{i-1}$. But transactions T_ℓ and T_i are disjoint-access in $\tilde{\pi}^{i-1} \cdot \beta^\ell \cdot \rho^i$ and by Lemma 1, they cannot contend on the same base object in this execution.

Consider the $(i - 1)$ different executions: $\pi^{i-1} \cdot \beta^1 \cdot \rho^i$, ..., $\pi^{i-1} \cdot \beta^{i-1} \cdot \rho^i$. For all $\ell, \ell' \le (i-1); \ell' \ne \ell$, M has an execution of the form $\pi^{i-1} \cdot \beta^\ell \cdot \rho^i \cdot \beta^{\ell'}$ in which transactions T_ℓ and $T_{\ell'}$ access mutually disjoint data sets. By weak invisible reads and Lemma 1, the pairs of transactions $T_{\ell'}$, T_i and $T_{\ell'}$, T_ℓ do not contend on any base object in this execution. This implies that $\pi^{i-1} \cdot \beta^\ell \cdot \beta^{\ell'} \cdot \rho^i$ is an execution of M in which transactions T_ℓ and $T_{\ell'}$ each apply nontrivial primitives to mutually disjoint sets of base objects in the execution fragments β^ℓ and $\beta^{\ell'}$ respectively (by Lemma 1).

This implies that for any $j \in \{1, 2\}$, $\ell \le (i - 1)$, the configuration C^i after E^i differs from the configurations after $\mathbb{E}_{j\ell}^i$ only in the states of the base objects that are accessed in the fragment β^ℓ. Consequently, transaction T_ϕ must access at least $i - 1$ different base objects in the execution fragment π_j^i to distinguish configuration C^i from the configurations that result after the $(i - 1)$ different executions $\pi^{i-1} \cdot \beta^1 \cdot \rho^i$, ..., $\pi^{i-1} \cdot \beta^{i-1} \cdot \rho^i$ respectively.

Thus, for all $i \in \{2, \ldots, m\}$, transaction T_ϕ must perform at least $i - 1$ steps while executing the ith t-read in π_j^i and T_ϕ itself must perform $\sum_{i=1}^{m-1} i = \frac{m(m-1)}{2}$ steps.

(2) Suppose that M is strictly serializable, but not opaque. Since M is strictly serializable, by Lemma 2, it has an execution of the form $E = \pi^{m-1} \cdot \rho^m \cdot \alpha_m$.

For each $\ell \le (i - 1)$, we prove that M has an execution of the form $E_\ell = \pi^{m-1} \cdot \beta^\ell \cdot \rho^m \cdot \bar{\alpha}^m$ where $\bar{\alpha}^m$ is the complete step contention-free execution fragment of $read_\phi(X_m)$ followed by the complete execution of $tryC_\phi$. Indeed, by weak invisible reads, π^{m-1} does not contain any nontrivial events and the execution $\pi^{m-1} \cdot \beta^\ell \cdot \rho^m$ is indistinguishable to transactions T_ℓ and T_m from the executions $\tilde{\pi}^{m-1} \cdot \beta^\ell$ and $\tilde{\pi}^{m-1} \cdot \beta^\ell \cdot \rho^m$ respectively, where $Rset(T_\phi) = \emptyset$ in $\tilde{\pi}^{m-1}$. Thus, applying Lemma 1, transactions $\beta^\ell \cdot \rho^m$ do not contend on any base object in the execution $\pi^{m-1} \cdot \beta^\ell \cdot \rho^m$. By ICF TM-liveness, $read_\phi(X_m)$ and $tryC_\phi$ must return matching responses in the execution fragment $\bar{\alpha}^m$ that extends $\pi^{m-1} \cdot \beta^\ell \cdot \rho^m$. Consequently, for each $\ell \le (i - 1)$, M has an execution of the form $E_\ell = \pi^{m-1} \cdot \beta^\ell \cdot \rho^m \cdot \bar{\alpha}^m$ such that transactions T_ℓ and T_m do not contend on any base object.

Strict serializability of M means that if $read_\phi(X_m) \to nv$ in the execution fragment $\bar{\alpha}^m$, then $tryC_\phi$ must return A_ϕ. Otherwise if $read_\phi(X_m) \to v$ (i.e. the initial value of X_m), then $tryC_\phi$ may return A_ϕ or C_ϕ.

Thus, as with (1), in the worst case, T_ϕ must access at least $m - 1$ distinct base objects during the executions of $read_\phi(X_m)$ and $tryC_\phi$ to distinguish the configuration C^i from the configurations after the $m - 1$ different executions $\pi^{m-1} \cdot \beta^1 \cdot \rho^m, \ldots, \pi^{m-1} \cdot \beta^{m-1} \cdot \rho^m$ respectively.

5   RMR Complexity of Strongly Progressive TMs

In this section, we prove every strongly progressive strictly serializable TM providing wait-free TM-liveness that uses only read, write and *conditional* RMW primitives has an execution in which in which n concurrent processes perform transactions on a single data item and incur $\Omega(\log n)$ *remote memory references* [2].

Remote Memory References(RMR) [3]. In the *cache-coherent (CC) shared memory*, each process maintains *local* copies of shared objects inside its cache, whose consistency is ensured by a coherence protocol. Informally, we say that an access to a base object b is *remote* to a process p and causes a *remote memory reference (RMR)* if p's cache contains a cached copy of the object that is out of date or *invalidated*; otherwise the access is *local*.

Algorithm 1. Mutual-exclusion object L from a strongly progressive, strict serializable TM M; code for process p_i; $1 \le i \le n$

```
1:  Local variables:
2:    bit face_i, for each process p_i

3:  Shared objects:
4:    strongly progressive, strictly
5:    serializable TM M
6:    t-object X, initially ⊥
7:    storing value v ∈ {[p_i,face_i]} ∪ {⊥}
8:    for each tuple [p_i,face_i]
9:    Done[p_i,face_i] ∈ {true,false}
10:   Succ[p_i,face_i] ∈ {p_1,...,p_n} ∪ {⊥}
11:   for each p_i and j ∈ {1,...,n} \ {i}
12:   Lock[p_i][p_j] ∈ {locked,unlocked}

13: Function: func():
14:   atomic using M
15:     value := tx-read(X)
16:     tx-write(X, [p_i,face_i])
17:   on abort Return false
18:   Return value

19: Entry:
20:   face_i := 1 − face_i
21:   Done[p_i,face_i].write(false)
22:   Succ[p_i,face_i].write(⊥)
23:   while (prev ← func) = false do
24:     no op
25:   end while
26:   if prev ≠ ⊥ then
27:     Lock[p_i][prev.pid].write(locked)
28:     Succ[prev].write(p_i)
29:     if Done[prev] = false then
30:       while Lock[p_i][prev.pid] = unlocked
      do
31:         no op
32:       end while
33:   Return ok
34:   // Critical section

35: Exit:
36:   Done[p_i,face_i].write(true)
37:   Lock[Succ[p_i,face_i]][p_i].write(unlocked)
38:   Return ok
```

In the *write-through (CC) protocol*, to read a base object b, process p must have a cached copy of b that has not been invalidated since its previous read. Otherwise, p incurs a RMR. To write to b, p causes a RMR that invalidates all cached copies of b and writes to the main memory.

In the *write-back (CC) protocol*, p reads a base object b without causing a RMR if it holds a cached copy of b in shared or exclusive mode; otherwise the access of b causes a RMR that (1) invalidates all copies of b held in exclusive mode, and writing b back to the main memory, (2) creates a cached copy of b in shared mode. Process p can write to b without causing a RMR if it holds a copy of b in exclusive mode; otherwise p causes a RMR that invalidates all cached copies of b and creates a cached copy of b in exclusive mode.

In the *distributed shared memory (DSM)*, each register is forever assigned to a single process and it *remote* to the others. Any access of a remote register causes a RMR.

Mutual Exclusion. The *mutex object* supports two operations: *Enter* and *Exit*, both of which return the response ok. We say that a process p_i *is in the critical section after an execution* π if π contains the invocation of Enter by p_i that returns ok, but does not contain a subsequent invocation of Exit by p_i in π.

A mutual exclusion implementation satisfies the following properties:

(Mutual-exclusion) After any execution π, there exists at most one process that is in the critical section.

(Deadlock-freedom) Let π be any execution that contains the invocation of Enter by process p_i. Then, in every extension of π in which every process takes infinitely many steps, some process is in the critical section.

(Finite-exit) Every process completes the Exit operation within a finite number of steps.

5.1 Mutual Exclusion from a Strongly Progressive TM

We describe an implementation of a mutex object $L(M)$ from a strictly serializable, strongly progressive TM implementation M providing wait-free TM-liveness (Algorithm 1). The algorithm is based on the mutex implementation in [15].

Given a sequential implementation, we use a TM to execute the sequential code in a concurrent environment by encapsulating each sequential operation within an *atomic* transaction that replaces each read and write of a t-object with the transactional read and write implementations, respectively. If the transaction commits, then the result of the operation is returned; otherwise if one of the transactional operations aborts. For instance, in Algorithm 1, we wish to atomically read a t-object X, write a new value to it and return the old value of X prior to this write. To achieve this, we employ a strictly serializable TM implementation M. Moreover, we assume that M is strongly progressive, *i.e.*, in every execution, at least one transaction successfully commits and the value of X is returned.

Shared Objects. We associate each process p_i with two alternating identities $[p_i, face_i]$; $face_i \in \{0, 1\}$. The strongly progressive TM implementation M is used to enqueue processes that attempt to enter the critical section within a single t-object X (initially $\perp$). For each $[p_i, face_i]$, $L(M)$ uses a register bit $Done[p_i, face_i]$ that indicates if this face of the process has left the critical section or is executing the **Entry** operation. Additionally, we use a register $Succ[p_i, face_i]$ that stores the process expected to succeed p_i in the critical section. If $Succ[p_i, face_i] = p_j$, we say that p_j is the *successor of* p_i (and p_i is the *predecessor* of p_j). Intuitively, this means that p_j is expected to enter the critical section immediately after p_i. Finally, $L(M)$ uses a 2-dimensional bit array $Lock$: for each process p_i, there are $n-1$ registers associated with the other processes. For all $j \in \{0, \ldots, n-1\} \setminus \{i\}$, the registers $Lock[p_i][p_j]$ are local to p_i and registers $Lock[p_j][p_i]$ are remote to p_i. Process p_i can only access registers in the $Lock$ array that are local or remote to it.

Entry Operation. A process p_i adopts a new identity $face_i$ and writes *false* to $Done(p_i, face_i)$ to indicate that p_i has started the **Entry** operation. Process p_i now initializes the successor of $[p_i, face_i]$ by writing $\perp$ to $Succ[p_i, face_i]$. Now, p_i uses a strongly progressive TM implementation M to atomically store its *pid* and identity i.e., $face_i$ to t-object X and returns the *pid* and identity of its *predecessor*, say $[p_j, face_j]$. Intuitively, this suggests that $[p_i, face_i]$ is scheduled to enter the critical section immediately after $[p_j, face_j]$ exits the critical section. Note that if p_i reads the initial value of t-object X, then it immediately enters the critical section. Otherwise it writes *locked* to the register $Lock[p_i, p_j]$ and sets itself to be the successor of $[p_j, face_j]$ by writing p_i to $Succ[p_j, face_j]$. Process p_i now checks if p_j has started the **Exit** operation by checking if $Done[p_j, face_j]$ is set. If it is, p_i enters the critical section; otherwise p_i spins on the register $Lock[p_i][p_j]$ until it is *unlocked*.

Exit Operation. Process p_i first indicates that it has exited the critical section by setting $Done[p_i, face_i]$, following which it *unlocks* the register $Lock[Succ[p_i, face_i]][p_i]$ to allow p_i's successor to enter the critical section.

5.2 Proof of Correctness

Lemma 3. *The implementation $L(M)$ (Algorithm 1) satisfies mutual exclusion.*

Proof. Let E be any execution of $L(M)$. We say that $[p_i, face_i]$ is the *successor* of $[p_j, face_j]$ if p_i reads the value of *prev* in Line 25 to be $[p_j, face_j]$ (and $[p_j, face_j]$ is the *predecessor* of $[p_i, face_i]$); otherwise if p_i reads the value to be $\perp$, we say that p_i has no predecessor.

Suppose by contradiction that there exist processes p_i and p_j that are both inside the critical section after E. Since p_i is inside the critical section, either (1) p_i read *prev* $= \perp$ in Line 23, or (2) p_i read that $Done[prev]$ is *true* (Line 29) or p_i reads that $Done[prev]$ is *false* and $Lock[p_i][prev.pid]$ is *unlocked* (Line 30).

(Case 1) Suppose that p_i read *prev* $= \perp$ and entered the critical section. Since in this case, p_i does not have any predecessor, some other process that returns successfully from the *while* loop in Line 25 must be successor of p_i in E. Since there exists $[p_j, face_j]$ also inside the critical section after E, p_j reads that either $[p_i, face_i]$ or some other process to be its predecessor. Observe that there must exist some such process $[p_k, face_k]$ whose predecessor is $[p_i, face_i]$. Hence, without loss of generality, we can assume that $[p_j, face_j]$ is the successor of $[p_i, face_i]$. By our assumption, $[p_j, face_j]$ is also inside the critical section. Thus, p_j *locked* the register $Lock[p_j, p_i]$ in Line 27 and set itself to be p_i's successor in Line 28. Then, p_j read that $Done[p_i, face_i]$ is *true* or read that $Done[p_i, face_i]$ is *false* and waited until $Lock[p_j, p_i]$ is *unlocked* and then entered the critical section. But this is possible only if p_i has left the critical section and updated the registers $Done[p_i, face_i]$ and $Lock[p_j, p_i]$ in Lines 36 and 37 respectively—contradiction to the assumption that $[p_i, face_i]$ is also inside the critical section after E.

(Case 2) Suppose that p_i did not read $prev = \perp$ and entered the critical section. Thus, p_i read that $Done[prev]$ is *false* in Line 29 and $Lock[p_i][prev.pid]$ is *unlocked* in Line 30, where $prev$ is the predecessor of $[p_i, face_i]$. As with case 1, without loss of generality, we can assume that $[p_j, face_j]$ is the successor of $[p_i, face_i]$ or $[p_j, face_j]$ is the predecessor of $[p_i, face_i]$.

Suppose that $[p_j, face_j]$ is the predecessor of $[p_i, face_i]$, *i.e.*, p_i writes the value $[p_i, face_i]$ to the register $Succ[p_j, face_j]$ in Line 28. Since $[p_j, face_j]$ is also inside the critical section after E, process p_i must read that $Done[p_j, face_j]$ is *true* in Line 29 and $Lock[p_i, p_j]$ is *locked* in Line 30. But then p_i could not have entered the critical section after E—contradiction.

Suppose that $[p_j, face_j]$ is the successor of $[p_i, face_i]$, *i.e.*, p_j writes the value $[p_j, face_j]$ to the register $Succ[p_i, face_i]$. Since both p_i and p_j are inside the critical section after E, process p_j must read that $Done[p_i, face_i]$ is *true* in Line 29 and $Lock[p_j, p_i]$ is *locked* in Line 30. Thus, p_j must spin on the register $Lock[p_j, p_i]$, waiting for it to be *unlocked* by p_i before entering the critical section—contradiction to the assumption that both p_i and p_j are inside the critical section.

Thus, $L(M)$ satisfies mutual-exclusion.

Lemma 4. *The implementation $L(M)$ (Algorithm 1) provides deadlock-freedom.*

Proof. Let E be any execution of $L(M)$. Observe that a process may be stuck indefinitely only in Lines 23 and 30 as it performs the *while* loop.

Since M is strongly progressive and provides wait-free TM-liveness, in every execution E that contains an invocation of **Enter** by process p_i, some process returns *true* from the invocation of *func()* in Line 23.

Now consider a process p_i that returns successfuly from the *while* loop in Line 23. Suppose that p_i is stuck indefinitely as it performs the *while* loop in Line 30. Thus, no process has *unlocked* the register $Lock[p_i][prev.pid]$ by writing to it in the **Exit** section. Recall that since $[p_i, face_i]$ has reached the *while* loop in Line 30, $[p_i, face_i]$ necessarily has a predecessor, say $[p_j, face_j]$, and has set itself to be p_j's successor by writing p_i to register $Succ[p_j, face_j]$ in Line 28. Consider the possible two cases: the predecessor of $[p_j, face_j]$ is some process $p_k; k \neq i$ or the predecessor of $[p_j, face_j]$ is the process p_i itself.

(Case 1) Since by assumption, process p_j takes infinitely many steps in E, the only reason that p_j is stuck without entering the critical section is that

$[p_k, face_k]$ is also stuck in the *while* loop in Line 30. Note that it is possible for us to iteratively extend this execution in which p_k's predecessor is a process that is not p_i or p_j that is also stuck in the *while* loop in Line 30. But then the last such process must eventually read the corresponding *Lock* to be *unlocked* and enter the critical section. Thus, in every extension of E in which every process takes infinitely many steps, some process will enter the critical section.

(Case 2) Suppose that the predecessor of $[p_j, face_j]$ is the process p_i itself. Thus, as $[p_i, face]$ is stuck in the *while* loop waiting for $Lock[p_i, p_j]$ to be *unlocked* by process p_j, p_j leaves the critical section, *unlocks* $Lock[p_i, p_j]$ in Line 37 and prior to the read of $Lock[p_i, p_j]$, p_j re-starts the Entry operation, writes *false* to $Done[p_j, 1 - face_j]$ and sets itself to be the successor of $[p_i, face_i]$ and spins on the register $Lock[p_j, p_i]$. However, observe that process p_i, which takes infinitely many steps by our assumption must eventually read that $Lock[p_i, p_j]$ is *unlocked* and enter the critical section, thus establishing deadlock-freedom.

We say that a TM implementation M *accesses a single t-object* if in every execution E of M and every transaction $T \in txns(E)$, $|Dset(T)| \leq 1$. We can now prove the following theorem:

Theorem 2. *Any strictly serializable, strongly progressive TM implementation M providing wait-free TM-liveness that accesses a single t-object implies a deadlock-free, finite exit mutual exclusion implementation $L(M)$ such that the RMR complexity of M is within a constant factor of the RMR complexity of $L(M)$.*

Proof. (Mutual-exclusion) Follows from Lemma 3.

(Finite-exit) The proof is immediate since the Exit operation contains no unbounded loops or waiting statements.

(Deadlock-freedom) Follows from Lemma 4.

(RMR complexity) First, let us consider the CC model. Observe that every event not on M performed by a process p_i as it performs the Entry or Exit operations incurs $O(1)$ RMR cost clearly, possibly barring the *while* loop executed in Line 30. During the execution of this *while* loop, process p_i spins on the register $Lock[p_i][p_j]$, where p_j is the predecessor of p_i. Observe that p_i's cached copy of $Lock[p_i][p_j]$ may be invalidated only by process p_j as it *unlocks* the register in Line 37. Since no other process may write to this register and p_i terminates the *while* loop immediately after the write to $Lock[p_i][p_j]$ by p_j, p_i incurs $O(1)$ RMR's. Thus, the overall RMR cost incurred by M is within a constant factor of the RMR cost of $L(M)$.

Now we consider the DSM model. As with the reasoning for the CC model, every event not on M performed by a process p_i as it performs the Entry or Exit operations incurs $O(1)$ RMR cost clearly, possibly barring the *while* loop executed in Line 30. During the execution of this *while* loop, process p_i spins on the register $Lock[p_i][p_j]$, where p_j is the predecessor of p_i. Recall that $Lock[p_i][p_j]$ is a register that is local to p_i and thus, p_i does not incur any RMR cost on account of executing this loop. It follows that p_i incurs $O(1)$ RMR cost in the DSM model. Thus, the overall RMR cost of M is within a constant factor of the RMR cost of $L(M)$ in the DSM model.

Theorem 3. *([3]) Any deadlock-free, finite-exit mutual exclusion implementation from read, write and conditional primitives has an execution whose RMR complexity is $\Omega(n \log n)$.*

Theorems 2 and 3 imply:

Theorem 4. *Any strictly serializable, strongly progressive TM implementation providing wait-free TM-liveness from read, write and conditional primitives that accesses a single t-object has an execution whose RMR complexity is $\Omega(n \log n)$.*

6   Related Work and Concluding Remarks

Theorem 1 improves the read-validation step-complexity lower bound [12,13] derived for *strict-data partitioning* (a very strong version of DAP) and (strong) invisible reads. In a *strict data partitioned* TM, the set of base objects used by the TM is split into disjoint sets, each storing information only about a single data item. Indeed, every TM implementation that is strict data-partitioned satisfies weak DAP, but not vice-versa. The definition of invisible reads assumed in [12,13] requires that a t-read operation does not apply nontrivial events in any execution. Theorem 1 however, assumes *weak* invisible reads, stipulating that t-read operations of a transaction T do not apply nontrivial events only when T is not concurrent with any other transaction.

The notion of weak DAP used in this paper was introduced by Attiya *et al.* [5].

Proving a lower bound for a concurrent object by reduction to a form of mutual exclusion has previously been used in [1,13]. Guerraoui and Kapalka [13] proved that it is impossible to implement strictly serializable strongly progressive TMs that provide *wait-free* TM-liveness (every t-operation returns a matching response within a finite number of steps) using only read and write primitives. Alistarh *et al.* proved a lower bound on RMR complexity of *renaming* problem [1]. Our reduction algorithm (Sect. 5) is inspired by the $O(1)$ RMR mutual exclusion algorithm by Hyonho [15].

To the best of our knowledge, the TM properties assumed for Theorem 1 cover all of the TM implementations that are subject to the validation step-complexity [6,7,14]. It is easy to see that the lower bound of Theorem 1 is tight for both strict serializability and opacity. We refer to the TM implementation in [17] or *DSTM* [14] for the matching upper bound.

Finally, we conjecture that the lower bound of Theorem 4 is tight. Proving this remains an interesting open question.

References

1. Alistarh, D., Aspnes, J., Gilbert, S., Guerraoui, R.: The complexity of renaming. In: IEEE 52nd Annual Symposium on Foundations of Computer Science, FOCS 2011, 22–25 October, 2011, pp. 718–727, Palm Springs, CA, USA (2011)

2. Anderson, T.E.: The performance of spin lock alternatives for shared-memory multiprocessors. IEEE Trans. Parallel Distrib. Syst. **1**(1), 6–16 (1990)
3. Attiya, H., Hendler, D., Woelfel, P.: Tight RMR lower bounds for mutual exclusion and other problems. In: Proceedings of the Twenty-seventh ACM Symposium on Principles of Distributed Computing, PODC 2008, pp. 447–447, New York, NY, USA. ACM (2008)
4. Attiya, H., Hillel, E.: The cost of privatization in software transactional memory. IEEE Trans. Comput. **62**(12), 2531–2543 (2013)
5. Attiya, H., Hillel, E., Milani, A.: Inherent limitations on disjoint-access parallel implementations of transactional memory. Theory Comput. Syst. **49**(4), 698–719 (2011)
6. Dalessandro, L., Spear, M.F., Scott, M.L.: Norec: streamlining STM by abolishing ownership records. SIGPLAN Not. **45**(5), 67–78 (2010)
7. Dice, D., Shalev, O., Shavit, N.N.: Transactional locking II. In: Dolev, S. (ed.) DISC 2006. LNCS, vol. 4167, pp. 194–208. Springer, Heidelberg (2006)
8. Dice, D., Shavit, N.: What really makes transactions fast? In: Transact (2006)
9. Dice, D., Shavit, N.: TLRW: return of the read-write lock. In: SPAA, pp. 284–293 (2010)
10. Ellen, F., Hendler, D., Shavit, N.: On the inherent sequentiality of concurrent objects. SIAM J. Comput. **41**(3), 519–536 (2012)
11. Fraser, K.: Practical lock-freedom. Technical report, Cambridge University Computer Laborotory (2003)
12. Guerraoui, R., Kapalka, M.: The semantics of progress in lock-based transactional memory. SIGPLAN Not. **44**(1), 404–415 (2009)
13. Guerraoui, R., Kapalka, M.: Principles of Transactional Memory. Synthesis Lectures on Distributed Computing Theory. Morgan and Claypool, San Rafael (2010)
14. Herlihy, M., Luchangco, V., Moir, M., Scherer III, W.N.: Software transactional memory for dynamic-sized data structures. In: Proceedings of the Twenty-Second Annual Symposium on Principles of Distributed Computing, PODC 2003, pp. 92–101, New York, NY, USA. ACM (2003)
15. Hyonho, L.: Local-spin mutual exclusion algorithms on the DSM model using fetch-and-store objects (2003). http://www.cs.toronto.edu/pub/hlee/thesis.ps
16. Israeli, A., Rappoport, L.: Disjoint-access-parallel implementations of strong shared memory primitives. In: PODC, pp. 151–160 (1994)
17. Kuznetsov, P., Ravi, S.: On the cost of concurrency in transactional memory. In: Fernàndez Anta, A., Lipari, G., Roy, M. (eds.) OPODIS 2011. LNCS, vol. 7109, pp. 112–127. Springer, Heidelberg (2011)
18. Kuznetsov, P., Ravi, S.: On partial wait-freedom in transactional memory. In: Proceedings of the 2015 International Conference on Distributed Computing and Networking, ICDCN 2015, Goa, India, p. 10, 4–7 Jan 2015
19. Papadimitriou, C.H.: The serializability of concurrent database updates. J. ACM **26**, 631–653 (1979)
20. Perelman, D., Fan, R., Keidar, I.: On maintaining multiple versions in STM. In: PODC, pp. 16–25 (2010)
21. Tabba, F., Moir, M., Goodman, J.R., Hay, A.W., Wang, C.: Nztm: nonblocking zero-indirection transactional memory. In: Proceedings of the Twenty-first Annual Symposium on Parallelism in Algorithms and Architectures, SPAA 2009, pp. 204–213, New York, NY, USA. ACM (2009)

Wavelet-Based Local Mesh Adaptation with Application to Gas Dynamics

Kirill Merkulov[(✉)]

ROSATOM Corp., Dukhov All-Russia Research Institute of Automatics,
Moscow, Russia
parovoz1991@yandex.ru

Abstract. The paper addresses a simple numerical method for calculating two-dimensional gas dynamics problems on Cartesian meshes with dynamic local refinement. For multilevel local adaptation, several mesh-related algorithms are proposed based on quadric trees and recursive functions. A global analyzer of the computed solution is developed on the wavelet-based decompositions. To project the numerical solution between different mesh levels a procedure is proposed for cell function reconstruction based on the WENO-approach. Different ways of the parallel realization for such dynamic mesh structures are discussed.

Keywords: Mesh refinement · Gas dynamics · Wavelet analysis · Multiscale calculations

1 Introduction

Numerical simulation in gas dynamics faces a problem when the flow of interest is spatially non-uniform and contains structures of very different scales that require different grid resolution. In this context, small scales might be so small that use of a uniform grid in the whole computational domain makes the problem to be computed impractical because of enormous computational resources required. One way to treat such problems is to employ dynamic locally adaptive grids that automatically increase space resolution (by dividing computational cells in smaller parts) in subdomains of small scale structures, and oppositely coarser grid resolution in regions of smooth distributions. Our research aims to develop and code a novel numerical technology that solves problems with different scales on simple Cartesian grids with local dynamic adaptation of the grid size to the solution computed. This technology is based on the combination of the idea of multiscaling in the description of flows realized by means of the wavelet analysis with the Method of Free Boundary (MFB) [1] that makes it possible to effectively model inner boundary conditions on an body unfitted Cartesian grid.

In the MFB, the compensating mass, momentum, and energy flux is introduced in the right-hand side of the governing equations so that the solution of the modified system of equations in the region off the body would exactly match the solution of the original boundary-value problem. In doing so, the problem is solved on a regular Cartesian grid, which ideally fits the procedure of local grid adaptation by means of tensor products of proper one-dimensional wavelets or multiwavelets.

© Springer International Publishing Switzerland 2015
V. Malyshkin (Ed.): PaCT 2015, LNCS 9251, pp. 426–435, 2015.
DOI: 10.1007/978-3-319-21909-7_41

The idea of multiscaling and adaptivity is not new and goes back to the ideas of A. Harten [2]. However, the appearance of new classes of multiwavelets orthogonal to polynomials and also having small compact supports in combination with the simple structure of computational cells in the method MFB offers fundamentally new possibilities for developing new wavelet adaptive algorithms on Cartesian grids for solution of hydro- and gas dynamic problems with the parallel computer systems of hybrid CPU/GPU architecture.

When solving hydro- and gas dynamic problems with the use of graphic accelerators, in most cases simple explicit time marching schemes are employed because of the specific architecture of the GPU. Program realizations, as a rule, simultaneously can operate several GPU of only one unit. In our work, we intend to develop a new local dynamic grid adaptive parallel algorithm that is based on the implicit time marching scheme, the realization of which is supposed to will have scalability near to the ideal one., we plan to create the new effective parallel computational technology.

2 Mathematical Model

In this paper, the method of mesh adaptation is considered for a simple two-dimensional gas dynamic model that is based on the system of inviscid Euler equations. The mathematical statement of the problem is reduced to solving the following system of equations:

$$
\begin{cases}
\frac{\partial \rho}{\partial t} + div(\rho \mathbf{U}) = 0, \\
\frac{\partial (\rho \mathbf{U})}{\partial t} + div(\rho \mathbf{U}\mathbf{U}) + grad(p) = 0, \\
\frac{\partial (\rho E)}{\partial t} + div(\rho \mathbf{U} H) = 0,
\end{cases}
\tag{1}
$$

in a region $V(\Gamma) \subset \mathbf{R}^2$ bounded by a given closed curve $\Gamma = \Gamma(x, y)$. Here we introduce the standard notation: $\mathbf{U}$ - velocity vector, ρ - density, p - pressure, $H = E + p/\rho$ - the total specific enthalpy, $E = e + 0.5\mathbf{U}^2$ - the total specific energy, e - specific internal energy.

These equations correspond to the laws of conservation of mass, momentum, and energy. To close these equations, the equation of state is used that functionally relates thermodynamical parameters. The gas is assumed to be ideal and calorifically perfect, i.e. $p = (\gamma - 1)\rho e$. The problem is completed with initial and boundary conditions imposed on the boundary Γ.

3 Numerical Method

The problem is solved in a rectangular computational domain with a Cartesian grid that embraces the flow field to be computed and also the boundary Γ. To model inner boundary conditions on Γ, we utilize the MFB. Following this method [2], the Eqs. (1) are modified by adding in the right-hand side properly defined compensating fluxes $\mathbf{F}_w$:

$$\frac{\partial \mathbf{Q}}{\partial t} + \left[\frac{\partial \mathbf{F}_x}{\partial x} + \frac{\partial \mathbf{F}_y}{\partial y}\right] = -\mathbf{F}_w \tag{2}$$

Here we introduce vector notations: $\mathbf{Q} = \left(\rho, \rho U_x, \rho U_y, \rho E\right)^T$ - the vector of conservative variables, $\mathbf{F}_k = \left(\rho U_k, \rho U_k U_x + p\delta_k, \rho U_k U_y + p\delta_k, \rho U_k H\right)^T, k = x, y$- the flux vectors in the x- and y-directions. The compensating flux $\mathbf{F}_w$ is applied only to those cells that cut by the boundary Γ. This flux depends on local gas dynamic parameters and also on the subcell geometry structure determined by three parameters – volume fraction of gas in the cut cell, the length of boundary Γ inside the cut cell, and the unit outward normal. Note that the geometry inside the cut cell is approximated by the linear reconstruction.

The integration of the system (2) over one time step is carried out at each computational cell of the Cartesian grid with the explicit finite-volume method. The numerical flux at the cell interface is calculated with the Godunov scheme on the base of the exact [3] or an approximate Riemann problem solution. We use a simple approximate Riemann solver with two-waves configuration proposed by Rusanov [4]. Because of the gas volume fraction that might be small rather small, the integration of the compensating flux in cut cells is accomplished with the implicit method. The latter is solved with local Newtonian iterations.

4   2D Adaptive Cartesian Mesh

When using grids with dynamic local adaptation, one of the most time-consuming and difficult tasks is the search for most appropriate and convenient format of data. To adapt the originally structured Cartesian grids where the cell partition is fulfilled with a pre-selected law, it is convenient to use different tree-type data structures. There are fast recursive algorithms that allows to fulfill effectively procedures of crawling and grid refinement or coarsening.

We use quadric trees for 2D adaptive grids similar to those used in [5]. An aexample of data performance you can see on Fig. 1. When refining the grid, each cell is divided into only four equal parts by halving each direction.

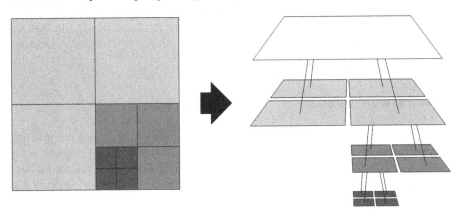

Fig. 1. Quadric tree as a format for adaptive mesh storage.

Each grid cell is described by the value of its level of adaptation (the base grid corresponds a level equal to 0), and the virtual position at this level - a couple of parameters: the integer coordinates of the cell in the virtual Cartesian grid corresponding the cell level. In addition, the cell possesses a dividing flag indicating whether this cell is the final settlement (the corresponding leaf of the tree) or it has 4 children cells of the next level. For divided cells we save pointers to all its children, for a leaf cell (we will call it physical) – the pointer to corresponding gas dynamic parameters.

This format is very useful for mesh storage because the creation, changing and traversal of quadric tree are describes by simple algorithms based on dichotomy. If you divide the row on current level of the instant cell evenly by two, you receive the row number of parent's cell on its current level.

Figure 2 just illustrates the importance of the grid adaptation. Here we show numerical results for the point explosion problem obtained on a two-level grid.

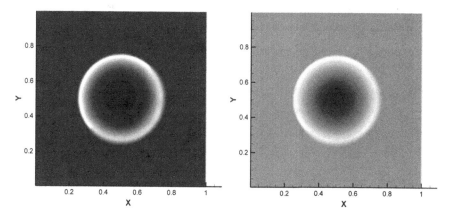

Fig. 2. Point explosion problem: pressure (left) and density (right) on the two-level grid. Black areas relates to low values and white to high values.

The base grid consists of 100×100 equally distributed cells. One level adaptation is then done only for the cells located above the main diagonal of the computational domain. One can see clear difference between solutions above and below the diagonal.

5   Wavelet-Based Analyzer of Numerical Solutions

It is very important to use simple unified procedure when analyzing big data on parallel system. We develop our method for computing problems of gas dynamics on the Cartesian grids of high resolution. So we need the simple operation, that can say us where we need fine grid and where we can stay it coarse. This operation should be homogeneously applicable to the each cell of non-structured mesh. So we decided to use global procedure based on wavelets that help us find subareas with singularities inside big arrays. Let us consider a discrete function (say the numerical density or pressure distribution) on a Cartesian grid, $\{g_{i,j}\}, 0 \leq i \leq 2N, 0 \leq j \leq 2N$. Introduce a threshold δ that is defined by the cell size and the order of approximation of the numerical method to be used. Then we can calculate the wavelet decomposition of a two-dimensional array of the data and define a set of flags:

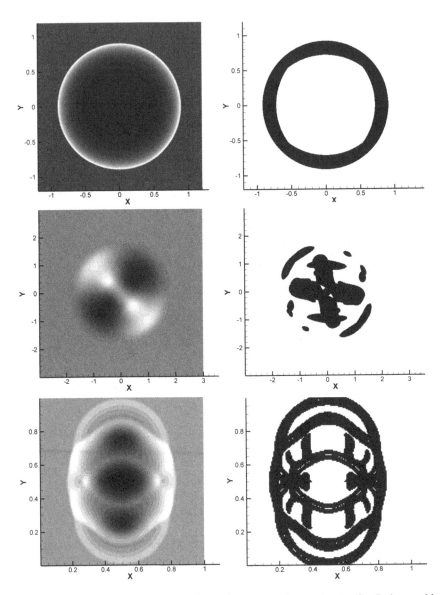

Fig. 3. Gas dynamic fields (left half of the picture, top down: density for Sedov problem numerical solution, pressure for the problem about the instability of isolated vortex and problem of interaction of three point explosures) and adaptation flags, corresponding to them. Black areas relates to low values and white to high values. For flags black area corresponds to the mesh part with refinement.

$$flag_{i,j} = \left(lh_{\left[\frac{i}{2}\right]\left[\frac{j}{2}\right]} > \delta \right) \vee \left(hl_{\left[\frac{i}{2}\right]\left[\frac{j}{2}\right]} > \delta \right) \vee \left(hh_{\left[\frac{i}{2}\right]\left[\frac{j}{2}\right]} > \delta \right),$$

$$0 \leq i \leq 2N, 0 \leq j \leq 2N$$

In this way, the flag points that area where the numerical solution dictates saving the fine grid; cells in other areas are assumed to coarsen [6]. Opportunities of this analyzer are illustrated in the Fig. 3, where we performance different gas dynamic fields and corresponding adaptation given by our analyzer.

This procedure can be easily adapted to multilevel case using sliding window approach or by recursive application of existing procedure, where analyzer says that the grid should be refined.

6 WENO-Reconstructions for Adaptive Cartesian Meshes

The calculation of the above problem on a dynamic locally adaptive grid requires a function to project the solution of the base grid onto the cells of the lower level grid. We propose a simple extension of the WENO interpolator [7] to adaptive 2D grids. Also we used the theoretical findings from [8] and [9]. Assuming that the difference in levels between two neighboring cells is not greater than 1, we use the regular eight-point pattern for each local mesh configuration. This allows us to construct efficient interpolation function that provides projected data in children cells on the base of the given data in the parent cell and its neighbors. In order to use WENO-technology for function reconstruction on non-regular mesh we should choose a group of linearly independent templates. Let's consider that all neighbors of the instant cell of the level l are of the level $l + 1$, i.e. instant cell has eight little neighbors. If in real situation configuration is not so, we use the data from the real cells that are physical and parent for the potential neighbors of the instant one. Let's give the indexes to each of these eight imaginary cells (see Fig. 4).

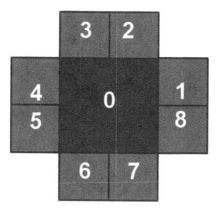

Fig. 4. Numeration and configuration of the imaginary neighbors of instant cell, used for the WENO-reconstruction of conservative data inside the instant cell.

Now we have nine values of conservative parameter: eight averaged for each of imaginary neighbor cells and one for the instant cell. We have chosen linearly eight independent 3-point templates: four triangular: 1-0-2, 3-0-4, 5-0-6, 7-0-8 and four right: 2-0-4,4-0-6,6-0-8,8-0-2 (we call it right, because they form right triangles). We can calculate the gradient of conservative parameter using each of eight templates. After that we need to weight influence of each template into final function reconstruction. Influence of each template of the group (triangular or right) will be proportional to the value $\frac{S}{d}$, where S is the square swept by the template and d is the distance between the centers of the instant cell and the template. All triangular templates has relative weight of $\frac{1}{2\sqrt{2}}$, and the right one – the weight of $\frac{3}{4\sqrt{5}}$. For each pattern we calculate the gradient, the resulting gradient we have as a weighted sum of gradients, received by the templates with weights $\frac{W_{triangular(right)}}{(\varepsilon+\beta)^2}$, where β is the smoothness (absolute value of the gradient), ε is small number that introduced to get rid of division by zero.

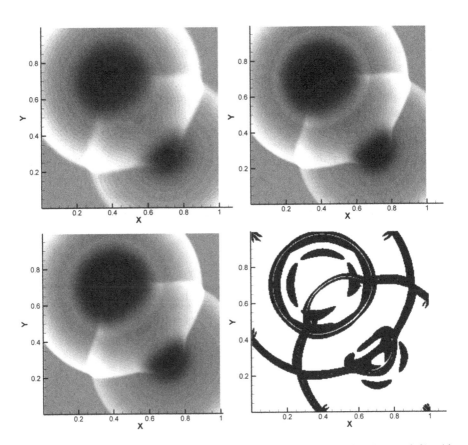

Fig. 5. Density field at the base (top left), shallow (top right) and adaptive (bottom left) grids, adaptation flags (bottom right) after calculation on dynamically adaptive grid for the problem of interaction of point explosions. Black areas relates to low values and white to high values. For flags black area corresponds to the mesh part with refinement.

7 Results of Numerical Tests

Figure 5 shows results of calculating two-points explosion problem with three grids: a base grid of 200 × 200 cells, a fine grid 400 × 400 cells, and a dynamic two-level adaptive grid. No difference can be seen between the fine and adaptive grid solutions. Adaptation flags are given for this problem in the bottom-right of Fig. 5. Black indicates cell of the fine grid, while white does the coarse grid. Figure 6 illustrates the numerical solution of Sedov problem [10], received by three modes of two-level grid (base, fine and dynamic adaptive) and adaptation flags for last variant of calculation.

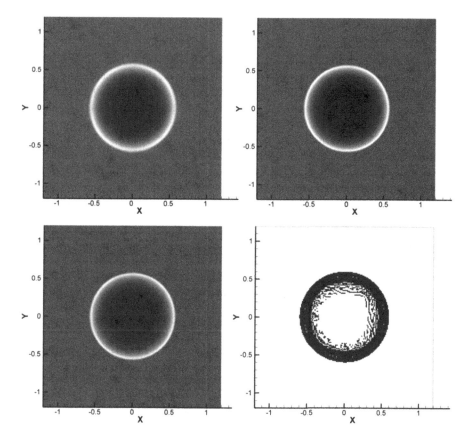

Fig. 6. Density field at the base (top left), shallow (top right) and adaptive (bottom left) grids, adaptation flags (bottom right) after calculation on dynamically adaptive grid for the Sedov problem. Black areas relates to low values and white to high values. For flags black area corresponds to the mesh part with refinement.

8 Specifics of Parallel Realization

The method above can be easily transferred on the parallel computer systems with shared memory. It can be implemented on the basis of dynamic parallelism (recursive functions in CUDA) or algorithms for the chains of different lengths. A challenging task is partitioning of dynamically changing adaptive grid and its optimal distribution in the computer to minimize calculation time and computing resources. Even developing the method for a system with common memory (for example, GPU), you need to optimize the memory access. For our problem localization of the each quadric tree (element of basic mesh) seems to be useful and accelerates calculation up to two times.

9 Conclusions

A simple numerical method has been proposed for calculating two-dimensional gas dynamics problems on Cartesian meshes with dynamic local refinement. All the mesh algorithms are based on quadric trees and recursive functions. We suggest the global analyzer for the different numerical solutions and the procedure of cell function reconstruction based on the WENO-approach. Also different ideas of the parallel realizations for such structures have been described.

This research was supported by the grant 14-11-00872 from Russian Scientific Fund. In the part of parallel implementation it is supported by the grant 14-01-31480 from the Russian Foundation for Basic Research.

References

1. Menshov, I., Kornev, M.: Free-boundary method for the numerical solution of gas-dynamic equations in domains with varying geometry. Math. Models Comput. Simul. **6**(6), 612–621 (2014)
2. Harten, A.: Multiresolution algorithms for the numerical solution of hyperbolic conservation laws. Comm. Pure Appl. Math. **48**(12), 1305–1342 (1995)
3. Godunov, S.K., et al.: Numerical Solution of Multidimensional Problems of Gas Dynamic. Nauka, Moscow (1976)
4. Rusanov, V.V.: The calculation of the interaction of non-stationary shock waves and obstacles. USSR Comput. Math. Math. Phys. **1**(2), 304–320 (1962)
5. Sukhinov, A.A.: Construction of cartesian meshes with dynamic adaptation to the solution. Matematicheskoe Modelirovanie **22**(1), 86–98 (2010)
6. Afendikov, A. L., Merkulov, K. D., Plenkin, A. V.: Local mesh adaptation in gas dynamic problems with the use of wavelet analysis (2014)
7. Semplice, M., Coco, A., Russo, G.: Adaptive Mesh Refinement for Hyperbolic Systems based on Third-Order Compact WENO Reconstruction (2014). arXiv:1407.4296
8. Kudryavtsev, A. N., Khotyanovsky, D. V.: Application of WENO schemes for numerical simulations of high-speed flows. In: the Abstract of International Conference on Computational Fluid Dynamics, Vol. 4 (2006)

9. Shu, C.W.: High order ENO and WENO schemes for computational fluid dynamics. In: Barth, T.J., Deconinck, H. (eds.) High-order methods for Computational Physics, pp. 439–582. Springer, Heidelberg (1999)

10. Sedov, L.I.: Similarity and Dimensional Methods in Mechanics. CRC Press, Boca Raton (1993)

On Implementation High-Scalable CFD Solvers for Hybrid Clusters with Massively-Parallel Architectures

Pavel Pavlukhin[1,2](✉) and Igor Menshov[1]

[1] Keldysh Institute of Applied Mathematics, Moscow 125047, Russia
{pavelpavlukhin,menshov}@kiam.ru
[2] Research and Development Institute "Kvant", Moscow 125438, Russia

Abstract. New approach for solving of compressible fluid dynamic problems with complex geometry on Cartesian grids is proposed. It leads to algorithmic uniformity for whole domain and structured memory accesses which are essential for effective implementations on massively-parallel architectures – GPUs. Methods used are based on implicit scheme and LU-SGS method. Novel parallel algorithm for last one is proposed. In-depth analysis of CUDA+MPI implementation (interoperability issues, libraries tuning) scalable up to hundreds GPUs is performed.

Keywords: CFD · CUDA · LU-SGS · Implicit schemes · Parallel algorithms

1 Introduction

Solving of compressible fluid dynamic problems with state-of-art computational systems is very challenging and face many difficulties. The first difficulty is due to discretization of the computational domain. The mesh is commonly unstructured if the domain is geometrically complex. Meshing procedure demands a huge amount of computations and can involve manual correction which also takes time. Unstructured mesh leads to irregular memory access which results in performance degradation due to rather memory-bound solver then compute-bound one. This concerns classical computational architecture and massive parallel processors in greater degree whose performance primarily depends on regularity of memory accesses. Consequently the favorite choice is to use structured meshes resulting in regular memory access patterns. But generation of grids fitted to domain boundaries is an exigent or even unresolvable problem.

Another difficulty rises from relationship between evolutions of numercal methods and processors architecture. To solve a problem on low performance computational systems one used a coarse mesh and therefore numerical methods were supposed to be sophisticated in order to obtain more accurate solution. From the other side processor cores became "heavier" at the same time: out-of-order execution, data prefetch, branch prediction, vector instructions allowed

© Springer International Publishing Switzerland 2015
V. Malyshkin (Ed.): PaCT 2015, LNCS 9251, pp. 436–444, 2015.
DOI: 10.1007/978-3-319-21909-7_42

effectively implement complex methods. But scalability of computational systems on such "heavy" cores was very limited and their development was operose thus led to emerging of new massive-parallel systems with high number of simple cores. A problem has appeared: accumulated amount of numerical methods weren't suited to new machines since their high performance was achieved by high number (about 1000) of cores simultaneously processing "light" threads not by low number (about 10) of "heavy" ones. In other words, from the one hand simple core structure of massive parallel units requires trivial numerical methods, from the other hand demands scalability and parallelizability much stronger than for "classical" computational systems. Explicit schemes can be well fitted to new processor architectures but they suffer from stability restrictions: in problems involving complex geometry with irregular grid the global time step will be defined by the smallest cell resulting in unreasonable computational costs. Implicit methods allow to overcome this restriction but they are more complicated from the abovementioned difficulties especially from parallelisation point of view.

Thus it's desirable to have a numerical method free of time step restricition, with capability of complex geometry usage by means of simple (Cartesian) grids for effective memory utilization and appropriate for massive-parallel compute systems.

Implicit schemes have no time step restrictions but their implementations on GPU are very complex since data dependency which is a common feature for scheme of such type especially in case of lack of GPU global synchronization. For example, the LU-SGS method [1] which is used in [2] for solving a linear system generated by implicit schemes can be regarded as robust for GPU but because of data dependency which complicates parallelisation another modification of LU-SGS method, DP-LUR [3], is chosen, this method is free of data dependence but suffers from higher computational cost. In our paper an original parallel algorithm for LU-SGS method is proposed. It exactly copies work of successive prototype and scales up to hundreds GPU.

As stated above, the type of spatial discretization also plays major role since GPU works with regular data structures which represent structured grid much more effectively in comparison with irregular ones typical for unstructured grids. But usage most common – body-fitted – grid type applies considerable restrictions on domain geometry complexity. The method of free boundaries [4] permits to solve problems in complex domains on plane non body-fitted Cartesian grids. This property makes it well suited candidate for GPU. This method is based on an alternative formulation of the problem in terms of which inner boundary conditions are modeled by a compensating flux a special right-hand side additive in governing equations thus permitting perform computations over all grid cells in a uniform way. In other words, the method of free boundaries possesses algorithmic uniformity quite important for massive-parallel architectures.

2   Numerical Method

Let us briefly describe methods used. Detailed description can be found in [1, 4–6]. Consider the Euler equations for compressible flow in conservation laws form:

$$\frac{\partial \boldsymbol{q}}{\partial t} + \frac{\partial \boldsymbol{f}_i}{\partial x_i} = -\boldsymbol{F}_w \ . \tag{1}$$

Non-penetrating boundary conditions are modeled by a properly defined right-hand additive – the compensating flux. After spacial discretization by the finite volume method implicit time integration scheme is applied which leads to the system of discrete equations. This one is solved by pseudo time relaxation method based on implicit discretization and Newtonian iterations. In this way, the linear system is consequently solved for finding iteration increment:

$$(D + L + U)\delta^s \boldsymbol{q}_i = -\boldsymbol{R}_i^{n+1,s} \ . \tag{2}$$

The LU-SGS method leads to approximate factorization of the left-hand side (1) and the equations are consequently splitted in two subsystems:

$$\begin{cases} (D + L)\delta^s \boldsymbol{q}_i^* = -\boldsymbol{R}_i^{n+1,s} \\ (D + U)\delta^s \boldsymbol{q}_i = D\delta^s \boldsymbol{q}_i^* \ . \end{cases} \tag{3}$$

Solving (3) is reduced to forward and backward sweeps over all grid cells without full matrix inverses.

3   Parallel LU-SGS Algorithm

The compensating flux addition (1) does not change structure of the linear equations (2) since that flux is local component relative to each grid cell. This property permits to construct parallel algorithm for LU-SGS method in the same way as for Cartesian grids without any inner boundaries. Solving (3) may be represented as forward and backward sweeps over all grid cells hence arising data dependency has local nature and is only defined by sweep order over all geometrical neighbors relative to current cell. In other words, depending on neighbor position ("before" or "after") in sweep order in relation to current cell various computing operations are performed. Sweep order may be chosen in different ways based not only on cell geometry neighborhood. This property will be exploited for parallel algorithm construction. There are two levels for this challenge – internode and intra-GPU parallelization.

First, let us consider the problem in internode level assuming each node equipped sequential computing unit only. Computational domain is divided into blocks with uniform cell distribution. The topology of blocks equals to grid cells one: "toe-to-toe" rectangles. Next, blocks are grouped in two sets in "chessboard" order: two neighbor blocks belong different ("black" and "white") sets. Inner and border cells are separated in each block. Global sweep is chosed in following way: first, inner cells in all "black" blocks are swept then half inner cells and whole border part in all "white" blocks, border cells in all "black" blocks and, finally, left half inner cells are swept (Fig. 1). This sweep permits to overlap parallel computing over inner cells and border cells exchanging between neighbor blocks preserving full method correctness.

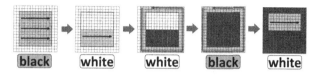

Fig. 1. Global sweep for internode parallel algorithm.

Consider now intra-GPU parallelization. Simultaneous computing over grid cells is possible only when they are not geometrical neighbors. Sweep chosing for GPU brings to graph coloring problem: any two neighbor vertex (cells) must be different colors. In case of structured, namely Cartesian grids in the method of free boundaries, only two colors are needed what leads to "chessboard" cells sweep similar to blocks one, i.e. computing on GPU is performed over all "black" cells (simultaneously) and then over all "white" cells (simultaneously).

Combining two levels of parallelism results in multinode multi-GPU algorithm: inner and border cells in blocks are splitted into "black" and "white" ones; in each part of block same color cells are computed simultaniously on GPU (Fig. 2). More details can be found in [7].

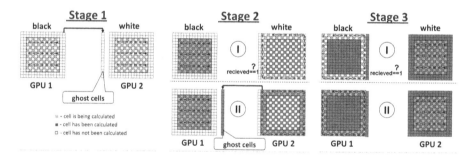

Fig. 2. Multinode multi-GPU parallel algorithm for LU-SGS method.

4   Numerical Experiment

Correctness of the proposed method of free boundaries was studied in flow over a NACA0012 airfoil problem. Angle of attack $\alpha = 1.25\,°$, Mach number $M = 0.8$. Two types of grid are used:

1. Body fitted C-grid with 400 nodes around aerofoil and total 400×200 resolution. Standard boundary conditions are applied.

2. Body non-fitted Cartesian grid with circumscribing around aerofoil 200 × 24 rectangle and total 650 × 324 resolution. Aerofoil body conditions were modeled by the compensating flux (1).

The problem was solved by using the hybrid explicit-implicit scheme [6] and parallel LU-SGS method described above with a Courant number of $C = 10$ on Tesla K20 GPU. Cp distribution for steady-state solution is presented on Fig. 3. One can observe that solutions for the two types of grid are very close. Discontinuity at trailing edge of aerofoil is caused by linear aproximation errors in this sharp part of NACA0012 on Cartesian grid. There is also characteristic local minimum in weak shock wave on lower aerofoil border for solution getting by the method of free boundaries; there is no one on C-grid, it can be only observed in higher resolution grids. That is explained by orthogonal property naturally inherenting for Cartesian grids and leading to more accurate solution is not valid for C-grid used in test problem.

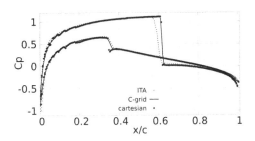

Fig. 3. Cp distribution, NACA0012, $M = 0.8, \alpha = 1.25\,°$, solution on C-grid is marked by solid line and on Cartesian grid – by rhombs.

5   Implementation

The key parallel solvers property for scalability is overlapping computing and data transfers between parts executing in parallel. In case of computer clusters equipped GPU accelerators interoperability of two API – CUDA and MPI – is exploited since interprocess data exchange consists of several stages: GPU RAM $\leftrightarrow$ CPU RAM transfers via pci-express bus, intranode CPU RAM transfers and internode communications via high speed low latency network (Infiniband). Nonblocking MPI calls and streams in CUDA API are used for efficient implementation of that exchanges. Kepler family specific features like Hyper-Q, GPUDirect RDMA are not utilized since modern supercomputers equipped Fermi generation accelerators are still prevalent.

First solver parallel implementation for 2D problems shows promising outlook of proposed methods and approaches. For example, strong scaling efficiency in 3.9 M cells problems is more 80 % on 64 Tesla C2050 GPUs (relative to 2 GPU), Fig. 4.

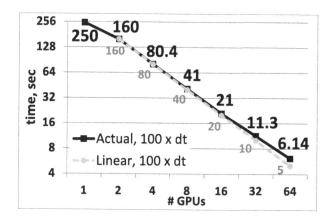

Fig. 4. Strong scaling in 2D problems, SC "K-100" [8].

One can pay attention at results for 1 and 2 GPU – performance increase only in 1.56 times although next values are with almost linear scalability. Analysis of stream tricks in CUDA [9] shows importance of order kernels and memory copies issued in streams. Scheme implemented by events and *cudaEventSynchronize* calls led to serializing these activities (and poor result on 2 GPU mentioned above) and therefore it was modified in 3D solver to work with *cudaStreamSynchronize* and without events. But another problem appears: amount of local per-thread memory is increased (up to around 2 KB for some kernels) due to increased code complexity (compared to 2D solver) and local memory resizes on every kernel launching what leads to serializing kernel execution with considerable launch overhead and blocking concurrent overlapped MPI activity (Fig. 5, up part, MPI calls marked by green color in "Markers and Ranges" line, kernels in "Compute" line).

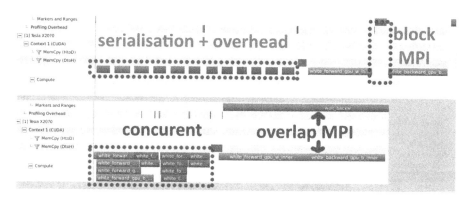

Fig. 5. Nvprof trace for default (up) and with *cudaDeviceLmemResizeToMax* flag (down) run, GPU X2070.

By setting $cudaDeviceLmemResizeToMax$ [10] CUDA GPU flag one able to eliminate local memory resizing on kernel launching what allows concurrent kernel execution without any overhead and overlapping MPI activity (Fig. 5, down part).

Still one interoperability feature of CUDA with MPI is discovered. There are two protocols in MPI library called *eager* and *rendezvous* for send/receive operations. *Eager* protocol requires more buffer memory but runs asynchronously as opposed to *rendezvous* not consuming extra memory and performing synchronization of sender and reciever. Switch from *eager* to *rendezvous* protocol happens when message size exceeds some predefined value depending on specific MPI library implementation (order around 10–100 KB). As it turned out, using *rendezvous* protocol has a negative performance impact (Cuda toolkit 5.*, 6.*, Intel MPI 4.*, 5.* is used for research). Grid size for test problem is 600000 cells; 2 Tesla X2070 is used (one GPU per process); message sizes is about 1 MB, therefore *rendezvous* protocol is used (by default, it is enabled in Intel MPI when message size exceeds 256 KB). Execution time for two MPI processes with default rendezvous protocol takes on 1 node 2.57 *sec*, on 2 nodes (1 process per node) 2.41 *sec* (Table 1). Forced switch-on *eager* protocol (via environment variables) reduces execution time to 2.17 *sec* for both cases.

Nvprof trace (Fig. 6) shows in detail that *MPI_wait* calls (green color, "markers and ranges" line) block kernel launching ("Compute" line) when *rendezvous* protocol is used whereas *eager* protocol does not limit overlapping kernels and MPI calls.

Table 1. CUDA + MPI: *rendezvous* vs *eager* protocol (times in *sec*)

2 GPU, intra-node, *rendezvous*	2 GPU, inter-node, *rendezvous*	2 GPU, intra-node, *eager*	2 GPU, inter-node, *eager*	1 GPU
2.57	2.41	2.17	2.17	4.13

Fig. 6. Nvprof trace: *rendezvous* (up) vs *eager* (down) (Color figure online).

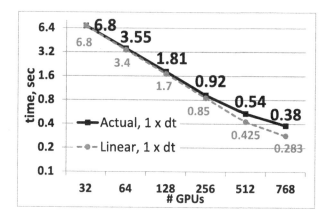

Fig. 7. Strong scaling on 150 M cells 3D problem, "Lomonosov" Supercomputer.

Consequently only by MPI library tuning one can reduce execution time up to 15 % for MPI+CUDA application. Performance on 2 GPUs increases almost 2 times compared to 1 GPU (Table 1). That demonstrates more effective scheme for CUDA streams compared to one in 2D solver (Fig. 4). Finally, 75 % strong scaling efficiency (Fig. 7) is achieved on 768 GPU (relative to 32 GPU) on "Lomonosov" Supercomputer [11] (150 M cells problem size). That result is obtained due to optimization approaches described above as well.

Acknowledgments. This work was supported by grant 14–01–31480 from the Russian Foundation for Basic Research.

References

1. Jameson, A., Turkel, E.: Implicit schemes and LU decomposition. Math.of Comp **37**, 385–397 (1981)
2. Lin, F., Zhenghong, G., Kan, H., Fang, X.: A multi-block viscous flow solver based on GPU parallel methodology. Comput. Fluids **95**, 19–39 (2014)
3. Wright, M.J., Candler, G.V.: A data-parallel LU rexation method for reacting viscous flows. In: Ecer, A., Satofuka, N., Periaux, J., Taylor, S. (eds.) Parallel Computational Fluid Dynamics 1995, pp. 67–74. North-Holland, Amsterdam (1996)
4. Menshov, I., Pavlukhin, P.: Numerical Solution of Gas Dynamics Problems on Cartesian Grids with the Use of Hybrid Computing Systems. Preprint of KIAM RAS. vol. 92 (2014)
5. Menshov, I., Nakamura, Y.: On implicit godunov's method with exactly linearized numerical flux. Comput. Fluids **29**(6), 595–616 (2000)
6. Menshov, I., Nakamura, Y.: Hybrid explicit-implicit, unconditionally stable scheme for unsteady compressible flows. AIAA J. **42**(3), 551–559 (2004)
7. Pavlukhin, P.: Parallel LU-SGS numerical method implementation for gas dynamics problems on GPU-accelerated computer systems. Vestn. of Lobachevsky State Univ. Nizhni novgorod **1**, 213–218 (2013)

8. "K-100" Supercomputer. http://www.kiam.ru/MVS/resourses/k100.html
9. CUDA C/C++ Streams and Concurrency. http://on-demand.gputechconf.com/gtc-express/2011/presentations/StreamsAndConcurrencyWebinar.pdf
10. CUDA Toolkit Documentation. http://docs.nvidia.com/cuda/cuda-runtime-api/group__CUDART__DEVICE.html
11. Lomonosov Supercomputer. http://hpc.msu.ru/?q=node/59

Parallelization of 3D MPDATA Algorithm Using Many Graphics Processors

Krzysztof Rojek and Roman Wyrzykowski$^{(\boxtimes)}$

Czestochowa University of Technology, Dabrowskiego 69,
42-201 Czestochowa, Poland
{krojek,roman}@icis.pcz.pl

Abstract. EULAG (Eulerian/semi-Lagrangian fluid solver) is an established numerical model for simulating thermo-fluid flows across a wide range of scales and physical scenarios. The multidimensional positive definite advection transport algorithm (MPDATA) is among the most time-consuming components of EULAG. In this study, we focus on adapting the 3D MPDATA computations to clusters with graphics processors. Our approach is based on a hierarchical decomposition including the level of cluster, as well as an optimized distribution of computations between GPU resources within each node. To implement the resulting computing scheme, the MPI standard is used across nodes, while CUDA is applied inside nodes. We present performance results for the 3D MPDATA code running on the NVIDIA GeForce GTX TITAN graphics card, as well as on the Piz Daint cluster equipped with NVIDIA Tesla K20x GPUs. In particular, the sustained performance of 138 Gflop/s is achieved for a single GPU, which scales up to more than 11 Tflop/s for 256 GPUs.

Keywords: Stencil computations · MPDATA algorithm · GPU · Cluster · Parallel programming · Algorithm adaptation · CUDA · MPI

1 Introduction

In recent years, there has been a rapid increase in using modern supercomputing architectures for modeling complex engineering systems. An important example is Numerical Weather Prediction (NWP) [8]. In NWP, the physical processes governing atmospheric flows are simulated by solving partial differential equations in three-dimensional space and time. Due to the high computational complexity of modeling mesoscale weather systems [19,20], it is necessary to employ efficient numerical algorithms and powerful computational resources. The new architectures based on modern multicore CPUs [1] and accelerators such as GPUs [12,13] and Intel Xeon Phi coprocessors [15,20] offer unique opportunities for modeling atmospheric processes significantly faster and with accuracy greater than ever before. The important goal is also the possibility to decrease the energy consumption when performing NWP. However, the traditional codes are bottlenecked by memory/communication bandwidth and cache performance. Therefore, to be able to effectively exploit the potential of new computing platforms, the structure of the traditional codes must be significantly redesigned.

© Springer International Publishing Switzerland 2015
V. Malyshkin (Ed.): PaCT 2015, LNCS 9251, pp. 445–457, 2015.
DOI: 10.1007/978-3-319-21909-7_43

EULAG (Eulerian/semi-Lagrangian fluid solver) is an established numerical model for simulating thermo-fluid flows across a wide range of scales and physical scenarios [9,16,19]. This model is an innovative solver in the field of numerical modeling of multiscale atmospheric flows. The multidimensional positive definite advection transport algorithm (MPDATA) [14] is among the most time-consuming components of EULAG. This algorithm represents a sequence of stencils [2], where each stencil depends on one or more others. The complex structure of data dependencies between stencils makes the efficient parallelization of MPDATA on modern computing platform a challenging problem.

Rewriting the EULAG code and replacing conventional HPC systems with heterogeneous clusters using accelerators such as GPUs was proposed in [6,11,17] to reduce the hardware cost and energy consumption. In particular, work [17] is focused on investigating aspects of an optimal parallel version of the 2D MPDATA algorithm on shared-memory hybrid architectures with GPU accelerators, where computations are distributed across both GPU and CPU components. At the same time, the parallelization of the 3D version of MPDATA requires a different approach. For modern GPU architectures, such an approach was proposed in paper [13], which provides the analysis of resource usage in GPU platforms and its influence on the overall system performance. The proposed approach to kernel processing with queues of data chunks placed in registers and shared memory increases the data locality significantly.

In this study, we focus on adapting the 3D MPDATA computations to clusters with GPU accelerators [4]. Our approach is based on a hierarchical decomposition including the level of cluster, as well as distribution of computations between GPU resources within each node. To implement the resulting computing scheme, the MPI standard is used across nodes, while CUDA is applied inside each node.

2   Overview of MPDATA

MPDATA solves continuity equation describing the advection of a nondiffusive quantity ϕ in a flow field, namely

$$\frac{\delta \psi}{\delta t} + div(\mathbf{V}\psi) = 0, \tag{1}$$

where $\mathbf{V}$ is the velocity vector. The spatial discretization of MPDATA is based on finite difference approximation. The algorithm is iterative and fast convergent. In the first substep, advection of a prognostic field ψ is computed with the standard donor-cell approximation [14]. This ensures the first order of accuracy only. In the subsequent substep, corrections are applied to make the scheme more accurate, i.e. second order in space and time. In the corrective substep, the donor-cell approximation is used again, but with new anti-diffusive velocities computed based on the advected fields.

The MPDATA scheme belongs to the group of nonoscilatory forward-in-time algorithms, and offers several options to model a wide range of complex geophysical flows. The number of required time steps depends on a type of simulated physical phenomenon, and can exceed few millions, especially when considering MPDATA as a part of the EULAG model.

Each MPDATA time step is determined by a set of 17 computational stages, where each stage is responsible for calculating elements of a certain array. These stages represent stencil codes which update grid elements according to different patterns. Figure 1 shows a part of the 3D MPDATA implementation, consisting of four stencils.

```
     pp(y)= amax1(0.,y)
     pn(y)=-amin1(0.,y)
     donor(y1,y2,a)=pp(a)*y1-pn(a)*y2
     do 111 k=1,n3m
     do 111 j=1,mp
     do 111 i=ilft,np
111  f1(i,j,k)=donor(c1*x(i-1,j,k)+c2,c1*x(i,j,k)+c2,v1(i,j,k))
     do 222 k=1,n3m
     do 222 j=jbot,mp
     do 222 i=1,np
222  f2(i,j,k)=donor(c1*x(i,j-1,k)+c2,c1*x(i,j,k)+c2,v2(i,j,k))
     do 333 k=2,n3m
     do 333 j=1,mp
     do 333 i=1,np
333  f3(i,j,k)=donor(c1*x(i,j,k-1)+c2,c1*x(i,j,k)+c2,v3(i,j,k))
     do 444 k=1,n3m
     do 444 j=1,mp
     do 444 i=1,np
555  x(i,j,k)=x(i,j,k)-( f1(i+1,j,k)-f1(i,j,k)
  .                +f2(i,j+1,k)-f2(i,j,k) + f3(i,j,k+1)-f3(i,j,k) )/h(i,j,k)
```

Fig. 1. Part of 3D MPDATA stencil-based implementation

The stages are dependent on each other: outcomes of prior stages are usually input data for the subsequent computations. Every stage reads a required set of arrays from the memory, and writes results to the memory after computation. In consequence, a significant memory traffic is generated, which mostly limits the performance of novel architectures [3]. A single MPDATA time step requires 5 input arrays, and returns one output array that is necessary for the next step. We assume that the size of the 3D MPDATA grid determined by coordinates i, j, and k is $n \times m \times l$.

3 Adaptation of MPDATA to a Single GPU Node

3.1 GPU Architecture and Software Environment

All the experiments performed in this work are performed using NVIDIA GPUs based on the Kepler architecture. The example is the NVIDIA GeForce GTX TITAN GPU [22]. It includes 14 streaming multiprocessors (SMX), each consisting of 64 double precision (DP) units with configurable size of 16/32/48 KB of shared memory and 48/32/16 KB of L1 cache. It gives the total number of $14 * 64 = 896$ DP units with the clock rate of 837 MHz, providing the peak DP performance of 1.5 TFlop/s. This graphics accelerator card includes 6 GB of global memory with the bandwidth of 288 GB/s. All the accesses to the global memory go through the L2 cache of size 1.5 MB. The number of load/store unit per SMX is 32, so it gives the possibility to load/store 256 bits at once per SMX.

To manage CPU and GPU components, we take advantage of using the CUDA programming standard [5]. CUDA is a scalable parallel programming model and a software environment for parallel computing. It allows for the utilization of a GPU as an application accelerator, when a part of an application is executed on a standard CPU processor, while another part is assigned to the GPU, as the so-called kernel. CUDA enables for the efficient management of GPU computing resources, beginning with GPU CUDA cores that are grouped into SMX. In the CUDA data parallel model, the same program (or kernel) runs concurrently on different pieces of data, and each invocation is called a *thread*. The set of threads is called a *block*. Each block is executed on a single SMX. Threads can be synchronized within a single block. However, there is no synchronization mechanism between blocks; they are executed independently.

Another key feature of modern GPUs is their hierarchical memory organization. In the CUDA memory model, all the GPU threads have access to the *global memory*, relatively large but rather slow. Within a particular block, all the threads share the fast *shared memory*. It is used for communication and synchronization among threads across the block. In addition, each thread has access to its *register file*. Furthermore, the L1 and L2 caches are applied to improve the data locality for memory accesses. In particular, all the accesses to the global memory go through L2, including copies to/from the CPU host.

3.2   Processing GPU Kernels

To increase the data locality within CUDA blocks, we employ a widely used method of 2.5D blocking [7,10,13], in which 2D blocks are responsible for computing $g_1 \times g_2$ data chunks, which correspond to sub-planes of a array, called here tiles. Between neighboring blocks, some extra computations take place on the borders. As a consequence, blocks have to be extended by adequate halo areas, both in vertical and horizontal directions. The loop inside a GPU kernel is used to traverse the grid in the dimension k. Because the MPDATA algorithm requires to store at most $3 \times (g_1 \times g_2)$ data chunks at the same time, we use a queue of data chunks placed in registers and shared memory. In this approach, we first copy data from the GPU global memory to registers, and then, for each iteration across the dimension k, we move data between registers and shared memory. This method is illustrated in Fig. 2.

The main advantage of this technique in relation to 3D blocking is the reduction of memory requirements. We need to store only three tiles of each array instead of the entire column of size l, to keep the same intensity of memory traffic between the global memory and shared memory or register file. It is par-

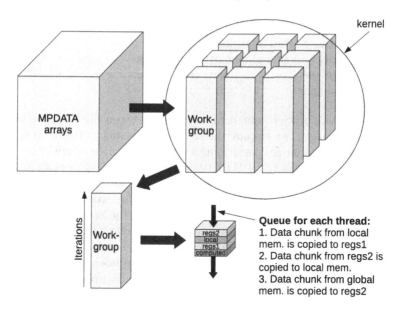

Fig. 2. GPU kernel processing

ticularly useful for GPUs, where the size of shared memory is too small to store 3D blocks of arrays.

3.3 Analysis of Stencils

The starting point of our considerations is when all the 17 stencils are distributed across 6 GPU kernels marked as: A, B, C, D, E, and F (this distribution takes into account synchronization points of MPDATA). Such a number of kernels is selected for the following reasons: (i) the stencils are distributed in such a way that for each kernel, the halo area from any side of a CUDA block does not exceed 1; (ii) the most memory- and register-consuming stencils are implemented in kernels B and C in order to increase the GPU occupancy, defined as the number of active threads per SMX divided by the maximum number of threads supported by SMX. This version of our implementation is called a naive one. In this version, we applied the most common techniques of optimization, including the usage of the shared memory, coalesced memory access, and 2.5D blocking. The kernels A and F are responsible for computing the donor-cell part of MPDATA, the kernels B and C compute anti-diffusive velocities, while the kernels D and E implement the non-oscillatory option for the MPDATA algorithm [17].

Our idea of efficient adaptation of MPDATA to GPU architectures is based on the detection of bottlenecks, and enables for reducing the most notable of them. We examine the following potential bottlenecks:

- data transfers between GPU global memory and host memory;
- instructions latency (stall analysis);
- arithmetic, logic, and shared memory operations;
- configuration of the algorithm taking into account the size of CUDA blocks, and GPU occupancy.

To overlap computation and data transfers between GPU global memory and CPU host memory, we employ the stream processing technique [11], where each stream is responsible for performing a sequence of three activities including: (i) data transfer from host to GPU that occurs only once (before computations); (ii) execution a sequence of six GPU kernels; (iii) data transfer from GPU to host memory (occurs after every time step). All the activities are processed synchronously within a single stream. However, all the streams are processed asynchronously. Thanks to that the activities from one stream are overlapped by the activities from another one. In consequence, the data transfer takes a relatively short time (about 18 %t of the total execution time). So we can conclude that data transfers between host and GPU are not a bottleneck for MPDATA.

The next step is devoted to the stall reasons analysis. It shows that the main reasons of stalls for kernels B and C includes: execution dependency, data requests, texture memory operations, synchronization, and instruction fetch. Among them the most important is the execution dependency caused by the complex structure of the MPDATA algorithm. The execution dependencies can be hidden in part by increasing the GPU occupancy. However, each of the kernels B and C uses about 47KB of shared memory for an active CUDA blocks, executing only 768 active threads per SMX (maximum is 2048 threads). It means that the GPU occupancy is only 37.5 % for both kernels. So the final conclusion is that the GPU utilization is limited by the shared memory usage. The main challenge is to find a solution where data transfers from the global to shared memory or register file are minimized.

3.4   Transformations of Stencils

Our idea of adaptation is based on an appropriate distribution of stencils across GPU kernels in order to minimize the number of memory transactions between shared and global memories. For this aim, we propose a method where a different number of kernels is considered. In each configuration, a single kernel processes a different number of stencils. We estimate a number of memory transactions for each configuration, and then select a configuration where the number of memory transactions is minimized.

The starting point for the optimization is a comprehensive analysis of data flows when executing MPDATA. The distribution of computational tasks is preceded by the estimation of the shared memory utilization, sizes of halo areas, as well as data dependencies between and within stencils. Based on such an extensive analysis, we are able to specify the most favorable number of kernels, as well as set an optimal distribution of stencils across kernels, and the sizes of CUDA blocks for each kernel. As a consequence, an efficient load balancing is preserved, and data communication is minimized and well structured [13].

The compression of stencils increases hardware requirements for CUDA blocks, and decreases the GPU occupancy. However, it allows for the reduction of the number of temporary arrays, and thereby it decreases the memory traffic. In consequence, the best configuration of the MPDATA algorithm is the compression of its 17 stencils into 4 GPU kernels [13]. Figure 3 presents a single GPU kernel which is obtained after compression of the four MPDATA stencils shown in Fig. 1. In this approach, the 2.5D blocking technique is used that allows us to reduce the global memory traffic and provide some subexpression elimination.

```
for(k=1; k<l-1; ++k) {
  q1 = fmax(N0(0.0),v3M(i,j,k+1))*xM(i,j,k)
       +fmin(N0(0.0),v3M(i,j,k+1))*xM(i,j,k+1);

  xP[ijk]=x[ijk] - ( fmax(N0(0.0),v1M(i+1,j,k))*xM(i,j,k)
                    +fmin(N0(0.0),v1M(i+1,j,k))*xM(i+1,j,k)
                    -fmax(N0(0.0),v1M(i,j,k)  )*xM(i-1,j,k)
                    -fmin(N0(0.0),v1M(i,j,k)  )*xM(i,j,k)
                    +fmax(N0(0.0),v2M(i,j+1,k))*xM(i,j,k)
                    +fmin(N0(0.0),v2M(i,j+1,k))*xM(i,j+1,k)
                    -fmax(N0(0.0),v2M(i,j,k)  )*xM(i,j-1,k)
                    -fmin(N0(0.0),v2M(i,j,k)  )*xM(i,j,k)
                    +q1
                    -q0
                    )/h[ijk];
  q0=q1;
  ijk+=M;
}
```

Fig. 3. Four MPDATA stencils compressed into a single GPU kernel

3.5 Performance Results

The performance results are achieved using a single node equipped with the Intel Core i7-3770 CPU clocked 3.40 GHz, NVIDIA GeForce GTX TITAN GPU (see Sect. 3.1), and 48 GB of host memory. The CUDA version used in these tests is V7.0. When testing the CPU version, we uses the original, parallel Fortran code [16] with -O2 flag (gfortran compiler version 4.8.2), which gives the better performance than -O3. This code was developed by the Institute of Meteorology and Water Management, Warsaw. It is parallelized using the MPI standard, without any manual vectorization.

In Tables 1 and 2, we can see two analysis of performance results achieved on the CPU and GPU. The first analysis corresponds to the naive GPU version, where the MPDATA stencils are distributed across 6 GPU kernels. This version uses the GPU shared memory, but it is not optimized for reducing GPU global memory transactions. In the improved version, we apply a set of optimizations. First of all, they include the compression of GPU kernels to 4 ones, which allow us for the reduction of GPU global memory transfers, and extensive elimination of common subexpressions.

Table 1. Execution time [s] and speedup for the naive version of the MPDATA algorithm

Grid size	CPU 1 core	CPU 4 cores	GPU	CPU 1 / GPU	CPU 4 / GPU
$16 \times 16 \times 16$	0.044	0.016	0.087	0.51	0.18
$32 \times 32 \times 16$	0.164	0.048	0.1	1.64	0.48
$64 \times 64 \times 16$	0.636	0.192	0.112	5.68	1.71
$64 \times 64 \times 64$	2.617	0.776	0.366	7.15	2.12
$128 \times 128 \times 64$	10.486	3.316	0.792	13.24	4.19
$128 \times 128 \times 128$	20.816	6.624	1.583	13.15	4.18
$256 \times 256 \times 64$	40.371	12.868	2.56	15.77	5.03

Table 2. Execution time [s] and speedup for the improved version of the MPDATA algorithm

Grid size	CPU 1 core	CPU 4 cores	GPU	CPU 1 / GPU	CPU 4 / GPU
$16 \times 16 \times 16$	0.044	0.016	0.044	1.00	0.36
$32 \times 32 \times 16$	0.164	0.048	0.046	3.57	1.04
$64 \times 64 \times 16$	0.636	0.192	0.055	11.56	3.49
$64 \times 64 \times 64$	2.617	0.776	0.164	15.96	4.73
$128 \times 128 \times 64$	10.486	3.316	0.344	30.48	9.64
$128 \times 128 \times 128$	20.816	6.624	0.678	30.70	9.77
$256 \times 256 \times 64$	40.371	12.868	1.268	31.84	10.15

Based on these results we can conclude that only for very small grids, the CPU implementation is faster than the GPU version. For grid sizes greater or equal $64 \times 64 \times 16$, that are interesting in practice, the GPU version outperforms the CPU one. Moreover, the performance gain is increasing with increasing the grid size. The important conclusion is a clear advantage of the improved GPU version against the naive one. The former outperforms the latter by a factor whose value is in the range from 2.02 to 2.34 (for grids greater or equal $64 \times 64 \times 16$). In consequence, the GPU allows us to speedup computations more than 10 times in comparison with the CPU code, for the largest grid of size $256 \times 256 \times 64$.

4   Adaptation of MPDATA to GPU-accelerated Clusters

4.1   MPDATA Decomposition

The performance results obtained in the previous subsection show that the GPU version of the 3D MPDATA code is profitable for mesh sizes greater or equal

$64 \times 64 \times 16$. To keep many GPUs busy we need to process even greater grids ($256 \times 256 \times 64$ or greater). When using the EULAG model for NWP purposes, typical simulations contain grids from $500 \times 250 \times 60$ to $2000 \times 2000 \times 120$. Moreover, the grid size l in the third dimension is much smaller than the first two grid sizes m and n, where usually $l \leq 128$. Therefore, to provide parallelization of 3D MPDATA on a cluster with GPU-accelerated nodes, it is sufficient to map the 3D MPDATA grid on a 2D mesh of size $r \times c$ (Fig. 4).

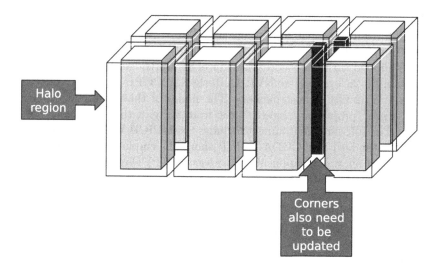

Fig. 4. 2D domain decomposition of MPDATA and communication model

In consequence, the MPDATA grid is partitioned into subdomains of size $n_p \times m_p \times l$, where each node is responsible for computing within a single subdomain, and:

$$n_p = \frac{n}{r}; \quad m_p = \frac{m}{c}. \qquad (2)$$

In EULAG computations, the MPDATA algorithm is interleaved with other algorithms in each time step, and particularly with the elliptic solver. So after every MPDATA call (one time step), we need to update halo regions between neighboring subdomains. The analysis shows that the halo regions are of size 3 on each side. The requirement to perform these updates generates a 2D model of communications between neighboring nodes, including the data exchange corresponding to corners of subdomains.

To implement the resulting computing scheme, the MPI standard is used across nodes, while CUDA is applied inside each node. In our current MPI implementation, only blocking communication routines are applied. In order to improve the performance and scalability of the resulting code, first of all we will use nonblocking MPI communication routines, and the double-buffering technique which permits programmers to overlap communication with computations [4]. An additional method is to adapt the GPUDirect RDMA technology [21], which allows for eliminating unnecessary memory copies, radically lowering CPU overhead, and reducing the communication latency. However, this method is not always applicable in practice, since this technology quite often is not supported by the software/hardware stack on a particular parallel platform.

4.2 Performance Results

The performance results are obtained for the Piz Daint supercomputer [23]. This machine is located in the Swiss National Supercomputing Centre. Currently it is ranked 6-th in the top500 list (November 2014 edition) [24]. Piz Daint is the largest Cray XC30 system that has been delivered and assembled so far. Each node is equipped with one 8-core 64-bit Intel SandyBridge CPU clocked 2.6 GHz (Intel Xeon E5-2670), and one NVIDIA Tesla K20X GPU with 6 GB of GDDR5 memory, and 32 GB of host memory. The nodes of this cluster are connected by the "Aries" proprietary interconnect from Cray, with a dragonfly network topology. The software environment includes the MPICH V6.2.2 implementation of the MPI standard, and CUDA V5.5. It should be emphasized that the current setup does not allow MPI applications to use the GPUDirect RDMA technology to speedup communications between nodes.

Table 3 presents results of weak scalability analysis, when the number of grid elements and number of GPUs increases twice in successive experiments. Here the R_p parameter corresponds to a ratio between the sustained performance for n nodes and sustained performance achieved in the previous experiment, using $n/2$ nodes. The last column shows the sustained performance that could be achieved if the MPDATA algorithm would be perfectly scalable.

Table 3. Weak scalability results of MPDATA on Piz Daint cluster

# nodes	# grid elements	time	sustained performance Gflop/s	R_p	perfect performance Gflop/s
1	2^{24}	4.165	138	-	138
2	2^{25}	4.211	273	1.98	276
4	2^{26}	4.288	537	1.97	552
8	2^{27}	4.430	1040	1.94	1104
16	2^{28}	4.719	1953	1.88	2208
32	2^{29}	5.244	3516	1.8	4416
64	2^{30}	6.361	5798	1.65	8832
128	2^{31}	8.935	8256	1.42	17664
256	2^{32}	13.057	11299	1.37	35328

Based on these results we can conclude that the current MPDATA implementation scales well up to 64 nodes, when the sustained performance is almost 5.8 Tflop/s, for the perfect performance of approximately 8.8 Tflop/s. Above this number of nodes, the scalability parameter R_p becomes smaller then 1.5.

However, this scalability drop is not very sharp, so even for 256 nodes we still have $R_p = 1.37$, with about 11.3 Tflop/s of the sustained performance. Therefore, it is reasonable to expect that the methods mentioned at the end of the previous subsection could provides good scalability results up to 256 nodes.

5 Conclusions and Further Work

Our approach to adapting the 3D MPDATA stencil-based algorithm to clusters with graphics processor is based on a hierarchical decomposition including the level of cluster, as well as an optimized distribution of computations between GPU resources within each node. In particular, our idea of adaptation to a single GPU node relies on an appropriate compression of GPU kernels, which first of all allows for minimizing the number of memory transactions between GPU shared and global memories. It should be noted here that the technique of stencil compression can be applied for the CPU code as well.

We present performance results for the 3D MPDATA code running on the NVIDIA GeForce GTX TITAN graphics card, as well as on the Piz Daint cluster installed in the Swiss Supercomputing Center, which is equipped with NVIDIA Tesla K20x GPUs. The sustained performance of 138 Gflop/s is achieved for a single GPU, which scales up to more than 11 Tflop/s for 256 GPUs.

In order to improve the performance and scalability of the resulting code, first of all we plan to use the non-blocking MPI communication routines, and double-buffering technique which permits programmers to overlap communication with computations. An additional method is to adapt the GPUDirect RDMA technology, which allows for eliminating unnecessary memory copies, lowering CPU overhead, and finally reducing the communication latency.

An important direction of our further work is also an efficient utilization of both components of a hybrid CPU-GPU cluster node, where computations are distributed across both GPU and CPU. Our previous experience shows that the main issue here is providing the numerical accuracy of the whole code when performing arithmetic operations on two components with quite different numerical properties. This research is directly related to a more general area of research which becomes more and more important – the management and optimization of energy costs required to perform complex numerical simulations.

Acknowledgments. This work was supported by the Polish National Science Centre under grant no. UMO-2011/03/B/ST6/03500, and National Centre for Research and Development under grant no. POIG.02.03.00-24-093/13-00, as well as by the grant from the Swiss National Supercomputing Centre (CSCS) under project ID d25.

References

1. Ciznicki, M., Kopta, P., Kulczewski, M., Kurowski, K., Gepner, P.: Elliptic solver performance evaluation on modern hardware architectures. In: Wyrzykowski, R., Dongarra, J., Karczewski, K., Waśniewski, J. (eds.) PPAM 2013, Part I. LNCS, vol. 8384, pp. 155–165. Springer, Heidelberg (2014)

2. Datta, K., Kamil, S., Williams, S., Oliker, L., Shalf, J., Yelick, K.: Optimization and performance modeling of stencil computations on modern microprocessors. SIAM Rev. **51**(1), 129–159 (2009)
3. Hager, G., Wellein, G.: Introduction to High Performance Computing for Science and Engineers. CRC Press, Boca Raton (2011)
4. Khajeh-Saeed, A., et al.: Computational fluid dynamics simulations using many graphics processors. Comput. Sci. Eng. **14**(3), 10–19 (2012)
5. Krotkiewicz, M., Dabrowski, M.: Efficient 3D stencil computations using CUDA. Parallel Comput. **39**, 533–548 (2013)
6. Kurowski, K., Kulczewski, M., Dobski, M.: Parallel and GPU based strategies for selected CFD and climate modeling models. Environ. Sci. Eng. **3**, 735–747 (2011)
7. Nguyen, A., Satish, N., Chhugani, J., Changkyu, K., Dubey, P.: 3.5-D blocking optimization for stencil computations on modern CPUs and GPUs. In: Proceedings of 2010 ACM/IEEE International Conference for High Performance Computing, Networking, Storage and Analysis, pp. 1–13 (2010)
8. Piotrowski, Z., Wyszogrodzki, A., Smolarkiewicz, P.: Towards petascale simulation of atmospheric circulations with soundproof equations. Acta Geophys. **59**, 1294–1311 (2011)
9. Prusa, J., Smolarkiewicz, P., Wyszogrodzki, A.: EULAG, a computational model for multiscale flows. Comput. Fluids **37**, 1193–1207 (2008)
10. Rivera, G., Tseng, Ch.-W.: Tiling optimizations for 3D scientific computations. In: SC 2000 Proceedings of ACM/IEEE Conference on Supercomputing (2000)
11. Rojek, K., Szustak, L.: Parallelization of EULAG model on multicore architectures with GPU accelerators. In: Wyrzykowski, R., Dongarra, J., Karczewski, K., Waśniewski, J. (eds.) PPAM 2011, Part II. LNCS, vol. 7204, pp. 391–400. Springer, Heidelberg (2012)
12. Rojek, K., Szustak, L., Wyrzykowski, R.: Performance analysis for stencil-based 3D MPDATA algorithm on GPU architecture. In: Wyrzykowski, R., Dongarra, J., Karczewski, K., Waśniewski, J. (eds.) PPAM 2013, Part I. LNCS, vol. 8384, pp. 145–154. Springer, Heidelberg (2014)
13. Rojek, K., Ciznicki, M., Rosa, B., Kopta, P., Kulczewski, M., Kurowski, K., Piotrowski, Z., Szustak, L., Wojcik, D., Wyrzykowski, R.: Adaptation of fluid model EULAG to graphics processing unit architecture. Concurrency Comput. Pract. Experience **27**(4), 937–957 (2015)
14. Smolarkiewicz, P.: Multidimensional positive definite advection transport algorithm: an overview. Int. J. Numer. Meth. Fluids **50**, 1123–1144 (2006)
15. Szustak, L., Rojek, K., Gepner, P.: Using intel xeon phi coprocessor to accelerate computations in MPDATA algorithm. In: Wyrzykowski, R., Dongarra, J., Karczewski, K., Waśniewski, J. (eds.) PPAM 2013, Part I. LNCS, vol. 8384, pp. 582–592. Springer, Heidelberg (2014)
16. Wójcik, D.K., Kurowski, M.J., Rosa, B., Ziemiański, M.Z.: A study on parallel performance of the EULAG F90/95 code. In: Wyrzykowski, R., Dongarra, J., Karczewski, K., Waśniewski, J. (eds.) PPAM 2011, Part II. LNCS, vol. 7204, pp. 419–428. Springer, Heidelberg (2012)
17. Wyrzykowski, R., Szustak, L., Rojek, K.: Parallelization of 2D MPDATA EULAG algorithm on hybrid architectures with GPU accelerators. Parallel Comput. **40**(8), 425–447 (2014)
18. Wyrzykowski, R., Szustak, L., Rojek, K., Tomas, A.: Towards efficient decomposition and parallelization of MPDATA on hybrid CPU-GPU cluster. In: Lirkov, I., Margenov, S., Waśniewski, J. (eds.) LSSC 2013. LNCS, vol. 8353, pp. 457–464. Springer, Heidelberg (2014)

19. Wyszogrodzki, A.A., Piotrowski, Z.P., Grabowski, W.W.: Parallel implementation and scalability of cloud resolving EULAG model. In: Wyrzykowski, R., Dongarra, J., Karczewski, K., Waśniewski, J. (eds.) PPAM 2011, Part II. LNCS, vol. 7204, pp. 252–261. Springer, Heidelberg (2012)
20. Xue, W., Yang, C., Fu, H., Xu, Y., Liao, J., Gan, L., Lu, Y., Ranjan, R., Wang, L.: Ultra-scalable CPU-MIC acceleration of mesoscale atmospheric modeling on Tianhe-2. IEEE Trans. Comput. (2014). doi:10.1109/TC.2014.2366754 (to appear)
21. GPUDirect RDMA. http://docs.nvidia.com/cuda/gpudirect-rdma/index.html
22. NVIDIA GeForce GTX TITAN Specification. http://www.geforce.com/hardware/desktop-gpus/geforce-gtx-titan/specifications
23. PizDaint & PizDora. http://www.cscs.ch/computers/piz_daint/index.html
24. Top 500 Supercomputing Sites. http://www.top500.org

Performance Evaluation of a Human Immune System Simulator on a GPU Cluster

Thiago M. Soares, Micael P. Xavier, Alexandre B. Pigozzo,
Ricardo Silva Campos, Rodrigo W. dos Santos, and Marcelo Lobosco[✉]

Graduate Program in Computational Modelling, UFJF, Juiz de Fora, Brazil
{thiagomarquesmg,micaelpx,alexbprr}@gmail.com,
{ricardo.campos,rodrigo.weber,marcelo.lobosco}@ufjf.edu.br

Abstract. The Human Immune System (HIS) is a complex system that protects the body against several diseases. Some aspects of such complex system can be better understand with the use of mathematical and computational tools. Huge computational resources are required to execute simulations of the HIS, so the use of parallel environments is mandatory. This work presents a parallel implementation of a 3D HIS simulator on a GPU cluster that uses CUDA, OpenMP and MPI to speedup the execution of the application. A performance evaluation is then carried out, and the impact of the use of InfiniBand, a low latency network, and GPU's Error-Correcting Code (ECC) are measured. Speedups up to 956 were obtained by the parallel version that uses Infiniband and turns off ECC.

1 Introduction

The immune system is of fundamental importance for several species of organisms. Its main function is to act in the recognition and elimination of any external pathogens that try to invade the body. In doing so, these pathogens can cause diseases that can take to death. The immune system also plays an important role in the maintenance of the body, removing dead and abnormal cells. To achieve these objectives, a complex network of cells, organs and substances work constantly to promote the proper functioning of the body [17]. Thus, due to the different relationships between their various components at varying levels of interaction, it is an extremely complex task to grasp how the immune system works. However, understanding it is of fundamental importance in the development of vaccines and drugs against many diseases. Mathematical and computational models can help in this task: in recent years, they have achieved some success in elucidating the mechanisms behind the immune response, being important, for example, in the definition of therapeutic strategies [3,4,10,16].

The high computational costs involved in the resolution of these mathematical and computational models impose the use of High Performance Computing (HPC) platforms. The massively parallel architecture of modern Graphics Processing units (GPUs) as well as their attractive performance-cost ratio make a GPU cluster a platform of choice for this kind of application [2]. This work presents a parallel implementation on a GPU cluster of a 3D version of a Human

© Springer International Publishing Switzerland 2015
V. Malyshkin (Ed.): PaCT 2015, LNCS 9251, pp. 458–468, 2015.
DOI: 10.1007/978-3-319-21909-7_44

Immune System (HIS) simulator. The simulator is based on previous works [13–15] and extends our previous implementation [18] in order to use a GPU cluster. Compared to the sequential version of the code, speedups up to 956 were achieved. The paper also presents the impacts of both InfiniBand and GPU's Error-Correcting Code (ECC) in performance.

This work is organized as follows. Section 2 presents an overview of the mathematical and computational model used in this work. Section 3 describes the implementation of the HIS simulator on a GPU cluster. The computational results obtained are presented in Sect. 4. Conclusions and future works are presented in the last section.

2 Mathematical and Computational Model

The mathematical model used in this work is based on previous models of the innate immune response that reproduces the spatial and temporal aspects of a bacterial infection [14,15] and the abscess formation [13]. The model simulates the spatial and temporal behavior of the bacteria (B), dead bacteria (BD), resting macrophages (MR), hyperactivated macrophages (AM), neutrophils (N), apoptotic neutrophils (ND), proinflammatory cytokine (CH), anti-inflammatory cytokine (CA), healthy tissue cells (HT) and dead tissue cells (TD). The relationship between the components are the following: neutrophils, resting macrophages, and active macrophages phagocyte bacteria. After that, neutrophils undergo apoptosis, which may or may not be induced by the phagocytosis. In this state, apoptotic neutrophils can not perform phagocytosis or produce proinflammatory cytokine. Consequently apoptotic neutrophils are eliminated from the body by activated macrophages. The apoptotic neutrophils will eventually die after a period of time, releasing cytotoxic granules and degrading enzymes in the body, causing damage in tissue by destroying healthy cells. Resting and active macrophages do the phagocytosis of dead tissue cells. Healthy tissue cells in contact with bacteria, neutrophils, and active macrophages produce proinflammatory cytokines that increase the permeability of the blood vessels. As a consequence, more cells, such as neutrophils and monocytes, leave the blood stream and enter the infected tissue. In addition, the proinflammatory cytokines act as a chemoattractant substance to the resting macrophages, active macrophages, and neutrophils [13,14].

A set of 10 partial differential equations are derived from the model [13] and implemented in 3D using the Finite Difference Method [9] for the spatial discretization and the explicit Euler method for the time evolution. The discretization of the chemotaxis term uses the First-Order Upwind scheme [6]. Therefore, the precision of our numerical implementation is first-order in time (explicit Euler) and first-order in space (upwind scheme). The upwind scheme discretizes the hyperbolic PDEs through the use of differences with bias in the direction given by the signal of the characteristics' speeds. The upwind scheme uses an adaptive or solution-sensitive stencil to numerically simulate more precisely the direction of information propagation. The details about the implementation of the upwind scheme in 3D can be found in [15].

3   GPU Cluster Programming

This work extends a previous version of our simulator [18], implemented in a shared-memory, multi-GPU platform, to a GPU cluster platform. The code was developed in C using CUDA [8]. To manage multiple GPUs, OpenMP [1] and MPI [11] are used. OpenMP is used to manage multiple GPUs in the same machine, so each host (or CPU) thread is responsible for invoking a CUDA kernel in a particular CUDA device. The use of multiple host threads is necessary to reduce the imbalance caused when a single thread launches all kernels in distinct GPUs located in the same machine. If a large number of GPUs are available on a machine, when the single host thread finishes to launch the last kernel in the last GPU, probably the first kernel launched in the first GPU has advanced a lot in its work, or even finished it. MPI is used to manage data movement and communication across processes located in distinct nodes.

To solve the system of Partial Differential Equations (PDEs) in a GPU cluster, the discretized space is divided among devices, so each one will operate on a specific slice of the original space such that the whole tissue is processed by the group of GPUs. Splitting was done by dividing the x dimension of a (N_x, N_y, N_z) mesh that describes the tissue by the number of GPUs available in the cluster, N_g, remaining a $(\frac{(N_x + N_g - 1)}{N_g}, N_y, N_z)$ mesh to be calculated by each device. The functions $cudaGetDeviceCount$ and MPI_Bcast were used in order to get the total number of GPUs available in the cluster.

To better explore GPU's memory bandwidth, the mesh is organized contiguously in the device memory using an unidimensional vector. Using this organization, access to the points in the tissue was done linearly. To correctly compute each point, each CUDA thread has to access its neighbor data, some of which can be located at distinct GPUs. Due to data splitting among GPUs, the data needed can be located in a distinct machine, as illustrated by Fig. 1. These parts of data, called boundaries, are necessary to execute the computation in two distinct GPUs.

Due to the division scheme used in this work, all data related to the neighbors of points in the y and z dimensions will always reside on the same GPU; only data related to the x dimension have to receive a distinct treatment. For this purpose, consider a thread that is responsible for computing data related to a point in the 3D space given by the coordinates (α, β, γ). In order to compute, this thread can need data from the following neighbors threads: $(\alpha - 1, \beta, \gamma)$ or $(\alpha + 1, \beta, \gamma)$. This data can be located in three distinct locations, as depicted by Fig. 2: (a) in the same GPU in which the thread is located; (b) in a distinct GPU located on the same computer; or (c) in a GPU located in another machine.

A global identifier is used to locate the appropriated neighbor and where data reside. If the neighbor is located in the same GPU (case a), data can be accessed through the GPU's global memory. If the neighbor is located on a distinct GPU (cases b and c), they are accessed through auxiliary vectors also located on GPU's global memory. These auxiliary vectors must be updated at each time-step to guarantee the correctness of the algorithm. This is done using a function called

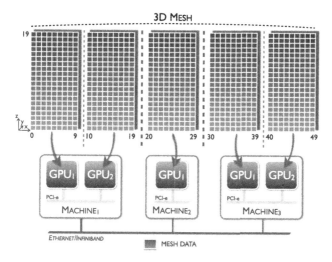

Fig. 1. Split example of a 3D mesh among 5 GPUs.

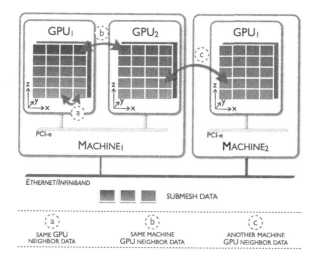

Fig. 2. Possible locations of a neighbor. Cases (b) and (c) illustrate the boundaries.

by the CPU at the end of each time-step. Unified Virtual Address (UVA) is used to copy data if the neighbor GPU is located in the same machine and MPI is used in the case the neighbor GPU is located in a distinct machine.

UVA implements the concept of unified address space among GPUs. To allow the access of a distinct GPU to its local memory space, it is necessary to call first the function *cudaEnablePeerAccess*. A previous work [18] has investigated two possible implementations using UVA. In this work, *cudaEnablePeerAccess* is called only once, during the initialization of the code, and then data is copied explicitly among GPUs, using *cudaMemcpy*.

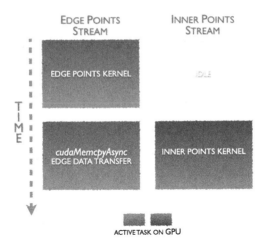

Fig. 3. Operations concurrence between two Independent CUDA Streams.

In order to reduce the communication cost due to data copy, specially when the network is used to complete data transfers, the kernel implementation was modified in order to overlap the boundary transfers with computation. To achieve this goal, computation of both the chemotaxis and the Laplacian operator for each point in the PDE were divided in two steps, the computation of interior points and boundary points, using for this purpose two CUDA kernels. Since the amount of boundaries points is smaller than interior points, its computation finishes first, so data copy can start early, while interior points computation is taking place. The computation and communication are executed concurrently with the use of CUDA streams, as illustrated by Fig. 3. A non-blocking call, *cudaMemcpyAsync*, is used in order to copy the boundary points while interior points are computed.

Since kernels are executed concurrently in each GPU, it is necessary to synchronize at each time-step all processes (located in distinct machines) and threads (located in the same machine) involved in computation. The *#pragma omp barrier* directive is used in order to synchronize threads running on the same machine, while MPI communication primitives implement an implicit synchronization by their blocking nature.

Also, a buffer was implemented to avoid race condition among CUDA threads. Its role is quite simple: two values at times $t - 1$ and t are stored for each point (α, β, γ) of a given population of cells. The value at time t is accessed only by the thread that is producing it, while the other one, $t - 1$, is accessed by threads in neighborhood that needs to read it. Thus, a thread at time t only gets access to data produced by its neighbors at time $t - 1$. These two buffer entries change their meaning at each time step to avoid data copy.

In CUDA, the execution configuration hugely impacts the performance of the application. In this work we choose a fixed block size, 128 threads, based on the memory demands of each thread. Then, a function was created to automatically generate the grid size. This function calculates the grid size taking into account that a thread computes a single point, and that the grid is unidimensional.

Algorithm 1 gives an overview of our GPU cluster implementation of the HIS simulator.

Algorithm 1. GPU cluster implementation of the HIS simulator

2: **main**

4: ... initialize MPI ...

6: ... verify the number of GPUs available in the cluster for computation ...

8: ... create one OpenMP thread for each GPU available in this machine ...

10: ... define the mesh slice to be computed by each GPU ...

12: ... initialize submesh according to their initial conditions ...

14: ... create two streams to deal with interior/boundary points computation and communication ...

16: **for** t **from** t_0 **to** t_f **do**

18: ... write the output files for each population ...

20: ... call the *kernel* that computes boundary points ...

22: ... call the *kernel* that computes interior points ...

24: ... call *sendRecievedBorders* to swap boundaries between GPUs as well as for synchronize them ...

26: ... synchronize all threads in this machine ...

28: **end-for**
    end-main

---

4   Numerical Results

This section presents the results obtained by the GPU cluster version of the code. The experiments were executed on a small cluster with 4 machines. Each machine has two AMD 6272 processors, with 32 GB of main memory, two Tesla M2075 GPUs, each one with 448 CUDA cores and 6 GB of global memory. Linux 2.6.32, CUDA driver version 6.0, OpenMPI version 1.6.2, *nvcc* release 6.0 and *gcc* version 4.4.7 were used to run and compile all versions of the code (sequential and parallel) with the usual optimizations flags(-O3 -march=bdver1, the last one to generate code optimized to the AMD 6272 processors). In order to evaluate the scalability of the solution, two distinct hardware configurations were used: two machines and four GPU's and four machines and eight GPUs.

There are two main communication bottlenecks in GPU cluster platforms [7]: accessing remote GPU memory and the communication between GPU and the host CPU. In order to evaluate the impacts of the first communication bottleneck for the HIS simulator, two distinct types of networks were used in the experiments: Gigabit Ethernet and InfiniBand [12]. Infiniband is a more expensive technology that provide low latency, high bandwidth, end-to-end communication between nodes in a cluster. InfiniBand is in average about 20 times faster then Ethernet. The latency to send a 0 byte message in Infiniband is $1.96\,\mu s$, while in Ethernet it takes $35.5\,\mu s$. The bandwidth to send $32,768$ bytes is $115.8\,\text{MB/s}$ on Gigabit Ethernet and $2,827.1\,\text{MB/s}$ on InfiniBand.

Another aspect in the GPU cluster that impacts performance is the use of the error-correcting code (ECC) support available in the GPU card. ECC uses Hamming code to check is memory contents are unaltered since random bit flips events can occur in memory during the execution of an application. When ECC is enabled, the effective maximum bandwidth is reduced due to the additional traffic for the memory checksums [5]. In this section we also evaluate the impacts of ECC on the simulator code.

In order to evaluate the performance gains obtained by the GPU cluster version of the code, experiments were performed using five distinct mesh sizes: $50 \times 50 \times 50$, $100 \times 100 \times 100$, $150 \times 150 \times 150$, $200 \times 200 \times 200$ and $250 \times 250 \times 250$ points. Since the execution time of each interaction is extremely regular, and the objective of this paper is to evaluate the techniques, nor the biological results, in this work we report the result for $10,000$ time steps. The execution times obtained by all versions of the code were measured 5 times and the standard deviation was lower than $1.52\,\%$. At each execution, Linux *time* application was used to measure the time spent in program execution.

The speedup factor (S_p), used to obtain the relative performance improvement due to the use the GPU cluster platform, is given by Eq. 1.

$$S_p = \frac{t_s}{t_p}, \tag{1}$$

where t_s is the sequential execution time and t_p is parallel execution time.

In order to analyze the impact of ECC and InfiniBand on application' performance, the HIS simulator has been evaluated using four distinct scenarios: (1) InfiniBand with ECC off (*Inf - Ecc Off*); (2) InfiniBand with ECC on(*Inf - Ecc On*); (3) Gigabit Ethernet with ECC off (*Eth - Ecc Off*) and (4) Gigabit Ethernet with ECC on (*Inf - Ecc On*). Table 1 presents the results for the two configurations: with two machines and four GPU's and with four machines and eight GPUs.

As could be expected, the results confirm that the best scenario, for all mesh sizes and configurations, is the one that uses InfiniBand for communication and that turns off ECC. The best speedups are obtained by the second configuration, using 4 machines and 8 GPUs: in this configuration, speedups increases when larger meshes are used. Larger meshes implies more data to be computed, which can reduce the idle time of the GPU multiprocessor during memory access. Also, the second configuration has more GPU cores available to handle this computation.

It also can be observed that InfiniBand has more impact in performance for smaller mesh sizes, compared with larger ones; and in the second configuration,

Table 1. Speedups obtained by running the code on 2 machines and 4 GPUs and on 4 machines and 8 GPUs. The versions are the following: (1) InfiniBand with ECC off; (2) InfiniBand with ECC on; (3) Gigabit Ethernet with ECC off and (4) Gigabit Ethernet with ECC on. The best speedups for each mesh size are marked in bold.

Mesh	Version	Average time(s)/ Speedups (4 GPUs)	Average time(s)/ Speedups (8 GPUs)
	Sequential	5,259.14	
	1	10.45 / 503.1	8.82 / **596.7**
50 x 50 x 50	2	10.62 / 495.3	9.91 / 530.8
	3	22.35 / 235.3	32.62 / 161.2
	4	22.40 / 234.7	33.47 / 157.1
	Sequential	42,096.63	
	1	84.72 / 496.8	52.44 / **802.6**
100 x 100 x 100	2	86.34 / 487.5	56.50 / 745.0
	3	129.37 / 325.3	142.73 / 294.9
	4	131.61 / 319.8	145.37 / 289.5
	Sequential	142,848.56	
	1	290.58 / 491.5	167.78 / **851.4**
150 x 150 x 150	2	299.48 / 476.9	171.50 / 832.9
	3	389.42 / 366.8	364.70 / 391.6
	4	398.85 / 358.1	365.51 / 390.8
	Sequential	334,907.98	
	1	670.86 / 499.2	361.79 / **925.6**
200 x 200 x 200	2	687.80 / 486.9	371.92 / 900.4
	3	842.68 / 397.4	716.43 / 467.4
	4	865.96 / 386.7	724.36 / 462.3
	Sequential	694,455.57	
	1	1,351.20 / 513.9	725.95 / **956.6**
250 x 250 x 250	2	1,385.88 / 501.0	763.53 / 909.9
	3	1,606.45 / 432.2	1,261.49 / 550.5
	4	1,650.89 / 420.6	1,293.77 / 536.7

compared with the first one. For example, in the first hardware configuration, InfiniBand is responsible for improving the performance for computing mesh $50 \times 50 \times 50$ by a factor of 2.1, while improves the performance for computing mesh $250 \times 250 \times 250$ by a factor of 1.19. Also, in the second configuration the impact in performance for computing mesh $50 \times 50 \times 50$ is 3.7. This is explained by the computation/communication ratio. The weight of communication is bigger for smaller mesh sizes. The same rule applies to the second configuration: in this case, more GPUs are available, so less computation per GPU is performed, reducing the computation/communication ratio.

The computation/communication ratio also helps to explain why in the first configuration a larger mesh size does not improve performance, till the largest mesh size, $250 \times 250 \times 250$. This is because in the first configuration more data has to be transfered per GPU compared with the second configuration, so their communication costs are higher. The largest mesh size also increases the computation cost, but now in a ratio higher than the increase imposed by the communication, so the speedup is a little better than the one obtained with the mesh $50 \times 50 \times 50$.

Compared to the gains obtained with the use of InfiniBand, the gains obtained by turning off ECC were modest. The results reveals gains ranging from 0 % to 12 %. Larger gains would be expected by turning off ECC if the GPU memory bandwidth is a bottleneck for the application performance. In the case of this specific application, the bottleneck is the network communication and synchronization, so reducing the network costs, as InfiniBand does, improves performance much more than ECC does.

Finally, figures in Table 1 show that the application nearly scales with the number of GPUs. That is, as the number of GPUs increases by a factor of 2, the speedup would be expected to increase by the same factor. The relative speedups computed for the best configurations of each mesh size are the following: 1.18, 1.61, 1.73, 1.85, and 1.86. As can be observed, the relative speedup increases as the mesh size increases, and for the largest mesh size an efficiency of 93 % is obtained.

5   Conclusion

This work extends a previous version of our simulator [18], implemented in a shared-memory, multi-GPU platform, to a GPU cluster platform. The impact of using a low latency network, InfiniBand, and turning off ECC was also evaluated.

The programming model proposed in this work was very effective in its purpose of speeding up the HIS simulator on a GPU cluster platform: speedups up to 957 were obtained in a cluster with 8×448 GPU cores, an efficiency of 27 %. It was observed that InfiniBand was responsible for improving performance from 1.19 to 3.7 times, compared to the same version that uses Gigabit Ethernet for communicating. Larger gains were observed in the cases where communication costs are higher, so it helps to improve the *computation/communication* ratio. The gains obtained by turning off ECC were more modest, ranging from 0 % to 12 %, but they could be bigger if the GPU memory bandwidth is a bottleneck for the application performance.

As future works, we plan to use all CPUs available in the cluster to perform part of the PDE computation. In the current implementation, the OpenMP threads that execute on the CPU call kernel functions that perform the computation on GPU. All CPU cores are idle while GPU solves the PDEs and they could be used to help in this task. A load balancing strategy will be implemented to distribute data with the objective of equalizing the load at each computational device, since GPU and CPU are heterogeneous. Finally, we would like to investigate better the scalability of our implementation using larger GPU clusters.

Acknowledgements. The financial supports provided by FAPEMIG, CAPES, UFJF and CNPq are greatly acknowledged.

References

1. Chandra, R., Dagum, L., Kohr, D., Maydan, D., MacDonald, J., Menon, R.: Parallel Programming in OpenMP, 1st edn. Morgan Kaufmann Publishers, San Francisco (2001)
2. Fan, Z., Qiu, F., Kaufman, A., Yoakum-Stover, S.: GPU cluster for high performance computing. In: Proceedings of the 2004 ACM/IEEE Conference on Supercomputing, SC 2004, pp. 47. IEEE Computer Society, Washington (2004)
3. Graw, F., Balagopal, A., Kandathil, A.J., Ray, S.C., Thomas, D.L., Ribeiro, R.M., Perelson, A.S.: Inferring viral dynamics in chronically hcv infected patients from the spatial distribution of infected hepatocytes. PLoS Comput. Biol. **10**(11), e1003934 (2014)
4. Guedj, J., Yu, J., Levi, M., Li, B., Kern, S., Naoumov, N.V., Perelson, A.S.: Modeling viral kinetics and treatment outcome during alisporivir interferon-free treatment in hepatitis c virus genotype 2 and 3 patients. Hepatology **59**(5), 1706–1714 (2014)
5. Habich, J., Feichtinger, C., Kastler, H., Hager, G., Wellein, G.: Performance engineering for the lattice boltzmann method on gpgpus: architectural requirements and performance results. Comput. Fluids **80**, 276–282 (2013)
6. Hafez, M.M., Chattot, J.J.: Innovative Methods for Numerical Solution of Partial Differential Equations. World Scientific Publishing Company, New Jersey (2002)
7. Kim, G., Lee, M., Jeong, J., Kim, J.: Multi-gpu system design with memory networks. In: 47th Annual IEEE/ACM International Symposium on Microarchitecture (MICRO), pp. 484–495, December 2014
8. Kirk, D., Hwu, W.: Massively Parallel Processors: A Hands-on Approach. Morgan Kaufmann, San Francisco (2010)
9. LeVeque, R.J.: Finite Difference Methods for Ordinary and Partial Differential Equations. Society for Industrial and Applied Mathematics, Philadelphia (2007)
10. Owen, M.R., Byrne, H.M., Lewis, C.E.: Mathematical modelling of the use of macrophages as vehicles for drug delivery to hypoxic tumour sites. J. Theo. Biol. **226**, 377–391 (2004)
11. Pacheco, P.S.: Parallel programming with MPI. Morgan Kaufmann Publishers Inc., San Francisco (1996)
12. Pfister, G.: Aspects of the infiniband architecture. In: Proceedings of IEEE International Conference on Cluster Computing, pp. 369–371, October 2001
13. Pigozzo, A.B., Macedo, G.C., Santos, R.W., Lobosco, M.: Computational modeling of microabscess formation. Comput. Math. Meth. Med. **2012**, 1–16 (2012)
14. Pigozzo, A.B., Macedo, G.C., Santos, R.W., Lobosco, M.: On the computational modeling of the innate immune system. BMC Bioinf. **14**, S7 (2013). Suppl. 6
15. Rocha, P.A.F., Xavier, M.P., Pigozzo, A.B., de M. Quintela, B., Macedo, G.C., dos Santos, R.W., Lobosco, M.: A three-dimensional computational model of the innate immune system. In: Murgante, B., Gervasi, O., Misra, S., Nedjah, N., Rocha, A.M.A.C., Taniar, D., Apduhan, B.O. (eds.) ICCSA 2012, Part I. LNCS, vol. 7333, pp. 691–706. Springer, Heidelberg (2012)

16. Rong, L., Guedj, J., Dahari, H., Perelson, A.S.: Treatment of hepatitis c with an interferon-based lead-in phase: a perspective from mathematical modeling. Antivir. Ther. **19**(5), 469–477 (2014)
17. Sompayrac, L.: How the Immune System Works. Wiley, New York (2011)
18. Xavier, M.P., do Nascimento, T.M., dos Santos, R.W., Lobosco, M.: Use of multiple gpus to speedup the execution of a three-dimensional computational model of the innate immune system. J. Phys. Conf. Ser. **490**(1), 012075 (2014)

HPC Hardware Efficiency for Quantum and Classical Molecular Dynamics

Vladimir V. Stegailov[1,2,3]([⊠]), Nikita D. Orekhov[1,2], and Grigory S. Smirnov[1,2]

[1] Joint Institute for High Temperatures of RAS, Moscow, Russia
stegailov@gmail.com
[2] Moscow Institute of Physics and Technology, Dolgoprudny, Russia
[3] National Research University Higher School of Economics, Moscow, Russia

Abstract. Development of new HPC architectures proceeds faster than the corresponding adjustment of the algorithms for such fundamental mathematical models as quantum and classical molecular dynamics. There is the need for clear guiding criteria for the computational efficiency of a particular model on a particular hardware. LINPACK benchmark alone can no longer serve this role. In this work we consider a practical metric of the time-to-solution versus the computational peak performance of a given hardware system. In this metric we compare different hardware for the CP2K and LAMMPS software packages widely used for atomistic modeling. The metric considered can serve as a universal unambiguous scale that ranges different types of supercomputers.

1 Introduction

The continuing rapid development of theoretical and computational methods of atomistic simulations during past decades provides a basis of analysis and prediction tools for chemistry, material science, condensed matter physics, molecular biology and nanotechnology. Nowadays molecular dynamics (MD) method that describes motion of individual atoms by the Newton's equations is a research tool of highest importance. The computational speed and the efficiency of parallelization are the main factors that pose limitations on the length and time scales accessible for MD models (the achievable extremes for classical MD are trillions of atoms [4] and milliseconds [12], a typical MD step being 1 fs).

A researcher working in the field of atomistic simulation is an end user of the complex and high performance software and hardware. The main technical question is to find a solution as fast as possible, that is to select appropriate HPC resources and to use them in a most efficient way [14].

In this work we consider a wide-spread type of supercomputer systems comprised of identical nodes and interconnected by a high speed network. Due to the rapid development of hardware, at the moment there is a wide spectrum of node types that can combine several CPUs and accelerators (e.g. GPU, MIC or FPGA). The interconnect architecture spectrum dominated previously by the fat tree and torus topologies has been enriched by the dragonfly and flattened butterfly topologies, the PERCS topology etc.

© Springer International Publishing Switzerland 2015
V. Malyshkin (Ed.): PaCT 2015, LNCS 9251, pp. 469–473, 2015.
DOI: 10.1007/978-3-319-21909-7_45

We can distinguish critical avenues in the development of high performance MD models. Quantum MD (QMD) models demonstrate much higher requirements to the data communication speed and hence to the interconnect properties [5,7]. The deployment of hybrid architectures for electronic structure calculations and quantum MD is not mature enough. Classical MD (CMD) models are less demanding with respect to data communication. The main limitation in CMD is the computational complexity of interatomic potentials (e.g. [10,11,13]) that is determined by the performance of supercomputer nodes. Therefore hybrid architectures of nodes are considered as a major perspective.

2 Problem Statement and Benchmarking Metric

Fundamental *mathematical models* (QMD and CMD) are well developed and practically not subjected to changes. HPC *hardware* architectures change quite quickly. Algorithms and *software* couple fundamental *mathematical models* with HPC *hardware*, however they can be adapted to new hardware quite slowly and therefore the role of legacy software is huge. Having in mind the criterion of the "time-to-solution" minimization for particular *mathematical models* we would like to answer the following questions: What *hardware* is more efficient if we use currently available *software*? What is the efficiency of emerging *software* designed for new *hardware*? And how complicated is this *software* development?

The LINPACK test can not serve as a tool for benchmarking atomistic models. More specialized tests have emerged [1,6,9]. Here we use CP2K and LAMMPS codes as representatives of the best HPC atomistic simulation software. Existing benchmarks suites (e.g. [9] and references therein) test the coupling of selected *software* with *hardware* and here we follow this route for QMD. But for CMD we would like to present a wider view: how efficiently *mathematical models* are coupled with *hardware* if we allow *software* to be tuned.

The "time-to-solution" criterion leads us to the evident choice of a time for one MD integration step as one parameter for the metric. The second parameter should characterize the hardware. Usually the number of some abstract processing elements (e.g. cores) is considered. However although this metric serves well in the *weak* and *strong* scaling benchmarks for the given system, it does not allow to compare essentially different hardware. In order to overcome this problem we consider the total peak performance R_{peak} as a second parameter for the metric that put on equal footing all HPC hardware under consideration. It is in favor of this metric that R_{peak} is a usual marketing aspect for novel hardware.

3 Comparison

Figure 1 shows the comparison for the standard H_2O benchmark for QMD (CP2K): IBM Regatta 690+ [8], Cray XT3 and XT5 [15], IBM BlueGene/P [2] and K-100 cluster of Keldysh Institute of Applied Mathematics in Moscow (64 nodes connected by Infiniband QDR, each node with 2 six-core Intel Xeon X5670 and 3 NVidia Fermi C2050).

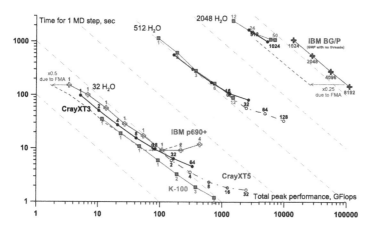

Fig. 1. Water model benchmarks with CP2K for various supercomputers (32-2048 water molecules). Numbers show how many nodes are used to run the benchmark. Dashed lines show ideal speed-up $t \sim R_{peak}^{-1}$.

For benchmarks with several nodes different supercomputers demonstrate close performance (in seconds per MD step). For large models this agreement is better. In the case of 512 molecules we see that the combination of hardware with compilers provides the same level of efficiency.

The role of the interconnect becomes evident in the multi-node cases where the speed-up worsens. Fat-tree systems show better performance for small model sizes. Torus interconnects of Cray XT3, XT5 and IBM BlueGene/P provides superior strong scaling for large system sizes (in accordance with the detailed analysis for another QMD code SIESTA [3]).

IBM systems show inferior performance in this metric because the fused multiply-add (FMA) operations supported by IBM PowerPC CPUs play no essential role for QMD algorithms.

Figure 2 shows the comparison for the standard Lennard-Jones benchmark for CMD (LAMMPS): pure CPU systems and hybrid systems with NVidia Fermi X5670, NVidia Kepler K40 and Intel Xeon Phi SE10X.

All the data (old benchmarks[1] including) for CPUs without vectorization follow the same trend (with the exception of IBM PowerPC 440 CPU due to the FMA issue mentioned above). Manual vectorization with the USER-INTEL package gives $\sim$ 2x speed-up. This is the most efficient way among all implemented in LAMMPS to deploy the total peak performance of hardware.

Hybrid nodes with GPUs show inferior timings with respect to CPU-only nodes when compared by the similar R_{peak}. There are three GPU-oriented versions of MD algorithms in LAMMPS implemented with NVidia CUDA technology (introduced in June 2007). The GPU package is the oldest one introduced in the 1st quarter 2010 and developed up to the 3rd quarter of 2013. The USER-CUDA

---

[1] http://lammps.sandia.gov/bench.html.

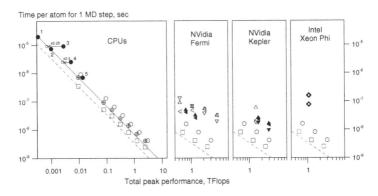

Fig. 2. Lennard-Jones liquid benchmarks with LAMMPS. Circles show CPU benchmarks without vectorization: open circles and crossed circles show Intel Xeon benchmarks on the "Lomonosov" cluster of Moscow State University and K-100 cluster (their discrepancy illustrate the precision of the metric deployed), black circles are the legacy data: 1 – Pentium II 333 MHz, 2 – DEC Alpha 500 MHz, 3 – PowerPC 440 700 MHz, 4 – Power4 1.3 GHz and 5 – Intel Xeon 3.47 GHz. Boxes correspond to Intel Xeon benchmarks with USER-INTEL. Triangles show the timings from the "Lomonosov" cluster using nodes with NVidia GPUs and different algorithms implemented in LAMMPS: △ – GPU, ▽ – USER-CUDA, ◁ – KOKKOS. Filled triangles are the benchmarks published on the LAMMPS web-site. The diamonds are the data for Intel Xeon Phi in the native mode (the lower diamond corresponds to the KOKKOS package).

package is a newer one introduced in the 3rd quarter 2011. The KOKKOS package is the most recent introduced in the 2nd quarter 2014 (and it performs essentially better on the novel NVidia Kepler K40).

Nodes with Intel Xeon Phi (an accelerator that became available in 2012–2013) in the *native mode* show more than $\sim$ 2x speed-up if LAMMPS is used with the KOKKOS package. However Intel Xeon Phi also shows inferior timings with respect to CPU-only nodes when compared by the similar R_{peak}.

4   Conclusions

We introduced a novel metric "time-to-solution (in seconds) vs R_{peak} (in Flops)" and applied it to representative examples of QMD and CMD. This metric allows us to compare existing HPC hardware, hybrid systems including.

CP2K shows better strong scaling on supercomputers with torus interconnects and especially on IBM BlueGene/P. LAMMPS performs with the best efficiency on Intel Xeon CPUs with manual vectorization of crucial routines. Since MD applications do not use FMA operations IBM PowerPC CPUs perform for these tasks at a fraction of R_{peak}.

The example of NVidia GPU shows that porting of an existing package on the new hardware takes several years (only after $\sim$ 7 years of development CUDA-based algorithms have approached CPU algorithms efficiency). After $\sim$ 3 years of development classical MD algorithms for Intel Xeon Phi are still not efficient.

Acknowledgment. The work is partially supported by the grant No. 14-50-00124 of the Russian Science Foundation.

References

1. Coral benchmark codes. https://asc.llnl.gov/CORAL-benchmarks/
2. Bethune, I., Carter, A., Guo, X., Korosoglou, P.: Million atom KS-DFT with CP2K. http://www.prace-project.eu/IMG/pdf/cp2k.pdf
3. Corsetti, F.: Performance analysis of electronic structure codes on HPC systems: a case study of SIESTA. PLoS ONE **9**(4), e95390 (2014)
4. Eckhardt, W., Heinecke, A., Bader, R., Brehm, M., Hammer, N., Huber, H., Kleinhenz, H.-G., Vrabec, J., Hasse, H., Horsch, M., Bernreuther, M., Glass, C.W., Niethammer, C., Bode, A., Bungartz, H.-J.: 591 TFLOPS multi-trillion particles simulation on SuperMUC. In: Kunkel, J.M., Ludwig, T., Meuer, H.W. (eds.) ISC 2013. LNCS, vol. 7905, pp. 1–12. Springer, Heidelberg (2013)
5. Gygi, F.: Large-scale first-principles molecular dynamics: moving from terascale to petascale computing. J. Phys. Conf. Ser. **46**(1), 268 (2006)
6. Heroux, M.A., Doerfler, D.W., Crozier, P.S., Willenbring, J.M., Edwards, H.C., Williams, A., Rajan, M., Keiter, E.R., Thornquist, H.K., Numrich, R.W.: Improving performance via mini-applications. Technical report, Sandia Nat. Laboratories (2009)
7. Hutter, J., Curioni, A.: Dual-level parallelism for ab initio molecular dynamics: reaching teraflop performance with the CPMD code. Parallel Comput. **31**(1), 1–17 (2005)
8. Krack, M., Parrinello, M.: Quickstep: make the atoms dance. High Perform. Comput. Chem. **25**, 29–51 (2004)
9. Muller, M.S., van Waveren, M., Lieberman, R., Whitney, B., Saito, H., Kumaran, K., Baron, J., Brantley, W.C., Parrott, C., Elken, T., Feng, H., Ponder, C.: SPEC MPI2007 – an application benchmark suite for parallel systems using MPI. Concurrency Comput. Pract. Experience **22**(2), 191–205 (2010)
10. Orekhov, N.D., Stegailov, V.V.: Graphite melting: atomistic kinetics bridges theory and experiment. Carbon **87**, 358–364 (2015)
11. Orekhov, N.D., Stegailov, V.V.: Molecular-dynamics based insights into the problem of graphite melting. J. Phys.: Conf. Ser. (2015)
12. Piana, S., Klepeis, J.L., Shaw, D.E.: Assessing the accuracy of physical models used in protein-folding simulations: quantitative evidence from long molecular dynamics simulations. Curr. Opin. Struct. Biol. **24**, 98–105 (2014)
13. Smirnov, G.S., Stegailov, V.V.: Toward determination of the new hydrogen hydrate clathrate structures. J. Phys. Chem. Lett. **4**(21), 3560–3564 (2013)
14. Stegailov, V.V., Norman, G.E.: Challenges to the supercomputer development in Russia: a HPC user perspective. Program Systems: Theory and Applications 5(1), 111–152 (2014). http://psta.psiras.ru/read/psta2014_1_111-152.pdf
15. VandeVondele, J.: CP2K: parallel algorithms. www.training.prace-ri.eu/uploads/tx_pracetmo/cpw09_cp2k_parallel.pdf

Automatic High-Level Programs Mapping onto Programmable Architectures

Boris Ya. Steinberg, Denis V. Dubrov$^{(\boxtimes)}$, Yury Mikhailuts,
Alexander S. Roshal, and Roman B. Steinberg

Southern Federal University, Rostov-on-Don, Russia
borsteinb@mail.ru, dubrov@sfedu.ru, aracks@yandex.ru,
teacplusplus@gmail.com, romanofficial@yandex.ru

Abstract. A technique for automatic high-level C program mapping onto compute systems with programmable pipeline architecture is presented in this article. An example of such a system could be a CPU with an FPGA accelerator or the corresponding system on a chip. The mapping is implemented on the base of Optimizing Parallelizing System (www.ops.rsu.ru) and C2HDL converter from C to the hardware description language (VHDL). HDL code generating from OPS internal representation would allow to utilize the user dialog to generate a family of equivalent chips, from which the user could select the most suitable one for various characteristics. The development of the current work would allow to create for the first time the C language compiler for the programmable pipeline architecture.

Keywords: Reconfigurable computing · Pipeline computing · High-level synthesis · Parallelizing compiler · High level internal representation · HDL · FPGA

1 Introduction

The use of programmable pipeline computing devices increases each year, while the development of high-level programming design tools for them still remains way behind. In this article a project of a compiler from C language to the computer with programmable architecture is considered as well as the current work on its implementation. The "target platform" here stands either for a system on a chip, which contains both the central processing core and the configurable logic block (CLB) matrix, or for a common CPU with an FPGA accelerator. Designing a C language compiler for programmable architectures could accelerate their application and development.

The pipeline computational systems stay apart from the well-known Flynn parallel computer classification: MIMD or SIMD. Pipeline computers are sometimes separated into MISD class (Multiple Instructions, Single Data flow). These computers are efficient for many those problems for which computers of MIMD or SIMD architectures are inefficient or poorly efficient.

© Springer International Publishing Switzerland 2015
V. Malyshkin (Ed.): PaCT 2015, LNCS 9251, pp. 474–485, 2015.
DOI: 10.1007/978-3-319-21909-7_46

The pipeline computers are used in many hardware-software systems and, for some problems, they show considerably higher performance (up to 2–3 orders) than the multi-purpose processors. The substantial progress in pipeline computing development is brought by the technologies of field-programmable gate arrays (FPGA). Conventionally, pipeline computers are used in hardware-software systems. The bottleneck of such computers is long time needed for an FPGA to be reprogrammed. The works of K. Bondalapati are dedicated to this problem [3]. To speed up FPGA reprogramming, a special buffer holding the next configuration, is used. In the Research Institute for Multiprocessor Computing Systems of the Southern Federal University the reconfigurable pipeline computers are designed with only the connections between the computing elements being able for reprogramming to achieve fast reconfiguration speed. Another architecture, which is being developed under V. Corneyev's direction, is also positioned as pipeline-based. In this architecture, it is possible to select pipeline configurations in the grid consisting of compute cores. The cluster with nodes equipped with programmable FPGA accelerators has been assembled under A. Lacis' direction. The language named Autocode-FPGA is suggested for software development for this cluster and is more high-level than VHDL but still is at the level close to Assembler.

The computing units with programmable and reprogrammable architectures are being developed for the wide range of applications and show high efficiency [9]. Particularly, multi-pipeline (or parallel pipeline) systems, which could be considered as a generalization of both the pipeline and the multi-core systems, are of great interest [13]. The algorithms for automatic mapping of a high-level language onto multi-pipeline systems are studied in [11]. In this work automatic generation of multi-pipeline circuits and mapping high-level programs onto them are discussed.

The compiler under development is based on Optimizing Parallelizing System (http://www.ops.rsu.ru/). It includes the converter from OPS program internal representation to the language of electronic circuits description (VHDL) [6]. This converter is able to accept input data ranges in addition to the source code as its input. It would be much more difficult to implement a pipeline system's VHDL code automatic generation using a low-level register-based internal representation, such as LLVM of Clang compiler or RTL and Gimple of GCC compiler family. Apart from OPS, a high-level internal representation (in which optimization analysis and transformations are performed) is used in ROSE compiler infrastructure [10] and SUIF parallelizing system [1].

2   Related Works Overview

High-level synthesis tools gain more interest in our days. The major electronic design automation (EDA) tool vendors incorporate converters from high-level language to electronic design into their products. These tools are targeted either to standalone FPGAs/ASICs, or to hybrid systems with FPGAs [5].

Two main approaches of the high level synthesis tools could be distinguished. The first one involves using traditional programming languages like C and C++

to express hardware algorithm implementation. Catapult C, Vivado Design Suite, Impulse CoDeveloper, Altium Designer, and HDL Coder for MATLAB are the examples of commercial products using such an approach. Academic research projects include PandA (http://panda.dei.polimi.it/), C-to-Verilog [2], TCE [7], and Parallel Intellectual Compiler. Sometimes the means of high-level languages used in these systems are not sufficient for expressing the reconfigurable schematics. For example, Trident Compiler [12] requires from the user to manually partition the code into software and hardware parts.

Other systems, namely Mitrion-C, Handel-C, and HaSCoL [4] use specially designed language constructs to express the hardware abstractions missing in traditional programming languages: manipulating with separate bits, parallel processes with data communications, synchronizations, etc. This makes the user to rewrite old programs and also makes these languages more low-level. However these languages still have advantages over HDL: many time-consuming routine tasks could be done automatically (for example, pipeline generating). On the other hand, tools that use traditional input languages try to incorporate more elaborate program analysis to extract the missing information. For example, a generalization of data-flow graph, called bit-flow graph, may help to generate efficient hardware implementations of the algorithms expressed with C bitwise operators (bit shfting, reversing, extraction, etc.) [14].

3   The Implementation

3.1   Structure of the Compiler from C to the Programmable Computational System

The task of compiling the C language source code into the program for the system with a programmable architecture consists of four subtasks, with the separate compilation module is dedicated to each of them (Fig. 1).

- The converter from C to VHDL (C2HDL). Loop pipelining and transformation of the initial program fragments into computational core descriptions in VHDL language takes place here.
- The configurable driver consisting of modules written in C and VHDL which provide:
  - Data and control commands transfer between a CPU and computational cores on FPGAs using the selected protocol;
  - Data exchange synchronization and the data transfer channel optimal bandwidth distribution;
  - Synchronization of FPGA computational threads with the CPU main control thread.
- The build manager. Using all the generated files, it assembles two projects in C and VHDL respectively, with compilation settings for each of them. It also manages the source code compilation process into the executable files for the target platforms.

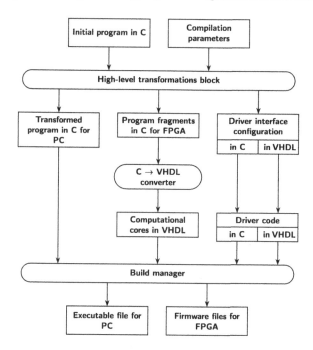

Fig. 1. The structure of the compiler from C to the programmable computational system.

The compiler takes a source code and the compilation settings as its input. The output is the transformed program in C language and the computational core description in VHDL.

3.2   Mapping Programs onto a Programmable Computer

Parallel computing, including the pipeline one, should be applied to the program fragments with long computational time. In this work we consider acceleration of code fragments which contain loop nests. Let us consider a nest of n nested loops:

```
for (I1 = L1; I1 <= R1; ++ I1)
  for (I2 = L2; I2 <= R2; ++ I2)
    ...
      for (In = Ln; In <= Rn; ++ In)
      {
        LOOPBODY(I1, I2, ..., In);
      }
```

The innermost nested loop is subject to be pipelined. The counters of the outer loops could be considered as parameters defining a cluster node with an accelerator that should execute the innermost loop (the same loop for different values of outer loops' counters should be executed on different nodes).

The main idea for the current project involves the system under development, which takes a C program with a loop nest as input. The loop nest is transformed into the form convenient for mapping it onto a pipeline architecture using the parallelizing system. The innermost loop of the nest is transformed into a VHDL description of a pipeline with C to VHDL converter (C2HDL). The programmable part of the compute node is then flashed with the obtained description. The original loop nest is transformed with the innermost loop being replaced with the call for the pipeline accelerator which will compute the given loop. The following techniques are used:

- The methods of loop nests mapping onto the multi-pipeline architecture described in [11] and partially implemented in OPS.
- C2HDL converter which adjusts the programmable part of the target platform according to the given code.

3.3   C2HDL Converter and a Multi-pipeline System Generating

Currently the converter supports the subset of input C programs with the following constraints:

- Integer arithmetics on int types.
- One-dimensional pipelineable loops with assignment expression statements.
- Variable occurrences containing regular linear index expressions.

As the result, the converter generates VHDL code for the synchronous pipeline computational circuit, which supports buffers at operations' inputs and initial pipeline loading stage, if needed. The circuit's input data may be sent to its input connectors from an external source, with the flow synchronized with the circuit operation. Likewise, the computation output flow may be read from the device's output connectors and sent to an external receiver (a memory storage, a control unit, etc.) Signed values of given bit widths (VHDL "`signed (N - 1 downto 0)`" type) are used as operands, standard VHDL operations ("+", "-", etc.) redefined for the given types in standard packages, are used as expression operations.

As an intermediate structure between the parallelizing system internal representation and a pipeline HDL description, the computational graph is used.

Future project development involves implementing the ability to generate several pipelines (for a loop nest) with synchronizing their functioning, if needed, with special delays between pipeline starts. Computing such delays involves information dependencies analysis between iteration space points. Such information dependencies are described with lattice graphs which are represented in memory as functions [8].

Example 1. Let us consider the following piece of code consisting of two nested loops:

```
for (I1 = L1; I1 <= R1; ++ I1)
  for (I2 = L2; I2 <= R2; ++ I2)
  {
    X[I1][I2] = X[I1 - 1][I2] + X[I1][I2 - 1];
  }
```

The data dependency graph between the iteration space points (lattice graph or algorithm graph) is presented on Fig. 2.

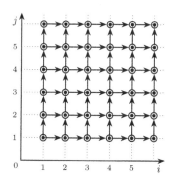

Fig. 2. The lattice graph of the program which represents dependencies between different points of the iteration space. Such information dependencies arise in some grid methods solving mathematical physics problems and in nucleotide sequences aligning.

It is possible to split the iteration space of this loop nest into stripes which have widths of two points. The iterations of each such a stripe could be computed with two pipelines with one being left behind by another (Fig. 3). The algorithms for delay computation in case of such a lag in time are presented in [11].

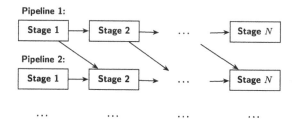

Fig. 3. Managing computations with information exchange between distinct pipelines.

3.4 Optimizing Parallelizing System (OPS)

OPS parallelizing system has the following advantages over the other programming systems:

- Dialogue-based optimization mode;
- Automatic data range analysis;
- Automatic delay computation for the multi-pipeline compute system.
- Optimized code generation not only for compute systems with shared memory, but also for the systems with distributed memory.
- Advanced methods of analysis and transformations of high-level programs which excel the analogs in the field of optimizing compilers and parallelizing systems:
  - lattice graphs;
  - SSA forms of arrays;
  - symbolic analysis;
  - recurrent loops parallelizing methods;
  - several new finer methods of loop parallelizing;
  - methods of program transformation testing.

The higher quality of HDL descriptions generating with C2HDL converter will be achieved with OPS utilization. Namely, it is beneficial to generate a whole family of equivalent HDL descriptions (instead of just one) for the given input C program. Then it would be possible to choose the one which will be optimal for each particular case (depending on restrictions of performance, chip area, etc).

For the currently existing systems of automatic program optimizing and parallelizing the user must set the optimization parameters before compiling a program (using compiler directives and options). This approach cannot give the satisfactory results in many cases, since the user does not know beforehand what can the system do automatically and for what aspects the system needs special indications. It is planned to use analysis (including symbolic one) to formulate questions for the user about those few particular variables, the value ranges of which would allow to apply required transformations or take certain engineering decisions. This approach is already partially implemented in OPS system and is considered to be fruitful for semi-automatic VHDL description construction for the given high-level program and semi-automatic program parallelization as well. No similar approaches are used worldwide yet.

Other advanced methods offered by OPS could be utilized by the programming systems. For example, lattice graphs is a useful tool for generating designs for multi-pipeline compute systems.

3.5 Chip Area Optimizing

When designing schematic implementation of the given functionality, the developer has always to take the hardware restrictions into account. In our work we use the following methods to achieve this:

- It is possible to set the ranges for input data. The user may also define the ranges for the output and intermediate data, the system uses this information to compute the bit widths of the generated arithmetic operations and communication lines. Optionally the system could profile the user code to compute the ranges for the intermediate and output data knowing initially only the input data ranges, using interval analysis. This technique is described in [6]. Knowing data ranges is essential for saving the FPGA resources. It gives an opportunity to get full advantage of this technology over general-purpose CPUs, where data registers and ALUs always have the fixed bit widths regardless the problem requirements.

- The system uses special arithmetic libraries where different schematic implementations of elementary operations (adders, multipliers, etc.) are stored. The user may have an option to select one for each arithmetic unit in the configuration to meet the particular requirements. For example, certain addition schemes may have better performance over the others at the cost of hardware resources use. The user may prefer the slower implementation to fit the particular computational algorithm into the given FPGA model. All arithmetic schemes are parametrized with the operands' bit widths.

- The system uses the heuristic procedure to estimate the required chip area for the given HDL description. The procedure traverses the HDL code internal representation tree to accumulate the user-defined weights for different language constructs (expression operators, variables, etc.) The language construct type is taken into consideration: for example, the effect of VHDL "generate" statement which duplicates its contents for the given number of times. We are planning to implement automatic selection of the alternative arithmetic implementations from the libraries according to their characteristics. It is possible to estimate the chip area for different implementations of the same operation using the above mentioned procedure.

3.6   Efficiency of Parallel Computing Use

It is well known that for modern processors main memory access takes tens of times longer than executing standard arithmetic operations. As to the distributed memory, these accesses take longer time than arithmetic operations almost more than two orders. Therefore, optimizing for program performance should be targeted, in the first place, to memory use.

Using programmable pipeline and parallel-pipeline architectures is considered to be rational only in case if the data for computations are able to be supplied in time. Therefore, in case of programmable computers, as well as of any other modern processors, data locality should be guaranteed first of all, afterwards it would be possible to process the data in parallel. For efficient data localization loop nest splitting into blocks is used. Additional acceleration could be achieved using array block placement in RAM.

4   A Running Example

Example 2. To keep the generated code size reasonable let us consider the following trivial program which implements summation of two arrays:

```
int main()
{
  int a[1000], b[1000], c[1000], i;
  for (i = 0; i < 1000; i ++)
  {
    c[i] = a[i] + b[i];
  }
  return 0;
}
```

To run this program we used the following hardware:

- Xilinx Virtex-4 family FPGA, model: XC4VLX25-FF668-10;
- Xilinx ML401 evaluation platform (used to communicate the FPGA with the PC via the Ethernet cable);
- A regular Intel Core2-based system running Windows 7 operating system.

We also used the following Xilinx development software and design components:

- ISE 14.7;
- A FIFO queue component generated with LogiCORE FIFO Generator;
- 1-Gigabit Ethernet MAC v8.5.

We used our components presented on Fig. 1. Some of the steps we still need to implement manually at the current stage of the compiler development: generating the control program code for the PC, adjusting communication of ports in the driver code (VHDL part) and executing the different external tools instead of using the build manager. The main steps for setting up the computational system are the following:

1. Running C2HDL converter for the above code, fixing possible errors. As the result the pipeline computational component code is generated (see below).
2. Adding the generated component to the Xilinx ISE project. The project already contains the reusable description of other components implementing Ethernet, IPv4, UDP protocols, and other primitives. The port communications are needed to be adjusted. Also the shift register component is adjusted according to the input/output data bit widths. We plan to automate this work in the future with a driver generator component.
3. Generating an FPGA firmware using ISE tools (iMPACT) with the project adjusted at the previous step.

4. Creating a C project for the PC control program which uses the PC part of the driver implemented in a library. The program basically copies the structure of the original program with the following differences:
   - Additional initialization of the used sockets and input/output buffers;
   - The computational loop is replaced with the library call which implements communications with an FPGA.
5. Flashing the FPGA with the firmware obtained at step 3.
6. Running the PC program created at step 4.

A fragment of the VHDL code generated at step 1 is presented below. The code is slightly reformatted to save space and improve readability.

```
architecture Sub1_synth of Sub1 is
  signal c_int: signed(15 downto 0);
  -- ...
  component adderN
    generic(n: integer := 16);
    port(a: in std_logic_vector((n - 1) downto 0);
      b: in std_logic_vector((n - 1) downto 0);
      sum: out std_logic_vector((n - 1) downto 0));
  end component;
  -- ...
begin
  uni0map: adderN port map(Rg2_conv, Rg1_conv, Rg3_conv);
  Rg1_conv <= Conv_std_logic_vector(Rg1, 16);
  Rg2_conv <= Conv_std_logic_vector(Rg2, 16);
  c_Out_Ready <= c_Out_Ready_int;
  c <= c_int;
  process (CLK) is
  begin
    if Rising_edge(CLK) then
      if (RST ='1' or Start ='1') then
        Rg1 <= x"0000";
      else
      if b_In_Ready = '1' then
        Rg1 <= b;
      end if;
      Rg1_Ready <= b_In_Ready;
      end if;
    end if;
  end process;
  process (CLK) is
  begin
    if Rising_edge(CLK) then
      if (RST = '1' or Start = '1') then
        Rg2 <= x"0000";
```

```
  else
  if a_In_Ready = '1' then
    Rg2 <= a;
  end if;
  Rg2_Ready <= a_In_Ready;
  end if;
  end if;
end process;
process (CLK) is
begin
  if Rising_edge(CLK) then
    if (RST = '1' or Start = '1') then
      c_int <= x"0000";
    else
    if Rg3_Ready = '1' then
      c_int <= signed(Rg3_conv);
    end if;
    c_Out_Ready_int <= Rg3_Ready;
    Rg3_Ready <= Rg1_Ready and Rg2_Ready;
    end if;
    end if;
  end if;
end process;
end Sub1_synth;
```

5 Conclusion

A method of mapping high-level programs onto compute systems with programmable architecture is presented in this work. A distinctive feature of this method is that the compute system is adjusted to the program automatically at compile time. Due to this feature, an optimizing parallelizing compiler from a high-level language to the compute system with programmable architecture could be developed. Not only such a compiler could optimize high-level program mapping onto the hardware, it could also optimize the hardware to run the given program. Developing such a compiler is possible on the base of a parallelizing system with high-level internal representation and HDL code generator from this representation.

The presented project aims the substantial simplification of the parallel-pipeline systems access. As the result of this the range of applied problems for this computational field should be expanded; the time needed for parallel-pipeline program developing should be decreased; the development of systems on a chip with programmable architecture should be encouraged.

References

1. Affine transformations for optimizing parallelism and locality. http://suif.stanford. edu/research/affine.html. Accessed: 13 February 2015
2. Ben-Asher, Y., Rotem, N., Shochat, E.: Finding the best compromise in compiling compound loops to Verilog. J. Syst. Archit. **56**(9), 474–486 (2010). http://dx.doi.org/10.1016/j.sysarc.2010.07.001
3. Bondalapati, K.: Modeling and mapping for dynamically reconfigurable hybrid architecture. Ph.D. thesis, University of Southern California, August 2001
4. Boulytchev, D., Medvedev, O.: Hardware description language based on message passing and implicit pipelining. In: Design Test Symposium (EWDTS), 2010 East-West, pp. 438–441, September 2010. http://dx.doi.org/10.1109/EWDTS.2010. 5742095
5. Cardoso, J.M.P., Diniz, P.C.: Compilation Techniques for Reconfigurable Architectures. Springer, US (2009). http://dx.doi.org/10.1007/978-0-387-09671-1
6. Dubrov, D., Roshal, A.: Generating Pipeline integrated circuits using C2HDL converter. In: East-West Design and Test Symposium, pp. 1–4, September 2013. http://dx.doi.org/10.1109/EWDTS.2013.6673108
7. Esko, O., Jääskeläinen, P., Huerta, P., de La Lama, C.S., Takala, J., Martinez, J.I.: Customized exposed datapath soft-core design flow with compiler support. In: Proceedings of the 2010 International Conference on Field Programmable Logic and Applications, FPL 2010, pp. 217–222. IEEE Computer Society, Washington, DC (2010). http://dx.doi.org/10.1109/FPL.2010.51
8. Feautrier, P.: Parametric integer programming. RAIRO Recherche Opérationnelle **22**, 243–268 (1988)
9. Gokhale, M., Graham, P.S.: Reconfigurable Computing. Accelerating Computation with Field-Programmable Gate Arrays. Springer US (2005). http://dx.doi.org/10. 1007/b136834
10. Liao, C., Quinlan, D.J., Willcock, J.J., Panas, T.: Semantic-aware automatic parallelization of modern applications using high-level abstractions. Int. J. Parallel Program. **38**(5–6), 361–378 (2010). http://dx.doi.org/10.1007/s10766-010-0139-0
11. Steinberg, R.B.: Mapping loop nests to multipipelined architecture. Program. Comput. Softw. **36**(3), 177–185 (2010). http://dx.doi.org/10.1134/S0361768810030060
12. Tripp, J.L., Gokhale, M.B., Peterson, K.D.: Trident: from high-level language to hardware circuitry. Computer **40**(3), 28–37 (2007). http://dx.doi.org/10.1109/ MC.2007.107
13. Yadzhak, M.S., Tyutyunnyk, M.I.: An optimal algorithm to solve digital filtering problem with the use of adaptive smoothing. Cybern. Syst. Anal. **49**(3), 449–456 (2013). http://dx.doi.org/10.1007/s10559-013-9528-x
14. Zhang, J., Zhang, Z., Zhou, S., Tan, M., Liu, X., Cheng, X., Cong, J.: Bit-level optimization for high-level synthesis and FPGA-based acceleration. In: Proceedings of the 18th Annual ACM/SIGDA International Symposium on Field Programmable Gate Arrays, FPGA 2010, pp. 59–68. ACM, New York (2010). http://doi.acm.org/ 10.1145/1723112.1723124

Applications

Implementation of a Three-Phase Fluid Flow ("Oil-Water-Gas") Numerical Model in the LuNA Fragmented Programming System

Darkhan Akhmed-Zaki[1], Danil Lebedev[1(✉)],
and Vladislav A. Perepelkin[2]

[1] Al-Farabi Kazakh National University, Almaty
Republic of Kazakhstan
Darhan.Ahmed-Zaki@kaznu.kz,
danil.lebedev.0881@gmail.com
[2] Institute of Computational Mathematics and Mathematical Geophysics
SB RAS, Novosibirsk, Russia
perepelkin@ssd.sscc.ru

Abstract. The fragmented programming technology and the language implementing it are briefly introduced as well as LuNA fragmented programming system, on the example of two-dimensional boundary value problem solution, for liquid filtration "oil-water-gas" system. For parallel implementation of the boundary value problem, the parallel longitudinal-transverse sweep algorithm was applied. Using this method, the fragmented program in the LuNA system has also been implemented. The calculations are made for different number of points in the spatial variables. To compare the quality of implementation the applied numerical algorithm has been implemented in several variations: the sequential program, the parallel program using MPI and the fragmented parallel program in LuNA language using LuNA programming system.

Keywords: Fragmented programming · LuNA · Numerical solution · MPI · Parallel program · Sweep method

1 Introduction

Implementation of large-scale numerical models on supercomputers is often challenging, especially for scientists, who are not experienced in system parallel programming. Consequently, of great importance are systems and tools of parallel programming. Such tools are aimed at reducing the complexity of parallel programming through its automation. In appropriate circumstances they reduce the complexity of parallel programs development, lower system parallel programming skill requirements, improve quality of resulting parallel programs, and so on.

© Springer International Publishing Switzerland 2015
V. Malyshkin (Ed.): PaCT 2015, LNCS 9251, pp. 489–497, 2015.
DOI: 10.1007/978-3-319-21909-7_47

Parallel programming automation is a subject for numerous research efforts, and its importance tends to increase. Worth mentioning are the following systems of parallel programming, which closely relate to scientific numerical modeling: PaRSEC [8], libgeodecomp [9], Charm++ [10], KeLP [11].

The LuNA programming system [6, 7, 12] is also a system of parallel programming, aimed at elimination of manual parallel programming from the process of parallel implementation of large-scale numerical models on supercomputers. In this paper we discuss the parallel implementation of two-dimensional boundary value problem solution for liquid filtration "oil-water-gas" system in LuNA programming system.

The paper is organized as follows. The next two sections describe the application problem and its mathematical formulation, Sect. 4 studies the parallel algorithm of the problem's solution. Sections 5 and 6 contain a brief description of LuNA system and how it executes fragmented algorithms. Section 7 represents the results of the performance tests.

2   The Problem of Filtration

This paper considers the two-dimensional problem of three-phase fluid filtration in the "oil-water-gas" system. The problem is a simulation of oil recovery process for secondary methods. Practical significance of the calculations for this class of problems is quite large. This is due to the fact that a very large part of oil production is associated with the use of secondary recovery techniques, such as displacement of oil by water or solvents, thermal impact on the field, etc. Considered model describes the secondary method of oil by water displacement. It has a number of specific features that make it difficult, and in some cases impossible to use standard numerical methods, proven to be efficient for other classes of problems. General formulation of the problem can be reduced to the following form: there is an oil reservoir, in which the water is pumped under pressure through the injection well, and it is necessary to calculate how much oil will be obtained from the production well. More information can be found in [1–4].

3   Definition of the Problem

Let us consider a two-dimensional boundary value problem for liquid filtration of "oil-water-gas" system. The problem is described by the following equations [1, 2] in dimensionless variables.

Given Eqs. (1−3) describe the change in pressure (P_l) and saturation (S_l) for each phase, on space-time coordinates. Influence of wells is accounted by the corresponding coefficients (q_l). The complexity of the solution is caused by presence of coefficients – functions of saturation inside of differential operator.

$$\begin{cases} \dfrac{\partial}{\partial x}\left[K_w\left(\dfrac{\partial P_w}{\partial x}-\gamma_w\dfrac{L}{P_H}\dfrac{\partial z}{\partial x}\right)\right]+\dfrac{\partial}{\partial y}\left[K_w\left(\dfrac{\partial P_w}{\partial y}-\gamma_w\dfrac{L}{P_H}\dfrac{\partial z}{\partial y}\right)\right]=\dfrac{\mu_w}{\mu_o}\dfrac{\partial S_w}{\partial \tau}+\dfrac{\mu_w L^2}{K\rho_w P_H}q_w \\[3mm]
\dfrac{\partial}{\partial x}\left[K_o(1+C_f P_H(P_o-1))\left(\dfrac{\partial P_o}{\partial x}-\gamma_o\dfrac{L}{P_H}\dfrac{\partial z}{\partial x}\right)\right]+\dfrac{\partial}{\partial y}\left[K_o(1+C_f P_H(P_o-1))\left(\dfrac{\partial P_o}{\partial y}-\gamma_o\dfrac{L}{P_H}\dfrac{\partial z}{\partial y}\right)\right]= \\[3mm]
=\dfrac{\partial}{\partial \tau}\left[S_o(1+C_f P_H(P_o-1))\right]+\dfrac{\mu_o L^2}{K\rho_H P_H}q_o \\[3mm]
+\dfrac{\partial}{\partial x}\left[R_s K_o(1+C_f P_H(P_o-1))\left(\dfrac{\partial P_g}{\partial x}-\dfrac{\partial P_{cog}}{\partial x}-\gamma_o\dfrac{L}{P_H}\dfrac{\partial z}{\partial x}\right)\right]+ \\[3mm]
\dfrac{\partial}{\partial y}\left[R_s K_o(1+C_f P_H(P_o-1))\left(\dfrac{\partial P_g}{\partial y}-\dfrac{\partial P_{cog}}{\partial y}-\gamma_o\dfrac{L}{P_H}\dfrac{\partial z}{\partial y}\right)\right]+ \\[3mm]
\dfrac{\mu_o P_H}{\mu_g \rho_H RTZ}\dfrac{\partial}{\partial x}\left[K_g P_g\left(\dfrac{\partial P_g}{\partial x}-\gamma_g\dfrac{L}{P_H}\dfrac{\partial z}{\partial x}\right)\right]+\dfrac{\mu_o P_H}{\mu_g \rho_H RTZ}\dfrac{\partial}{\partial y}\left[K_g P_g\left(\dfrac{\partial P_g}{\partial y}-\gamma_g\dfrac{L}{P_H}\dfrac{\partial z}{\partial y}\right)\right]= \\[3mm]
=(1-C_f P_H)\dfrac{\partial}{\partial \tau}(R_s S_o)+C_f P_H\dfrac{\partial}{\partial \tau}(S_o)+\dfrac{P_H}{\rho_H RTZ}\dfrac{\partial}{\partial \tau}(P_g S_g)+\dfrac{\mu_o L^2}{K\rho_H P_H}(R_s q_o+q_g) \\[3mm]
P_o-P_w=P_{cow} \\[2mm]
P_g-P_o=P_{cog} \\[2mm]
S_w+S_o+S_g=1 \end{cases}$$

$$\tag{1}$$

Initial and boundary conditions

$$P_0(x,y,0)=P_o^H(x,y),P_w(x,y,0)=P_w^H(x,y),P_g(x,y,0)=P_g^H(x,y)$$
$$S_0(x,y,0)=S_o^H(x,y),S_w(x,y,0)=S_w^H(x,y),S_g(x,y,0)=S_g^H(x,y) \tag{2}$$

$$\left.\dfrac{\partial P_o}{\partial n}\right|_\Gamma=0,\left.\dfrac{\partial P_w}{\partial n}\right|_\Gamma=0,\left.\dfrac{\partial P_g}{\partial n}\right|_\Gamma=0 \tag{3}$$

4 Algorithm of the Solution

To solve the system (1) an iterative method with implicit pressure and explicit saturation is used. In order to do this, first three equations of the system are summarized and using fourth and fifth ratios of the system (1). Obtained equation is solved relative to the gas pressure. Resulting equation is reduced to the implicit difference scheme, then solved by the longitudinal-transverse sweep method [5] at each iteration layer using saturation values from the previous iteration layer. The idea of the method is to calculate values for next time step through intermediary time step, where at intermediary time step the function value is calculated as derivative relative to one spatial variable, and in the second spatial variable takes the value from previous time step (longitudinal direction). In the transverse direction, the value of the function at next time step is calculated as derivative by second spatial variable and first spatial variable takes the value calculated at intermediary time step. Calculated values of gas pressure are used to find other pressure values. Then, we find saturation of oil and water, using the first and second equation of the system (1), and gas saturation value – using the last ratio of the system. Iterative process stops when the convergence condition is satisfied.

$$max\left|S_{g,i,j}^{r} - S_{g,i,j}^{r-1}\right| \le \varepsilon_1 \tag{4}$$

Convergence of the process is affected by number of wells, because gradients of pressure and saturation phases are rapidly changing around the wells and the rest of the area is changing smoothly and therefore the convergence condition is reached faster.

We suggest the following parallel algorithm. We make decomposition of the spatial area into rectangles, as shown in the following figure.

Figure 1 shows that direction j has fragment size M/fgcnt, where fgcnt is the number of fragments. Direction i has fragment size N/fgcnt. Therefore, we have $fgcnt^2$ data fragments of size $N/fgcnt \times M/fgcnt$. Then, the original algorithm can be represented as sequence of following steps:

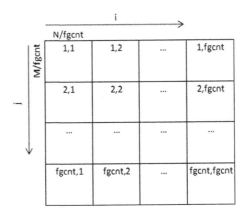

Fig. 1. Decomposition area.

Step 1. Generate variable matrices $P_o, P_w, P_g, S_w, S_o, S_g$ of size Nx(M/fgcnt + 1) for first and last data fragment and Nx(M/fgcnt + 2) for others. The values of these matrices are determined from the initial conditions (2).

Step 2. Solve the equation for the gas pressure obtained from addition of the system (1) at intermediate time step for $j = \overline{1, M/fgcnt}$, $i = \overline{1, N-1}$ and all $k = \overline{1, fgcnt}$

Step 3. Calculate values for pressure of oil and water, using ratios 4 and 5 of the system (1) for all $k = \overline{1, fgcnt}1$.

Step 4. Calculate values for saturation of oil and water using Eqs. 1 and 2 of the system (1), gas saturation can be found from ratio 6 of the system (1) for all $k = \overline{1, fgcnt}$.

Step 5. Check convergence condition (4), if it is satisfied for all $k = \overline{1, fgcnt}$ then go to step 6, otherwise override values of variables matrix $P_o, P_w, P_g, S_w, S_o, S_g$ with values calculated in steps 2−4.

Step 6. Generate variables matrix $P_o, P_w, P_g, S_w, S_o, S_g$ of size *(N/fgcnt + 1)* × *M* for first and last data fragment and *(N/fgcnt + 2)* × *M* for others.

Step 7. We define the values of these matrices according to the following scheme: fragments k send all l fragments submatrices of their variables P_o, P_w, P_g, S_w, S_o, S_g, of size (N/fgcnt + 2) × (M/fgcnt + 2) for $l \neq 1$ or $l \neq fgcnt$ where $i = \overline{(l-1)*N/fgcnt - 1, l*N/fgcnt + 1}$ and size of (N/fgcnt + 1) x (M/fgcnt + 1) otherwise, where $i = \overline{(l-1)*N/fgcnt, l*}$ $N/fgcnt + 1$ $by\, l = 0$ and $i = \overline{(l-1)*N/fgcnt, l*N/fgcnt}$ $by\, l = fgcnt$.

Step 8. Solve the equation for gas pressure obtained from addition of the system (1) at intermediary time step for $i = \overline{1, N/fgcnt}$, $j = \overline{1, M-1}$ and all $l = \overline{1, fgcnt}$.

Step 9. Similar to steps 3 and 4, find the pressure of oil and water and saturation of oil, water and gas for all $l = \overline{1, fgcnt}$.

Step 10. Check convergence condition (4), if it is satisfied for all $l = \overline{1, fgcnt}$ then go to step 11, otherwise override values of variable matrices $P_o, P_w, P_g, S_w, S_o, S_g$ to values calculated in step 9.

Step 11. Fill variable matrices $P_o, P_w, P_g, S_w, S_o, S_g$, for longitudinal direction by the following scheme: l fragment sends all k fragments submatrix of their variables $P_o, P_w, P_g, S_w, S_o, S_g$ of size $\overline{(N/fgcnt + 2)}$ x $\overline{(M/fgcnt + 2)}$ for $k \neq 1 or\, k \neq fgcnt$ where $i = \overline{(k-1)*N/fgcnt - 1, k*N/fgcnt + 1}$ and of size *(N/fgcnt + 1)*x*(M/fgcnt + 1)*, otherwise, where $i = \overline{(k-1)*}$ $N/fgcnt, k*N/fgcnt + 1$ $by\, k = 0$ and $i = \overline{(k-1)*N/fgcnt, k*N/fgcnt}$ $by\, k = fgcnt$ and go to step 2.

The process stops when it reaches specified time.

A feature of this parallel algorithm is the absence of data exchange within the computations by directions. Only when the iterative process converged in one direction, the calculated values will be transferred to other processes. This reduces amount of communications, while checking convergence conditions of the iterative method, but leads to sending the entire array, obtained by each process as a result of calculation, to all other processes, in order to start computation in other direction.

To compare implementation quality the applied numerical algorithm has been implemented in several forms: sequential Java program, parallel C++ program using the MPI standard, parallel fragmented program in LuNA language [6, 7] using LuNA programming system.

5 LuNA Language and System of Fragmented Programming

LuNA (Language for Numerical Algorithms) – is a language and a system of parallel programming, aimed at automation of parallel programming of large-scale numerical models on supercomputers. LuNA language and system are being developed in the Supercomputer software department of the Institute of computational mathematics and mathematical geophysics of the Siberian branch of Russian academy of sciences.

The theoretical basis of LuNA is the theory of structured synthesis of parallel programs on the basis of computational models [13]. The main approach of LuNA can be described as follows. A user reformulates the application algorithm into an explicitly parallel form, called fragmented algorithm (FA). The FA is automatically transformed by LuNA compiler into a parallel program, executable by LuNA run-time system on a multicomputer. LuNA compiler and run-time system take care of such problems as performing communications, data access synchronization, distributed memory alloca- tion, dynamic load balancing, and so on.

An important peculiarity of the approach is the possibility to improve the quality of FA execution by specifying "recommendations" – partial decisions on such problems as: how to distribute data and operations of the FA among computing nodes, what kind of workload scheduling to choose, and so on. The recommendations allow manual tuning the FA execution in order to achieve better performance without the user having to drive into complex system parallel programming.

Main advantages, offered by LuNA system, are reduction of programmer qualifi- cation requirements, reduction of parallel program development laboriousness, auto- mation of provision of such properties of parallel program execution as dynamic load balancing, performing communications in parallel with computations, and so on. The programmer does not do parallel programming as such, his role is limited to algorithm decomposition, sequential programming and declaration of recommendations in a domain specific language (DSL). So, the parallel programming as such is eliminated form the process of implementation of numerical models for multicomputers.

Currently, LuNA system is implemented as a prototype. FA, described in LuNA language, is interpreted by the run-time system, which invokes user sequential pro- cedures, encapsulated in a traditional dynamic load library, according to the FA. In such way, implementation of coarse-grained parallel algorithm mainly consists of native code execution and minor system overhead. On the contrary, fine-grained FA is likely to have poor performance due to run-time system overhead. The LuNA run-time system is designed to be scalable to supercomputers of any size, therefore it only employs scalable distributed system algorithms with localized communications.

6   Fragmented Algorithm Execution

In this section we consider the fragmented algorithm (FA) representation and its execution by LuNA run-time system. In particular, we consider, how it provides a number of important properties of FA execution, such as resources distribution, exe- cution scheduling, dynamic load balancing, data access synchronization, garbage collection, etc.

FA is basically a bipartite directed graph. One part is a set of computational fragments, the other part is a set of data fragments. Data fragment (DF) is a single assignment variable, which value is an aggregate of values. A submatrix is an example of a DF. Computational Fragment (CF) is a pure-functional (with no side effects) operation on DFs. Each CF has a number if input DFs and a number of output DFs.

If at given moment of FA execution all input DFs for a CF are computed, then the CF may be executed. Execution of the CF produces values for its output DFs. Execution of FA is execution of all of its CFs. In such a way, it is a dataflow model with single assignment aggregated values.

In general, the graph of the FA is potentially infinite (due to potentially infinite iteration processes and single assignment). Thus the FA is represented as an enumerating algorithm, capable of enumerating all entities (CFs and DFs) of the FA. The enumeration is performed in run-time. As a result of the enumeration CFs and DFs are initiated.

A peculiarity of such representation is, that each CF may be implemented as a sequential ("native") procedure call, provided all input DFs are present on current Processing Element (PE, e.g., computing node). Being serializable, DFs can be moved by the run-time system automatically from one PE to another (through network message passing). Thus the run-time system is able to provide DFs migration and CFs execution in order to execute FA. The PEs to store DFs or execute CFs can be assigned automatically and dynamically. The FA is, therefore, highly portable, because various CFs and DFs distributions among PEs can be chosen, according to the multicomputer configuration.

If a load imbalance occurs, the overloaded PEs may have some of their CFs migrated to underloaded PEs automatically, resulting in load balancing.

The run-time system may employ different scheduling, distribution and load balancing strategies in order to improve efficiency of FA execution.

To make this possible, each DF must be small enough to fit into memory of any PE, and each CF must be small enough (in terms of execution time) to provide granularity for load balancing.

Efficient FA execution is a hard problem, because such problems as CFs scheduling, CFs and DFs distribution and dynamic redistribution are computationally hard, so heuristics must be used (see [6] for details).

FA execution comprises the following activities.

1. CFs and DFs are generated by enumerating algorithm
2. Ready for execution CFs are found through data dependencies tracking
3. Ready CFs are executed, producing new DFs
4. No longer needed DFs are disposed through garbage collection
5. PEs workload is monitored to detect load imbalances
6. CFs and DFs are transferred to other PEs to balance PEs workload
7. CFs execution is monitored to detect system full stop.

Current implementation of LuNA system has moderate overhead, since it is only a prototype, but system algorithms can be improved. The system is designed to be scalable, therefore all the system algorithms are distributed and with localized communications (only allowed communications are between neighbor PEs).

7 Performance Tests

Experiments were conducted for different number of mesh points in the spatial coordinates and different number of processors for parallel implementations with fixed number of iterations. Parallel version and LuNA version were run on a cluster of

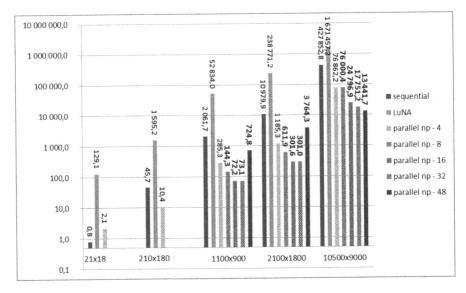

Fig. 2. The computation time (sec.) depending on the number of mesh points.

Siberian Supercomputer Center. Time step and number of wells were not changed. The purpose of the test was to determine the efficiency of various implementations of the numerical algorithm. Test results are shown in Fig. 2.

Figure 2 shows that at a small number of mesh points the minimum execution time belongs to the sequential implementation of the algorithm.

With increase of the number of points, parallel implementation is at first catching up with and then overtakes the sequential implementation. This is due to the fact that, with the growth of computation load benefit from parallelism begins to outweigh the overheads arising from communications. Execution time of the LuNA program exceeds the other execution times at about half order. The reason for that is relatively small size of the problem and big number of iterations, which results in fine granularity of the fragmented algorithm. Such fine-grained fragmented algorithms cause large run-time system execution overhead.

Despite lower quality of the implementation, the LuNA program is easier to develop, because the LuNA program only contains application computations and algorithm scheme description. No communications, synchronization or resources allocation code is present.

8   Conclusion

An implementation of a three-phase fluid flow ("oil-water-gas") numerical model is considered, parallel algorithm is suggested and its implementation with LuNA fragmented programming system is presented. Peculiarities of fragmented algorithms execution by LuNA run-time system are shown. Comparative performance tests were presented. The study has shown that LuNA system simplifies algorithm's implementation, but the efficiency is poor due to small problem size.

Acknowledgements. This work was supported by grant funding of scientific and technical programs and projects by the Committee of Science, Ministry of Education and Science RK, grant No. 528/GF2; and Russian Foundation for Basic Research (grants No. 14-07-00381-a and 14-01-31328-mol_a).

References

1. Aziz, K., Settari, A.: Petroleum Reservoir Simulation, p. 407. Applied Science Publishers Ltd, London (1979)
2. Crichlow, H.B.: Modern Reservoir Engineering – A Simualation Approach, p. 303. Prentice-Hall, Inc., Englewood Cliffs (1977)
3. Konovalov, A.N.: The Problem of Filtration Multiphase Incompressible Fluid. Novosibirsk: science. 166p (1988)
4. Akhmed-Zaki, D.Z., Danaev, N.T., Mukhambetzhanov, S.T., Imankulov, T.: Analysis and evaluation of heat and mass transfer processes in porous media based on darcy - stefan's model. In: ECMOR XIII (2012). http://dx.doi.org/10.3997/2214-4609.20143274
5. Janenko, N.N.: The method of fractional steps for solving multidimensional problems of mathematical physics. 197 p. (1967)
6. Malyshkin, V., Perepelkin, V.: Optimization methods of parallel execution of numerical programs in the LuNA fragmented programming system. J. Supercomput. **61**(1), 235–248 (2012). doi:10.1007/s11227-011-0649-6
7. Kireev, S., Malyshkin, V., Fujita, H.: The LuNA library of parallel numerical fragmented subroutines. In: Malyshkin, V. (ed.) PaCT 2011. LNCS, vol. 6873, pp. 290–301. Springer, Heidelberg (2011)
8. ParSEC: Parallel Runtime Scheduling and Execution Controller. http://icl.cs.utk.edu/parsec/index.html accessed on 15 January 2015
9. Schäfer, A., Fey, D.: LibGeoDecomp: a grid-enabled library for geometric decomposition codes. In: Lastovetsky, A., Kechadi, T., Dongarra, J. (eds.) EuroPVM/MPI 2008. LNCS, vol. 5205, pp. 285–294. Springer, Heidelberg (2008)
10. Kale, L.V., Krishnan, S.: CHARM++: a portable concurrent object oriented system based on C++. In: OOPSLA 1993 Proceedings of the Eighth Annual Conference on Object-Oriented Programming Systems, Languages, and Applications, ACM, New York, NY, USA. pp 91–108 (1993). doi:10.1145/165854.165874
11. Gershon, E., Shaked, U.: Applications. In: Gershon, E., Shaked, U. (eds.) Advanced Topics in Control and Estimation of State-multiplicative Noisy Systems. LNCIS, vol. 439, pp. 201–216. Springer, Heidelberg (2013)
12. Malyshkin, V.E., Perepelkin, V.A.: LuNA fragmented programming system, main functions and peculiarities of run-time subsystem. In: Malyshkin, V. (ed.) PaCT 2011. LNCS, vol. 6873, pp. 53–61. Springer, Heidelberg (2011)
13. Valkovsky, V.A., Malyshkin, V.E.: Synthesis if parallel programs and system on the basis if computational models. Nauka, Novosibirsk, 1988, 128 p (1988). (In Russian)

Development of a Distributed Parallel Algorithm of 3D Hydrodynamic Calculation of Oil Production on the Basis of MapReduce Hadoop and MPI Technologies

Darkhan Akhmed-Zaki, Madina Mansurova[✉], Timur Imankulov,
Bazargul Matkerim, and Ekaterina Dadykina

al-Farabi Kazakh National University, Almaty, Republic of Kazakhstan
darhan.ahmed-zaki@kaznu.kz,
{mansurova01,imankulov_ts}@mail.ru,
bazargulmm@gmail.com

Abstract. The developed hybrid model of high performance computing and the realized applications on the basis of MapReduce Hadoop and MPI technologies allow to solve effectively the different classes of oil production problems.

Investigations of high performance computing to solve the problem of 3D hydrodynamic calculation of oil production resulted in proposition of a constructive approach to organization of distributed parallel computing using MapReduce Hadoop and MPI technologies for which a general scheme of the iteration infrastructure of MapReduce model is designed; the structure for fulfillment of *map* and *reduce* methods and organization of decomposition of the computational area at different Map/Reduce stages are presented; a computational experiment on specific infrastructure is carried out. As the result of this work the architecture of the system realized on the advantages of MapReduce Hadoop and MPI technologies is constructed.

Keywords: MapReduce Hadoop · MPI · Oil production problems · Parallel computing · Distributed computing

1 Physical and Mathematical Models of 3D Problem of Hydrodynamic Calculation of Oil Production

In this section, physical and mathematical models of the isothermal process of oil displacement based on Buckley-Leverett model are considered.

Water of the pre-determined temperature is pumped through the injection well of the oil-gas deposit. In the injection and production wells, either pressures P_{ijk}^n and $P_{prod}\left(P_{ijk}^n > P_{prod}\right)$ or volumes of the water being pumped or discharges of wells are specified. The pumped water displaces the oil left in the pool which goes to the production well.

A mathematical model of a two-phase filtration consists of equations of oil and water balance in the flow and the generalized Darcy's law with the following

© Springer International Publishing Switzerland 2015
V. Malyshkin (Ed.): PaCT 2015, LNCS 9251, pp. 498–504, 2015.
DOI: 10.1007/978-3-319-21909-7_48

assumptions: fluids are incompressible; the capillary effects can be neglected; gravitational forces are not taken into account; the flow obeys the Darcy's law.

The set of equations for region Ω with the boundary $\partial\Omega$ can be written in the following way (1)–(3) [1]:

$$m\frac{\partial s}{\partial t} + div(\vec{v}_1) = q_1 \tag{1}$$

$$-m\frac{\partial s}{\partial t} + div(\vec{v}_2) = q_2 \tag{2}$$

$$\vec{v}_i = -K_0\frac{f_i(s)}{\mu_i}\nabla P \tag{3}$$

where m, μ_i, f_i, K_0, q_i are porosity of the medium, viscosity of fluids, relative phase permeabilities, absolute permeability of the medium, operational characteristics of the wells, respectively.

It is necessary to find functions {P, s, V} – pressure, saturation of water, the flow rate, respectively, satisfying ratios (1)–(3) with initial and boundary conditions (4)–(5):

$$S|_{t=0} = S_0(x) \tag{4}$$

$$\frac{\partial P}{\partial n}\bigg|_{\partial\Omega} = 0, \frac{\partial s}{\partial n}\bigg|_{\partial\Omega} = 0 \tag{5}$$

2 Numerical Model of 3D Problem of Hydrodynamic Calculation of Oil Production

For a numerical solution of the problem, the algorithm of separate determination of pressure and saturation fields is used. By the pre-determined distribution of saturation on the n-th time layer, pressure on this layer P_{ijk}^n is obtained, the use of which allows finding s_{ijk}^{n+1}. Then computations are repeated in the same sequence. To check up the accuracy of the results, coincidence of discharges of the production and injection wells is controlled. The results of numerical solution of oil displacement problem are presented in [2, 3]. Adding (1) and (2), we will have the equation for pressure and after reducing it to difference form, the equation is solved by the Jacobi method [4]. Using several arithmetical operations, it is easy to obtain the P_{ijk}^{n+1}. To solve the equation of saturation, the value of pressure in this layer is substituted into (1). The difference analogs for the equation of saturation are written according to [4] in the form which can be reduced to the difference form.

3   The Distributed Parallel Algorithm on the Basis of MapReduce Hadoop and MPI Technologies

The analysis of hybrid solutions of MapReduce Hadoop and MPI technologies showed that the use of MPI technology for development of MapReduce applications increases performance of the applications [5–10]. Hybrid solutions are created with the aim of using the advantages of separate technologies. As is known, in MapReduce Hadoop, the tasks important for distributed systems such as fault tolerance, load balancing, distributed storage and processing of large data volumes are solved at the level of MapReduce technology itself. A flexible organization of communications and exchange of data between parallel processes can be referred to the advantages of MPI technology. It is to unification of advantages of the technologies of parallel, distributed and cloud computing that the majority of works [11, 12] are devoted owing to their special practical importance. But the questions of effective organization of computations, scalability of algorithms and their adaptation to a wide range of scientific problems remain open.

Thus, the problem of development of applications for high performance computing where the tasks of storage and distribution of data are supported by MapReduce Hadoop technology and the computational part performed by MPI technology is actual.

Usually when there is a huge computational task one has to choose between performance (provided by MPI) and fault tolerance (main feature of Hadoop). In our case, when solving this specific task, we use MPI and Hadoop technologies in combination. Practical unification of the two considered technologies requires solution of the following tasks:

- Development of the interaction mechanism of MapReduce Hadoop and MPI environments (Fig. 1);
- Development of a hybrid algorithm using MapReduce and MPI technologies.

The mechanism of MapReduce Hadoop and MPI interaction via files and calling up MPI processes in the MapReduce Hadoop environment proposed by the authors was successfully used for solution of a number of problems [13], in particular, Dirichlet problem for Poisson equation [14], 3D problem of fluid dynamics in anisotropic elastic porous medium [15, 16]. This section presents the results of further investigations in this direction, namely, the use of a constructive approach of a hybrid unification of MapReduce Hadoop and MPI for the problem of 3D hydrodynamic calculation of oil production.

To solve the problem (1)–(3), a scheme of iteration infrastructure shown in Fig. 1 is used. Implementation of distributed parallel algorithms begins with starting of the initialization stage presented by MapReduce task *Initial* at the main node of the cluster. Then, a cycle of iterations on time is started in which, firstly, a MapReduce task MR0 is fulfilled which computes auxiliary coefficients for calculation of pressure; secondly, an MPI is compiled and called up in which pressure is calculated; thirdly, a MapReduce task MR1 is fulfilled for calculation of saturation.

Thus, the algorithm consists of two main stages: the initialization stage and the iteration stage which, in its turn, contains internal iteration cycles. A more detailed description of the algorithm stages is presented below.

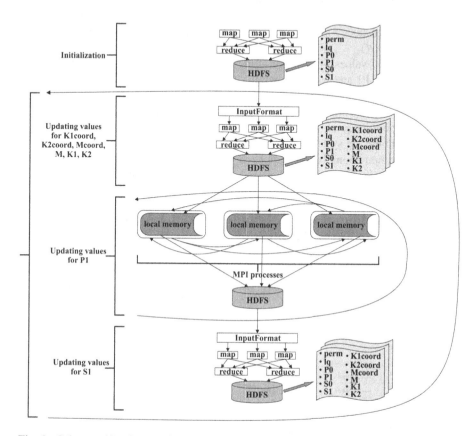

Fig. 1. Scheme of implementation of MapReduce MPI algorithm for oil production problem

Initialization stage is presented by the MapReduce task *Initial* in which the parameters of the oil reservoir necessary for calculations are given. The function *map* reads the task numbers and writes them down in the output file. In the function *reduce* there takes place assignment of initial values to the problem characteristics such as permeability *perm*, porosity of the medium *m*, coordinates of the wells location *lq*. Then, initial values of pressure P_1, initial values of saturation s_0, s_1 are calculated. All the new values obtained at the stage *Initial* are written down into the distributed file system of Hadoop HDFS. Thus, at the stage *Initial* there proceeds preparation of the data necessary for the main computational part of the algorithm.

Iteration stage begins with starting of the MapReduce task MR0 in which the function *map* reads the numbers of tasks in the form of a key and writes them down the output file. The function *reduce* reads the data from the distributed file system HDFS, calculates the values of auxiliary arrays – $K[0][1][i][j][k]$ of the size 5, M $[1][i][j][k]$ of the size 4 – and writes them down in the form of output data into HDFS. Output files in MR0 task are read as input files by MPI processes. Auxiliary arrays K and M from HDFS are used for calculation of pressure.

In the MPI part of the algorithm, 1D decomposition of the data is used. The exchange between nodes is implemented with the help of primitives MPI.COMM_-WORLD.Sendrecv().

The calculated values of pressure at the current iteration of the external cycle are stored in the array P1 and written down into HDFS as shown in Fig. 1. Here, the work of the MPI task is completed. Then, the next MapReduce task MR1 is started. The following data – the current values of saturation s_1, permeability $perm$, pressure P_1, porosity m, coordinates of wells location lq and the elements of the auxiliary array K are read from the file system HDFS. A cycle of computations of the saturation value begins. The computations being completed, the new values of saturation are written down into the file system HDFS. Here, the work of MR1 is completed.

4   Implementation of the Distributed Parallel Algorithm and Analysis of the Results

We have configured Apache Hadoop 2.6.0 cluster which consists of one master node and 8 slaves (Figs. 2, 3). All slave nodes have Ubuntu 14.04 on board; the master node has Ubuntu Server 14.04.

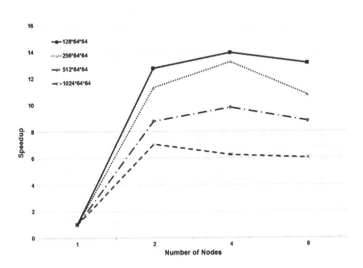

Fig. 2. Speedup of the distributed parallel program

All slave nodes are configured to work with MPICH-3.1.4. To share the data between MPICH nodes, we use NFS and NFS server is configured on the master node. For experimental design the hybrid of Hadoop 2.6.0 MapReduce (Yarn) and MPICH

3.1.4 technologies is used. YARN schedules the MPI tasks and delivers input output files to MPI nodes via HDFS. This provides the reliability and maximum computing performance.

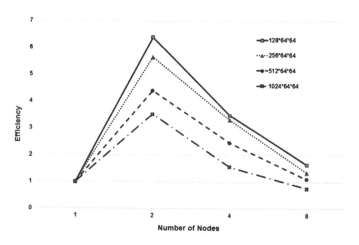

Fig. 3. Efficiency of the distributed parallel program

5 Conclusion

Investigations on high performance computing for solution of the problem of 3D hydrodynamic computation of oil production (1)–(3) resulted in proposition of a constructive approach to organization of distributed parallel computations using MapReduce Hadoop and MPI technologies for which: a general scheme of the iteration infrastructure of MapReduce model of problem (1)–(3) computation was designed; the structure for fulfillment of *map* and *reduce* methods and organization of decomposition of the computation area at different Map/Reduce stages were presented; a computational experiments were carried out on specific infrastructure. The practical result of investigations is adaptation of high performance technologies to solve of actual oil and gas industry problems of Kazakhstan.

References

1. Aziz, K., Settari, A.: Petroleum Reservoir Simulation, p. 407. Applied Science Publishers Ltd., London (1979)
2. Imankulov, T.S., Mukhambetzhanov, S.T., Akhmed-Zaki, D.Z.: Simulation of generalized plane fluid filtration in a deformable environment. In: Computational Technologies, part 1, vol. 1, pp. 183–191 (2013). Bulletin of the East-Kazakhstan State Technical University Named by D. Serikbaev

3. Zhumagulov, B.T., Mukhambetzhanov, S.T., Akhmed-Zaki, D., Imankulov, T.S.: Computer modeling of nonisothermal oil displacement at gelepolimer flooding. Bull. NEA RK. 4(50), 14–22 (2013)

4. Samarskiy, A.A., Gulin, A.V.: Numerical Methods. Nauka, Moscow (1989). (in Russian)

5. Jin, H., He, X.: Sun performance comparison under failures of MPI and MapReduce. An analytical approach. J. Future Gener. Comput. Syst. 7(29), 1808–1815 (2013)

6. Lu, X., Wang, B., Zha, L., Xu, Z.: Can MPI benefit Hadoop and MapReduce applications? In: Proceedings of the 2011 40th International Conference on Parallel Processing Workshops, ICPPW 2011, pp. 371–379 (2011)

7. Mohamed, H., Marchand-Maillet, S.: Enhancing MapReduce using MPI and an optimized data exchange policy. In: Proceedings of the 2012 41st International Conference on Parallel Processing Workshops, ICPPW 2012, pp. 11–18 (2012)

8. Steven, J., Plimpton, K.D.: Devine MapReduce in MPI for large-scale graph algorithms. J. Parallel Comput. 9(37), 610–632 (2011)

9. Matsunaga, A., Tsugawa, M., Fortes, J.: CloudBLAST: combining MapReduce and virtualization on distributed resources for bioinformatics applications. In: IEEE 4th International Conference on eScience 2008. IEEE (2008)

10. Sul, S., Tovchigrechko, A.: Parallelizing BLAST and SOM algorithms with MapReduce-MPI library. In: Proceedings of the 2011 IEEE International Symposium on Parallel and Distributed Processing Workshops and PhD Forum, IPDPSW 2011, pp. 481–489 (2011)

11. Hoefler, T., Lumsdaine, A., Dongarra, J.: Towards efficient MapReduce using MPI. In: Ropo, M., Westerholm, J., Dongarra, J. (eds.) PVM/MPI. LNCS, vol. 5759, pp. 240–249. Springer, Heidelberg (2009)

12. Slawinski, V.S.: Adapting MPI to MapReduce PaaS clouds: an experiment in cross-paradigm execution. In: 2012 IEEE/ACM Fifth International Conference on Utility and Cloud Computing, pp. 199–203 (2012)

13. Akhmed-Zaki, D., Danaev, N., Matkerim, B., Bektemessov, A.: Design of distributed parallel computing using by MapReduce/MPI technology. In: Malyshkin, V. (ed.) PaCT 2013. LNCS, vol. 7979, pp. 139–148. Springer, Heidelberg (2013)

14. Kumalakov, B., Shomanov, A., Dadykina, Y., Ikhsanov, S., Tulepbergenov, B.: Solving Dirichlet problem for Poisson's equation using MapReduce Hadoop. In: Theoretical and Applied Aspects of Cybernetics, Proceedings of the III International Scientific Conference of Students and Young Scientists, pp. 224–227. Kyiv (2013)

15. Akhmed-Zaki, D., Mansurova, M., Matkerim, B., Kumalakov, B., Shomanov, A.: Iterative MapReduce oil reservoir simulator. Applying MDA and Hadoop to solve hydrodynamics problems. In: Extended Abstracts of the International Conference Electronics, Telecommunications and Computers, Lisbon, Portugal (2013)

16. Mansurova, M., Akhmed-Zaki, D., Kumalakov, B., Matkerim, B.: Distributed parallel algorithm for numerical solving of 3D problem of fluid dynamics in anisotropic elastic porous medium using MapReduce and MPI technologies. In: Proceedings of 9th International Joint Conference on Software Technologies ICSOFT 2014, Vienna, Austria, pp. 525–528 (2014)

A Two-Level Parallel Global Search Algorithm for Solution of Computationally Intensive Multiextremal Optimization Problems

Victor Gergel$^{(\boxtimes)}$ and Sergey Sidorov

Lobachevsky State University of Nizhni Novgorod, Nizhni Novgorod, Russia
gergel@unn.ru

Abstract. The work considers a new parallel global search algorithm developed within the framework of the information-statistical approach to multiextremal optimization. The proposed algorithm is intended for maximum possible use of the potential of state-of-the-art high-performance computing systems, in particular, for solving the most computationally intensive problems of multiextremal optimization. The key feature of the algorithm is organization of parallel calculations through the use of multiple mappings based on Peano curves for dimensionality reduction. This approach enables effective use of supercomputers with shared and distributed memory and a large number of processors for solving global search problems.

Keywords: Global optimization · Information-statistical theory · Parallel computations · High-performance computing systems · Supercomputer technologies

1 Introduction

Problems of global or multiextremal optimization currently belong to one of the key research areas in the theory and practice of decision-making. In problems of such kind, it is assumed that the optimized criterion has several local extrema with different values in the search domain. The existence of several local extrema significantly complicates the search for the global optimum as it requires complete analysis of the whole search domain.

Such problems involve a large number of variables, multiextremality of optimized functions, existence of nonlinear constraints, multicriteriality, and, most importantly, a high computational complexity of the functions that form the basis of the optimized criteria and constraints. Thus, it is only possible to solve such problems with realistic resource costs when using high computing capacities of state-of-the-art supercomputers with the use of parallel global optimization algorithms.

The present state of the art of global optimization is adequately presented in a number of key monographs [1–5] et al. With regard to parallel computation, comparatively obvious parallel extensions of the exhaustive search method, Monte Carlo [6] and local multistart schemes [7] can be distinguished. There are considerable efforts to develop parallel genetic algorithms [8] and parallel algorithms in interval global

© Springer International Publishing Switzerland 2015
V. Malyshkin (Ed.): PaCT 2015, LNCS 9251, pp. 505–515, 2015.
DOI: 10.1007/978-3-319-21909-7_49

optimization [9]. An interesting approach to parallel algorithms based on the non-uniform space-covering technique is given in [10]. The well-known algorithm of global optimization DIRECT and its modification LBDIRECT (locally-biased DIRECT) is presented in [11, 12].

This work continues to further develop effective parallel global search algorithms within the framework of information-statistical theory of multiextremal optimization [5].

The key feature of the proposed algorithm is that the parallel computations is organized through the use of multiple mappings based on Peano curves for dimensionality reduction. As a result, multidimensional optimization problems are reduced to problems of one-dimensional global search, thus enabling the construction of effective computation schemes for multiprocessor computation systems with both distributed and shared memory. This approach allows the maximum possible use of the potential of modern supercomputers for solving the most computationally intensive problems of multiextremal optimization.

The structure of the paper is as follows. The problem statement is presented in Sect. 2, the scheme of dimensionality reduction on the basis of multiple mappings based on Peano curves used further for organization of parallel computation is considered. The computation scheme of the constructed two-level global search algorithm is provided in Sect. 3. The results of the performed computation experiments are considered in Sect. 4. At the end of the paper, general conclusions are given and prospects for continued research are discussed.

2 Problem Statement

Let us consider a problem of global optimization of the form

$$\varphi^* = \varphi(y^*) = \min\{\varphi(y) : y \in D\}, \tag{1}$$

$$D = \{y \in R^N : a_i \leq y_i \leq b_i, \ 1 \leq i \leq N\},$$

where the objective function $\varphi(y)$ meets the Lipschitz condition

$$|\varphi(y') - \varphi(y'')| \leq L\|y' - y''\|, \ y', y'' \in D \tag{2}$$

where $L > 0$ is the Lipschitz constant, and $\|*\|$ denotes the Euclidean norm in the space R^N.

Let us further assume that the value of a minimized function $\varphi(y)$ can be calculated for any vector value $y \in D$ by some computational procedure (hereinafter, obtaining the value of a minimized function is called a *trial*). As a rule, this procedure is computationally intensive, i.e. computation costs of the optimization problem solution (1) are determined first of all by the number of executed trials of function values.

By using *Peano curves* or *evolvents* $y(x)$, which are single-valued mappings of the interval [0,1], on the N-dimensional hypercube D, the initial problem (1) can be reduced to the following one-dimensional problem [5, 16]:

$$\varphi(y(x^*)) = \min\{\varphi(y(x)) : x \in [0,1]\}. \tag{3}$$

The considered scheme of dimensionality reduction allows a multidimensional problem with a Lipschitz minimized function to be transformed to a one-dimensional problem, where corresponding functions meet the uniform Hölder condition, i.e.

$$|f(y(x')) - f(y(x''))| \le \widehat{L}|x' - x''|^{1/N}, \; x', x'' \in [0,1] \tag{4}$$

where the constant $\widehat{L}$ is determined by the formula $\widehat{L} = 4L\sqrt{N}$, L is the Lipschitz constant from (2), and N is the dimensionality of the optimization problem (1).

Algorithms for numerical construction of Peano curve approximations are given in [5]. For the purpose of illustration, Fig. 1 shows a Peano curve approximation for the third level of density. The curve in Fig. 1 shows the procedure for bypassing a two-dimensional domain; the accuracy of the Peano curve approximation is determined by the implemented construction density level.

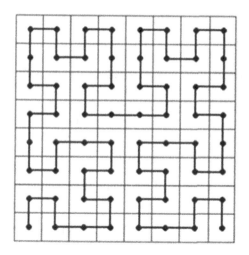

Fig. 1. Peano curve approximation for the third level of density

By reducing multidimensional problems (1) to optimization of one-dimensional functions (3) using mappings based on Peano curves it is possible to preserve uniform boundedness of differences between function values in the case of limited argument variation, i.e. the feasibility of (4). However, in this case a partial loss of information on closeness of points in multidimensional space occurs, as point $x \in [0,1]$ has only neighbors on the left and on the right, and its corresponding point $y(x) \in R^N$ has neighbors in 2^N directions. Thus, when using Peano-type mappings quite distant preimages x', x'' on the interval $[0,1]$ can correspond to images y', y'' in R^N, which are close in N-dimensional space. As a result, several (up to 2^N) local extrema in a one-dimensional problem can correspond to the only global minimum point in a multidimensional problem. This fact significantly complicates the solution of the one-dimensional problem (3) since global search algorithms, as a rule, provide convergence to all global minimum points.

To reduce this effect it was proposed in [5] to use a set of mappings

$$Y_s(x) = \{y^1(x), \ldots, y^s(x)\}. \tag{5}$$

instead of using one Peano curve $y(x)$.

To construct the set $Y_s(x)$, several different approaches can be used. For example, in the work [5] a scheme was implemented, according to which each transformation $y^i(x)$ from $Y_s(x)$ is constructed as a result of shifting along the main diagonal of the hyperinterval D. The set of Peano curves thus constructed allows one to obtain y', y'' from D for any close multidimensional images, which differ only in one coordinate, close preimages x', x'' from the interval [0,1] for a transformation $y^k(x)$, $1 \leq k \leq s$.

The set of transformations $Y_s(x)$ from (5) generates (for the multidimensional problem (1)) s auxiliary information-linked one-dimensional problems (3) of the form:

$$\varphi\big(y^k(x^*)\big) = \min\big\{\varphi\big(y^k(x)\big) : x \in [0, 1]\big\}, \ 1 \leq k \leq s. \tag{6}$$

It must be noted that the family of one-dimensional problems $\varphi(y^k(x))$, $1 \leq k \leq s$, obtained as a result of dimensionality reduction is information-linked: a function value calculated for any problem $\varphi(y^k(x))$ of the family (6) can be used for all other problems of this family without any complicated computations (see Sect. 3.2 of this paper).

Information compatibility of problems of the family (6) provides a natural way to organize parallel computations. Each separate problem can be solved on a separate processor of a computing system. In the course of computations, it is necessary to provide exchange of obtained search information between the processors. In more detail, organization of parallel computations based on simultaneous solution of one-dimensional reduced problems of the family (6) is considered in Sect. 3.

3 Parallel Two-Level Global Optimization Algorithm

It should be noted that it is not easy to parallelize solving global optimization problems. In this case the "classical" approach of domain decomposition can not be applied because only one subdomain can contain global optimum point. Also it is not reasonable to parallelize global algorithm itself because the most time consuming computations are executed for evaluating minimized function values. So to parallelize solving global optimization problems a new perspective way can be proposed – this is computing function values at several points simultaneously. The question is how to select these several points to provide high efficiency of parallel computations? The parallel two-level global optimization algorithm described in the paper applies for that the original approach for reducing multidimensional optimization problems to one-dimensional ones by using Peano curves.

The information-statistical theory of global search formulated in [5] has provided the basis for developing a large number of effective methods of multiextremal optimization – see, for example, [13–15, 17] and some other works.

The key feature of the algorithm proposed in this paper is the use of a unified approach to organize parallel computations for multiprocessor systems with both distributed and shared memory. With this integrated approach, one can achieve the maximum possible use of the potential of modern-day supercomputers for solving the most computationally intensive multiextremal optimization problems. According with the proposed scheme there is no any leading processor and all processors are equivalent. At any time some processors can be added (or excluded) for calculations. Data that should be sent between distributed nodes have a relatively small size and message passing can not decelerate parallel calculations. Moreover there are no any data exchange between computational cores with shared memory.

The description of the proposed parallel algorithm is divided into two parts. The first part contains the rules of the algorithm with regard to computing nodes of supercomputers with shared memory; the scheme of organization of parallel computation for multiprocessor systems with distributed memory is given in the second part of the description.

3.1   Parallel Computations for Nodes with Shared Memory

State-of-the-art supercomputers consist of a large number of computing nodes including several multicore processors. The random access memory of computing nodes is shared: the value of any element of the memory can be read (written) by any of the available computing cores at any random moment. In most cases, shared memory is uniform, i.e. the memory access time characteristics are identical for all computing cores and for all elements of the memory.

For the sake of simplicity of the description of the proposed approach, let us assume that each computing node of a supercomputer is used for solving one of the one-dimensional reduced problems of the family (6). Let us denote the problem being solved on a computing node in a simpler way:

$$f(x) = \varphi(y(x)) : x \in [0, 1]. \tag{7}$$

The parallel global search algorithm on multiprocessor multicore computing nodes is based on the results presented in [5] and can be described as follows [18].

The initial iteration is carried out in some point $x^1 \in (0,1)$. Then assume that τ, $\tau > 1$, iterations of global search are carried out. Selection of trial points $\tau + 1$ of the next iteration is regulated by the following rules.

Rule 1. Renumber points of the set of trial points using subscripts in the increasing order of coordinate values

$$0 = x_0 < x_1 < \ldots < x_i < \ldots < x_k < x_{k+1} = 1, \tag{8}$$

where points x_0, x_{k+1} are used additionally in order to simplify the description below, the values of the minimized function z_0, z_{k+1} at these points are not defined.

Rule 2. Calculate the current estimate of the Hölder constant from (4)

$$m = \begin{cases} rM, & M > 0, \\ 1, & M = 0, \end{cases} \quad M = \max_{1 < i \le k} \frac{|z_i - z_{i-1}|}{\rho_i}, \tag{9}$$

as relative differences of the minimized functions $f(x)$ from (7) on the set of executed trial points x_i, $1 \le i \le k$ from (8). Hereinafter, $\rho_i = (x_i - x_{i-1})^{1/N}$, $1 \le i \le k + 1$. Constant r, $r > 1$, is a *parameter* of the algorithm.

Rule 3. For each interval (x_{i-1}, x_i), $1 \le i \le k + 1$, calculate the characteristic $R(i)$, where

$$R(i) = \rho_i + \frac{(z_i - z_{i-1})^2}{m^2 \rho_i} - 2\frac{(z_i + z_{i-1})}{m}, \, 1 < i \le k,$$

$$R(i) = 2\rho_i - 4\frac{z_i}{m}, \, i = 1, \tag{10}$$

$$R(i) = 2\rho_i - 4\frac{z_{i-1}}{m}, \, i = k + 1.$$

Rule 4. Locate the interval characteristics obtained in (10) in the decreasing order

$$R(t_1) \ge R(t_2) \ge \ldots \ge R(t_{k-1}) \ge R(t_{k+1}) \tag{11}$$

and select p intervals with numbers t_j, $1 \le j \le p$, with maximum values of characteristics (p is the number of processors (cores) used for parallel computations).

Rule 5. Execute new trials (computation of values of the minimized function $f(x)$) at points x^{k+j}, $1 \le j \le p$, located in the intervals with maximum characteristics from (11)

$$x^{k+j} = \frac{x_{t_j} + x_{t_j-1}}{2} - \text{sign}(z_{t_j} - z_{t_j-1})\frac{1}{2r}\left[\frac{|z_{t_j} - z_{t_j-1}|}{m}\right]^N, \, 1 < t_j < k + 1,$$

$$x^{k+j} = \frac{x_{t_j} + x_{t_j-1}}{2}, \, t_j = 1, \, t_j = k + 1. \tag{12}$$

The stop condition, according to which calculations are terminated, is defined by the condition

$$\rho_t \le \varepsilon,$$

that should be fulfilled at least for one of the numbers t_j, $1 \le j \le p$, from (11) and $\varepsilon > 0$ is the specified coordinate-wise accuracy of the problem solution.

If the stop condition is not fulfilled, then the number of iteration τ increases by one and a new iteration of global search is executed.

To provide some explanations of the given algorithm let us note the following.

Remark 1. Because of the high computational complexity of trials (calculation of values of the minimized function), the parallel algorithm is based on the approach whereby parallel computing is achieved by running multiple trials simultaneously at different points of the search domain.

Remark 2. Characteristics $R(i)$, $1 \le i \le k + 1$, calculated in (10) can be interpreted as some measures of the importance of intervals in terms of the point of the global minimum they may contain. Then the scheme of interval selection for parallel computations used in (11) becomes evident: the points of subsequent trials are selected in the intervals with maximum values of characteristics.

Remark 3. It is assumed that parallel computations are carried out simultaneously. First, p points for carrying out trials are defined, which are then distributed between processors (cores). Each processor (core) executes a trial at one of the trial points calculated in (12). The global search algorithm continues calculations only after all the trials of the current search iteration have been completed.

Remark 4. The number s of used evolvents in (6) may be arbitrary. If $s > 1$ the set of points of executed search trials for different evolvents may be either integrated in (8) or the rules of the considered algorithm may be applied separately to each of the available evolvents.

Remark 5. At the very beginning of the global search, the number of available points of executed trials may be less than the number of processors (cores) used, i.e. $k < p$. In this case, some computing elements will be idle due to the insufficient number of trial points.

The conditions of convergence of the proposed algorithm and nonredundancy of the parallel computations are considered in [5]. Thus, in the case of an appropriate estimation of the Hölder constant ($m > 2^{2-1/N} \hat{L}$, m from (9)) the algorithm converges to all global minimum points. Besides, the conditions are defined under which parallel computations are nonredundant in comparison with the serial method when using less than 2^N processors (cores).

3.2 Parallel Computations for Systems with Distributed Memory

The following level of organization of parallel computations in high-performance systems consists in the use of several computing nodes. Thus, each computing node has its individual memory and, in this case, interaction between different nodes can be provided only by means of data transmission via the communication network of the computing system.

The unified approach to organization of parallel computations for multiprocessor computers with distributed and shared common memory is as follows.

1. The family of one-dimensional reduced information-linked problems (6) is distributed between computing nodes of the multiprocessor system. One or several problems of the family can be allocated for each computing node.

2. For solving allocated problems of the family (6) on each computing node, the parallel global search algorithm considered in Sect. 3.1 is used. The algorithm is supplemented with the rules of information exchange given below.

(a) Prior to a new trial for any problem $\varphi(y^k(x))$, $1 \leq k \leq s$, at any point $x' \in [0, 1]$ the following must be performed:

   (i) Calculate the image $y' \in D$ for the point $x' \in [0, 1]$ according to mapping $y^k(x)$,
   (ii) Calculate preimages x'_i, $1 \leq i \leq s$, for every problem of the family (6),
   (iii) Transmit the obtained images x'_i, $1 \leq i \leq s$, to all used computing nodes to exclude repeated selection of intervals that contain received preimages.

To organize data transmission, a queue of received messages is formed at each computing node. This queue keeps the transmitted data on trial points and values of the minimized function at these points.

(b) After completion of any trial for any problem $\varphi(y^k(x))$, $1 \leq k \leq s$, the following must be executed at a point $x' \in [0, 1]$:

   (i) As in item 2.a, calculate all preimages for the point of the executed trial x'_i, $1 \leq i \leq s$, for each problem of the family (6),
   (ii) Transmit the preimages x'_i, $1 \leq i \leq s$, with the trial result $z' = \varphi(y_k(x'))$ to all used computing nodes for including the obtained data into the search information in the rules of the global search algorithm.

(c) Prior to the start of the next iteration of global search the algorithm has to check the queue of received messages; if data is present in the queue, the received data have to be included into the search information.

The key feature of this scheme of parallel computations is the possibility of asynchronous data transmission (computing nodes process the received data as they become available). Besides, there is no single control node in this scheme. The number of computing nodes can change in the process of global search, but the exclusion (loss) of any node does not lead to the loss of the sought global minimum of the minimized multiextremal function.

4 Results of Computational Experiments

Computational experiments were executed on a high-performance cluster of the University of Nizhni Novgorod. Each computing node of this cluster contains two dual-core Intel Xeon 2.13 MHz processors with 24 Gb memory. Nodes are connected with the Gigabit Ethernet network. The cluster is operated under Microsoft Windows HPC Server 2008. The development environment was Microsoft Visual Studio 2008 with Microsoft 32-bit C/C++ Optimizing Compiler. The MPI library was MPICH 1.2.5.

In the first series of experiments, we used the problems of multiextremal optimization generated by a GKLS test functions generator [19] with a priori known properties (the point and the value of the global minimum, the number of local minima, etc.).

Computational experiments were executed for three-dimensional functions of the classes Simple and Hard. A stop rule

$$\|y - y^*\| \leq \Delta$$

was used to terminate computations, i.e. it was assumed that estimation of the global minimum was evaluated with the specified accuracy, if any point of the executed trial y^k got into the neighborhood Δ of point y^*.

In our experiments, the evolvent density was $\gamma = 10$, the reliability parameter r took on the values $r = 2.5$ and 2.7, the number of evolvents was $s = 5$, the accuracy in the stopping conditions was defined as $\Delta = 0.01\sqrt{N}$.

In each computation experiment, a total of 100 problems generated by the GKLS generator had to be solved. For the cases when not all 100 problems were solved, the number of unresolved problems is given in brackets.

First of all the three serial algorithms – DIRECT [11], LBDIRECT [12] and global search algorithm (GSA) from Sect. 3 (a serial version) – are compared. The results of numerical experiments are provided in Table 1 (number of the executed trials is presented, results of the first two algorithms are given in paper [19]). Numerical comparison was carried out on the function classes Simple and Hard of dimension 3.

Table 1. Numerical results of the three serial algorithms

Problem class	DIRECT	LBDIRECT	GSA	
			$r = 2.5$	$r = 2.7$
Simple	1117	1785	397	487(1)
Hard	42322(4)	4858	1567(5)	1980(2)

As it can be noted Table 1 shows that GSA outperforms the DIRECT and LBDIRECT methods on all problems in terms of average number of iterations.

Then a computing node with two dual-core processors was used for experiments. The number of trials executed by the parallel method was taken to be equal to the maximum value among all used computing cores.

The average number of trials by the serial (SA) and parallel (PA) methods to achieve global optimum with the specified accuracy is given in Table 2.

In the second series the GKLS problems with increased dimension (6 and 8) have been solved. In this experiments, the evolvent density was $\gamma = 10$, the reliability parameter was $r = 4$, the number of evolvents was equal to the number of cores, the accuracy in the stopping conditions was defined as $\Delta = 0.000001$.

To provide the GKLS problems be more comparable with real applied time-consuming optimization problems some additional delays have been added for calculations of minimized function values (the delay value is presented in the column "Delay" in Table 3).

Table 2. Average number of trials for solving 100 test problems

Problem class	Reliability parameter $r = 2.5$			Reliability parameter $r = 2.7$		
	SA	PA	Speedup	SA	PA	Speedup
Simple	397	312	1,27	487(1)	204	2,38
Hard	1567(5)	911	1,72	1980(2)	809	2,44

The numerical results of this series are given in Table 3.

Table 3. Numerical results for the second series of experiments

N	Delay (ms)	Cores	SA		PA		Speedup (time)	Speedup (iters)
			Iters	Time, sec.	Iters	Time, sec.		
6	100	40	500000	54500	16620	2370	23	30.1
8	200	60	1000000	203000	17008	5049	40.4	58.8

As it appears from the results, the speedup demonstrated by the developed parallel two-level global search algorithm can provide some opportunities for solving serious multidimensional global optimization problems.

5   Conclusion

This work considers a parallel global search algorithm developed within the framework of the information-statistical approach to multiextremal optimization. The developed algorithm offers a uniform integrated scheme for organizing parallel computations using multiprocessor systems with distributed and shared memory. With this approach, it is possible to achieve maximum use of the potential of state-of-the-art high-performance computing systems for solving the most computationally intensive multiextremal optimization problems. Computational experiments confirm the effectiveness of the proposed approach.

More research is envisaged in this promising area. Some additional experiments using larger numbers of computing elements (up to several tens of thousands processors) are necessary. Another important direction for future research is to extend the approach for solving multiextremal optimization problems with additional nonlinear constraints.

Acknowledgements. The research is supported by the grant of the Ministry of education and science of the Russian Federation (the agreement of August 27, 2013, № 02.B.49.21.0003).

References

1. Törn, A., Žilinskas, A.: Global Optimization. Lecture Notes in Computer Science, vol. 350. Springer, Heidelberg (1989)
2. Horst, R., Tuy, H.: Global Optimization: Deterministic Approaches. Springer, Heidelberg (1990)
3. Zhigljavsky, A.A.: Theory of Global Random Search. Kluwer Academic Publishers, Dordrecht (1991)
4. Pintér, J.D.: Global Optimization in Action (Continuous and Lipschitz Optimization: Algorithms, Implementations and Applications). Kluwer Academic Publishers, Dordrecht (1996)
5. Strongin, R.G., Sergeyev, Y.D.: Global Optimization with Non-convex Constraints: Sequential and Parallel Algorithms. Kluwer Academic Publishers, Dordrecht (2000)
6. Byrd, R.H., Dert, C.L., Rinnoy Kan, H.G., Schnabel, R.B.: Concurrent stochastic methods for global optimization. Math. Program. **46**, 1–29 (1990)
7. Lootsma, F.A., Ragsdell, K.M.: State of the art in parallel nonlinear optimization. Parallel Comput. **6**, 133–155 (1988)
8. Luque, G., Alba, E.: Parallel Genetic Algorithms: Theory and Real World Applications. Springer, Heidelberg (2011)
9. Eriksson, J., Lindström, P.: A parallel interval method implementation for global optimization using dynamic load balancing. Reliable Comput. **1**(1), 77–91 (1995)
10. Evtushenko, Y., Posypkin, M.: A deterministic approach to global box-constrained optimization. Optim. Lett. **7**(4), 819–829 (2013)
11. Jones, D.R., Perttunen, C.D., Stuckman, B.E.: Lipschitzian optimization without the Lipschitz constant. J. Optim. Theory Appl. **79**(1), 157–181 (1993)
12. Gablonsky, J.M., Kelley, C.T.: A locally-biased form of the DIRECT algorithm. J. Global Optim. **21**(1), 27–37 (2001)
13. Gergel, V.P.: A method of using derivatives in the minimization of multiextremum functions. Comput. Math. Math. Phys. **36**(6), 729–742 (1996)
14. Gergel, V.P., Strongin, R.G.: Parallel computing for globally optimal decision making. In: Malyshkin, V.E. (ed.) PaCT 2003. LNCS, vol. 2763, pp. 76–88. Springer, Heidelberg (2003)
15. Sergeyev, Y.D.: An information global optimization algorithm with local tuning. SIAM J. Optim. **5**(4), 858–870 (1995)
16. Sergeyev, Y.D., Strongin, R.G., Lera, D.: Introduction to Global Optimization Exploiting Space-Filling Curves. Springer, New York (2013)
17. Barkalov K.A., Gergel V.P.: Multilevel scheme of dimensionality reduction for parallel global search algorithms. In: Proceedings of the 1st International Conference on Engineering and Applied Sciences Optimization, pp. 2111–2124 (2014)
18. Grishagin, V.A., Sergeyev, Y.D., Strongin, R.G.: Parallel characteristical global optimization algorithms. J. Global Optim. **10**(2), 185–206 (1997)
19. Gaviano, M., Lera, D., Kvasov, D.E., Sergeyev, Y.D.: Software for generation of classes of test functions with known local and global minima for global optimization. ACM Trans. Math. Softw. **29**, 469–480 (2003)

Efficient Parallel Implementation of Coherent Stacking Algorithms in Seismic Data Processing

Maxim Gorodnichev[1,2,3]($\boxtimes$), Anton Duchkov[2,4], and Alexander Kupchishin[3,4]

[1] Institute of Computational Mathematics and Mathematical Geophysics SB RAS, Novosibirsk, Russia
maxim@ssd.sscc.ru
[2] Novosibirsk State University, Novosibirsk, Russia
DuchkovAA@ipgg.sbras.ru
[3] Novosibirsk State Technical University, Novosibirsk, Russia
vsegdatrezv@mail.ru
[4] Chinakal Institute of Mining SB RAS, Novosibirsk, Russia

Abstract. We discuss efficient parallel implementation of coherent stacking algorithms which form basis for a class of seismic processing procedures. In detail we address the problem of processing data of microseismic monitoring for localizing seismic events in space and time. Continuous data recording by seismic array quickly generates terabytes of data to be processed in a timely manner, including real-time analysis in some cases. Thus processing requires efficient parallel implementation with a special attention to data partitioning between nodes, and using computations to mask data reading from disk. Efforts were taken to minimize cache misses and vectorize loops. Sequential version of the code demonstrates 8x speed up compared to a naive implementation of the algorithm; parallel code scales almost linearly.

Keywords: Coherent stacking · Microseismic monitoring · Parallel computing · Xeon Phi

1 Introduction

Seismic studies is the main source of our knowledge about the Earth structure including mineral resources exploration, and seismic risk assessment. Seismic data processing requires high performance computing due to extremely large volumes of acquired data [1]. In particular, coherent stacking (summation) operation is used in a number of processing procedures [2]. These include velocity analysis, stacking, Kirchhoff-type migration in time and depth, location of microseismic events, and many others.

Keeping in mind this broad list of perspective applications here we will consider coherent summation algorithm with application to microseismic data processing. Microseismic monitoring is widely used to monitor hydraulic fracturing, subsidence related to depletion, cap rock integrity, mapping fluid migration, detection of casing failure [3]. Such monitoring requires continuous recording of

© Springer International Publishing Switzerland 2015
V. Malyshkin (Ed.): PaCT 2015, LNCS 9251, pp. 516–521, 2015.
DOI: 10.1007/978-3-319-21909-7_50

seismicity by a network of stations. This results in large data volumes which preferably need to be processes in a real time manner.

Microseismic data processing results in localisation of seismic events in space and time. For surface arrays the coherent stacking (Kirchhoff-type stacking) is the most popular processing method producing coherency images [4]. Maximums in these images indicate hypocenters and origin time of microseismic events which characterize associated with subsurface geomechanical processes.

In this paper we develop parallel implementation of the coherent stacking method for the analysis of large microseismic data volumes (terabytes) optimised for modern high performance computational platforms.

2   Coherent Summation Method

The method takes microseismic monitoring data and the parameters of the volume being studied as an input and solves the inverse problem discovering spatio-temporal location of a series of seismic events. Microseismic monitoring data are represented as a set of traces, were a trace $G(t, D_k), t = 0, .., T - 1$ — is a signal sampled by seismometer $D_k, k = 0..(Q-1)$. A surface location of the seismometer D_k is given as (x_k, y_k). The studied volume is described by its spatial dimentions.

The method starts with building a regular mesh over the volume. Let us denote the set of mesh nodes as U. Each node $u \in U$ is then tested as a possible event location.

In order to do so, for each $k = 0..(Q-1)$ and $u \in U$, we compute the time t_{uk} required for a wave front to reach the seismometer D_k from the location u. Let us denote $Tmin_u = \min_k t_{uk}$, $Tmax_u = \max_k t_{uk}$, and $Td_u = Tmax_u - Tmin_u$. The length of the signal will be denoted as L. The coherent summation algorithm can be then defined as follows:

1: **for** $u \in U$ **do**

2:     **for** $t \in 0..(T - Td_u - L)$ **do** $S_{u,t} = \sum_{l=0}^{L-1} \sum_{k=0}^{Q-1} G(t_{uk} - Tmin_u + t + l, D_k)$

3:         **end for**

4: **end for**

5: find u_{max} and t_{max} such that $S_{u_{max}, t_{max}} = \max_{u,t} S_{u,t}$

If there is such a mesh node u_e that corresponds to the location of the seismic event, this u_e is expected to be found as $u_e = u_{max}$, and t_{max} will be the time of the event. If we are interested to find a series of events, we will choose several S_{ut} that exceed a certain treshold value.

In practice, there can be variations to this method, but the general scheme remains and it is this simple. However, straightforward implementation of this scheme results in poor computational performance. The following sections describe the steps to efficient implementation of this method Fig. 1.

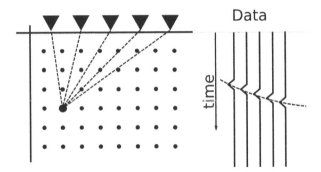

Fig. 1. Coherent summation. Seismic sensors are shown as triangles. The grid represents the set of probe points U. Summation of seismic data is done along the dotted line for the given point.

3 Mesh Refinement

The process starts [5] with a coarse mesh U^0 covering the whole volume V^0 being studied. The coherent summation method is then applied to this mesh and, in this way, we find a mesh node u^0_{max} that is close to the seismic event. Then we define a volume $V^1 \in V^0$ with a center in u^0_{max} and build a successive finer mesh U^1 in V^1. These steps are repeated until the volume V^j is small enough thus determing an event location with appropriate accuracy.

4 Hiding Disk Access Operations Behind Computation

Typical volumes of seismic data can be measured in hundreds of gigabytes or terabytes. Thus, it is necessary for the program to access low performance external memory (local disks, networked filesystems). In order to confront this issue, the double buffering approach is taken. The new portion of data is loaded from external memory asynchronously by a special thread while the previously loaded portion is being processed.

5 Elimination of Recomputing

One can notice that $S_{u,t+1} = S_{u,t} - \sum_{k=0}^{Q-1} G(t_{ku} - Tmin_u + t, D_k) + \sum_{k=0}^{Q-1} G(t_{ku} - Tmin_u + t + L, D_k)$. Thus, naive implementations of the method will require multiple computation of the same values.

Elimination of recomputing made computing time independent of the signal length L (Fig. 2).

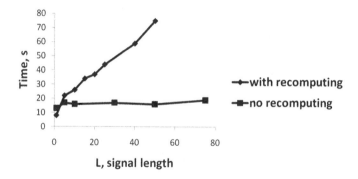

Fig. 2. Computation time, s, dependence on signal length. U: a grid of 151515 points, $Q = 100$, $T = 2000$.

6 Loop Vectorization and Avoiding Cache Misses

Processor caches and hardware prefetching work efficiently when an application traverse data in memory in consequitive way [6]. In our case, the data is stored in memory in such a way that $G(t + 1, D_k)$ follows $G(t, D_k)$ immediately. So, in order to force consequitive memory traversal, the loops should be organized in the following order:

1: **for** $k = 0..(Q - 1)$ **do**
2:     **for** $t = 0..(T_{end} - Td_u - L)$ **do**
3:         $S_{u,t} = S_{u,t} + \sum\limits_{l=0}^{L-1} G(t_{k,u} - Tmin_u + t + l, D_k)$
4:     **end for**
5: **end for**

It is important to vectorize the inner loop manually or write it in a form that would permit a compiler to do this job.

Changing the loop order to the correct one reduces the time of computation by 3.8 times on Intel® Xeon® E5-2690 and by 9.8 times on Intel® Xeon Phi™ 7110X. Vectorization of this loop speeded-up the sequential program by a factor of 2 for E5-2690 and by a factor of 2.8 for 7110X. The Intel® C++ Compiler ver. 14.0.1 20131008 with -O3 was used for all the tests.

7 Parallel Implementation

The method can be easily paralleled. Within a computing node and the shared memory processing paradigm, the computation can be divided among computing cores by partitioning the nodes of a mesh U. The nodes also can be processed independently of each other, thus the only problem with parallel implementation can be a memory bottleneck. This problem can be addressed by improving data access locality (see Sect. 6). The performance results of a shared memory parallel program obtained on Intel®Xeon Phi™7110 are presented in the Table 1.

Table 1. Computation time, s, of shared memory parallel program execution for different number of threads on Intel® Xeon® E5-2690 and on Intel® Xeon Phi™ 7110X.

Number of OMP threads	1	4	8	16	30	61	244
Xeon E5-2690	28	7	4	3			
Xeon Phi 7110X	779	193			27	21	12

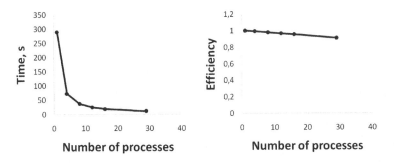

Fig. 3. Computation time, s, and efficiency of the program execution on the Intel®Xeon Phi™ 7110 co-processors depending on the number of co-processors used. Only one co-processor per computing node was used and 244 OMP threads were running on each co-processor.

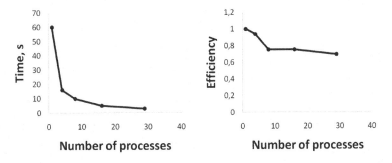

Fig. 4. Computation time, s, and efficiency of the program execution on the computing nodes with 2 Intel® Xeon® E5-2690 each depending on the number of computing nodes. Each node was running 16 OMP threads.

In order to distribute the work between the nodes of a supercomputer we partition seismic data in chunks by time. Most of the processing can be done independently and the nodes will have to compare their maximal sums only in the end of computation. The program was implemented with Intel® MPI Library. The performance results are presented in Figs. 3 and 4.

The problem characteristics for all the tests in this Sections are: U is a grid of 404010 points, Q = 2000, and T = 1200.

8 Conclusion

A number of optimization steps to naive implementation of coherent summation algorithm were taken in order to obtain efficient parallel program. The overal speed-up of the sequential computations is over 7x on Intel® Xeon® E5-2690 and over 26x on Intel® Xeon Phi™ 7110X. The efficiency characteristics measured on Intel(R) Core(TM) i5-3550 processor are as following: CPI (clocks per instruction) = 0.55, LLC (last-level cache) miss = 0.026 (a ratio of cycles with outstanding LLC misses to all cycles). A parallel program was implemented with OpenMP and MPI. The program is capable of processing large data sets on computational clusters and scales well over distributed memory computing nodes. Memory subsystem seems to be a bottleneck for a shared-memory program scalability and more efforts on memory access optimization are required.

The algorithm is to be included into different procedures for seismic data processing. It should be enhanced by introduction of workload balancing for heterogenous computing systems so that it could evenly distribute workload among all the available processing units, including CPU cores and co-processors. We also plan to integrate it into a popular open-source seismic processing package Madagascar [7].

Acknowledgements. Work was supported by the Russian Ministry of Education and Science (project # RFMEFI60414X0047).

References

1. Camp, W., Thierry, P.: Trends for high-performance scientific computing. Lead. Edge **29**(1), 44–47 (2010)
2. Rückemann, C.: Comparison of stacking methods regarding processing and computing of geoscientific depth data. GEOProcessing **7**(27), 35–40 (2012). http://dx.doi.org/10.12988/ces.2014.410187
3. Warpinski, N.: Microseismic monitoring: inside and out. J. Pet. Technol. **61**(11), 80–85 (2009)
4. Chambers, K., Kendall, J., Brandsberg-Dahl, S., Rueda, J.: Testing the ability of surface arrays to monitor microseismic activity. J. Pet. Technol. **58**(5), 821–830 (2010)
5. Lemeshko, B.: Optimization Methods. NSTU Publishing, Novosibirsk (2009). (in Russian)
6. Cherkasov, A., Gorodnichev, M., Kireev, S., Markova, V., Artyom, M.: On optimization of numerical simulation programs. In: Proceedings of the 9th Russian-Korean International Symposium on Science and Technology, KORUS 2005, pp. 584–589. IEEE (2005)
7. Fomel, S., Sava, P., Vlad, I., Liu, Y., Bashkardin, V.: Madagascar: open-source software project for multidimensional data analysis and reproducible computational experiments. J. Open Res. Softw. **1**(1), e8 (2013)

Accurate Parallel Algorithm for Tracking Inertial Particles in Large-Scale Direct Numerical Simulations of Turbulence

Takashi Ishihara[1]([✉]), Kei Enohata[2], Koji Morishita[3], Mitsuo Yokokawa[3], and Katsuya Ishii[4]

[1] JST CREST, Center for Computational Science, Graduate School of Engineering, Nagoya University, Nagoya 464-8603, Japan
ishihara@cse.nagoya-u.ac.jp
[2] Computational Science and Engineering, Graduate School of Engineering, Nagoya University, Nagoya 464-8603, Japan
enohata@fluid.cse.nagoya-u.ac.jp
[3] Education Center on Computational Science and Engineering, Kobe University, Kobe 650-0047, Japan
{morishita,yokokawa}@port.kobe-u.ac.jp
[4] Information Technology Center, Nagoya University, Nagoya 464-8601, Japan
ishii@cc.nagoya-u.ac.jp

Abstract. Statistics on the motion of small heavy (inertial) particles in turbulent flows with a high Reynolds number are physically fundamental to understanding realistic turbulent diffusion phenomena. An accurate parallel algorithm for tracking particles in large-scale direct numerical simulations (DNSs) of turbulence in a periodic box has been developed to extract accurate statistics on the motion of inertial particles. The tracking accuracy of the particle motion is known to primarily depend on the spatial resolution of the DNS for the turbulence and the accuracy of the interpolation scheme used to calculate the fluid velocity at the particle position. In this study, a DNS code based on the Fourier spectral method and two-dimensional domain decomposition method was developed and optimised for the K computer. An interpolation scheme based on cubic splines is implemented by solving tridiagonal matrix problems in parallel.

Keywords: Large-scale DNS of turbulence · Particle tracking · Cubic spline interpolation · Parallel computation

1 Introduction

The turbulent diffusion phenomena of small heavy (inertial) particles such as PM2.5, yellow sand, dust, and pollen in the air relate to our daily life and are of public concern. To predict such turbulent diffusion phenomena, the fundamental physics and statistics of the motion of particles in highly nonlinear turbulent flows need to be understand. Direct numerical simulations (DNSs) of turbulence

© Springer International Publishing Switzerland 2015
V. Malyshkin (Ed.): PaCT 2015, LNCS 9251, pp. 522–527, 2015.
DOI: 10.1007/978-3-319-21909-7_51

provide detailed data free from uncertainties and are one of the most effective tools for studying the physics and statistics of particle motion (e.g. [1]). Recently, there have been extensive studies on the statistics of the motion of inertial particles based on turbulence DNSs. In these simulations, particles are tracked in a simulated velocity field by using linear interpolation (e.g. [2,3]) or a piecewise cubic Lagrangian interpolation (e.g. [4]). Such interpolations are attractive because they have a simple implementation and require no global communication for data transfer.

The tracking accuracy of the particle motion is known to primarily depend on the spatial resolution of the turbulence DNS and the accuracy of the interpolation scheme used to calculate the fluid velocity at the particle position (e.g. [5,6]). Yeung and Pope [5] showed that cubic splines give higher interpolation accuracy. However, cubic spline interpolation requires the solving of tridiagonal matrix problems. Usually, the implementation of cubic splines in parallel codes requires global communication for data transposition. However, the method developed by Mattor et al. [7] minimises the amount of data transfer.

In this study, we developed an accurate and effective parallel algorithm for tracking inertial particles in a large-scale DNS to obtain accurate statistics on the motion of inertial particles in turbulent flows with a high Reynolds number. The DNS code is based on a Fourier spectral method and two-dimensional domain decomposition method and was developed and optimised for the K computer, which has a peak performance of 10.6 PFlops by 82,944 nodes (663,552 cores). The interpolation scheme is based on cubic splines and is implemented by solving tridiagonal matrix problems in parallel with the method developed by Mattor et al. [7].

2   Equations and Numerical Methods

Consider an incompressible fluid of unit density under periodic boundary conditions which obeys the following Navier–Stokes equations:

$$\partial \mathbf{u}/\partial t + (\mathbf{u} \cdot \nabla)\mathbf{u} = -\nabla p + (1/Re)\nabla^2 \mathbf{u} + \mathbf{f}, \quad \nabla \cdot \mathbf{u} = 0, \tag{1}$$

where $\mathbf{u}, p$, and $\mathbf{f}$ are the velocity, pressure, and external force, respectively, and $Re = UL/\nu$ is the Reynolds number, which measures the nonlinearity of turbulence. Here, U and L are the characteristic velocity and length scale in turbulence, and ν is the kinematic viscosity. The external force is used to obtain a statistically steady state of turbulence.

According to the classical turbulence theory [8], the degree of freedom in turbulence is proportional to $Re^{9/4}$. Therefore to obtain high Re turbulence in a DNS, large-scale simulations are necessary. In the DNS, Eq. (1) is solved by using a Fourier spectral method for spatial discretisation and a fourth-order Runge–Kutta method for time marching. The numerical methods are the same as those used for the Earth simulator [9].

Consider the inertial particles which obey the following equations:

$$d\mathbf{X}/dt = \mathbf{V}, \quad d\mathbf{V}/dt = (1/St)(\mathbf{u}(\mathbf{X},t) - \mathbf{V}), \tag{2}$$

where $\mathbf{X}$ and $\mathbf{V}$ are the position and velocity of the particles and St is the Stokes number, which is the ratio of the characteristic time of a particle to the characteristic time of the turbulent flow. When $St = 0$, Eq. (2) becomes $d\mathbf{X}/dt = \mathbf{u}(\mathbf{X}, t)$, i.e. the equation for fluid particles. Equation (2) is solved with the fourth-order Runge–Kutta method, where the time interval is set to twice as large as that used for solving Eq. (1). The velocity $\mathbf{u}$ at the particle position is evaluated by using an interpolation method. To obtain highly accurate statistics, we used cubic spline interpolation (see [5]).

3 Implementation

The DNS code based on the spectral method was optimised for the K computer by using parallel processing with Message Passing Interface (MPI) and OpenMP and utilising the FFTW library to implement the 3D-FFT. We used a two-dimensional domain decomposition method (see Fig. 1) for the data distribution in MPI, which is suited to massively parallel computers like the K computer. High peak performance efficiencies of 3.835 %, 3.143 %, and 2.242 % were obtained in double-precision DNSs with $N^3 = 6144^3$, 8192^3, and 12288^3 grid points using 96×64, 128×64, and 192×128 nodes, respectively, of the K computer. The details of the DNS code are described in [10].

Particles are tracked by solving Eqs. (1) and (2), where the velocity $\mathbf{u}$ at the particle positions in the *real* space of the turbulence field must be evaluated. The DNS code uses the velocity components in the *wavenumber* space as dependent variables. For each time step of the spectral method, the velocity components in the *real* space are calculated to evaluate the nonlinear terms. Therefore, no additional 3D-FFTs are needed. However, to use cubic spline interpolation, the tridiagonal matrix problems need to be solved to obtain the second derivatives u_i'' $(i = 0, ..., N-1)$:

$$\frac{1}{6}u_{i-1}'' + \frac{2}{3}u_i'' + \frac{1}{6}u_{i+1}'' = \frac{1}{\Delta^2}(u_{i+1} - 2u_i + u_{i-1}), \quad i = 0, 1, ..., N-1, \quad (3)$$

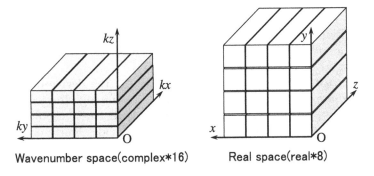

Wavenumber space(complex*16)   Real space(real*8)

Fig. 1. Two-dimensional domain decomposition used in spectral DNS of turbulence

where $u_i = u(\alpha_i)$ is the value of the velocity component u at an equal-interval grid α_i, $\Delta(= \alpha_{i+1} - \alpha_i)$ is the grid interval, and $u''_{-1} = u''_{n-1}$ and $u_{-1} = u_{n-1}$ from the periodicity. Once the second derivatives (i.e. spline coefficients) are obtained, the value of $u(\alpha)$ for $\alpha \in [\alpha_i, \alpha_{i+1})$ is calculated as

$$u(\alpha) = A_\alpha u_i + B_\alpha u_{i+1} + C_\alpha u''_i + D_\alpha u''_{i+1}, \tag{4}$$

where

$$A_\alpha = \frac{\alpha_{i+1} - \alpha}{\Delta}, \; B_\alpha = 1 - A_\alpha, \; C_\alpha = \frac{\Delta^2}{6}(A_\alpha^3 - A_\alpha), \; D_\alpha = \frac{\Delta^2}{6}(B_\alpha^3 - B_\alpha).$$

The cubic spline interpolation of u at the point $\mathbf{x} = (x, y, z)$ in the three-dimensional space can be expressed as follows:

$$
\begin{aligned}
u = A_z \{ A_y (& A_x u_{i,j,k} + B_x u_{i+1,j,k} + C_x u_{i,j,k}^{(2,0,0)} + D_x u_{i+1,j,k}^{(2,0,0)}) \\
+ B_y (& A_x u_{i,j+1,k} + B_x u_{i+1,j+1,k} + C_x u_{i,j+1,k}^{(2,0,0)} + D_x u_{i+1,j+1,k}^{(2,0,0)}) \\
+ C_y (& A_x u_{i,j,k}^{(0,2,0)} + B_x u_{i+1,j,k}^{(0,2,0)}) \\
+ D_y (& A_x u_{i,j+1,k}^{(0,2,0)} + B_x u_{i+1,j+1,k}^{(0,2,0)}) \} \\
+ B_z \{ A_y (& A_x u_{i,j,k+1} + B_x u_{i+1,j,k+1} + C_x u_{i,j,k+1}^{(2,0,0)} + D_x u_{i+1,j,k+1}^{(2,0,0)}) \\
+ B_y (& A_x u_{i,j+1,k+1} + B_x u_{i+1,j+1,k+1} + C_x u_{i,j+1,k+1}^{(2,0,0)} + D_x u_{i+1,j+1,k+1}^{(2,0,0)}) \\
+ C_y (& A_x u_{i,j,k+1}^{(0,2,0)} + B_x u_{i+1,j,k+1}^{(0,2,0)}) \\
+ D_y (& A_x u_{i,j+1,k+1}^{(0,2,0)} + B_x u_{i+1,j+1,k+1}^{(0,2,0)}) \} \\
+ C_z \{ A_y (& A_x u_{i,j,k}^{(0,0,2)} + B_x u_{i+1,j,k}^{(0,0,2)}) \\
+ B_y (& A_x u_{i,j+1,k}^{(0,0,2)} + B_x u_{i+1,j+1,k}^{(0,0,2)}) \} \\
+ D_z \{ A_y (& A_x u_{i,j,k+1}^{(0,0,2)} + B_x u_{i+1,j,k+1}^{(0,0,2)}) \\
+ B_y (& A_x u_{i,j+1,k+1}^{(0,0,2)} + B_x u_{i+1,j+1,k+1}^{(0,0,2)}) \},
\end{aligned}
\tag{5}
$$

where the point (x, y, z) is assumed to be in $[x_i, x_{i+1}) \times [y_i, y_{i+1}) \times [z_i, z_{i+1})$; the coefficients $A_{x,y,z}$, $B_{x,y,z}$, $C_{x,y,z}$, and $D_{x,y,z}$ are defined in the same manner as A_α, B_α, C_α, and D_α, respectively, in the corresponding x, y, and z directions; and the superscripts $(2,0,0)$, $(0,2,0)$, $(0,0,2)$ denote the second partial derivative with respect to x, y, and z, respectively. Note that expressions different from Eq. (5) are possible (see [5]). Equation (5) uses four arrays, but Eq. (5) uses the same number of stencils as the linear interpolation.

When the computational domain is decomposed in two dimensions as in Fig. 1, Eq. (3) can be solved in the z direction without message passing. However, the corresponding equations in the x and y directions cannot be solved without message passing. Therefore, we use the method developed by Mattor et al. [7] to solve the corresponding equations in the x and y directions in parallel. This method minimizes the amount of data transfer, which is proportional to the number of nodes used in the decomposed directions.

Fig. 2. Particle distributions for different values of St in subdomain of size $(1/8)^3$ for turbulence DNS in periodic box $(N^3 = 256^3)$: $St = 0, 0.2, 1.0$ and 5.0 from the left.

4   Results and Discussion

Figure 2 shows the St-dependence of the distribution of inertial particles under the same initial conditions and in the same turbulent flow fields. The results for $St \neq 0$ were consistent with the previous studies, which showed a preferential concentration of the inertial particles (see, e.g., [11]). The approximately homogeneous distribution of particles for $St = 0$ was consistent with the motion of fluid particles, which tracked the motion of the fluid completely. Based on the results (figure omitted) for the time dependence of the average vorticity at the particle positions, we confirmed that the code developed in this study can be used to accurately track the motion of particles.

At present, we have performed DNSs of turbulence in a periodic box with up to 2048^3 grid points and tracked up to 8×128^3 particles by using a parallel implementation of cubic spline interpolation on $16 \times 16(= 256)$ nodes of the K computer. The computational time for tracking 8×128^3 particles was less than 16% of the total computational time. This implies that we can obtain accurate statistics on the inertial particles with eight different values of St by using less than 20% additional computational time.

To obtain accurate statistics on the inertial particles for much higher values of Re, we need to increase not only the size of the DNS but also the number of particles. To do so, we need to optimise the memory access in the computations of Eq. (5) for the particles and the data-transfer of particle position information between the adjacent nodes as the next steps.

Acknowledgement. This research used computational resources of the K computer provided by the RIKEN Advanced Institute for Computational Science through the HPCI System Research Project (Project ID: hp150174) and the supercomputer system at Nagoya University. The work is partially supported by "Joint Usage/Research Center for Interdisciplinary Large-scale Information Infrastructures" in Japan. This research was partly supported by KAKENHI, Grant Numbers: (B) 15H03603, and (C) 26390130.

References

1. Ishihara, T., Kaneda, Y.: Relative diffusion of a pair of fluid particles in the inertial subrange of turbulence. Phys. Fluids **14**, L69–L72 (2002)

2. Bec, J., Biferale, L., Cencini, M., Lanotte, A.S., Toschi, F.: Intermittency in the velocity distribution of heavy particles in turbulence. J. Fluid Mech. **646**, 527–536 (2010)

3. Onishi, R., Takahashi, K., Vassilicos, J.C.: An efficient parallel simulation of interacting inertial particles in homogeneous isotropic turbulence. J. Comput. Phys. **242**, 809–827 (2013)

4. Sundaram, S., Collins, L.R.: Numerical considerations in simulating a turbulent suspension of finite-volume particles. J. Comput. Phys. **124**, 337–350 (1996)

5. Yeung, P.K., Pope, S.B.: An algorithm for tracking fluid particles in numerical simulations of homogeneous turbulence. J. Comput. Phys. **79**, 373–416 (1988)

6. Balachandar, S., Maxey, M.R.: Methods for evaluating fluid velocities in spectral simulations of turbulence. J. Comput. Phys. **83**, 96–125 (1989)

7. Mattor, N., Williams, T.J., Hewett, D.W.: Algorithm for solving tridiagonal matrix problems in parallel. Parallel Comput. **21**, 1769–1782 (1995)

8. Kolmogorov, A.N.: The local structure of turbulence in incompressible viscous fluid for very large reynolds number. C. R. Acad. Sci. URSS **30**, 299–303 (1941)

9. Yokokawa, M., Itakura, K., Uno, A., Ishihara, T., Kaneda, Y.: 16.4-tflops direct numerical simulation of turbulence by a fourier spectral method on the earth simulator. In: Proceeding of the IEEE/ACM SC2002 Conference, p. 50 (2002)

10. Morishita, K., Yokokawa, M., Uno, A., Ishihara, T., Kaneda, Y.: Highly-efficient direct numerical simulation of turbulence by a fourier spectral method on the K computer. In: Parallel CFD 2015 (submitted)

11. Toschi, F., Bodenschatz, E.: Lagrangian properties of particles in turbulence. Annu. Rev. Fluid Mech. **41**, 375–404 (2009)

Treating Complex Geometries with Cartesian Grids in Problems for Fluid Dynamics

Igor Menshov[(✉)]

Keldysh Institute for Applied Mathematics, Russian Academy of Sciences,
Moscow, Russia
menshov@kiam.ru

Abstract. The paper addresses an efficient and simple numerical approach to simulating the 3D unsteady gas dynamics problems in geometrically complex domains discretized with the use of plane Cartesian grids. The key point of our consideration is to equivalently substitute the solution of the boundary value problem for the homogeneous system of equations with the solution in the whole space of non-homogeneous equations with a properly designed right-hand side term. The geometry is represented by the set of cut cells where additional (compensating) fluxes of mass, momentum, and energy are introduced to model inner boundary conditions. This approach leads to algorithmically transparent codes, simple for computer realization on parallel multiprocessor computing systems.

Keywords: Cartesian grid · Unsteady gas dynamics · Immersed boundary

1 Introduction

The development of supercomputers today and seemingly in perspective of the near future follows the way of hybrid SIMD architectures with massive multithreading systems (GPGPU, PHI). Effectiveness of using such kind of systems depends on the algorithmical complexity of numerical methods employed to solve the problem. Modern computing systems are based on very large number of so-called light compute kernels that able to proceed lightweight threads of execution. This architecture requires developing special solution numerical methods that respond the condition of computational primitivism.

Explicit Cartesian grid schemes best fit this condition. They result in algorithms that work with simple local data and free from logical operators. Spatial discretization is performed in a straightforward way. The main drawback of this approach is the treatment of geometrically complex computational domains and corresponding boundary conditions.

In the present paper we deal with this issue and show one technique for implementing Cartesian grids in problems with complex geometries. We consider numerical solution of the unsteady gas dynamics equations in the domain external to a solid object that in general case may move in space.

The method we consider can be related to the class of so-called immersed boundary (IB) approaches where the computational domain that includes both gas and solid

© Springer International Publishing Switzerland 2015
V. Malyshkin (Ed.): PaCT 2015, LNCS 9251, pp. 528–535, 2015.
DOI: 10.1007/978-3-319-21909-7_52

regions is discretized with a grid so that the solid surface intersects computational cells. The main advantage of the IB method compared with the body fitted grid method is the simplicity of grid generation, and the possibility to avoid regridding needed to adapt changes in geometry due to solid motion. The penalty is the problem of cut cell calculations – how to calculate flow parameters in cells that are cut by the solid surfaces?

Since pioneering works by C. Peskin (1977) [1] there were many efforts to cope with this problem, which can be classified into two groups. One is based on the finite volume formulation applied to the fraction of the cut cell occupied by fluid. In this way one may face many difficulties, such as the problem of "small cell", arising new fluid cells and collapsing cells because of solid motion, etc. (Pember et al. (1995) [2].

Other methods are based on the finite-difference discretization and employ rather sophisticated interpolation schemes to treat the boundary conditions at solid surfaces. These methods are mostly developed for incompressible flow problems. Their generalization to compressible flows is in somewhat trickish problem.

We propose a novel approach to treat the boundary conditions in the framework of the IB method. It uses the finite volume formulation. However, in contrast to the conventional procedure we apply it to the whole computational cell and introduce properly defined additional fluxes that model the effect of boundary conditions. We name this approach as free boundary method. Its basics are given in [3].

2 Method of Free Boundaries

A key point of the method of free boundaries is an alternative mathematical formulation of the problem. For example, the conventional mathematical statement of a fluid flow problem is the following. Let Ω be a domain occupied by the solid, and Γ its boundary surface. Then, one need to solve the system of Euler equations exterior to the solid, $R^3 \backslash \Omega$, with the impermeability condition boundary conditions imposed at the solid surface Γ:

$$\partial_t \mathbf{q} + \partial_k \mathbf{f}_k = 0, \ \mathbf{x} \in R^3 \backslash \Omega, (\mathbf{u} - \mathbf{U}_s, \mathbf{n}) = 0, \ \mathbf{x} \in \Gamma \tag{1}$$

where q is the conservative vector, $\mathbf{f}_k$ is the flux vectors, u is the fluid velocity vector, $\mathbf{U}_s$ is the solid velocity vector, n is the outward normal to the solid surface.

Giving an alternative formulation of the problem we aim to replace the solution of the boundary value problem in a part of the space with the solution of an initial value problem in the whole space. With this end, we modify the original system of partial differential Eq. (1) by adding in the right-hand side a vector-function $\mathbf{F}_c$ that we refer in what follows as the compensating flux:

$$\partial_t \mathbf{q} + \partial_k \mathbf{f}_k = -\mathbf{F}_w, \ \mathbf{x} \in R^3 \tag{2}$$

We want to choose the compensating flux so that the reduction of the solution to the problem (2) onto the domain $R^3 \backslash \Omega$ exactly matches the solution to the original boundary value problem (1). In our previous paper [3] we suggest an expression for the compensating flux that warranties the above condition:

$$\mathbf{F}_w = \begin{pmatrix} \rho u_k n_k \\ \rho u_k u_m n_k + (p - p_w) n_m \\ \rho u_k n_k H \end{pmatrix} \delta(\mathbf{x}, \Gamma) \tag{3}$$

where $\delta(\mathbf{x}, \Gamma)$ denotes the generalized Dirac's function of the surface Γ defined by the following relation:

$$\int_V \delta(\mathbf{x}, S) \varphi(\mathbf{x}) dV = \int_{V \cap \Gamma} \varphi(\mathbf{x}) dS \quad \forall V \in \Re^3 \tag{4}$$

with $\varphi(\mathbf{x})$ being any integrable function. The value p_w in (3) represents the instantaneous reaction (pressure) of the solid wall on forcing from the side of the fluid flow. It is analytically calculated by local values of flow parameters [3]. The expression (3) has a simple mechanical interpretation. The compensating flux is modeled by two parts. The first accounts for convection of mass, momentum, and energy through the solid surface Γ, while the second one represents the solid reaction and the corresponding work.

As for programming and realization of calculations, the alternative formulation of the problem seems to be much easier than its conventional counterpart. Calculations are performed in a plain cube-like domain with a simple Cartesian grid. The computation algorithm is homogeneous with minimal logical branching and simple for implementation on multiprocessor computer systems, in particular with the SIMD architecture.

The discrete model is developed by implementing the principle of splitting in physics. Updating the solution from one time level to another is performed in two steps. First we integrate the system of equations without compensating fluxes on the time step Δt in the whole computational domain including the part occupied by solid and cut cells. The solution obtained is then once more updated in cut cells, only by integrating the compensating fluxes over the time step Δt.

The first step is executed with the finite-volume method discretization of the governing equations, which results in the following discrete equations for calculating predict-values:

$$\tilde{\mathbf{q}} = \mathbf{q}^n - \frac{\Delta t}{Vol} \sum_\sigma \mathbf{f}_{k,\sigma} n_{k,\sigma} S_\sigma \tag{5}$$

where σ denotes the cell interface, and the summation is fulfilled over all interfaces surrounded the cell. Parameters Vol and S_σ are the cell volume and the interface area, respectively.

The vectors $\mathbf{f}_{k,\sigma}$ in (5) are numerical fluxes at the interface that can be approximated with any of known gas dynamics numerical schemes. For example, in our calculations we use the hybrid explicit-implicit second-order accurate MUSCL-type scheme presented in [4]. As the flux function the Godunov approach is employed, and the flux at the interface is approximated on the basis of the exact Riemann problem solution.

The solution is calculated in the whole domain including solid cells. However, only part of this solution, in fluid and cut cells is of interest. The solid cells solution is complementary, which should have no relation to the actual one and therefore should not affect the latter. In the case of steady geometry, there is even no need to calculate this solution. At the same time the unsteady geometry results in the situation when the status of the computational cell may change - the solid cell is becoming a cut and then a fluid one. In this transition, emerging cut and fluid cells will inherit the non-actual solid cells solution. In other words, non-physical solution inside the solid domain will affect the solution of interest.

To avoid this phenomenon, the solutions in and outside the solid should have smooth conjunction, i.e., the solution in the solid domain near the boundary surface should be smooth continuation of the actual solution. This may be assured with the following way.

We introduce doubled fluxes at interfaces that separate solid and cut (or fluid) cells. The flux for the solid cell is computed in the standard manner, for example with the Godunov method through the Riemann problem solution. In this way the solution outside the solid plays the role of boundary conditions providing smooth extension of the solution in the solid domain. In order to prevent the backward effect on the actual solution, the flux for the real (cut or fluid) cell is merely calculated by the cell values of flow parameters.

The predict-solution $\tilde{\mathbf{q}}$ is then modified in the cut cells by the compensating flux. We use the linear approximation for the subcell geometry. Therefore, the geometry inside the cut cell is represented by three parameters - the volume fraction of fluid ω_f, the area of the geometry cut off by the cell S_f, and the outward unit normal $\mathbf{n}_f$. Taking $\tilde{\mathbf{q}}$ as initial data we integrate over the time step Δt the following system of equations:

$$\omega_f Vol \frac{d\mathbf{q}}{dt} = -\sum_{\sigma \in f} \mathbf{f}_k n_k S_\sigma + \mathbf{F}_p S_f \tag{6}$$

where the summation in the right-hand side is performed only over the interfaces that are in the fluid, and $\mathbf{F}_p = \left(0, p_w \mathbf{n}_f, p_w(\mathbf{U}_s, \mathbf{n}_f)\right)^T$. Because of the compensating flux effect is taken into account at this step, one can simply approximate the interface flux by the cell values, $\mathbf{f}_k = \mathbf{f}_k(\mathbf{q})$. Then the Eq. (6) can be recast as

$$\omega_f Vol \frac{d\mathbf{q}}{dt} = -\mathbf{F}_c S_f + \mathbf{F}_p S_f = -\mathbf{F}_w S_f \tag{7}$$

where $\mathbf{F}_c$ we denote the convective part of the flux.

The parameter ω_f is in generally changed between 0 and 1, and may take very small values. Therefore the integration of (7) is carried out with the implicit scheme to maintain stability. Combining with (5), finally we come to the following discrete equations describing the solution update at one time step:

$$\mathbf{q}^{n+1} = \mathbf{q}^n - \frac{\Delta t}{Vol}\sum_{\sigma}\mathbf{f}_{k,\sigma}n_{k,\sigma}S_{\sigma} - \frac{\Delta t}{\omega_f Vol}\mathbf{F}_w(\mathbf{q}^{n+1})S_f \qquad (8)$$

This non-linear equation is solved with the Newton iterative method. At each iteration, the matrix of the linear system is block-diagonal if the flux approximation employs the explicit scheme. In this case, one time step computations are performed in a single run over the cells for calculating residuals (total fluxes) followed by local Newtonian iterations for compensating fluxes in cut cells. The computational work is parallelized fairly straightforward.

For implicit flux approximation the system of Eq. (8) becomes more involved. Newton iterations results in a linear system with a sparse block matrix, $A\delta\mathbf{q} = \mathbf{R}$, where $\delta\mathbf{q}$ is the iterative residual. The solution of this linear equations is executed with the Lower-Upper Symmetric Gauss-Seidel approximate factorization method. Splitting the matrix in the diagonal, low-, and upper-triangle parts, $A = D + L + U$, we implement an approximate factorization by replacing A with $(D + L)D^{-1}(D + U)$, and consider solution of the factorized equations, $(D + L)D^{-1}(D + U)\delta\mathbf{q} = \mathbf{R}$, which can be executed in forward and backward sweeps.

It should be noted that the order cells are swept in this method can be chosen arbitrarily. This gives us in fact a set of algorithms. The sweep order can be decided in a way that would best fit for parallel executions. One such a choice has been proposed in [5]. It is based on the checkerboard order. The computational cells are sorted in "black" and "white" in accordance with chess coloring. The computational loop is performed over first the black and then white cells. This allows us to separate neighboring cells in the loop ordering; the forward sweep over cell is then represented by the black loop while the white loop serves to execute the backward sweep.

3   Numerical Results

In this section we show several test calculations that demonstrate accuracy and effectiveness of the proposed methodology. The first problem has 1D formulation. Its sketch is illustrated in Fig. 1. This is a piston problem. The piston initially is located in the interval between X_1 and X_2 and moves with a velocity U_s. A gas with constant flow parameters is outside the piston. Depending of this initial data the piston movement produces shock or rarefaction waves on the both sides of the piston.

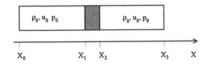

Fig. 1. The sketch of the 1D piston problem.

The problem has the analytical solution that is shown in Fig. 2. along with the numerical solution obtained on a Cartesian grid with the method of free boundary (BIC-solution). The numerical solution is calculated in the whole domain including the solid piston. One can see that the solution outside the solid region well matches the analytical solution.

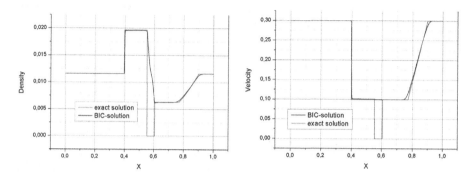

Fig. 2. The piston problem: $X_1 = 0.4$ $X_2 = 0.5$ $U_S = 0.3$ $p_0 = 0.001 \rho_0 = 0.016$ grid = 1000 cells, $\Delta t = 0.0009$, steps = 600.

The second problem is the calculation of supersonic flow around a wedge. The problem serves to verify the method of free boundaries on 2D flows with shock and rarefaction waves. The wedge angle is $10°$. The inflow Mach number $M = 3$. The angle of attack is $0°$ and $20°$ that corresponds the formation of a shock wave and a rarefaction wave, respectively. Numerical solutions obtained with a Cartesian grid 1200x480 cells are shown in Fig. 3.

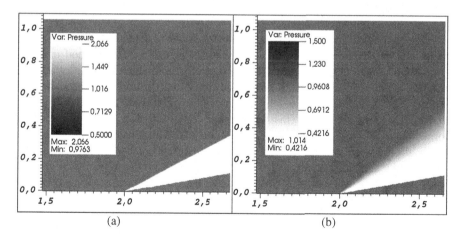

Fig. 3. Numerical simulation of supersonic flow around a wedge: shock wave (a) and rarefaction wave (b).

The numerical data gives the shock angle of 17.4° and the rarefaction fan angle of 13.2° that well agree with analytical predictions - 17.383° and 13.24°, respectively. Next example demonstrates the work of the method on computing complex unsteady flows. This is diffraction of a plane shock wave on a set of cylinders. A Cartesian grid of 1024x1024 cells is used in a rectangular computational domain. Calculations are executed with 32 GPU. Numerical results are presented in Fig. 4. For comparison we also show a reference numerical solution obtained with an alternative numerical method - the penalty function method [5].

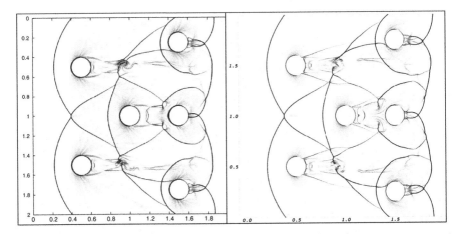

Fig. 4. Shock diffraction on a set of cylinders: the present method (left), the penalty function method [5] (right).

4   Conclusions

A simple numerical method has been proposed for calculating unsteady three-dimensional gas dynamics problems with complex geometry. The method treats the geometry on a plain Cartesian grid. The geometry is represented by a set of cut cells separating actual (fluid) cells from non-actual (those out the flow domain). The inner boundary conditions are realized by means of an additional (compensating) flux introduced in only cut cells. This flux depends on local flow parameters and the information about the geometry subcell structure – the volume fraction of fluid in the cut cell, the outward unit normal, and the area of the geometry inside the cell. The method works similarly for all the cells of the Cartesian grid, and results in algorithmically transparent and uniform codes, simple for computer realization on parallel multiprocessor computing systems.

Acknowledgments. This research was supported by the grant No 14-11-00872 from Russian Scientific Fund.

References

1. Peskin, C.S.: Numerical analysis of blood flow in the heart. J. Comput. Phys. **25**, 220–252 (1977)
2. Pember, R.B., Bell, J.B., Colella, P., Crutchfield, W.Y., Welcome, M.L.: An adaptive cartesian grid method for unsteady compressible flow in irregular regions. J. Comput. Phys. **120**, 278–304 (1995)
3. Menshov, I., Kornev, M.: Free_boundary method for the numerical solution of gas_dynamic equations in domains with varying geometry. Math. Models Comput. Simulations. **6**(6), 612–621 (2014)
4. Menshov, I., Nakamura, Y.: Hybrid explicit-implicit, unconditionally stable scheme for unsteady compressible flows. AIAA J. **42**(3), 551–559 (2004)
5. Boiron, O., Chiavassa, G., Donat, R.: A high-resolution penalization method for large mach number flows in the presence of obstacles. Comput. Fluids **38**, 703–714 (2009)

Architecture, Implementation and Performance Optimization in Organizing Parallel Computations for Simulation Environment

Maria Nasyrova[1(✉)], Yury Shornikov[1,2], and Dmitry Dostovalov[1,2]

[1] Novosibirsk State Technical University, Novosibirsk, Russia
maria_myssak@mail.ru, shornikov@inbox.ru,
d.dostovalov@corp.nstu.ru
[2] Design Technological Institute of Digital Techniques SB RAS,
Novosibirsk, Russia

Abstract. This paper discusses architectural concepts, implementation details and performance optimization techniques in the context of instrumental environment ISMA2015 supporting parallel computations for hybrid models. The paper considers the approach of organizing computations so that the user can work with the environment in the terms of the application field omitting the complex implementation details and to simply running models in a suitable mode: sequential, parallel on a multi-core machine or a cluster. The technology of the remote class loading is proposed. The framework for extending the library by new numerical methods is considered. The results of performance optimization are given. The technology of optimizing communication between cluster nodes is described. Simulation results are presented on the example of generated reaction-diffusion problems.

Keywords: Simulation · Distributed memory · Performance optimization · MPI · Parallel computations · Hybrid systems · Explicit methods · Accuracy control · Stability control

1 Introduction

Hybrid system (HS) is a mathematical model for convenient description of systems that can be both in continuous and discrete state simultaneously [1, 2]. Continuous behavior of HS is usually described by a system of differential equations and discrete behavior is determined by the instantaneous transition from one state to another.

Analytical research of HS is difficult and often impossible due to gaps in the system behavior. Therefore the research of HS dynamics is performed in special instrumental environments such as Charon, AnyLogic, Scicos, Rand Model Designer, Hybrid Toolbox, HyVisual, Dymola, OpenMVLShell, ISMA, etc.

Recent trends in simulation show a shift towards modeling of high-dimensional systems. Among the environments listed above the support of parallel computations is provided only by commercial software (AnyLogic, Scicos, Dymola, etc.). Despite of a rich set of tools and methods designed to simulate a variety of systems the specifics of

© Springer International Publishing Switzerland 2015
V. Malyshkin (Ed.): PaCT 2015, LNCS 9251, pp. 536–545, 2015.
DOI: 10.1007/978-3-319-21909-7_53

HS imposes the limitations to parallelization. Often the use of an environment assumes that the user is familiar with the programming and parallel programming in particular. Therefore, their use for research or educational purposes may not be available for a wide range of users. The approach implemented in the ISMA2015 environment allows applying the original methods and tools designed for hybrid model analysis and hides the complexity of parallel programming.

This paper also discusses issues of performance optimization important for any modern software and describes the technologies implemented in ISMA2015. These techniques can be applied to systems using MPI as a technology for computing nodes communication. In addition, the class caching approach is suitable for distributed systems regardless of the interaction way.

2 Computing Core Architecture

ISMA is a simulation environment of complex dynamical and hybrid systems developed at the department of Automated control systems of Novosibirsk state technical university (NSTU) (Russia) in 2007 [3]. ISMA mainly focuses on analysis of HS characterized by stiff modes. Enhanced specification tools such as structural schemes, dynamically typed language LISMA and electricity scheme editor are the system content of the instrumental environment. Analytical content is presented by a library of original numerical methods and event detection algorithms that allow accurately handling the moments when the system has switched to another state.

Today HS theory is extensively studying and, as a result, the requirements of easy adapting the environment to a new application fields become more and more urgent. Due to the high coupling of ISMA components, the addition of new objects is accompanied by considerable efforts and sometimes even impossible. Furthermore, the existing architecture cannot be modified to provide calculations of distributed models. Therefore, the new flexible architecture based on loosely coupled modules and supported parallel computations is proposed.

The architecture of the core is presented on the Fig. 1. Integration API consisting of public classes and interfaces used by other components to interact with the implemented solvers and to create a new ones. Integration core provides the implementation of solvers and integration strategies for sequential and parallel modes. Method library is an independent and easily extendable library of numerical methods. Integration server modules include the server API declaring the protocol of the client-server communication, the implementation of the server itself supports a multi-core or a cluster configuration and the server client respectively, which is responsible for sending models from ISMA2015 to the integration server.

3 Organizing Computations

The specifics of simulation environment supposes that the model is obtained during the runtime. If the environment is designed for execution on the shared memory architectures it does not impose any restrictions.

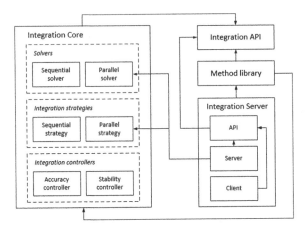

Fig. 1. The architecture of ISMA computing core.

In Java-based applications the program is running on the Java virtual machine (JVM). JVM is loading classes only on startup. When the model is generated in ISMA2015 the integration server and its nodes do not know anything about the class of the model and as a result cannot handle it. To solve the problem the technology of loading classes to JVM in runtime based on RMI is proposed. The scheme of components interaction is shown on the Fig. 2. The client built in ISMA2015 publishes the implementation of the specific class provider to the registry of objects based on RMI. The server uses the class provider to load classes of generated models from client. In turn, the server publishes in the registry the implementation of the methods. The client uses the implementation to invoke the remote server. After loading the classes to JVM the server allocates simulation tasks between child nodes as shown on the Fig. 3.

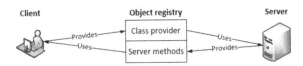

Fig. 2. The scheme of client-server communication.

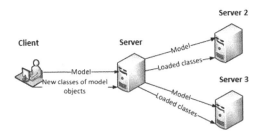

Fig. 3. Dynamic class loading in runtime.

4 Performance Optimization

First performance evaluations for the given approach shows that the most of the time is occupied by operations related to transferring and deserialization of objects participated in calculations. In this paragraph we consider the techniques implemented in ISMA2015 to improve the performance.

4.1 Caching

RMI technology based on the direct transfer of serialized classes. Deserialization and serialization processes are well-known bottleneck of program performance. Serialization is used to persist the state of an object so that the object can be saved and then regenerated by deserialization mechanism later. The default serialization behavior can easily lead to unnecessary overheads.

According with the technology described in the previous paragraph we obtain a significant performance degradation due to many deserialization processes caused by intensive class loading on each integration step. However, for the most problems a system of equations included in the model remains unchanged from step to step. Thus the mechanism of class caching is proposed to decrease the deserialization time on the server side. If the server knows the class then it is already in the cache and can be easily retrieved. Otherwise, it should be loaded to the cache.

4.2 Optimizing MPI Data Transfer

Many scientific applications use objects for storing data or utilize multi-dimensional arrays for these purposes. The comparative analysis of MPI data type performance [4, 5] shows that it is preferable to use an array of primitives over an array of objects.

In ISMA2015 solver cluster nodes send to each other values calculated on the step including approximate solution, function values or stage results for equations calculated on the node. These values has double type and each value corresponds to the equation index. Thus, the transferring of objects can be replaced with the transferring of one-dimensional array of specific format. Each value is represented by two parameters: index of equation and the value itself. Figure 4 shows the algorithm implemented in ISMA 2015 on the example of two nodes sending and receiving the array of five elements. The first node calculates three elements and the second node calculates the remaining. To collect the results we use MPI.COMM_WORLD.Allgather operation. The initial buffer received by the node shown on the Fig. 5.

5 Method Library

Particular attention should be paid to the choice of the integration method. Fully implicit methods cannot be used for HS because they require the calculation of the function at a potentially dangerous area, where the model is not defined [6].

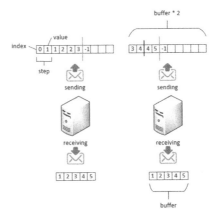

Fig. 4. The algorithm of sending values each of which is represented by two parameters in one-dimensional array.

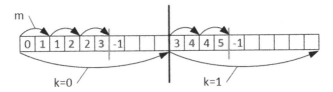

Fig. 5. The buffer received by all ranks in MPI.COMM_WORLD.Allgather operation.

Therefore, ISMA2015 primarily uses the explicit methods with accuracy and stability control [7] to limit the size of the integration step. The mathematical description of the used integration methods can be found in [8].

5.1 Integration Algorithms

Sequential Algorithm. Let the approximate solution y_n is known at the moment t_n with the step h_n. Then to obtain the approximate solution y_{n+1} at moment t_{n+1} we have the following common algorithm:

1. Calculate the approximate solution y_{n+1} at the moment t_{n+1} with the step h_n according to the performing method.
2. Calculate approximate function value $f(y_{n+1})$.
3. Obtain the accuracy characteristics of the integration step.
4. If the solution is accurate then go to 5, else set the integration step h_n equals to the step h^{ac} corrected by accuracy according to the performing method and go to 1.
5. Obtain the stability characteristics of the integration step.
6. If the solution is stable then go to 7, else set the integration step h_n equals to the step h^{st} corrected by stability according to the performing method and go to 1.

7. Get size of the next integration step using formula (5).
8. Perform the next integration step.

Parallel Algorithm. The developed parallel algorithm is based on the presented above sequential algorithm with the following differences.

For definiteness, we assume that the computer system consists of p processors, $N > p$ and let k is a number of equations per rank. Then all of N equations should be evenly allocated between computing nodes. Taking into consideration assumptions about beginning of the sequential method base parallel algorithms can be defined in the following way:

1. Allocate equations evenly between ranks using Round-Robin algorithm.
2. Calculate in each rank the approximate solution y_{n+1}^j, $0 \le j \le k$ at the moment t_n with the step h_n according to the performing method.
3. Send the obtained y_{n+1}^j from each rank to others.
4. Calculate in each rank the approximate function value $f^j(y_{n+1})$, $0 \le j \le k$.
5. Execute for each rank the sequential algorithm starting from the step 3 of the previous section.

5.2 New Methods Implementation

One of the main advantages of ISMA2015 is that it hides all complex logic inside allowing user to work with the environment in the terms of an application field. On the other side, it requires to add new features as fast as possible for providing the acceptable functionality that is enough to solve new problems. ISMA2015 supports the library of original numerical methods that is constantly growing. This paragraph considers the framework of ISMA2015 for developers that allows to timely extending the library designed so that the developer can focus on the implementation of the specific method avoiding unnecessary details of the common numerical algorithm.

The library of numerical methods is delivered as an independent module. It contains the enumeration of available methods and speaking in terms of the object-oriented patterns the factory of these methods. To add a new numerical method to the library the following sequence of actions should be done:

1. Create a class for new integration method that implements the IntgMethod interface for single-stage methods and the StagedIntgMethod for methods that have two or more stages.
2. If the method is staged then implement the required stage calculators, otherwise got to 3.
3. Register the method in the library:
   (a) Add the new type to the IntgMethodType enumeration.
   (b) Add the method of the new method creation to the IntgMethodFactory.
4. If the method is accurate then provide the implementation of the AcurateMethod interface:
   (a) Create the accuracy controller.
   (b) Register the accuracy controller in the class of the method.

5. If the method is stable then provide the implementation of the StableMethod
interface:
   (a) Create the stability controller.
   (b) Register the stability controller in the class of the method.

It is worth noting that the created method can be run in sequential on in parallel
mode without any further implementation costs. It is the responsibility of the environment computing core.

6   Reaction-Diffusion Problem

Let us consider the use of the proposed approaches on the simulation example of
reaction-diffusion problem in two-dimensional space, which is associated with competition model of Lotka-Volterra [9].

Two kinds of variables $c^1(x, z, t)$ and $c^2(x, z, t)$ represent density of competing
species in the habitat area $\Omega = \{(x, z) : 0 \leq x \leq 1, 0 \leq z \leq 1\}$ and in time $0 \leq t \leq 3$:

$$\frac{\partial c^i}{\partial t} = d_i \left(\frac{\partial^2 c^i}{\partial x^2} + \frac{\partial^2 c^i}{\partial z^2} \right) + f^i(c^1, c^2), i = 1, 2 \tag{1}$$

where $d_1 = 0.05$, $d_2 = 1.0$, $f^1(c^1, c^2) = c^1(b_1 - a_{12}c^2)$, $b_1 = 1$, $a_{12} = 0.1$,
$f^2(c^1, c^2) = c^2(-b + b_{21}c^1)$, $a_{21} = 100$, $b_2 = 1000$.   Boundary   conditions   are
$\partial c^i/\partial x = 0$ at $x = 0$, $x = 1$ and $\partial c^i/\partial z = 0$ at $z = 0$, $z = 1$ Initial conditions are
$c^1(x, z, 0) = 10 - 5\cos(\pi x)\cos(10\pi z)$ and $c^2(x, z, 0) = 17 + 5\cos(10\pi x)\cos(\pi z)$.

At $t \to \infty$ solution becomes spatially homogeneous and tend to periodically solve
ODE system of Lotka-Volterra. This ODE system is alternately stiff and non-stiff
depending on the solution position in the phase space. Turning to the grid of size $J \times K$
by x and z respectively we obtain $\Delta x = 1/(J-1)$ and $\Delta z = 1/(K-1)$ are grid steps
by x and z coordinates, c^i_{jk} is approximation of $c^i(x_j, z_k, t)$, where $x_j = (j-1)\Delta x$,
$z_k = (k-1)\Delta z$, $1 \leq j \leq J$, $1 \leq k \leq K$. Thus we obtain differential equations system of
$N = 2JK$ dimension:

$$\dot{c}^i_{jk} = \frac{d_i}{\Delta x^2} \left(c^i_{j+1,k} - 2c^i_{jk} + c^i_{j-1,k} \right) + \frac{d_i}{\Delta z^2} \left(c^i_{j,k+1} - 2c^i_{jk} + c^i_{j,k-1} \right) + f^i_{jk}, \tag{2}$$

where $1 \leq i \leq 2$, $1 \leq j \leq J$, $1 \leq k \leq K$, $f^i_{jk} = f^i\left(c^1_{jk}, c^2_{jk} \right)$. Boundary conditions on the
grid are the following: $c^i_{0,k} = c^i_{2,k}$, $c^i_{J+1,k} = c^i_{J-1,k}$ for $1 \leq k \leq K$ and $c^i_{j,0} = c^i_{j,2}$, $c^i_{j,K+1} = c^i_{j,K-1}$ for $1 \leq j \leq J$.

Figure 6 shows that model created in ISMA2015 is maximally close to the original
mathematical description. Simulation settings are presented on the Fig. 7. Two staged
Runge-Kutta method with accuracy and stability control enabled is chosen for the
model solving. The host and the port of the remote simulation server are specified. The
number of computing processes is configured on the integration server.

Fig. 6. The model of the reaction-diffusion problem in ISMA2015.

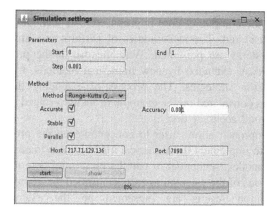

Fig. 7. Simulation settings window in ISMA2015.

Measurements are performed on the grid system of NSTU consisting of four computing nodes with hyper-threading. Each node has Intel Xeon X5355 2.66 HGz processor, 28 GByte of main memory and 8 MByte of L2 cache. The computing speed of the grid equals to 0.7 TFlops. Figure 8 presents the results of the comparative analysis of the performance changing according with the approaches described in the paragraph 4. The presented calculations are performed on the Ethernet network. The figure shows that the described approaches greatly improve performance, especially replacing the transferring of objects by transferring of primitives. Further, it is planned to configure the integration server to obtain result on InfiniBand and improve the MPI collective communication performance.

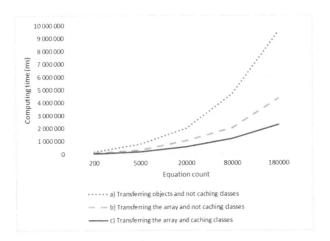

Fig. 8. The results of the performance improvements in ISMA 2015 for parallel computing mode.

7 Conclusion

The architecture of the computing core is proposed. The technology of the dynamic class loading is considered. The technologies of performance improvement for distributed memory systems are presented. With the use of the presented framework ISMA2015 can be easy extended by new numerical methods, hybrid model types, solvers and integration strategies. The process of extending the library by new numerical methods is considered on the example of the two-staged stable Runge-Kutta algorithm of the second accuracy order.

The presented approach allows user to concentrate on the problem solution and eliminates the need to implement the numerical methods. Any method from the library can be run in three modes: sequential or parallel on a multi-core machine and on the cluster. The evaluations of the calculation costs confirmed the viability of the proposed approaches. Simulation results are presented on the example of generated reaction-diffusion problems based on Lotka-Volterra model.

Acknowledgements. This work was supported by grant 14-01-00047-a from the Russian Foundation for Basic Research, RAS Presidium project № 15.4 "Mathematical modeling, analysis and optimization of hybrid systems".

References

1. Novikov, E.A., Shornikov, Y.V.: Computer Simulation of Stiff Hybrid Systems: Monograph. Publishing House of NSTU, Novosibirsk (2012). (in Russian)
2. Esposito, J.M., Kumar, V., Pappas, G.J.: Accurate event detection for simulating hybrid systems. In: Di Benedetto, M.D., Sangiovanni-Vincentelli, A.L. (eds.) HSCC 2001. LNCS, vol. 2034, pp. 204–217. Springer, Heidelberg (2001)

3. Shornikov, Y.V.: Instrumental tools of computerized analysis (ISMA). In: Shornikov, Y.V., Druzhinin, V.S., Makarov, N.A., Omelchenko, K.V., Tomilov, I.N. (eds.) Official Registration License for Computers No 2005610126. Rospatent, Moscow (2005)

4. Ross, R.J., Miller, N., Gropp, W.D.: Implementing fast and reusable datatype processing. In: Dongarra, J., Laforenza, D., Orlando, S. (eds.) EuroPVM/MPI 2003. LNCS, vol. 2840, pp. 404–413. Springer, Heidelberg (2003)

5. Carpenter, B., Fax, G., Ko, S.H., Lim, S.: Object serialization for marshalling data in a java interface to MPI. In: Proceedings of the ACM 1999 Conference on Java Grande, pp. 66–71, New York (1999)

6. Hairer, E., Vanner, G.: Solving Ordinary Differential Equations. Stiff and differential-algebraic problems. Mir, Moscow (1999). (in Russian)

7. Novikov, E.A., Vashchenko, G.V.: Parallel Explicit Runge-Kutta Method 2nd Order: Accuracy and Stability Control. Int. J. Appl. Fundam. Res. Phys. Math. Sci. 1, 101–102 (2011)

8. Shornikov, Y.V., Myssak, (Nasyrova) M.S., Dostovalov D.N.: Computer simulation of hybrid systems by ISMA instrumental facilities. In: Proceedings of the 2014 International Conference on Mathematical Models and Methods in Applied Sciences (MMMAS 2014), pp. 257–262, Saint Petersburg, Russia (2014)

9. Brown, P.N., Hindmarsh, A.C.: Matrix free methods in the solution of stiff systems of ODEs, p. 38. Lawrence Livermore National Laboratory, San Francisco (1983)

Author Index

Printed in the United States
By Bookmasters